Free Radicals in Synthesis and Biology

NATO ASI Series

Advanced Science Institutes Series

A Series presenting the results of activities sponsored by the NATO Science Committee, which aims at the dissemination of advanced scientific and technological knowledge, with a view to strengthening links between scientific communities.

The Series is published by an international board of publishers in conjunction with the NATO Scientific Affairs Division

A	Life Sciences	Plenum Publishing Corporation
B	Physics	London and New York
C	Mathematical and Physical Sciences	Kluwer Academic Publishers
D	Behavioural and Social Sciences	Dordrecht, Boston and London
E	Applied Sciences	
F	Computer and Systems Sciences	Springer-Verlag
G	Ecological Sciences	Berlin, Heidelberg, New York, London,
H	Cell Biology	Paris and Tokyo

Free Radicals in Synthesis and Biology

edited by

Francesco Minisci

Department of Chemistry,
Politecnico di Milano, Milano, Italy

Kluwer Academic Publishers

Dordrecht / Boston / London

Published in cooperation with NATO Scientific Affairs Division

Proceedings of the NATO Advanced Research Workshop on
Free Radicals in Synthesis and Biology
Bardolino (VR), Italy
May 8–13, 1988

Library of Congress Cataloging in Publication Data

NATO Advanced Research Workshop on Free Radicals in
    Synthesis and Biology (1988 : Bardolino, Italy)
    Free radicals in synthesis and biology.

    (NATO ASI series. Series C, Mathematical and
physical sciences ; vol. 260)
    "Published in cooperation with NATO Scientific
Affairs Division."
    Includes index.
    1. Free radicals (Chemistry)--Physiological effect--
Congresses. 2. Free radical reactions--Congresses.
3. Chemistry, Organic--Synthesis--Congresses.
I. Minisci, Francesco. II. North Atlantic Treaty
Organization. Scientific Affairs Division. III. Title.
IV. Series: NATO ASI series. Series C, Mathematical and
physical sciences ; no. 260.
QP527.N38 1988   547'.2   88-27298

ISBN-13: 978-94-010-6892-5   e-ISBN-13: 978-94-009-0897-0
DOI: 10.1007/ 978-94-009-0897-0

Published by Kluwer Academic Publishers,
P.O. Box 17, 3300 AA Dordrecht, The Netherlands.

Kluwer Academic Publishers incorporates the publishing programmes of
D. Reidel, Martinus Nijhoff, Dr W. Junk, and MTP Press.

Sold and distributed in the U.S.A. and Canada
by Kluwer Academic Publishers,
101 Philip Drive, Norwell, MA 02061, U.S.A.

In all other countries, sold and distributed
by Kluwer Academic Publishers Group,
P.O. Box 322, 3300 AH Dordrecht, The Netherlands.

This book contains the proceedings of a NATO Advanced Research Workshop held within the programme of activities of the NATO Special Programme on Selective Activation of Molecules running from 1983 to 1988 as part of the activities of the NATO Science Committee.

Other books previously published as a result of the activities of the Special Programme are

BOSNICH, B. (Ed.) - *Asymmetric Catalysis* (E103), 1986

PELIZZETTI, E. and SERPONE, N. (Eds.) - *Homogeneous and Heterogeneous Photocatalysis* (C174) 1986

SCHNEIDER, M. P. (Ed.) - *Enzymes as Catalysts in Organic Synthesis* (C178) 1986

SETTON, R. (Ed.) - *Chemical Reactions in Organic and Inorganic Constrained Systems* (C165) 1986

VIEHE, H. G., JANOUSEK, Z. and MERÉNYI, R. (Eds.) - *Substituent Effects in Radical Chemistry* (C189) 1986

BALZANI, V. (Ed.) - *Supramolecular Photochemistry* (C214) 1987

FONTANILLE, M. and GUYOT, A. (Eds.) - *Recent Advances in Mechanistic and Synthetic Aspects of Polymerization* (C215) 1987

LAINE, R. M. (Ed.) - *Transformation of Organometallics into Common and Exotic Materials: Design and Activation* (E141) 1988

BASSET, J.-M., et al. (Eds.) - *Surface Organometallic Chemistry: Molecular Approaches to Surface Catalysis* (C231) 1988

WHITEHEAD, J. C. (Ed.) - *Selectivity in Chemical Reactions* (C245) 1988

CHANON, M., JULLIARD, M. and POITE, J. C. (Eds.) - *Paramagnetic Organometallic Species in Activation/Selectivity, Catalysis* (C257) 1988

CONTENTS

PREFACE

Free-radical reactions have been for long time the domain of the physical chemists and the basic chemical industry (polymerization of vinyl monomers, oxidation by molecular oxygen, chlorination of methane etc.), where the use of simple molecules and the possibility of partial conversions without heavy problems of product separations makes less dramatic the aspects of regio and chemoselectivity.

As synonym of unselectivity, free radical reactions were considered of poor use in the synthesis of fine chemicals or sophisticated molecules, where a high selectivity is an essential condition for the success, or in the involvement of biological processes.

Within the last 15 years, however, an authentic explosion of synthetic applications of free radical reactions occurred; they have gained a remarkable position among the selective methods of synthesis. At the same time the great importance of free radical reactions in fundamental biological processes and in the metabolism of drugs has been recognized. Thus a specialized meeting on these topics was generally felt appropriate.

I had the honour and the onus to organize this workshop because for more than 30 years I have been involved in the research of free radical reactions.

Thanks to a generous grant from NATO Scientific Affairs Division and the financial support of CNR and chemical industry (Montedison, Enichem, Zambon) such a meeting among almost 50 specialists and 15 observers of sufficient standard to take advantage of the discussion, became possible at Bardolino (Italy).

The two following main aspects of the general subjects were among the purposes of the workshop:
i) The frontier of two areas of research
ii) The Chemistry-Biochemistry interface in free radical reactions, being convinced that synergystic effects could arise from the impact of the two cultural areas, both involved in free radical chemistry.

A variety of topics have been faced by the contributors:
- Captodative and polar effects in synthesis
- Regio, chemo and stereoselectivity in cyclization
- Iodine atom transfer in inter and intramolecular reactions
- Silyl and stannyl radicals in organic synthesis
- Homolytic aromatic substitution
- Organometallic complexes in biomimetic synthesis
- Synthetic applications by electron-transfer processes

- Hydroperoxides and peroxyl radicals in biological processes
- Free radicals with polynucleotides and DNA
- Enzimic stereospecificity
- Fluorescent pigment in biological tissues
- Free radicals in metabolism of drugs
- Detection of free radical metabolites

I hope that the wide discussion (a whole session has been devoted to the general discussion of Chemistry-Biochemistry interface) has stimulated the participants and primed synergystic mechanisms of research. At the same time I also hope that the readers of this book can take advantage of its contents, as incentive for further developments of selective synthetic methods and for the understanding of complex biological processes.

F. Minisci

Milano, 1988.

CONTRIBUTORS

Bachi M., The Weizmann Institut, Department of Chemistry, 76100 Rehovot, Israel

Baciocchi E., Dipartimento di Chimica, Università di Roma P.le A. Moro 5, Roma, Italy

Barclay L.R.C., Department of Chemistry, Mount Allison University, Sackville, New Brunswick E0A 3C0, Canada

Caronna T. Dipartimento di Chimica del Politecnico. P. L. da Vinci 32, 20133 Milano, Italy

Chanon M., Laboratoire de Chimie Inorganique Moléculaire, Université d'Aix-Marseille III, Rue H. Poincaré, 13397 Marseille, France

Chatgilialoglu C., Istituto di Chimica del CNR, Via Tolara di Sotto 89, 40064 Ozzano Emilia (BO), Italy

Citterio C., Dipartimento di Chimica del Politecnico, P. L. da Vinci 32, 20133 Milano, Italy

Correa C., Departamento de Quimica, Faculdade de Ciencias do Porto, Portugal

Crick D., Department of Chemistry, University College London, 20 Gordon Street, London WCIH OAJ, England

Crozet M.P., Laboratoire de Chimie Organique B, Université d'Aix-Marseille, Rue H. Poincaré, 1339 Marseille Cedex 13, France

Curran D.P., Department of Chemistry, University of Pittsburg, Chemistry Building, Pittsburg, PA 15260, USA

Fischer H., Physikalisch-Chemisches Institut der Universität Zürich, Wintherthurer Str., 8057 Zürich, Switzerland

Fontana F., Dipartimento di Chimica del Politecnico, P. L. da Vinci 32, 20133 Milano, Italy

Giese B., Institut für Organische Chemie, Technische Hochschule Darmstadt, Petersenstr. 22, 6100 Darmstadt, Germany

Griller D., Division of Chemistry, National Research Council, 100 Sussex Drv., KIA OR6 Ottawa, Canada

Heinisch G., Institut für Pharmazeutische Chemie der Universität Wien, Wahringer Str.10, A-1090 Wien, Austria

Koster J.F., Erasmus Universität Rotterdam, 3000 DR Rotterdam, The Netherlands

Louw R., Department of Chemistry, Leiden University, 5 Einsteinweg, The Netherlands

Maillard B., Laboratoire de Chimie Organique, Université de Bordeaux I, France

Malatesta V., Istituto di Ricerca Farmitalia-Carlo Erba, Via dei Gracchi, Milano, Italy

Maples K.R., National Institutes of Health, Research Triangle Park, N.C. 27709, USA

Minisci F., Dipartimento di Chimica del Politecnico, P. L. da Vinci 32, 20133 Milano, Italy

Neumann W.P., Lehrstuhl für Organische Chemie I, Universität Dortmund, Otto-Hahn-Str., Postfach 500500, 4600 Dortmund, Germany

Porter N., Department of Chemistry, Duke University, Durham, N.C. 27706, USA

Retey J., Institut für Organische Chemie der Universität Karlsruhe, Germany

Rotilio G., Dipartimento di Biologia, Università di Roma, Roma, Italy

Rüchardt C., Chemisches Laboratorium, Albert-Ludwigs-Universität, Albertstr. 21, 7800 Freiburg, Germany

Schulte-Frohlinde D., Max-Plank-Institut für Strahlenchemie, 4330 Mülheim/Ruhr, Stifstrasse 34-36, Germany

Stella L., Laboratoire de Chimie Organique B, Université d'Aix-Marseille III, 13397 Marseille, France

Steenken S., Max-Plank-Institut für Strahlenchemie, Stifstr.34-36, 4330 Mülheim, Germany

Tiecco M., Istituto di Chimica Organica, Facoltà di Farmacia, Perugia, Via del Liceo, 06100 Perugia, Italy

Viehe H.G., Laboratoire de Chimie Organique, Université de Louvain, Place L. Pasteur 1, 1348 Louvain-Neuve, Belgium

Vismara E., Dipartimento di Chimica del Politecnico, P. L. da Vinci 32, Milano 20133, Italy

Vittimberga B., Department of Chemistry, The University of Rhode Island, Kingston, RI02881-0801, USA

Zard S.Z., Laboratoire de Synthèse Organique, Ecole Polytechnique, 91128 Palaiseau Cedex, France

PARTICIPANTS

Bachi M., The Weizmann Institut, Department of Chemistry
76100 Rehovot, Israel

Baciocchi E., Dipartimento di Chimica, Università di Roma
P.le A. Moro 5, Roma, Italy

Barclay L.R.C., Department of Chemistry, Mount Allison
University, Sackville, New Brunswick EOA 3CO, Canada

Bayrakceken F., Department of Engineering Physics, Ankara
University, 06100 Tandagan, Ankara, Turkey

Caronna T., Dipartimento di Chimica del Politecnico, P. L.
da Vinci 32, 20133 Milano, Italy

Chanon M., Laboratoire de Chimie Inorganique Moléculaire,
Université d'Aix-Marseille III, Rue H.Poincaré, 13397
Marseille, France

Chatgilialoglu C., Istituto di Chimica del CNR, Via Tolara
di Sotto 89, 40064 Ozzano Emilia (BO), Italy

Citterio C., Dipartimento di Chimiea del Politecnico, P.L.
da Vinci 32, 20133 Milano, Italy

Correa C., Departamento de Quimica, Faculdade de Ciencias
do Porto, Portugal

Crich D., Department of Chemistry, University College
London, 20 Gordon Street, London WCIH OAJ, England

Crozet M.P., Laboratoire de Chimie Organique B, Université
d'Aix-Marseille, Rue H.Poincaré, 1339 Marseille Cedex 13,
France

Curran D.P., Department of Chemistry, University of Pittsburg, Chemistry Building, Pittsburg, PA 15360, USA

Fischer H., Physikalisch-Chemisches Institut der Universität Zürich, Wintherthurer Str., 8057 Zürich, Switzerland

Fontana F., Dipartimento di Chimica del Politecnico, P. L. da Vinci 32, 20133 Milano, Italy

Giese B., Institut für Organische Chemie, Technische Hochschule Darmstadt, Petersenstr. 22, 6100 Darmstadt, Germany

Greci L., Dipartimento di Scienze dei Materiali, Università di Ancona, Via Brecce Bianche, 60131 Ancona, Italy

Griller D., Division of Chemistry, National Research Council, 100 Sussex Drv., KIA OR6 Ottawa, Canada

Heinisch G., Institut für Pharmazeutische Chemie der Universität Wien, Wahringer Str. 10, A-1090 Wien, Austria

Kaptan Y., Department of Physics Engineering, Hacettepe University Ankara, 06532 Beytepe, Ankara, Turkey

Koster J.F., Erasmus Universität Rotterdam, 3000 DR Rotterdam, The Netherlands

Louw R., Department of Chemistry, Leiden University, 5 Einsteinweg, The Netherlands

Lusztyk Jaunsz, Division of Chemistry, National Research Council, 100 Sussex Drv., K1A OR6 Ottawa, Canada

Maillard B., Laboratoire de Chimie Organique, Université de Bordeaux I, France

Malatesta V., Istituto di Ricerca Farmitalia-Carlo Erba, Via dei Gracchi, Milano, Italy

Manitto P., Dipartimento di Chimica Organica, Università di Milano, Via Venezian 19, Milano, Italy

Maples K.R., National Institutes of Health, Research Triangle Park, N.C. 27709, USA

Mezenyi R., Laboratoire de Chimie Organique, Université de Louvain, Place L. Pasteur 1, 1348 Louvain-la-Neuve, Belgium

Minisci F., Dipartimento di Chimica del Politecnico, Piazza L. da Vinci 32, 20133 Milano, Italy

Neumann W.P., Lehrstuhl für Organische Chemie I, Universität Dortmund, Otto-Hahn-Str., Postfach 500500, 4600 Dortmund, Germany

Pedulli, G., Dipartimento di Chimica Organica, Università di Bologna, Bologna, Italy

Penco S., Farmitalia-Carlo Erba, Via dei Gracchi, Milano, Italy

Platone E., ENIRICERCHE, Via Maritano 26, S. Donato Milanese (MI), Italy

Porter N., Department of Chemistry, Duke University, Durham, N.C. 27706, USA

Retey J., Institut für Organische Chemie der Universität Karlsruhe, Germany

Romano U., ENICHEM Sintesi, Via Maritano 26, S. Donato Milanese (MI), Italy

Rotilio G., Dipartimento di Biologia, Università di Roma, Roma, Italy

Rüchardt C., Chemisches Laboratorium, Albert-Ludwigs-Universität, Albertstr. 21, 7800 Freiburg, Germany

Santi R., Istituto Donegani, Montedison, Novara, Italy

Schulte-Frohlinde S., Max-Plank-Institut für Strahlenchemie, 4330 Mülheim/Ruhr Stifstrasse 34-36, Germany

Stella L., Laboratoire de Chimie Organique B, Université d'Aix-Marseille III, 13397 Marseille, France

Screttas C.G., The National Hellenic Research Foundation S 48 Vas Constantinon Av., Athens 106-75, Greece

Steenken S., Max-Plank-Institut für Strahlenchemie, Stifstr. 34-36, 4330 Mülheim, Germany

Surzur J.M., Laboratoire de Chimie Organique, Université d'Aix-Marseille, 13397 Marseille, France

Tiecco M., Istituto di Chimica Organica, Facoltà di Farmacia, Perugia, Via del Liceo, 06100 Perugia, Italy

Tordo P., Université de Provence, Rue H. Poincaré, 13397 Marseille, France

Viehe H.G., Laboratoire de Chimie Organique, Université de Louvain, Place L. Pasteur 1, 1348 Louvain-la-Neuve, Belgium

Vismara E., Dipartimento di Chimica del Politecnico, Piazza L. da Vinci 32, 20133 Milano, Italy

Vittimberga B., Department of Chemistry, The University of Rhode Island, Kingston, RI02881-0801, USA

Zard S.Z., Laboratoire de Synthèse Organique, Ecole Polytechnique, 91128 Palaiseau Cedex, France

CAPTODATIVE SUBSTITUENT EFFECTS IN SYNTHESIS[1]

H.G. Viehe, Z. Janousek and R. Merényi
Université Catholique de Louvain
Laboratoire de Chimie Organique
Place L. Pasteur 1
1348 Louvain-la-Neuve
Belgium

ABSTRACT. In contrast to the effect of two substituents with like polarity, captodative (cd) substitution leads to a synergy of the individual substituent effects. In synthetic organic chemistry this results in preferential recombination of cd radicals, increased reactivity in thermal (2+2) and (4+2) cycloadditions. A number of rearrangements are facilitated by such substitution such as vinyl-cyclopropane - to cyclopentene and especially oxazines to epoxy-epimines. Thus cyclodienes are stereospecifically transformed via chiral oxazines to epoxy-epimines with four chiral centers at all the original four dienic sp2 carbon atoms.

10 years ago we found a rearrangement[2a] reaction of captor substituted iminium salts which, together with other results, led us to the formulation of the captodative effect : "The combined action of a captor (electron withdrawing) and a dative (electron releasing) substituent on a radical center leads to enhanced stabilisation"[3]. Similar ideas have been formulated earlier[4], but today it appears that especially this postulate of the cd-effect together with experimental support which we have provided has stimulated many other theoretical and experimental studies[5]. More generally the role of polar effects on the ground state of molecules and on the transition state of their homolysis became of interest.

Let us first remind the experimental facts which were one starting point of recognition of the cd-effect in our group :

The 1.3-Cl-H- rearrangement is a monomolecular reaction which, with negligibly low solvent effects, is not ionic and appears to be triggered by the presence of the captodative substitution on CCl_2-group (Fig. 1).

The captor group may be electron-withdrawing σ-substituents, e.g. trifluoromethyl and trichloromethyl groups[2b]. The rates are lowered ($\Delta\Delta E_a$ = 7.6 Kcal) as compared to an ester group, but the reaction is very clean (Fig. 2).

The rearrangement of oxalyl amide chloride leads to follow-up reactions thereby giving a trichloroimidazolium salt[6] (Fig. 3).

1

F. Minisci (ed.), Free Radicals in Synthesis and Biology, 1–26.
© 1989 by Kluwer Academic Publishers.

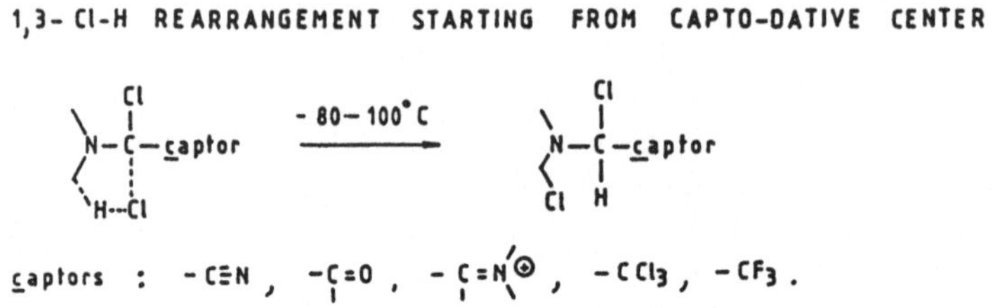
1,3- Cl-H REARRANGEMENT STARTING FROM CAPTO-DATIVE CENTER
- 80 – 100°C
captor
captors : -C≡N , -C=O , -C=N⊕ , -CCl3 , -CF3 .

Fig. 1

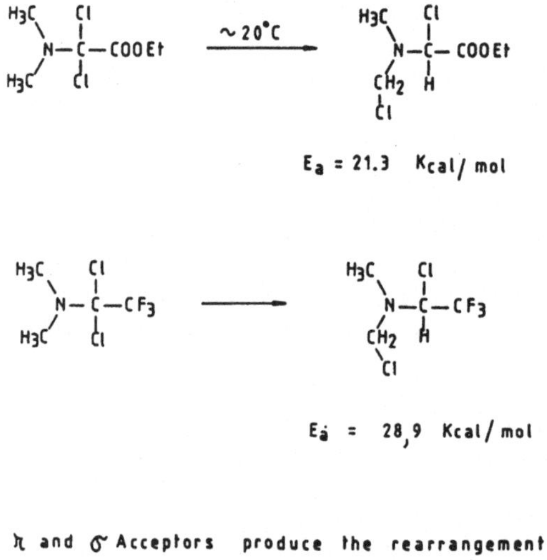
~20°C
COOEt
Ea = 21.3 Kcal/mol
CF3
Ea = 28,9 Kcal/mol
π and σ Acceptors produce the rearrangement

Fig. 2

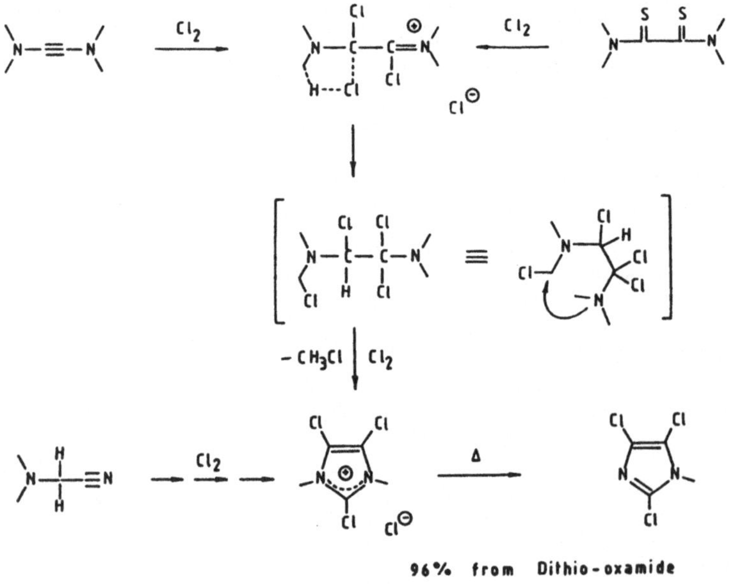
Cl2
Cl2
- CH3Cl
Cl2
Cl2
Δ
96% from Dithio-oxamide

Fig. 3

The CF_3-substituent in this series is of interest because this molecule contains a potential carbocation next to the CF_3 group and this feature can be exploited before and after the 1,3-Cl-H-rearrangement[7] (Fig. 4-5).

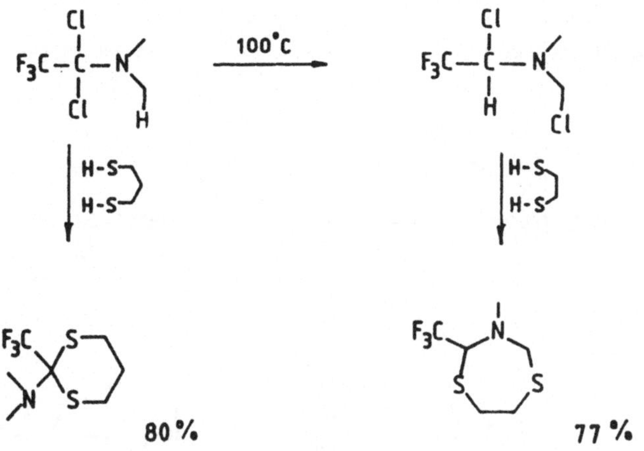

Fig. 4

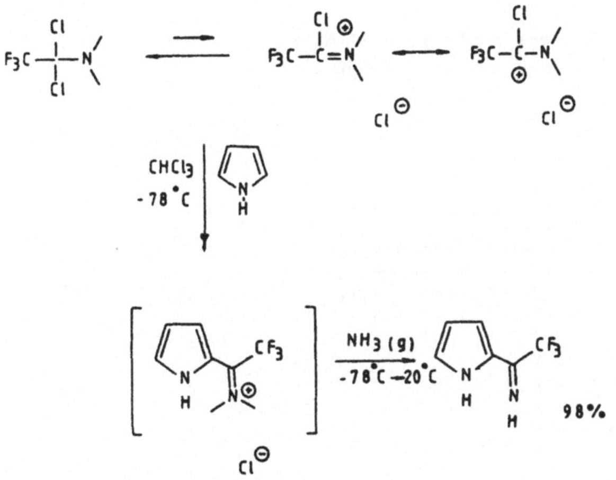

Fig. 5

Based on the cd-effect we reported on selective dehydrodimerisations[8] (Fig. 6).

Radical trapping by cd-olefins was established as a useful synthesis in which cd-stabilised adducts generally coupled to symmetrical dimers[9]. Thus both dehydrodimers or bridged dehydrodimers of crown ethers could be synthetised by this method[10] (Fig. 7-8).

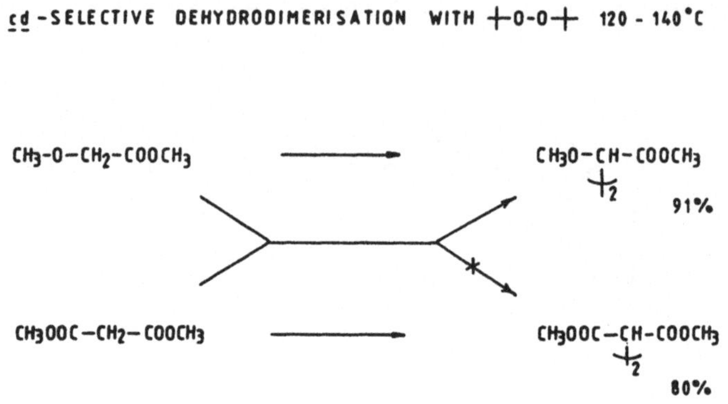

4

CH₃-O-CH₂-COOCH₃ ⟶ CH₃O-CH-COOCH₃ ⊢₂ 91%

CH₃OOC-CH₂-COOCH₃ ⟶ CH₃OOC-CH-COOCH₃ ⊢₂ 80%

Fig. 6

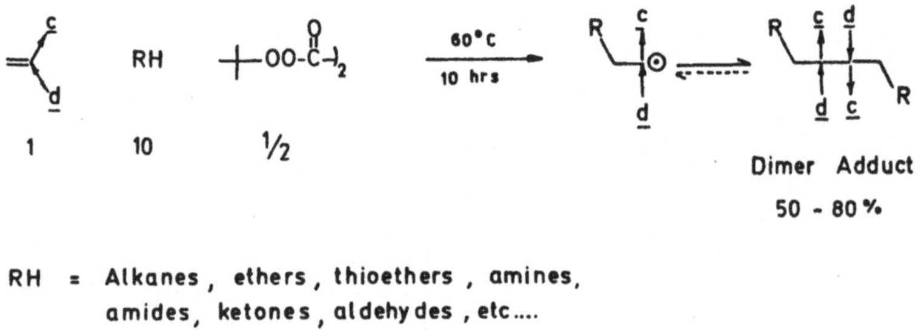

RH   ⊢OO-C⊣₂   60°C / 10 hrs   Dimer Adduct 50 - 80 %

1   10   ½

RH = Alkanes , ethers , thioethers , amines, amides , ketones , aldehydes , etc....

Fig. 7

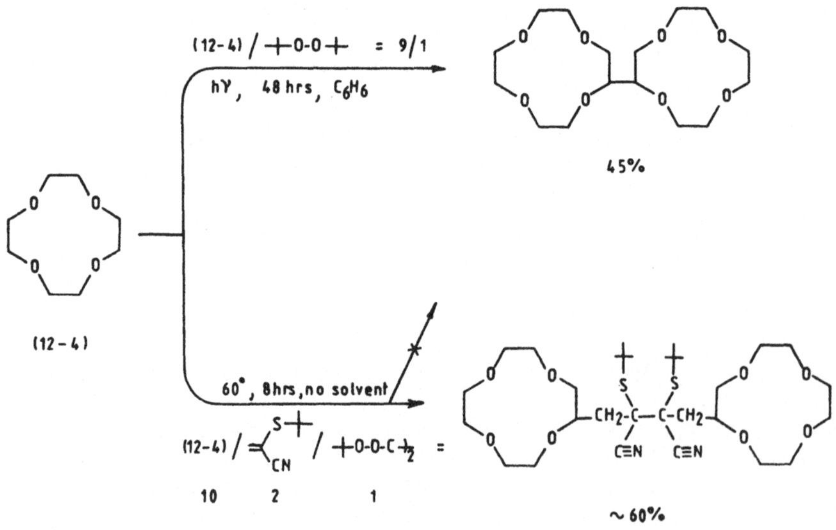

(12-4) / ⊢O-O⊣ = 9/1

hγ , 48 hrs , C₆H₆

45%

(12-4)

60° , 8hrs , no solvent

(12-4) / C=S⊣(CN) / ⊢O-O-C⊣₂ =

10   2   1

~60%

Fig. 8

cd-olefins were also recognised as good partners in (2+2)[11] and in (2+4)[12] cycloaddition reactions as summarised in the following formulation (Fig. 9).

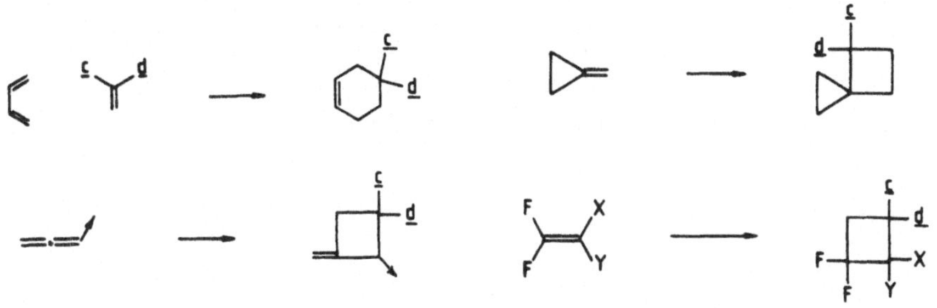

Fig. 9

Before developing in more detail the topic on cd-substitution effects in synthesis a glance at theory and on physico-chemical evaluation appears necessary. Substituent effects on radical stabilisation are generally much smaller than on ion stabilisation[13]. The stabilisation enthalpy of polar substituents varies approximately from 3 to 10 Kcal/mol[3c,14]. The effect of two substituents of same polarity is generally not additive. In contrast, captodative substitution may lead to additivity or even to synergism with values which are comparable to benzylic, propargylic or allylic stabilisation. The importance, however, of all these effects does not appear so much in the amount of the radical stabilisation energies but in the resulting reactivity and in selectivity for synthesis when radical intermediates are involved. The cd-effect is, however, still under discussion for its magnitude compared to dicapto (cc) or didative (dd) radical stabilisation. The following considerations shall throw more light on this question.

Quantification of substituent effects presents a general problem in chemistry because experimental values refer always only to the difference of states. For example, since the effect of polar substituents is different on the ground state of radical dimers and on the corresponding radicals there is no possibility to determine the absolute values of the cumulative influence of substituents by bond dissociation energy measurements. Furthermore, different methods of evaluation of radical stabilisation are expected to produce different quantities. In the best case only relative values can be obtained in adequate series.

The groups of Sustmann[15] and Walton[16] have shown that measuring barriers of rotation in allylic and propargylic radicals represents a reliable method for the evaluation of substituent effects. The transition state in the allylic rotation is a more localised system with greater influence of substituents than in the ground state, however, the influence is in the same direction for both states. These measurements support the cd-effect (Fig. 10).

Another more direct way of the evaluation of the effect of substituents on radicals derives from spin delocalisation measurements in their ground state by ESR. This method, however, does not furnish

6

	Kcal/mol E A	Δ EA

$$\overset{\odot}{H_2C=CH-CH}-D \qquad\qquad 15.7$$

$$\overset{\odot}{H_2C=CH-CH}-C\equiv N \qquad\qquad 10.2 \qquad\qquad 5.5$$

$$\overset{\odot}{H_2C=CH-CH}-O-CH_3 \qquad\qquad 14.5 \qquad\qquad 1.2$$

$$H_2C=CH-\overset{\odot}{\underset{\underset{C\equiv N}{|}}{C}}-O-CH_3 \qquad\qquad 6.0 \qquad\qquad 9.7$$

A barrier of 6.0 Kcal is found

Additivity would require 9.0 Kcal

Conclusion : There is a synergetic c d effect
of ∼3 Kcal / mol

Fig. 10

energy but delocalisation values. Still, the trend permits an estimation of substituent effects on radical stabilisation. These spectroscopic measurements support the postulate of the cd-effect with the conclusion: there is synergism of cd-substituents whereas for substituents with like polarity antagonism is observed[16b,17,18] (Fig. 11).

Sequence of radical stabilisation by disubstitution of a carbon centered radical

ΔH of dissociation : Allylic radical dimers

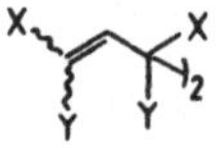

$$X,Y = NMe_2, CN > St\text{-}But, CN = SMe, CN > OEt, CN = OMe, CN \gg COOMe, COOMe$$

(This work).

Synergy on delocalisation parameters (E S R) :

Benzylic radicals

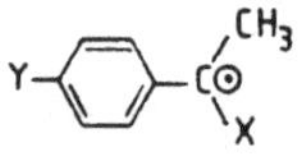

$$X,Y = CN, OMe > OMe, CN \gg CN, CN \geqslant OMe, OMe$$

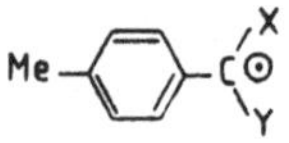

$$X,Y = OMe, CN \gg CN, CN > OMe, OMe$$

Allylic radicals

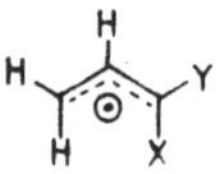

$$X,Y = OMe, CN \geqslant CN, OMe \gg CN, CN > OMe, OMe$$

Fig. 11

Complementary to spectroscopy we measured together with other groups bond-dissociation energies and kinetics for the evaluation of the cd-effect being well aware of the qualitative character of the conclusions therefrom. First, steric effects were of our concern. In cyclopropane these are minimised. Therefore we synthetised the then unknown bis-cd-cyclopropanes and compared their ease of cis-trans isomerisation with other cyclopropanes carrying polar substituents : cd-substitution leads to lower energies of isomerisation than cc-substitution[19] (Fig. 12).

CYCLOPROPANE CIS-TRANS ISOMERISATIONS

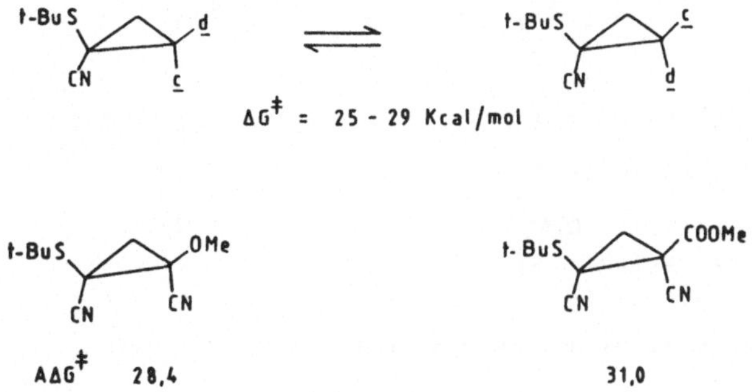

$\Delta G^{\ddagger}$ = 25 - 29 Kcal/mol

$\Delta \Delta G^{\ddagger}$ 28,4   31,0

Fig. 12

As another means of reducing steric effects we synthetised dimers of allylic radicals with bis-cd-substitution. These radicals enjoy by allylic conjugation the influence of two cd-substituent pairs on their radical center. Compared to captor substituted radical dimers the bis-cd-allyl compounds homolyse easier[20] (Fig. 13).

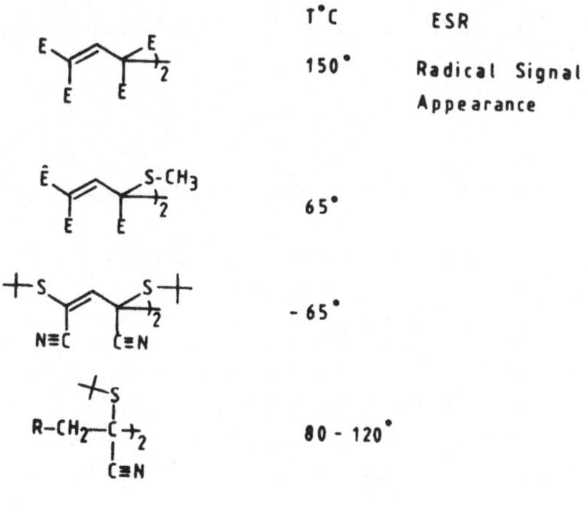

Fig. 13

Besides steric effects the influence of σ-interactions on the energy level of a dimer under study for homolysis is important. It leads to stabilisation or to destabilisation. Thus the anomeric effect of two gem-alkoxy pairs in the tetramethoxyethane[21] is in the order of 12-14 Kcal/mol stabilisation whereas tetracyanoethane could be destabilised by up to 30 Kcal/mol[13a]. Leroy and his group have done computations concerning this problem[13a,14] and also discuss the example of persistency of the amino-dicyanomethyl radical as deriving not only from dicapto-dative thermodynamic radical stabilisation but especially from the destabilisation of its dimer by steric and polar effects (Fig. 14).

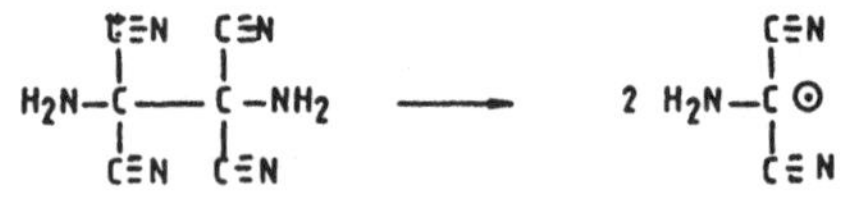

$\Delta H_f^\bullet (R^\bullet)$ 89.55   BDE $(c-c)$ = 16.36

$\Delta H_f^\bullet (R-R)$ 162.34 $(E_a)r$ = 6.39

SE $(R^\bullet)$ 12.08 ΔH_d = 22.75

SE $(R-R)$ -44 53

Radical-Persistancy because Destabilisation of Dimer

G. LEROY 1986

Fig. 14

With such proximity effects in mind we studied the dl-meso isomerisation of benzylic radical[22] dimers where only polar substituents in the para position are varied (Fig. 15).

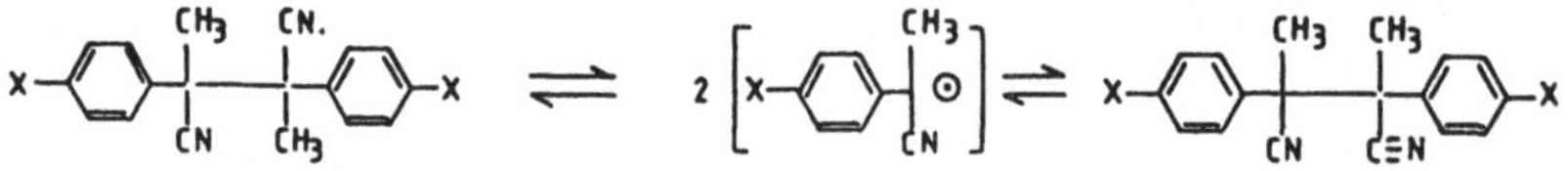

Para-Substituent Variation keeps steric effects constant and

minimises polar proximity and groundstate effects.

Fig. 15

This "para variation" system remains sterically constant in the dimer and at the arising benzylic radical center and avoids as much as possible disturbing polar proximity effects which arise unavoidably when varying substituents in benzylic positions. This is furthermore influenced by the phenyl group as the third stabilising and interfering

substituent which should preclude the reliable evaluation of polar substituent effects.
  This last system was chosen by Rúchardt and his group[23a] presuming
additivity of substituent effects (Fig. 16). The then observed high

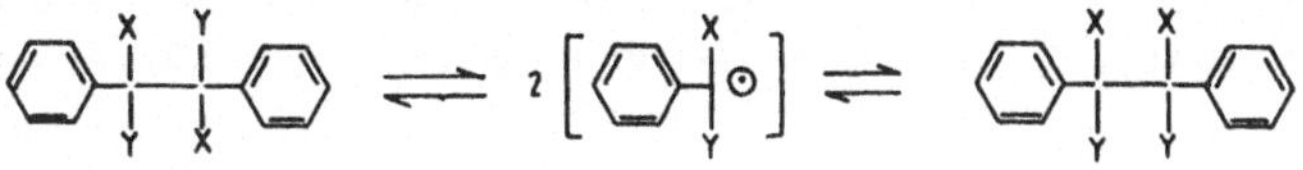

Gem-disubstituted Benzylic Radical Dimers and their
Radicals are under the direct varying steric and polar
influence of changing substituent.

Fig. 16

enthalpy of dissociation (51.4 Kcal/mol) of the gem-dimethoxy benzylic
dimer could be explained by the anomeric effect. As a further
consequence we must consider the influence of four cyano substituents on
the ground state of the benzylic dimer. One expects a strong
destabilisation and therefore easier dissociation ($\Delta H^{\ddagger}$ = 26 Kcal/mol)[23b]:
if indeed the cd-dimer ($OCH_3/C \equiv N$; $\Delta H^{\ddagger}$ = 36.6 Kcal/mol) requires a
higher enthalpy of activation for dissociation this can be explained as
only a seeming exception to the cd-effect. In the para substituted
benzylic system the cd-effect is confirmed : the phenylogous cd-com-
pounds isomerise faster than the cc-derivatives (Fig. 17).

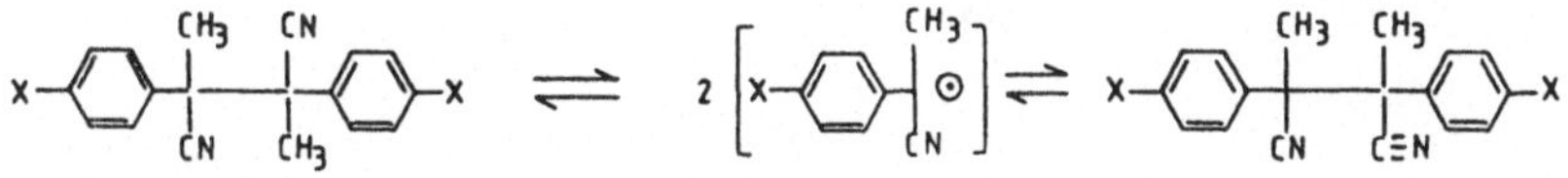

Para-Substituent Variation keeps steric effects constant and

minimises polar proximity and groundstate effects.

X	$\Delta G^{\ddagger a}$	$t\frac{1}{2}^{b}$	X	$\Delta G^{\ddagger a}$	$t\frac{1}{2}^{b}$
H	33.3	86	OMe	32.3	26
Cl	32.9	50	SMe	31.8	15
CN	32.8	45	NH$_2$	30.7	4
COOMe	32.8	45	NMe$_2$	30.0	2

a) Kcal/mol (160°C)   b) Minutes (160°C)

Fig. 17

As the table shows the effect is neat although it is expectedly small because attenuated by the phenylogous arrangement of substituents. Here also, as proposed by Roduner et al.[18] for spin-density measurements, a "substituent interaction parameter" Δ_{XY} can be defined. It is calculated from the G_X/G_H ratio taking account of the effect of monosubstitution (Δ_X). The Δ_{XY} parameter is negative in case of antagonism and positive if there is synergy (Fig. 18-19).

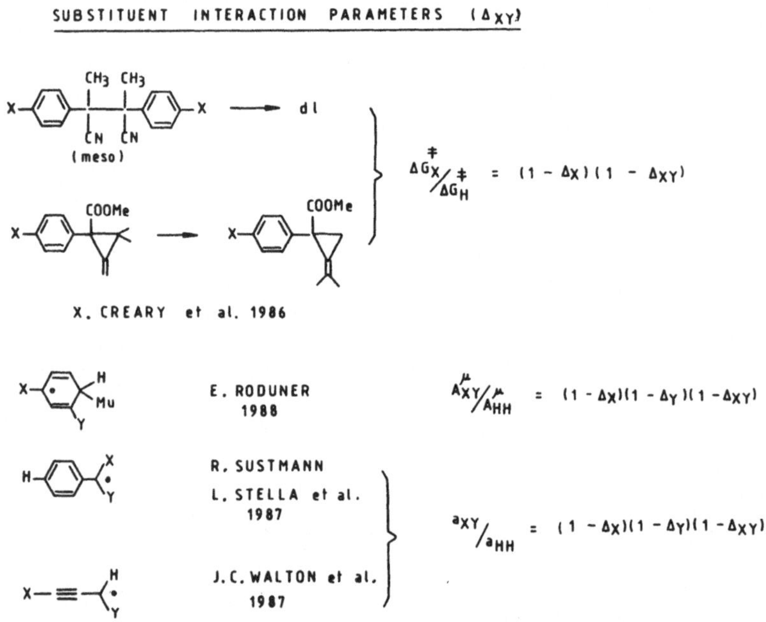

SUBSTITUENT INTERACTION PARAMETERS (Δ_{XY})

Fig. 18

For comparison the same Δ_{XY} parameter can be derived from the kinetics of the methylenecyclopropane rearrangement[24] and the different ESR and Mu-resonance studies[16-18].

Without exception two substituents of the same polarity show antagonism and practically all cd-couples give positive Δ_{XY} values meaning synergy.

It is not our purpose to go into further details of quantification of the cd-effect here. It must again be underlined that generally all substituent effects in thermodynamic radical stabilisations are small. Together with all the other effects they may be important and useful in synthesis which is the real topic here.

Another example for the cd-effect derives from cd-cyclopropane chemistry. We have already reported that cd-substitution makes cyclopropanes cis-trans isomerise more readily than with cc-substitution. In agreement we have found a more facile rearrangement of a vinyl-cd-cyclo-propane to cd-cyclopentenes if compared to dicapto-analogs[25-28] (Fig. 20).

This example is related to the question of the homo-allylic

SUBSTITUENT INTERACTION PARAMETER (Δxy·100)

X/Y	X—⟨Ph⟩—Ċ(Me)(CN)	X—⟨Ph⟩—Ċ(COOEt)	X—Ċ(H)(Mu)	H—⟨Ph⟩—Ċ(X)(Y)	X—≡—Ċ(H)(Y)
CN / CN	- 2.4			- 12.5	
COOMe / CN	- 2.1			- 12.0	
COOR / COOR		~ 0.8		- 12.0	
SOMe / COOEt		- 0.6			
SO₂Me / COOEt		- 0.6			
OMe / OMe			- 3.4	-13.6	
Me / NH₂					- 7.7
Me / OH					- 3.3
OMe / CN	+ 1.0		+ 5.17	+ 16.0	
CN / OMe			+ 6.98		
SR / CN	- 0.3			+ 13.0	
NH₂ / CN	+ 4.2			+ 9.6	
COOEt / N(SiMe)₃					+ 12.8
COOEt / OH					+ 7.4
OMe / COOEt		+ 2.6			
SMe / COOEt		+ 0.8			

Fig. 19

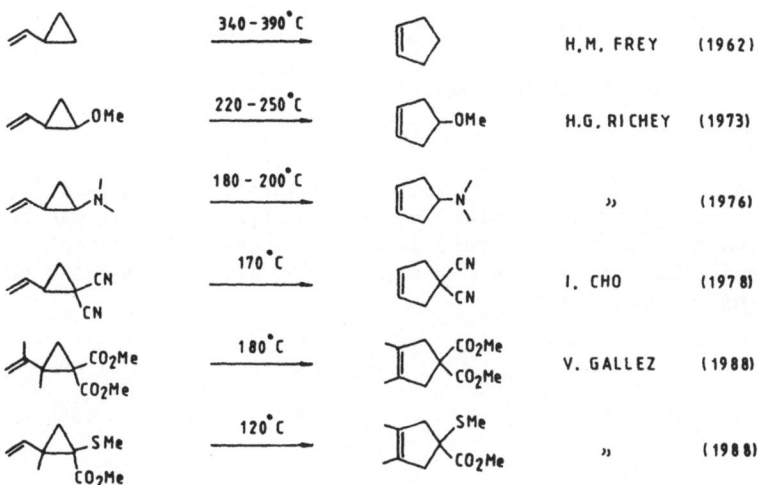

	340 - 390°C		H.M. FREY (1962)
-OMe	220 - 250°C	-OMe	H.G. RICHEY (1973)
-N<	180 - 200°C	-N<	" (1976)
-CN, CN	170°C	-CN, CN	I. CHO (1978)
-CO₂Me, CO₂Me	180°C	-CO₂Me, CO₂Me	V. GALLEZ (1988)
-SMe, CO₂Me	120°C	-SMe, CO₂Me	" (1988)

Fig. 20

rearrangement of cyclopropyl-methyl radicals under the influence of cd-substituents.

Thus cyclopropyl-methyl radicals are well known for their isomerisation to homo-allylic radicals at rather low temperatures[29] (Fig. 21).

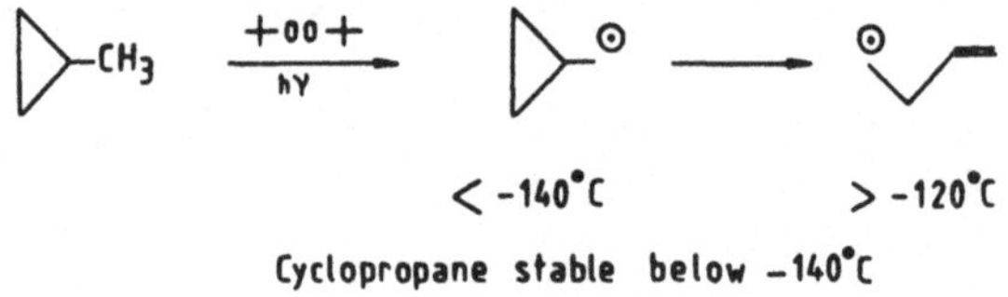

$< -140°C$ $> -120°C$

Cyclopropane stable below −140°C

Fig. 21

J. Walton et al.[30] and independently our group[28] have shown that cd-substitution on the cylopropyl-methyl radical shows according to ESR-Spectroscopy a remarkable stabilising effect. This observation invites for preparative control (Fig. 22).

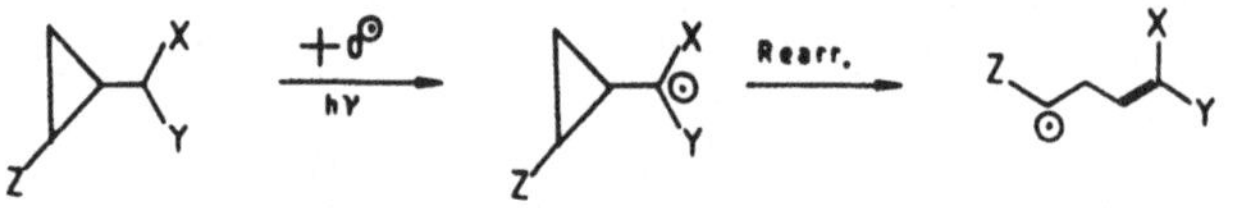

X	Y	Z	Cyclopropane stable until
(CH₃)₃SiO	CN	H	+ 32°C ∗
(CH₃)₃SiO	CO₂Me	H	+ 135°C ∗
Et₂N	CN	CO₂Me	+ 20°C
Et₂N	CO₂Me	CO₂Me	+ 20°C
+S	CN	CO₂Me	+ 20°C
+S	CO₂Me	CO₂Me	+ 20°C

∗   John C. WALTON (1986)

Fig. 22

In one interesting case we found that in spite of cd-stabilisation on the cyclopropyl methyl radical the homo-allylic radical can be obtained again if further cd-substitution is conferred to the cyclopropane ring. It becomes thereby proradical and undergoes the ring-opening[31] (Fig. 23).

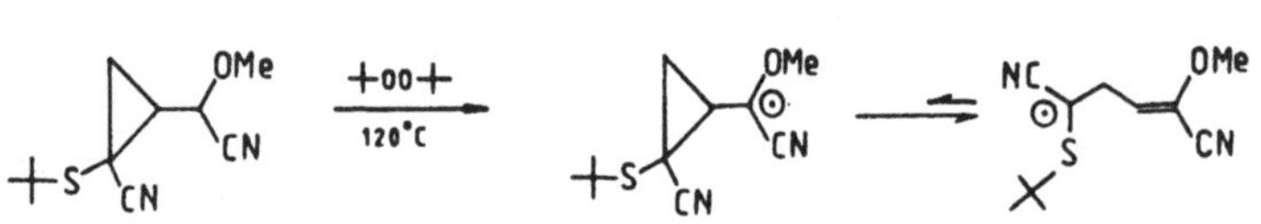

Fig. 23

A report[32] on a system which behaves rather as a cd-stabilised 1.3 diradical than as a 1.3 dipole may be of interest here : Instead of the expected intramolecular dipolar cycloaddition a head-to-head cyclo-dimerisation rationalises the result (Fig. 24).

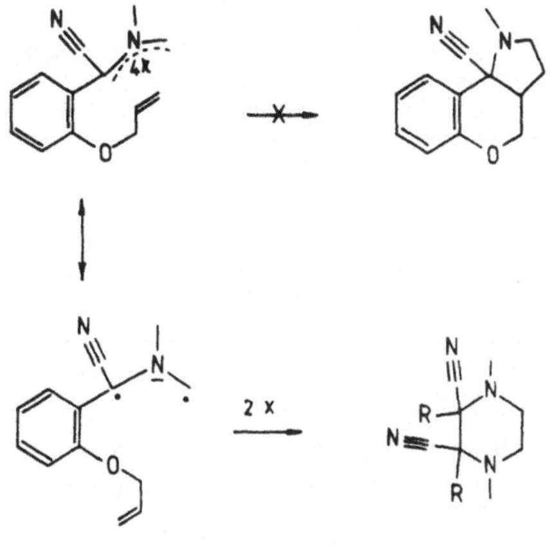

PADWA 1986

Fig. 24

Of the many cycloaddition types, which have been mentioned above for cd-olefins, the remarkable head-to-head cyclodimerisation of acrylonitrile α-thioethers represents the longest known case[33] (Fig. 25).

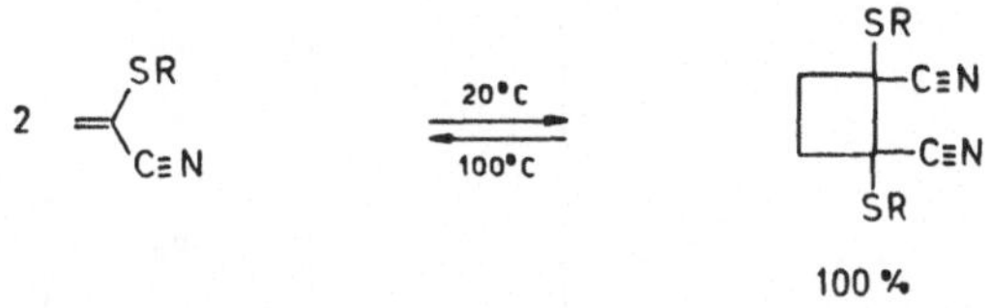

100 %

K.D. GUNDERMANN 1956

Fig. 25

Intramolecular (2+2) cycloadditions have been reported by Alder and Bellus to occur the easiest with cd-olefins[34] (Fig. 26).

We have studied (2+2) cycloadditions of various fluoro-ethenes to cd-olefins. A comparison shows the beneficial cd-effect[11a] (Fig. 27).

Since we introduced 1,1-difluoro-2,2-dithioether ethenes[35] as handy reagents for such reactions, they have also been used for comparing the influence of cd-substituents on acrylonitrile[11] (Fig. 28).

cd substituted allenes have been prepared in our group and their behaviour in various cycloadditions was examined[36]. As expected they cyclodimerise already at room temperature but can be intercepted with cd-olefins, which in many variations become synthetically more useful in cycloadditions with allene itself or with its more accessible derivatives (Fig. 29-30).

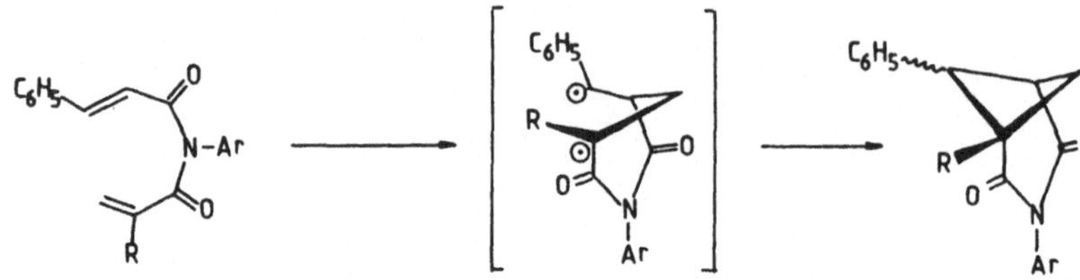

R	Cdts	yield (%)
CH$_3$	140°C , 4 hrs	75
SCH$_3$	140°C , 5 min	91
SCH$_3$	40°C , 3 days	90

Fig. 26

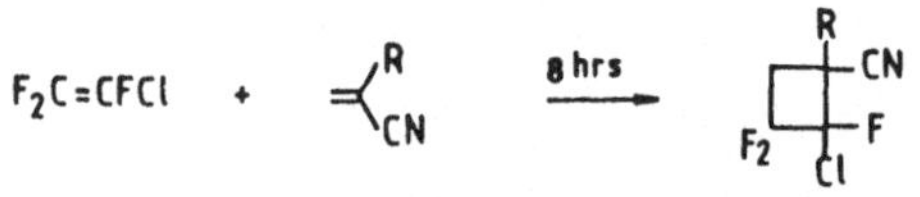

R	Temp °C	Yield %
H	160	65
Cl	120	72
SePh	120	70
SEt	120	83
SPh	120	85
S十	120	90

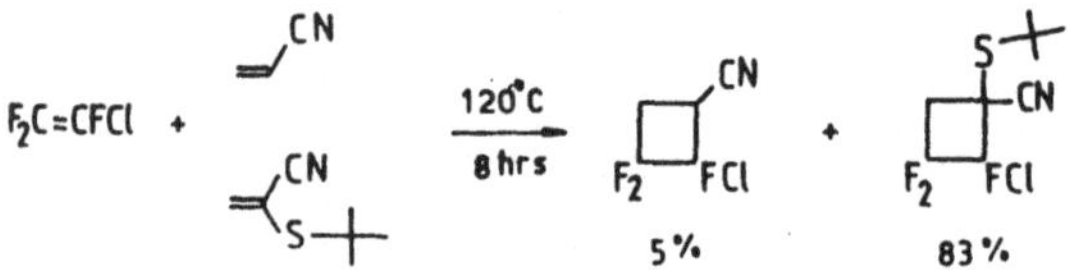

Fig. 27

The head-to-head adducts may rearrange thermally with varying ease depending on the efficiency of cd-substituent couples (Fig. 31).

Using the adduct of allene to α-chloro acrylonitrile β-cyano-cyclo-butenone has been obtained with remarkably temperature dependent react-ivity either as a dienophile at room temperature or as cyano-vinylketene at 80°[37] (Fig. 32).

α-Thioalkyl-acrylonitriles show in polar additions to ynediamines

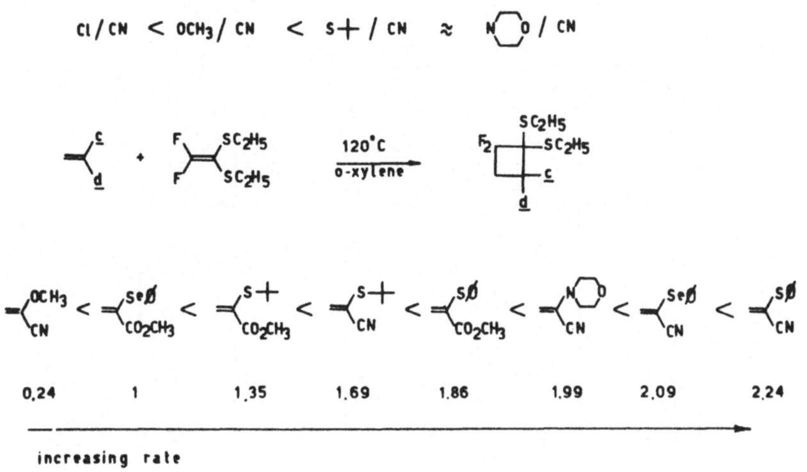

$Cl/CN \; < \; OCH_3/CN \; < \; S{+}/CN \; \approx \; N\!\!\bigcirc\!O \, /CN$

(reaction scheme: 120°C, o-xylene)

$\cdots OCH_3 / CN < \cdots Se\phi / CO_2CH_3 < \cdots S{+} / CO_2CH_3 < \cdots S{+} / CN < \cdots S\phi / CO_2CH_3 < \cdots N\bigcirc O / CN < \cdots Se\phi / CN < \cdots S\phi / CN$

| 0,24 | 1 | 1,35 | 1,69 | 1,86 | 1,99 | 2,09 | 2,24 |

increasing rate

Fig. 28

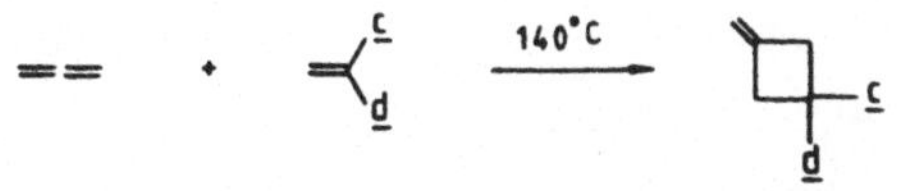

(reaction scheme: 140°C)

c	d	yield %	time (days)
-CN	-S$+$	86	2
-CN	-Cl	66	6
-CO$_2$CH$_3$	-S$+$	42	6
-CO$_2$CH$_3$	-Cl	39	6

Fig. 29

higher reactivity than acrylonitrile itself. The cyclobutene adduct
itself does not undergo ring-opening to the cd-diene but rather isomer-
ises first to another cyclobutene which apparently is much more prone to
opening because of its favorable influence of substituents on presumably
dipolar or diradical transition states[38] (Fig. 33).

The remarkable substituent influence on cyclobutene-butadiene isomerisations is furthermore demonstrated by the analogous reaction where
the thiotert-butyl ether group is replaced by a phenylthio group. The
primary adduct undergoes again a 1,3-thioether rearrangement to a
push-pull stabilised cyclobutene which does not open. It can be either
isolated as such or hydrolysed to the amino substituted cyanocyclobute-
none. This product has a much greater thermal stability than the analog-

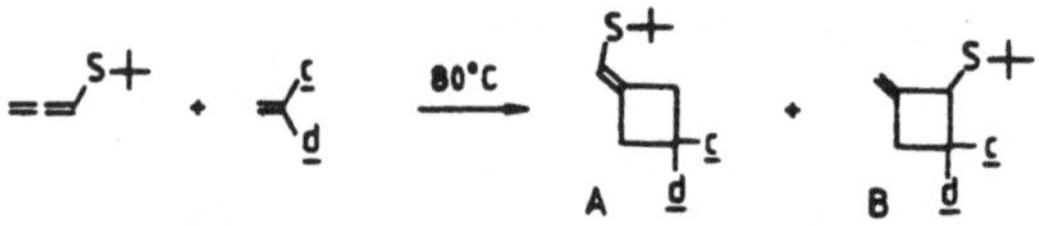

c	d (R)	global yield %	A %	B %
-CN	-N<	49	82	18
-CN	-N O (morpholine)	75	88	12
-CN	-S+	95	70	30
-CN	-OCH$_3$	73	70	30
-CN	-Cl	48	72	28
-CO$_2$CH$_3$	-S+	60	65	35
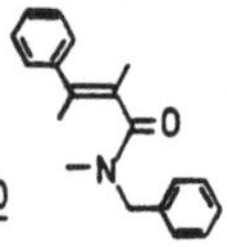 **10**	(structure)	44	100	0
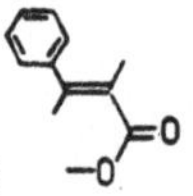 **11**	(structure)	47	88	12
-CN	-CH$_3$	38	86	14
-CN	-H	59	75	25

Fig. 30

c	d	$\Delta H^{\neq}$ kcal / mol	$\Delta S^{\neq}$ cal / K.mol	$\Delta G^{\neq}_{300}$ kcal / mol
-CN	-N O	28.6 ± 1.8	+ 7.9 ± 6.9	26.2 ± 3.8
-CN	-S+	24.9 ± 1.5	- 5.1 ± 5.8	26.4 ± 3.2
-CN	-OCH$_3$	44.2 ± 0.6	+ 36.1 ± 2.2	33.4 ± 2.8

Fig. 31

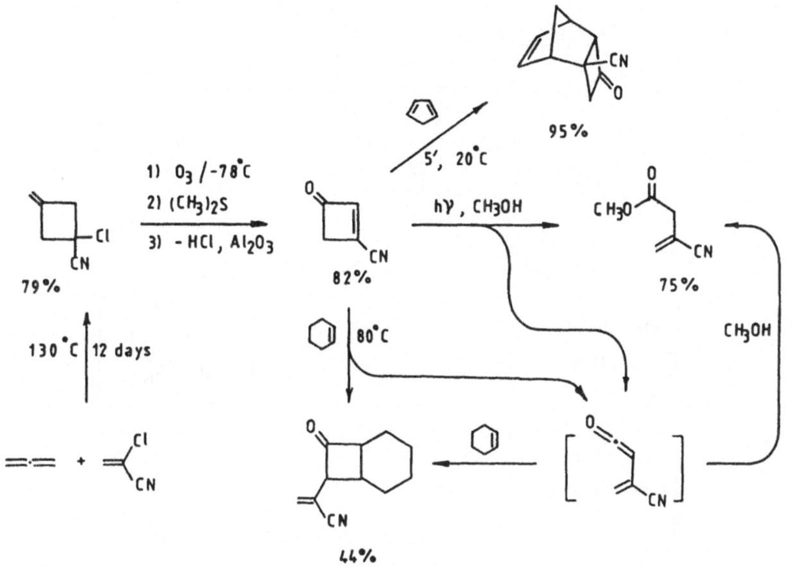

Fig. 32

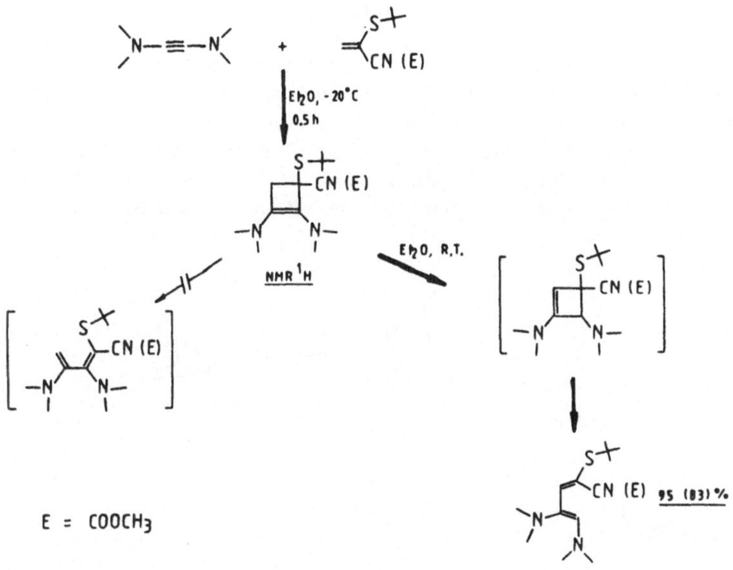

E = COOCH₃

Fig. 33

ous compound without the amino group perhaps because of the push-pull stabilisation[38] (Fig. 34).

Before the cd-postulate for radical stabilisation nitrosoolefins were a topic of interest in our group within the general framework of

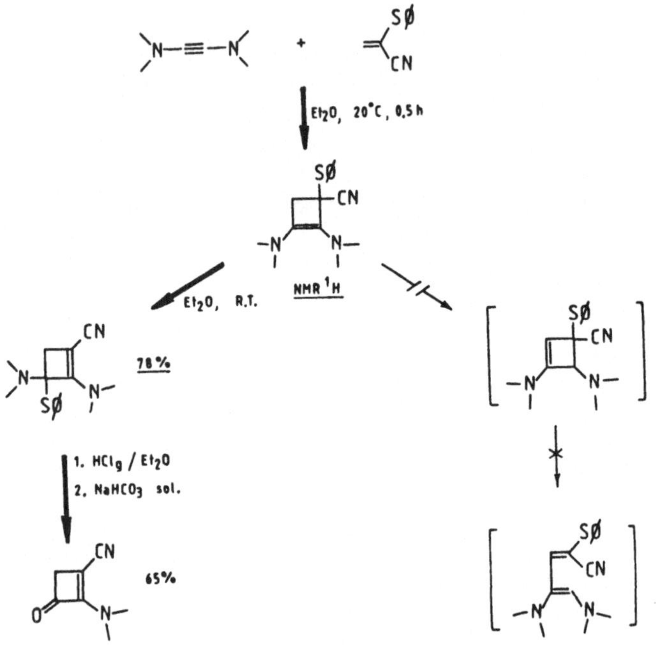

Fig. 34

studies on polar substituent effects on multiple bonds. The comparison
of the nitroso group with singlet-oxygen in (4+2) cycloadditions led to
the discovery of the oxazine adduct rearrangement to epoxy-epimines[39].
Their structure and the stereospecificity of this reaction were proven
by x-ray analysis[40]. The cyclopentadiene adduct rearranged at room
temperatures whereas the cyclohexadiene adduct required 60°[41] (Fig.
35-36).

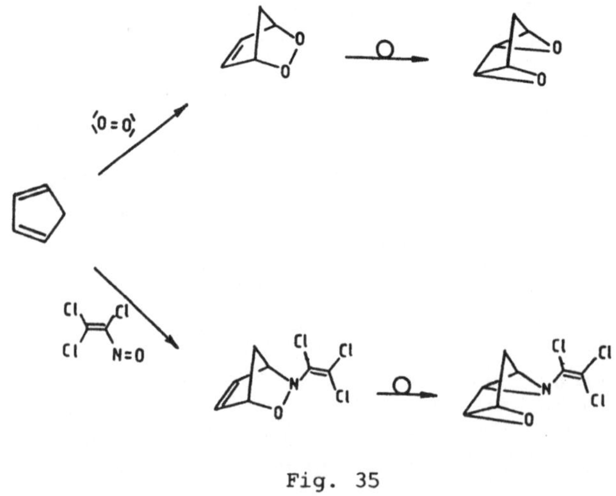

Fig. 35

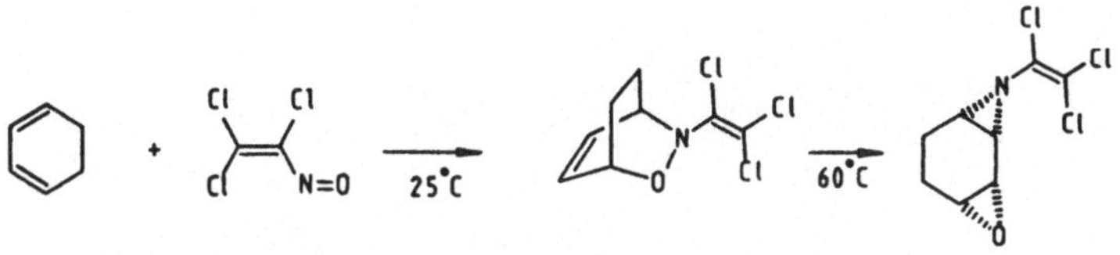

FRANCOTTE , VIEHE 1978

Fig. 36

Studying the scope of the oxazine-epoxy-epimine rearrangement limitations were found first in the difficulties for access to nitroso ethenes with appropriate substitution[42] ; furthermore these fragile compounds had to be trapped before their decomposition. The following table shows the cases where the vinyl group carries only phenyl and/or halogen substituents and the rearrangement proceed with the yields indicated[42a,b] (Fig. 37).

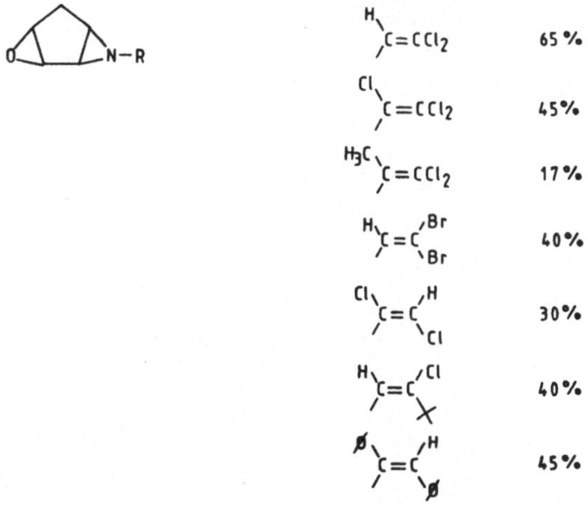

$H\backslash C=CCl_2$	65 %
$Cl\backslash C=CCl_2$	45%
$H_3C\backslash C=CCl_2$	17%
$H\backslash C=C\,^{Br}_{Br}$	40%
$Cl\backslash C=C\,^{H}_{Cl}$	30%
$H\backslash C=C\,^{Cl}$	40%
$\varnothing\backslash C=C\,^{H}_{\varnothing}$	45%

Fig. 37

The mechanistic rationale for the N-O-bondbreaking under the influence of substituents necessitates either ground state destabilisation of the oxazine derivative or stabilisation of intermediates (Fig. 38).

Thus more varied substitution of oxazines could be achieved by avoiding the limitations of the approach via nitroso-olefins. N-substituted oxazines have proven to be much more versatile[42,43] starting materials (Fig. 39).

These are indeed easily accessible through a one-pot reaction between cyclodienes and α-chloro-nitroso-cyclohexene - the latter being available by chlorination of cyclohexanone oxime. The oxazine adducts cleave to the hydrochlorides of bicyclic N-H-oxazines in ethanol[44] (Fig. 40).

20

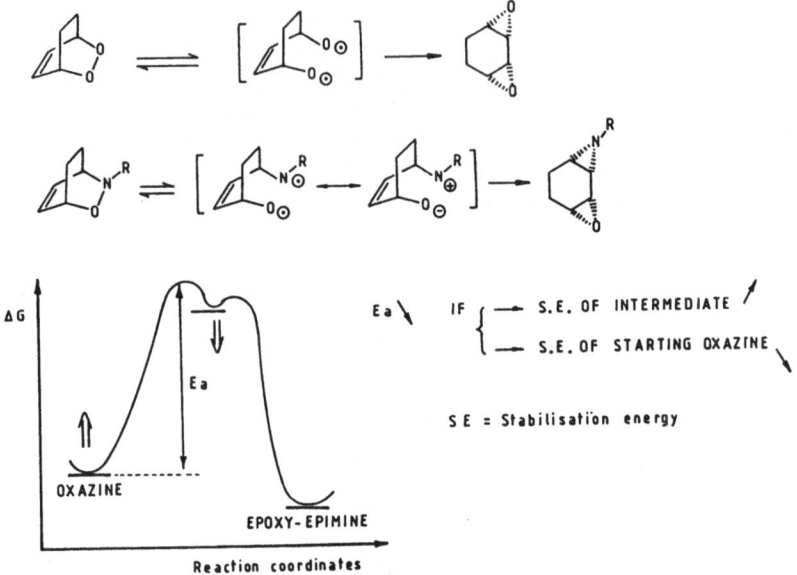

Fig. 38

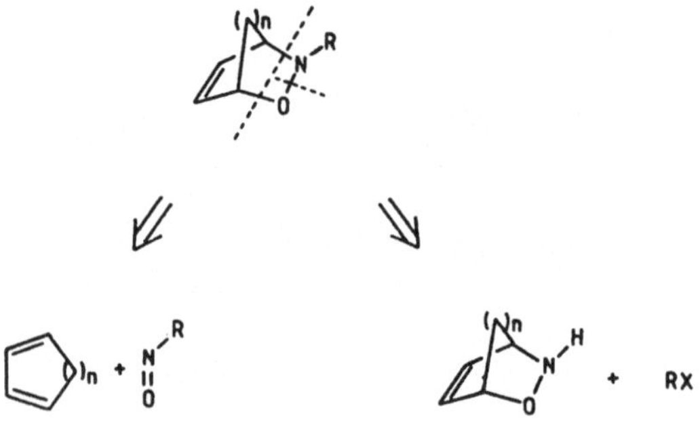

Fig. 39

These compounds permit facile derivatisation via nucleophilic substitution. The arising compounds permit many other transformations besides the desired rearrangement which leads to epoxy-epimines. Thus finally the application of the cd-concept brought success. Indeed cd-vinyl substitution with β-chloro-α-thio-tert-butyl-acrylonitrile permitted the desired smooth rearrangement at 100°C[43,45] (Fig. 41).

The contrast in particularly with the analogous β,β-dicyano-vinyl and β,β-diester derivatives is remarkable : these dicapto derivatives do not rearrange[43]. The activation parameters and the negligible solvent effect indicate that the rearrangement may proceed via radicals (Fig. 42).

There are other facts which support a mechanism with diradical intermediates. Since the epoxy-epimine rearrangement made radical

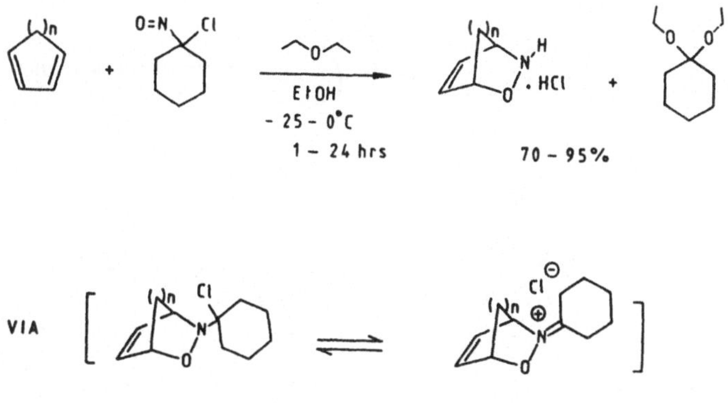

Fig. 40

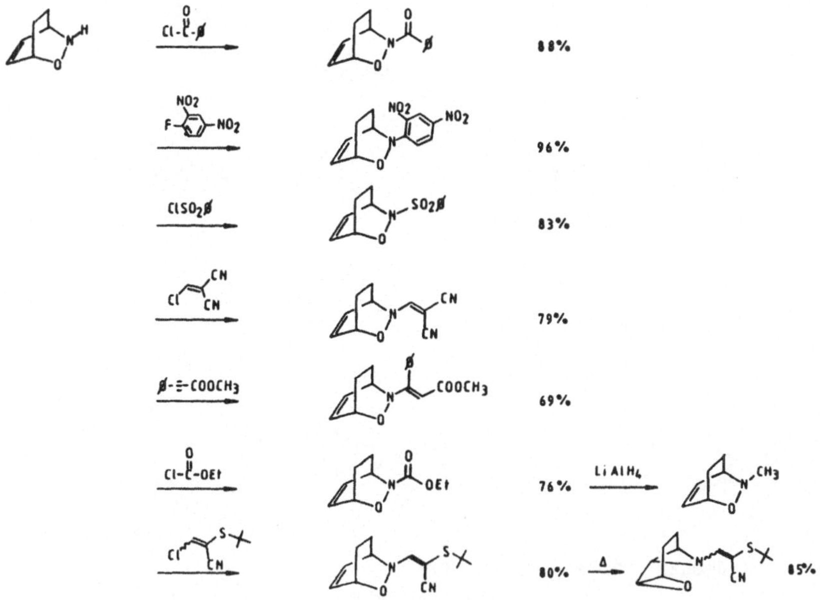

Fig. 41

trapping rather difficult, by diimine reduction the dihydrooxazine was prepared which cannot rearrange. It could be shown by photo-electron spectroscopy of both the reduced and the original unsaturated oxazine that N-O bond and double bond do not interact[46]. Then it was found that heating the cd-vinyl substituted dihydro-oxazine in xylene still led to homolysis and to subsequent hydrogenation of the N-O bond. Without cd-vinyl substitution the acetylated dihydro-oxazine did not homolyse (Fig. 43).

The cd-vinyl group promoted oxazine rearrangement to epoxy-epimines succeeds with bicyclic adducts derived from cyclopentadiene, cyclohexa-

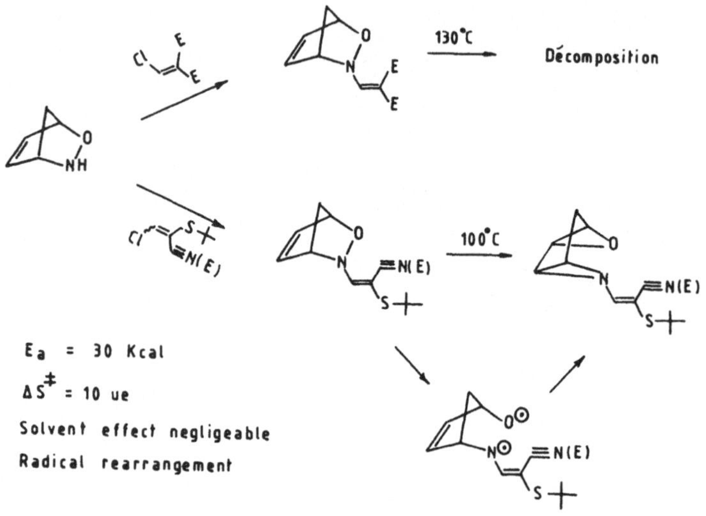

22

Fig. 42

HYDROGEN DONORS = XYLENE
XYLENE / Ø₃CH
XYLENE / ØSH

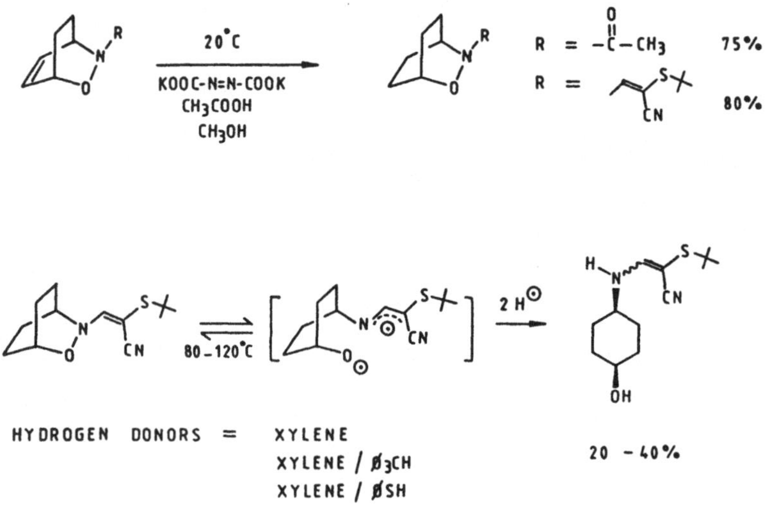

Fig. 43

diene and with cycloheptadiene but not from cyclo-octadiene or from
butadienes[45]. The reaction obviously follows the endoperoxide analogy[47]
and is favoured by molecular strain (Fig. 44).

N-unsubstituted **chiral** oxazines with two chiral centers are readily
available from cyclodienes[48] and permit by cd-vinylation their
stereospecific epoxy-epimine rearrangement to cyclanes with now four
chiral centers at the originally four sp2-carbon atoms of the diene. It
could be shown by chiral shift reagents that the chiral tranfer attains
at least ∿ 96 % ee[43] (Fig. 45).

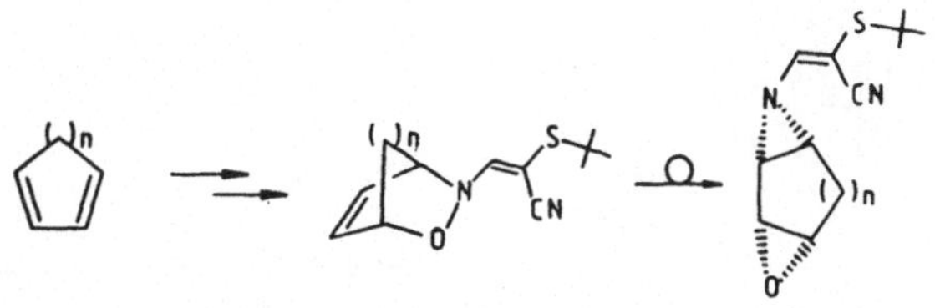

n	T (°C)	Yield (%)**
1	0	40
2	80	85
3	140	75
4	140 or F.V.P.*	0

* FLASH VACUUM PYROLYSIS UP TO 500°C

** ISOLATED PRODUCT

Fig. 44

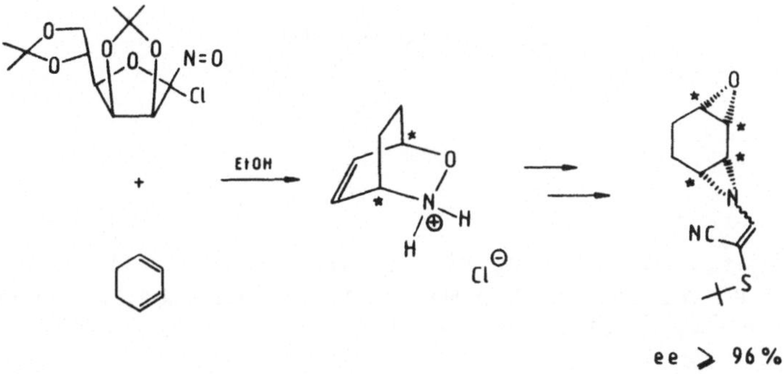

ee ≥ 96%

Fig. 45

A collaboration[49] with the late G. Kresze and his group has shown, that when using 1,2 dioxy-cyclohexadiene derivatives a new efficient stereocontrolled synthesis of amino cyclitols and of inosamine becomes possible. The latter compounds are building blocks for certain antibiotics (Fig. 46).

While this work is still under way it is rewarding to conclude that the cd-postulate has been very stimulating and useful also in organic synthesis.

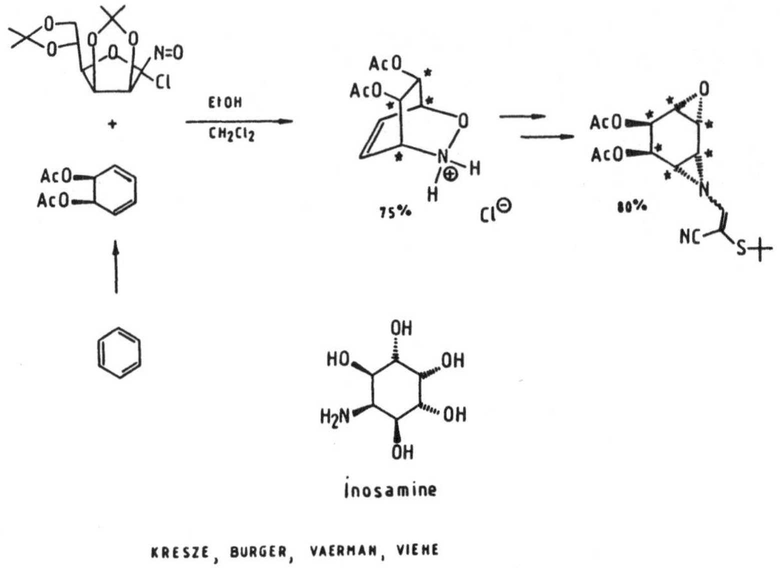

24

inosamine

KRESZE, BURGER, VAERMAN, VIEHE

Fig. 46

REFERENCES

1. Captodative Substituent Effects Part 44; Part 43 see B. Tinant, S. Wu, J.P. Declerq, M. Van Meerssche, W. Masamba, A. de Mesmaeker and H.G. Viehe, J. Chem. Soc. Perkin Trans. II accepted for publication.
2a. F. Huys, R. Merényi, Z. Janousek, L. Stella, H.G. Viehe, Angew. Chem. 91, 650 (1979); Internat. Edit. English 18, 615 (1979).
2b. M. Rover-Kevers, L. Vertommen, F. Huys, R. Merényi, L. Stella, Z. Janousek, H.G. Viehe, Angew. Chem. 93, 1091 (1981); Internat. Ed. English 20, 1023 (1981).
3a. L. Stella, Z. Janousek, R. Merényi and H.G. Viehe, Angew. Chem. 90, 741 (1978); Internat. Ed. Engl. 17, 691 (1978).
3b. H.G. Viehe, Z. Janousek, R. Merényi, L. Stella, Acc. Chem. Res. 18, 148 (1985).
3c. R. Merényi, Z. Janousek, H.G. Viehe in ref. 5, page 301.
4. M.J.S. Dewar, J. Am. Chem. Soc. 74, 3353 (1952); A.T. Balaban, Rev. Roum. Chim. 16, 725 (1971); R.W. Baldock, P. Hudson, A.R. Katritzky, F. Soti, J. Chem. Soc. Perkin Trans. 1, 1422 (1974).
5. Substituent Effects in Radical Chemistry, H.G. Viehe, Z. Janousek, R. Merényi, Editors, NATO ASI Series C189, D. Reidel Publisher, Dordrecht (1986).
6. Z. Janousek, F. Huys, L. René, M. Masquelier, L. Stella, R. Merényi, H.G. Viehe, Angew. Chem. 91, 651 (1979); Internat. Ed. Engl. 18, 615 (1979).
7a. M. Rover-Kevers, Ph.D. Thesis, Louvain-la-Neuve (1980).
7b. Ph. Dellis, Ph.D. Thesis, Louvain-la-Neuve (1987).
7c. M.A. Decock-Plancquaert, Ph.D. Thesis in preparation.

8.   M. Beaujean, Ph.D. Thesis in preparation.
9a.   S. Mignani, M. Beaujean, Z. Janousek, R. Merényi, H.G. Viehe,
      Tetrahedron 37, Suppl. 1, 111 (1981).
9b.   S. Mignani, Z. Janousek, R. Merényi, H.G. Viehe, Bull. Soc. Chim.
      France ,1267 (1985).
10.   M. Beaujean, S. Mignani, R. Merényi, Z. Janousek, H.G. Viehe, M.
      Kirch, J.M. Lehn, Tetrahedron 40, 4395 (1984).
11a. Ch. De Cock, S. Piettre, F. Lahousse, Z. Janousek, R. Merényi, H.G.
      Viehe, Tetrahedron 41, 4183 (1985).
11b. A. De Meijere, H. Wenck, F. Seyed-Mahdavi, H.G. Viehe, V. Gallez,
      I. Erden, Tetrahedron 42, 1291 (1986).
12a. J.L. Boucher, L. Stella, Tetrahedron, 3871 (1986).
12b. L. Stella in ref. 5, p. 361.
13a. G. Leroy, Adv. Quantum Chem. 17, 1, (1985).
13b. G. Leroy, C. Wilante, D. Peeters, M.M. Uyewa, J. of Molecular
      Struct. (theochem) 124, 107 (1985).
14.   G. Leroy, D. Peeters, M.Sana, C. Wilante in ref. 5, p. 1.
15a. H.G. Korth, P. Lommes, R. Sustmann, J. Am. Chem. Soc. 106, 663
      (1984).
15b. R. Sustmann in ref. 5, p. 143.
16.   I. MacInnes, J.C. Walton, D.C. Nonhebel, J. Chem. Soc. Chem.
      Commun. 712 (1985); I. MacInnes, J.C. Walton, J. Chem. Soc. Perkin
      Trans. II, 1077 (1987).
17a. L. Sylvander, L. Stella, H.G. Korth, R. Sustmann, Tetrahedron Lett.
      26, 749 (1985).
17b. H.G. Korth, P. Lommes, R. Sustmann, L. Sylvander, L. Stella, New
      Journal of Chemistry 11, 365 (1987).
18.   Ch.J. Rhodes, E. Roduner, Tetrahedron Lett. 29, 1437 (1988).
19a. A. De Mesmaeker, L. Vertommen, R. Merényi, H.G. Viehe, Tetrahedron
      Lett. 23, 69 (1982).
19b. R. Merényi, A. De Mesmaeker, H.G. Viehe, Tetrahedron Lett. 24, 2765
      (1983).
20.   M. Van Hoecke, A. Borghese, J. Penelle, R. Merényi, H.G. Viehe,
      Tetrahedron Lett. 27, 4569 (1986).
21.   N.M. Gutner, N.D. Lebedeva, S.L. Dobychin, N.N. Kiseleva, Zhur.
      Prikl. Khim. 53, 2061 (1980); Engl. Transl. p. 1523.
22a. R. Merényi, V. Daffe, J. Klein, W. Masamba, H.G. Viehe, Bull. Soc.
      Chim. Belge 91, 456 (1982).
22b. R. Merényi, J. Klein, H.G. Viehe, in preparation.
23a. H. Birkofer, J. Hadrich, N.D. Beckhaus, C. Ruchardt, Angew. Chem.
      99, 592 (1987); Internat. Edit. Engl. 26, 573 (1987).
23b. H.A.P. de Jongh, C.R.H.I. de Jonge, H.J.M. Sinnige, W.J. de Klein,
      W.G.B. Huysmans, W.J. Mijs, M.W.J. van den Hoeck, J. Smidt, J. Org.
      Chem. 37, 1960 (1972).
24a. X. Creary in ref. 5, p. 245.
24b. X. Creary, M.E. Mehrheikh-Mohammadi, J. Org. Chem. 51, 2664 (1986).
25.   H.M. Frey, Trans. Faraday Soc. 58, 516 (1962).
26a. J.M. Simpson and H.G. Richey, Tetrahedron Lett. 27, 2575 (1973).
26b. H.G. Richey and D.W. Shull, Tetrahedron Lett. 1976, 575.
27.   D. Cho and K.D. Ahn, J. Korean Chem. Soc. 22, 158 (1978); C.A. 90,
      38368 (1979).

26

28. V. Gallez, Ph.D. Thesis in preparation.
29. J.K. Kochi, D.J. Krusic, D.R. Eaton, J. Am. Chem. Soc. 91, 1877 (1969).
30. D. Laurie, E. Lucas, D.C. Nonhebel, C.J. Suckling, J.C. Walton, Tetrahedron 42, 1035 (1986).
31. W. Masamba, Ph.D. Thesis, Louvain-la-Neuve (1987).
32. A. Padwa, W. Dent, H. Nimmesgern, M.K. Venkatramanan, G.S.K. Wong, Chem. Ber. 119, 813 (1986).
33. K.D. Gundermann, A. Losler, Liebigs Ann. Chem. 758, 155 (1972).
34. A. Alder, D. Bellus, J. Am. Chem. Soc. 105, 6712 (1983).
35. S. Piettre, Ch. De Cock, R. Merényi, H.G. Viehe, Tetrahedron 43, 4309 (1987).
36a. G. Coppe-Motte, Ph.D. Thesis, Louvain-la-Neuve (1987).
36b. G. Coppe-Motte, A. Borghese, Z. Janousek, R. Merényi, H.G. Viehe in ref. 5, . 371.
37. B. Bienfait, Ph.D. Thesis in preparation.
38. M. Goffin-Vermander, Ph.D. Thesis in preparation.
39. H.G. Viehe, R. Merényi, E. Francotte, M. Van Meerssche, G. Germain, J.P. Declercq, J. Am. Chem. Soc. 99, 2340 (1977).
40. M. Van Meerssche, G. Germain, J.P. Declercq, J. Bodart-Gilmont, H.G. Viehe, R. Merényi, E. Francotte, Acta Cryst. B33, 3553 (1977).
41. E. Francotte, R. Merényi, H.G. Viehe, Angew. Chem. 90, 911 (1978).
42a. E. Francotte, Ph.D. Thesis, Louvain-la-Neuve (1978).
42b. B. Vandenbulcke-Coyette, Ph.D. Thesis, Louvain-la-Neuve (1982).
42c. R. Faragher, T.L. Gilchrist, J. Chem. Soc. Perkin I, 249 (1979).
43. J.L. Vaerman, Ph.D. Thesis, Louvain-la-Neuve (1988).
44a. Y.A. Arbusov, A. Markovskaya, Izv. Akad. Nauk. SSSR, Otdel., Khim. Nauk., 363 (1952); C.A. 47, 33168 (1953).
44b. K. Iida, Y. Watanabe, C. Kibayashi, J. Org. Chem. 50, 1818 (1981).
44c. D. Ranganathan, S. Ranganathan, C.B. Rao, K. Raman, Tetrahedron 37, 629 (1981).
45. H.G. Viehe, J.L. Vaerman, J. Prakt. Chem., submitted.
46. R. Sustmann, unpublished results.
47. For a review see M. Balci, Chem. Rev. 81, 91 (1981).
48. H. Felber, G. Kresze, R. Prewo, A. Vasella, Helv. Chim. Acta 69, 1137 (1986).
49. W. Burger, Ph. Work, Munich and Louvain-la-Neuve, in preparation.

H. Birkhofer, J. Hädrich, J. Pakusch, H.-D. Beckhaus and C. Rüchardt*

Institut für Organische Chemie und Biochemie der Universität Freiburg, Albertstr. 21, D-7800 Freiburg,FRG

K. Peters and H.-G. v. Schnering

Max-Planck-Institut für Festkörperforschung, Heisenbergstr. 1, D-7000 Stuttgart 80, FRG

ABSTRACT   The thermodynamic stabilisation of radicals by captodative substitution is little influenced by the polarity of solvents. The high thermal stability of Tetramethoxy-1.2-diphenylethane as compared to 1.2-dimethoxy-1.2-diphenyl-succinonitrile is due to the anomeric effect of gem-dialkoxy substitution and not to synergetic stabilisation of captodative substituted radical centers. This interpretation is in qualitative agreement with structural data.

INTRODUCTION

In previous work[1] it was shown that bond dissociation energies of CC-bonds are strongly dependent on the substitution pattern.

1. On the relaxation of strain on bond cleavage

2. On the stabilisation of the radicals generated, by the attached substituents

Fig. 1   Reaction Coordinate of a CC-Bond Cleavage

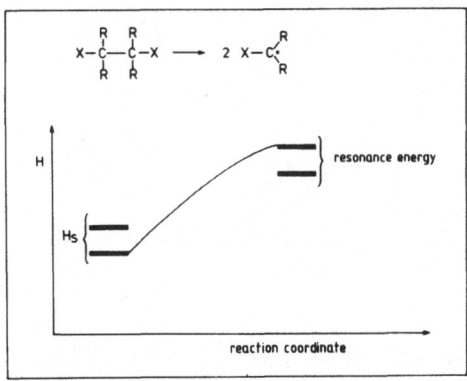

F. Minisci (ed.), Free Radicals in Synthesis and Biology, 27–36.
© 1989 by Kluwer Academic Publishers.

A quantitative separation of the two factors was possible. Within series of compounds carrying identical substituents X but differing in bulk of the attached side chains R and R'.

$$
\begin{array}{ccc}
R' & & R' \\
| & & | \\
X - C & - & C - X \\
| & & | \\
R & & R \\
\end{array}
$$

linear correlations between the free enthalpies of activation $\Delta G^{\#}(300^\circ C)$ and the changes in strain on bond dissociation were observed (see for example Fig. 2).

Fig. 2   Thermolysis of α-Phenyl-C_q-C_q-Alkanes[1]

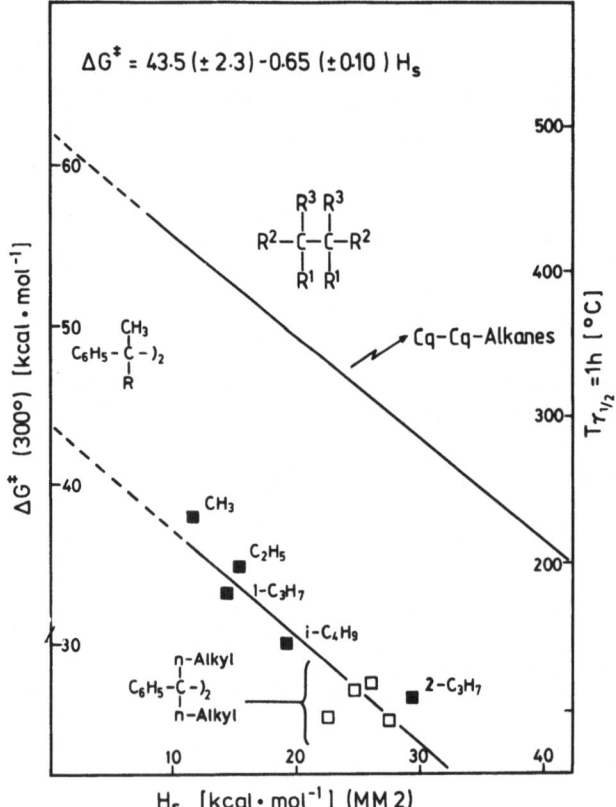

From these data, by proper analysis[1] stabilisation energies of alkyl radicals substituted by X were obtained. Selected results are shown in Table 1. Note, that these stabilisation energies are defined by exchanging X for methyl.

Table 1:[1)]   Selected Stabilisation Energies of Alkyl
                Radicals.

	H_r[kcal·mol^{-1}]
$CH_3-C\overset{R}{\underset{R}{\cdot}}$	≡0
$C_6H_5-C\overset{R}{\underset{R}{\cdot}}$	8,4
$R-\overset{\underset{\parallel}{O}}{C}-C\overset{R}{\underset{R}{\cdot}}$	6,5
$N≡C-C\overset{R}{\underset{R}{\cdot}}$	5,5
$CH_3O-\overset{\underset{\parallel}{O}}{C}-C\overset{R}{\underset{R}{\cdot}}$	3,5
$CH_3O-C\overset{R}{\underset{R}{\cdot}}$	1,3

In a similar manner, but on a smaller basis of data,
the stabilisation of alkyl radicals by gem. disubstitution was analysed. Selected results are reproduced
in table 2.

Table 2:[1]   Selected Stabilisation Energies of gem.-
Substituted Radicals. [H_r(calc. =
stabilisation energy expected for an
additive substituent effect]

Radical	H_r(exp)	H_r(calc)
	[kcal·mol^{-1}]	
$(CH_3)_3C\cdot$	$\equiv 0$	$\equiv 0$
$(C_6H_5)_2\dot{C}-R$	12	16.8
$(C_6H_5)_3C\cdot$	19	25.2
$C_6H_5-\dot{C}\begin{smallmatrix}/CH_3\\\backslash CN\end{smallmatrix}$	15	14
$C_6H_5-\dot{C}\begin{smallmatrix}/CH_3\\\backslash OCH_3\end{smallmatrix}$	9.4	9.7
$C_6H_5-\dot{C}\begin{smallmatrix}/OCH_3\\\backslash CN\end{smallmatrix}$	14	15.2

From these data the following conclusion can be
drawn:

1. The stabilisation by the first, second and third
   phenyl is not additive. This is due to the non
   planarity of the 1.1-diphenylalkyl and trityl
   radicals.

2. The stabilising effects of phenyl and cyano;
   phenyl and methoxy; and phenyl, methoxy and cyano,
   i.e. a captodative substituted radical are additive.

The captodative substituted system was investigated
in particular detail in order to be sure of the
reliability of this result.

Fig. 3 Reaction Coordinate for the Dissociation of
2.3-Diphenyl 2.3-Dimethoxy Succinonitriles[2)]

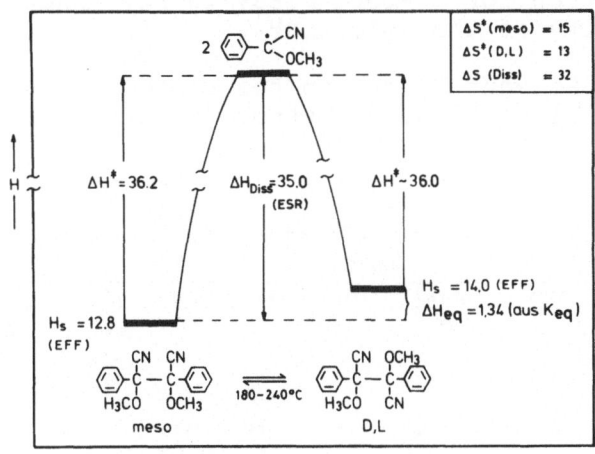

All data in kcal/mol or e.u.

The enthalpy of the equilibrium between the meso and
the racemic diastereomer was verified by force field
calculations. In addition, the activation enthalpies
for decomposition compared favourably with the
enthalpy of dissociation obtained by ESR experiments.
The entropy of dissociation, as expected, is much
larger than the entropy of activation is. This indicates that a radical pair, still restricted in its
motions, is generated primarily when the central CC-
bond is broken. Similar results were obtained for
other captodative substituted radicals, which were
not substituted by phenyl (see table 4).

This was an unexpected result, because it had been
predicted that captodative substitution has a strong
synergetic stabilising effect on radicals[3,4)]. The
consequences of this assumption for synthesis have
been discussed in detail[3,4)].

In order to get a better understanding of captodative
substitution the following questions were proposed:

1. Is synergetic captodative stabilisation dependent
   on a polar solvent as claimed in the recent literature on the basis of calculations[5)]?

2. How does captodative stabilisation compare with
   the stabilisation by two capto or by two dative
   substituents?

SOLVENT EFFECT ON CAPTODATIVE RADICALS?

The results in table 3 and 4 show that the enthalpies of activation for CC-bond cleavage are independent of the polarity of the solvents[6].

Table 3:   Solvent Effect on the Dissociation of 2.3-Diphenyl 2.3-Dimethoxy Succinonitriles[6]

Konfig.	solvent	$\epsilon(T°C)$	method	ΔH_R	$\Delta H^{\ddagger}$	ΔS_R	$\Delta S^{\ddagger}$
meso			nmr		36.6		15.1
D.L			nmr		36.0		13.2
meso/D.L		~3	esr	35.0		31.8	
meso/D.L	HOCH₂–CH₂OH	26(100)	esr	35.3		29.8	
meso/D.L		50(60)	esr	37.2		33.6	

ΔH: <±1(kcal/mol)
ΔS: ±1,5(cal/mol·k)

Table 4:   Solvent Effect on the Dissociation of 2.3-Di-t.Butyl 2.3-Dimethoxy Succinonitriles[6]

Konfig.	solvent	$\epsilon(T°C)$	method	ΔH_R	$\Delta H^{\ddagger}$	ΔS_R	$\Delta S^{\ddagger}$
meso			DSC		34.8		10.9
meso	CH₃–C–NHCH₃	70(150)	DSC		34.8		10.9
D.L			DSC		34.2		12.6
D.L	CH₃–C–NHCH₃	70(150)	DSC		34.2		11.0
meso/D.L		~3	ESR	34.3		41.2	

ΔH: <±1 [kcal/mol]
ΔS: ±1.5 [cal/mol·k]

The same is true for the enthalpies of dissociation
obtained from ESR-equilibrium studies. The identity
of the two types of results confirms the assumption
that the dimerisation of these radicals occurs without activation enthalpy. The typical differences
between the entropies of activation and the entropies
of dissociation are again found in these examples.

gem.-DIALKOXY- and gem.-DICYANO SUBSTITUTION.

In order to get an answer for the second question,
mentioned above, tetramethoxy-1.2-diphenylethane and
tetracyano-1.2-di(p-tolyl)ethane were investigated.
The compounds listed in table 5 have similar strain
enthalpies.

Table 5: Thermolysis of gem.-Substituted 1.2-
Diphenyl-ethanes[7]

no	X	Y	$H_S^{a)}$ [kcal/mol]	$\Delta G^{\ddagger}(300°C)$ [kcal/mol]	$\Delta H^{\ddagger}$	$\Delta S^{\ddagger}$ [e.u.]	$a_H^{b)}$ $[m^T]$
1[c]	CH_3	CH_3	12	39.9	46.0	14	0.567[d]
2[c]	C_2H_5	CH_3[e]	15	34.3	48.4	24	—
3[c]	C_2H_5	OCH_3[e]	16	34.5	44.0	17	0.625[f]
4	OCH_3	OCH_3	16[g]	43.6	51.4	14	0.603
5[h]	CN	CH_3[e]	11	30.9	36.1	9	0.470
6[i]	CN	OCH_3[e]	13	27.9	36.6 [35.0][k]	15 [32][k]	0.409
7[l]	CN	CN		18.0	26.0 [21.8][k]	14 [35][k]	0.503

[a] strain enthalpy as defined by P.v.R.Schleyer[1],
[b] ESR-hyperfine coupling constant of p-H, [c] cf.ref.1,
[d] cf.ref.7, [e] meso-configuration, [f] for X = CH_3, [g] preliminary value, [h] cf.ref.1,7. [i] cf.ref.1, [k] enthalpy-
and entropy of dissociation determined by ESR, [l] for
$4-CH_3-C_6H_4$ instead of C_6H_5 from ref. 10.

Data for two hydrocarbons of similar strain are
listed as no. 1 and 2 in the table. It is seen, that
exchanging one methoxyl for methyl (no. 3) on each
carbon results in 4.4 kcal/mol decrease in $\Delta H^{\ddagger}$. Exchanging, in additon, two ethyls for methoxyl (no.4),
in contrast, results in an increase of $\Delta H^{\ddagger}$ by 7.4
kcal/mol. This can hardly be interpreted by a destabilisation of a 1.1-dimethoxyalkyl radical as compared to a 1-methoxyl-1-alkyl-radical. It is, therefore, convincing evidence for an extra stabilisation

of the starting material by the geminal methoxyl substitution. The well known anomeric effect, apparently, has a stabilising function in saturated hydrocarbon structures but not in radical centers. The anomeric effect, therefore, should retard also hydrogen abstractions from acetal positions if the transition state of the reaction is late[8] and if it is not overshadowed by a strong rate enhancing polar effect[8,9].

The effect of gem. cyano substitution (no. 7) in table 5 is harder to interpret, because the strain enthalpy of the model compound and the importance of dipolar interactions are not safely known. The enthalpy of dissociation (21.8 kcal/mol) from ESR-experiments is believed to be more reliable than the enthalpy of activation[10] which was obtained by a special nmr-technique[10]. While the introduction of the first pair of cyano groups results in a decrease in $\Delta H^{\#}$ by 10 kcal/mol (cf. no.1 and 5) the second pair of cyano's decreases the bond energy even more (by ~ 15 kcal/mol).

These results clearly show that a comparison of the reactivity of capto-capto, captodative and dative-dative substituted model compounds must not rely on the differences in resonance stabilisation of the various types of radicals alone. Stabilisation by σ-type orbital interaction or dipolar interaction in the ground state and the radicals must not be neglected.

We have previously[1] shown that the lengths of the central CC-bonds in substituted bisquaternary ethanes are influenced only by the ground state strain and not depending on the radical stabilizing effects of the substituents attached. It is interesting to find that, as a consequence of the anomeric effect, in the tetramethoxy substituted model compound, a rather short central CC-bond is observed.

Fig. 4   Crystal Structure Analysis of Tetramethoxy 1.2-diphenylethane[11]

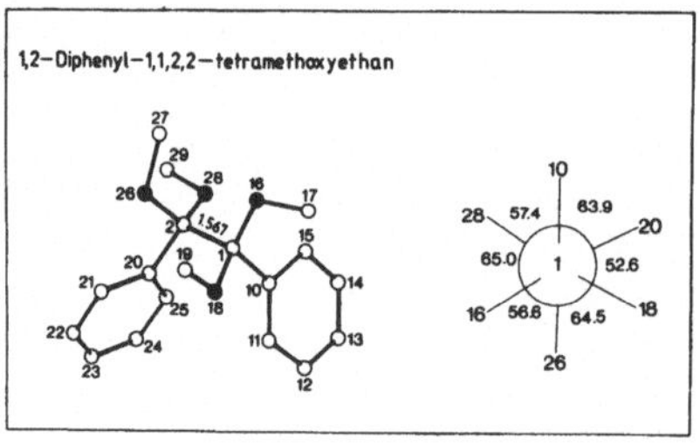

The structures of two compounds of this type have central bond lengths of 1.567 Å (Fig.4) and.1.544 Å (Fig.5). The captodative disubstituted analog has a longer bond of 1.578 Å (Fig.6).

Fig. 5   Crystal Structure Analysis of 2.2'-Diphenyl-bi(1,3-dioxolan-2-yl)[11].

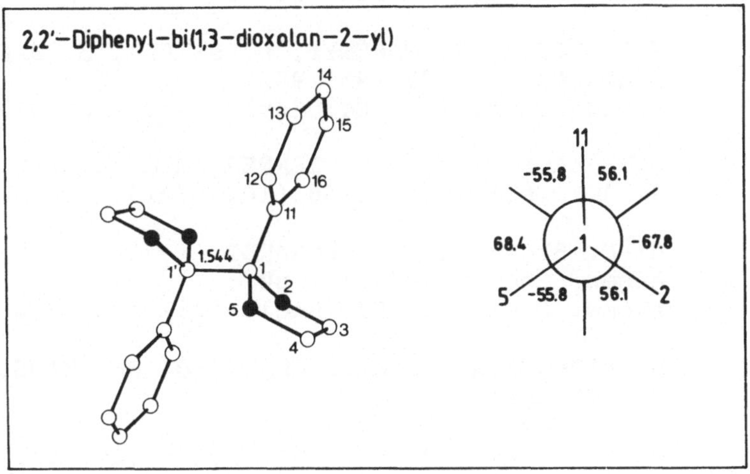

2,2'−Diphenyl−bi(1,3−dioxolan−2−yl)

Fig. 6   Crystal Structure Analysis of meso-1.2-Dimethoxy 1.2-Diphenyl Succinonitrile[11]

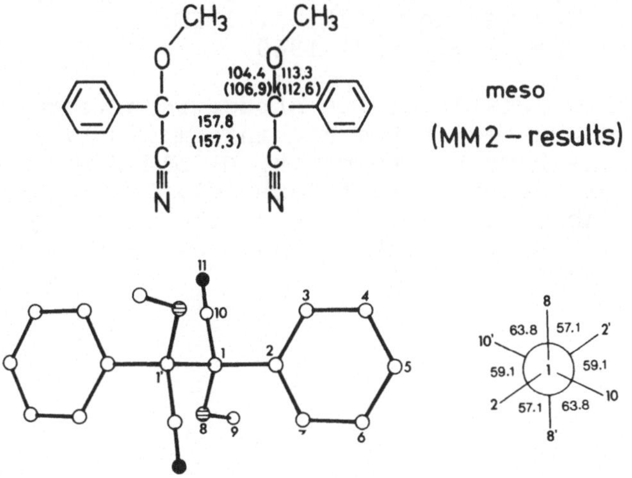

meso

(MM2−results)

36

References:

1. C. RÜCHARDT, H.-D. BECKHAUS, Topics in Current
   Chem. 1985 (130) 1 and references cited.

2. H. BIRKHOFER, H.-D. BECKHAUS, C. RÜCHARDT
   Substituent Effects on the CC-Bond Strength
   NATO-ASI Series 1986 (189) 199.

3. H.G. VIEHE, Z. JANOUSEK, R. MERÉNYI
   Acc. Chem. Res. 1985 (18) 148.

4. H.G. VIEHE, R. MERÉNYI, L. STELLA, Z. JANOUSEK
   Angew.Chem. 1979 (91) 982;
   Angew. Chem. Int. Ed. Engl. 1979 (18) 917.

5. A.R. KATRITZKY, M.C. ZERNER, M.M. KARELSON
   J. Am. Chem. Soc. 1986 (108) 7213.

6. H.-D. BECKHAUS, C. RÜCHARDT
   Angew. Chem. 1987 (99) 807;
   Angew. Chem. Int.Ed. Engl. 1987 (26) 770.

7. H. BIRKHOFER, J. HÄDRICH, H.-D. BECKHAUS,
   C. RÜCHARDT
   Angew.Chem. 1987 (99) 592;
   Angew. Chem. Int.Ed. Engl. 1987 (26) 573.

8. C. RÜCHARDT
   Angew. Chem. 1970 (82) 845;
   Angew. Chem. Int.Ed. Engl. 1970 (9) 830

9. J. FLEMING
   Frontier Orbitals and Organic Reactions
   Wiley N.Y. 1975, Chapt. 5

10. H.A.P. de JONGH, C.R.H.I. de JONGE, H.J.M.
    SINNIGE, W.J.de KLEIN, W.G.B. HUYSMANS, W.J.
    MIJS, W.J.van den HOEK, J. SMIDT,
    J.Org.Chem. 1972 (37) 1960.

11. Further details of the structure determinations
    are deposited at the Fachinformationszentrum
    Energie Physik Mathematik, D-7514 Eggenstein-
    Leopoldshafen 2 (West Germany). These data are
    available with quotation of the registry number
    CSD 53084, the authors, and the eference to this
    publication.

IODINE ATOM TRANSFER REACTIONS IN ORGANIC SYNTHESIS

Dennis P. Curran,
Department of Chemistry
University of Pittsburgh
Pittsburgh, PA 15260
USA

ABSTRACT. Sequential free radical reactions have been used with considerable success to rapidly assemble functionalized organic molecules. Often, such reactions are carried out with a trialkyltin hydride reagent. However, the use of tin hydride is not always desirable (or possible), particularly when one or more slow steps are involved in a given sequence. The unique capability of iodine atom transfer reactions to control free radical sequences will be discussed. In this "atom transfer" method, the starting iodide substitutes for the tin hydride in the key chain transfer step. With proper design, intermediate radicals are permitted maximum lifetimes (to undergo desired reactions) while final radicals are permitted very short lifetimes. The development of addition, cyclization, and annulation reactions controlled by iodine atom transfer is described.

1. INTRODUCTION

The dramatic advances in the application of free radical reactions to problems in organic synthesis can be attributed in good measure to the the versatility of trialkyltin hydride reagents.[1] Tin hydrides mediate radical chains by providing a means for site selective radical generation and a means for chain transfer. Between generation and chain transfer, the radical is permitted a variable lifetime (depending on the tin hydride concentration) in which useful chemistry can occur. Carbon–carbon bonds can be formed by addition or cyclization reactions, or cleaved by fragmentation reactions. In addition, sequences of reactions in which multiple carbon–carbon bonds are formed can be conducted, provided that certain selectivity requirements are met.[1d]

Like all reagents, tin hydrides have limitations which are important to recognize for synthetic planning. These include the difficulty to conduct slow reactions due to competing reduction by tin hydride and the need to separate reaction products from stoichiometric amounts of organotin compounds. Various experimental techniques address both of these problems. To conduct slow reactions, low concentrations of tin hydride are required; however, this dilution technique is limited because the rates of propagation steps can drop below the critical limits that are needed to maintain the chain. One potential limitation cannot be addressed by experimental techniques; it is inherent. Since the tin hydride method is reductive, a sequence of radical reactions must always be terminated by hydrogen atom transfer. From the synthetic perspective, valuable functionality is lost.

The use of allyl- and vinyltin reagents (pioneered by Keck, Russell, Baldwin, and others[1]) provides an alternative to the tin hydride method which bypasses some of the above problems. A completely different approach is available by conducting chains based on the

37

F. Minisci (ed.), Free Radicals in Synthesis and Biology, 37–51.

chemistry of thiohydroxamate esters (pioneered by Barton).[2] Many other chain and non-chain methods are useful for synthesis.

Kharasch reactions (atom transfer reactions) are among the fundamental reactions of organic free radicals and the development of a detailed understanding of such reactions predates all of the above methods. The key chain transfer step in Kharasch reactions is the abstraction of a univalent atom (hydrogen or halogen). In contrast to tin hydride chemistry, slow reactions can be conducted, stoichoimetric quantities of tin are not required, and reaction products are in the same oxidation state as the starting materials (functional groups are not lost). Of course, atom transfer reactions have their own set of limitations (for example, only exothermic atom transfer reactions are rapid enough for chain propagation) and the result is a strong complementary relationship with tin hydride chemistry.

Despite this knowledge, such reactions have not been widely applied in synthesis. Hydrogen atom transfer reactions between carbon radicals are relatively slow (that is, viable chains are difficult to design) and this severely restricts the use of bimolecular hydrogen atom transfer reactions in complex synthetic applications (where the presence of more than one "C–H" bond is probable). Chlorine and bromine atoms are transferred more rapidly and valuable synthetic methods have been developed by using inherently weak heteroatom–halogen bonds and by weakening carbon–halogen bonds with adjacent radical-stabilizing substituents (such as other halogens).[1d]

Despite these important developments, the vast potential use of atom transfer reactions to mediate synthetic transformations by free radical chains has not been recognized. For the past three years, our research group has been involved in the development of new transformations based on halogen atom transfer. The simple expedient of introducing iodine as the atom donor provides a dramatic acceleration of the atom transfer steps. Otherwise useless reactions can now be conducted with ease. This paper will provide an overview of our research concerning the development of new synthetic methods based on iodine atom transfer.

2. ATOM TRANSFER ADDITION AND CYCLIZATION REACTIONS OF ALKYL IODIDES

The development of iodine atom transfer reactions in our laboratories was ignited by an accidental discovery during a synthesis of the important linear triquinane capnellene (Scheme 1).[3] In an application of our tandem radical cyclization strategy to triquinanes, bromide 1a was reduced with tri-n-butyltin hydride under the usual conditions. As expected, capnellene (2) was formed in good yield (>80% by GC) in less than 30 min. Reduction of the corresponding iodide 1b under identical conditions required an even shorter time for disappearance of starting material. To our surprise, capnellene was formed in a much lower yield (<20%). However, if the reaction mixture was heated further, the yield of capnellene continued to increase! After eight hours, the yield of capnellene from iodide 1b reached the same value as the bromide 1a had given in 30 min.

From these experiments, we concluded that the iodide 1b was first converted to an intermediate product, and that this intermediate product was then converted slowly to capnellene. In contrast, no intermediate product was formed in the reduction of the bromide. Given this, we were soon able to isolate the intermediate, capnellene vinyl iodide 3. A series of mechanistic experiments allowed us to conclude that 3 was formed by the expected tandem radical cyclization to give a vinyl radical and that this vinyl radical abstracted an iodine atom from the starting material (to give the starting radical) more rapidly than it abstracted a hydrogen atom from tin hydride! (The mechanism is discussed in detail below.) Vinyl iodide 3 was then reduced in a slower reaction by tin hydride to produce 2. When the vinyl radical was generated from the bromide, bromine atom abstraction could not compete with direct reaction with tin hydride and capnellene was formed directly by the normal mechanism.

Scheme 1

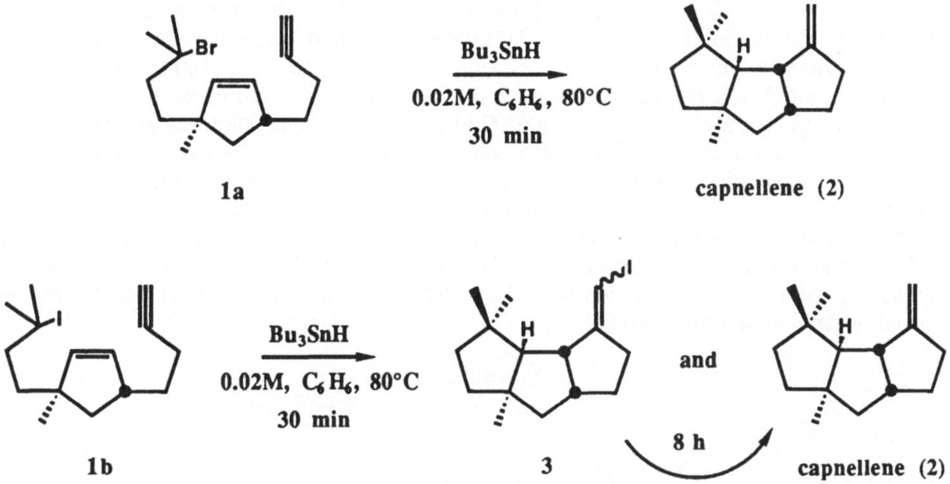
Br
Bu₃SnH
0.02M, C₆H₆, 80°C
30 min
H
1a
capnellene (2)
I
Bu₃SnH
0.02M, C₆H₆, 80°C
30 min
I
H
and
H
8 h
1b
3
capnellene (2)

The results of these experiments implied that it was possible to isomerize hexenyl iodides to cyclic vinyl iodides. A screen of possible experimental conditions resulted in the development of a method that used a catalytic amount of hexaalkylditin.[4] Typically, a solution of iodide (0.3 M in benzene) and hexaalkylditin are irradiated for 30 – 120 min with a 275w sunlamp. The reaction temperature generally reaches 80 °C. Good yields of (iodomethylene)cyclopentanes are produced and several representative examples of this isomerization are listed in Figure 1. Primary, secondary, and tertiary iodides all isomerize to the cyclic vinyl iodides without difficulty. The method is particularly valuable since vinyl iodides are very useful in a variety of synthetic transformations.

Figure 1

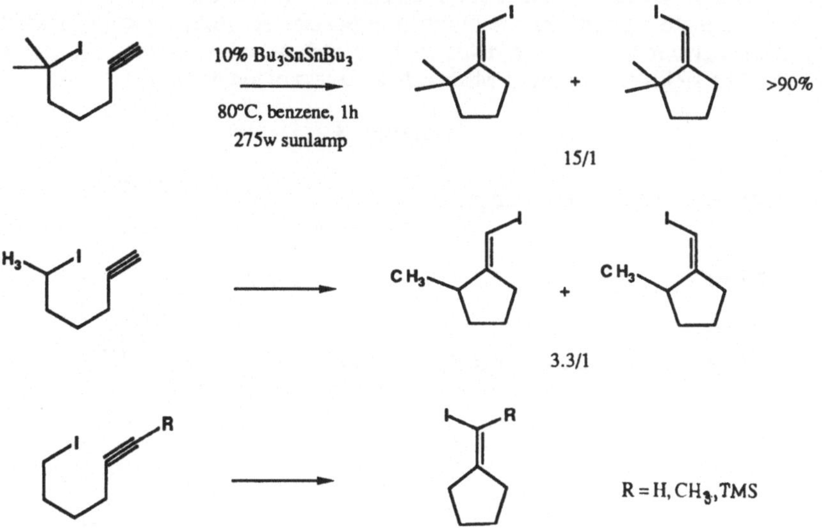
I
10% Bu₃SnSnBu₃
80°C, benzene, 1h
275w sunlamp
I
+
I
>90%
15/1
CH₃
I
CH₃
I
+
CH₃
I
3.3/1
I
R
I
R
R = H, CH₃, TMS

The precise nature of the initiation step(s) is unclear. Photolytic cleavage of the hexaalkylditin to trialkyltin radical is one possible initiation step. Another, perhaps more likely, initiation is photolytic cleavage of the carbon–iodine bond. Here, the role of ditin may be as a scavenger for iodine atoms or molecules, both of which would suppress radical chains. The propagation steps are shown in Scheme 2. Hexenyl radical cyclization is exothermic despite the formation of a vinyl radical from an alkyl radical. This is in large measure because a C–C sigma bond has been gained at the expense of a C–C pi bond. The iodine atom transfer step is also significantly exothermic and is kinetically rapid due to the low strength of the C–I bond (experiments indicate that k_I in Scheme 2 is about 10^9 M^{-1} sec^{-1}). This reaction, named atom transfer cyclization because of the two propagation steps, has the same requirement as any Kharasch reaction: all the propagation steps must be rapid and exothermic. Because iodine atom transfers are so rapid, any exothermic addition, cyclization, or fragmentation reaction which converts a more resonance stabilized radical to a less resonance stabilized counterpart will be followed by rapid iodine atom transfer.

Scheme 2

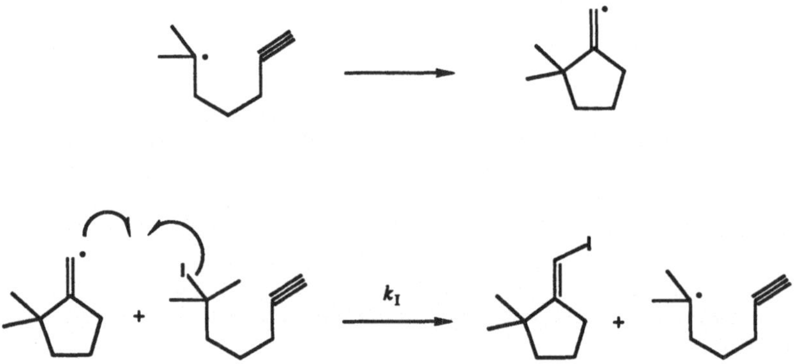

Although the yields are not as high, bimolecular atom transfer addition reactions can be conducted based on the conversion of alkyl radicals to vinyl radicals. The additions of several alkyl iodides to methyl propiolate are outlined in Scheme 3. As is typical of such reactions, primary iodides give relatively poor yields while secondary and tertiary iodides give modest to good yields. The yields are improved when a slight excess of iodide is used.

Scheme 3

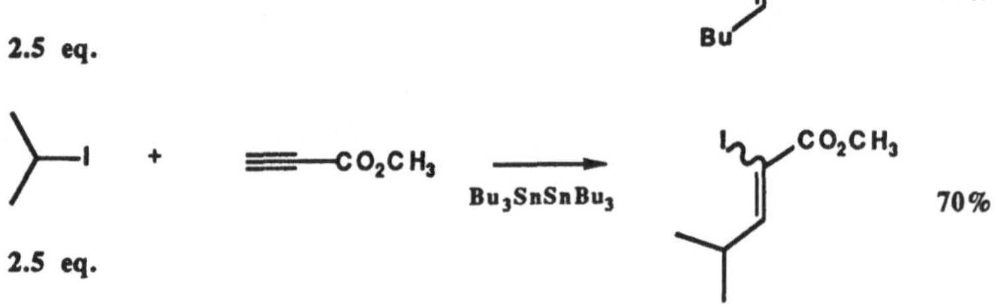

Scheme 3 (cont.)

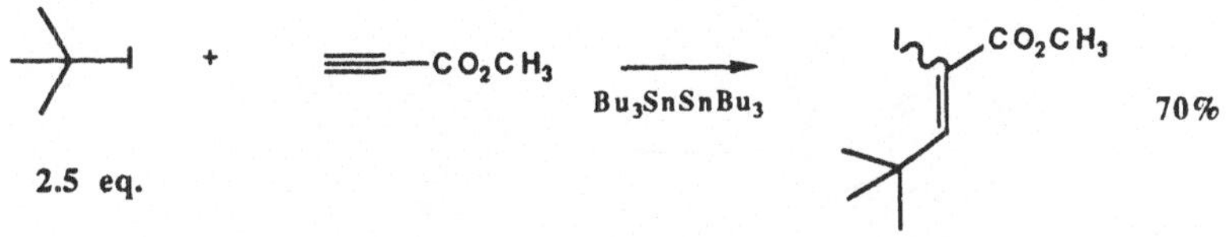

2.5 eq.   70%

In this addition reaction, an activating group on the alkyne is required to raise the rate of bimolecular addition to a sufficient level for a chain reaction. However, to the extent that this activating group stabilizes the adduct radical, it reduces the exothermicity (and hence, the rate) of the atom transfer step. Thus, a delicate balance must be maintained. Examples of the addition of isopropyl iodide and *t*-butyl iodide to dimethylacetylene dicarboxylate, phenylacetylene, and phenylsulfonyl acetylene are contained in Figure 2. The last acceptor is particularly interesting for synthetic applications because it exhibits a good selectivity for formation of the Z-alkene.

Figure 2

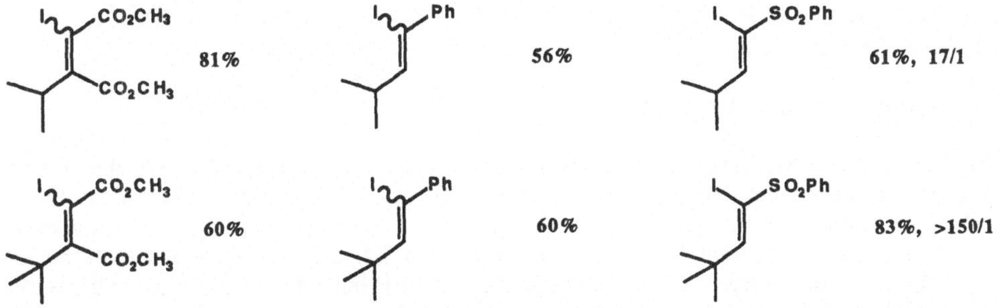

81%   56%   61%, 17/1

60%   60%   83%, >150/1

Alkyl iodides are such good atom donors that they can even be transferred to other alkyl radicals (near-thermoneutral reactions) with sufficient rates to propagate chains.[5] Some simple isomerizations which have been conducted in our group are listed in Scheme 4.[6] In addition, a nice example from the work of Crimmins and Mascarella shows that the atom transfer method can also mediate fragmentation reactions.[7] Direct reductive cleavage of the cyclobutylmethyl iodide requires very slow syringe pump addition of tin hydride to keep a low concentration. Otherwise, significant amounts of the reduced-unfragmented product are formed. However, the ditin-promoted isomerization can be conducted at high concentration, as can the subsequent reductive deiodination with tin hydride.

Scheme 4

46%

Scheme 4 (cont.)

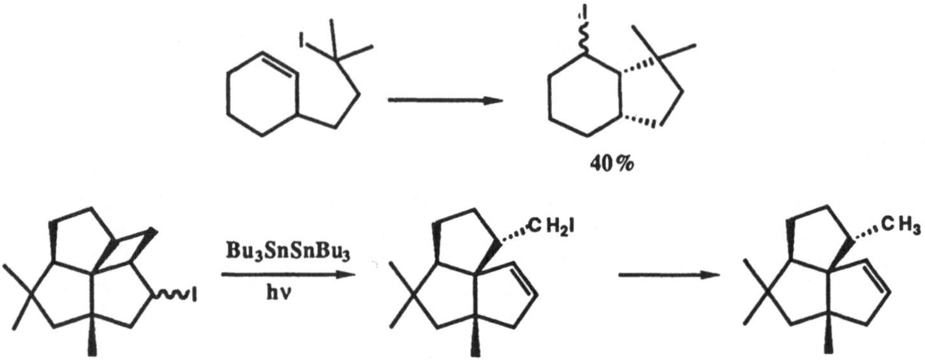
I
40%
Bu3SnSnBu3
hv
CH2I
CH3

Crimmins and Mascarella

3. ATOM TRANSFER ADDITION AND CYCLIZATION REACTIONS OF α-IODOCARBONYLS

3.1 α-Iodoesters and Ketones

If an exothermic cyclization or addition reaction converts a resonance stabilized radical to a simple alkyl radical, subsequent transfer of an iodine atom should again be rapid and exothermic. We have investigated the cyclization of a variety of α-iodoesters and ketones with both alkenes and alkynes and several representative examples are given in Scheme 5.[8] Ester-substituted radicals usually show a good preference for 5-exo cyclization (>10/1) but ketone-substituted radicals often give significant amounts of the 6-endo product (up to 25%).

Scheme 5

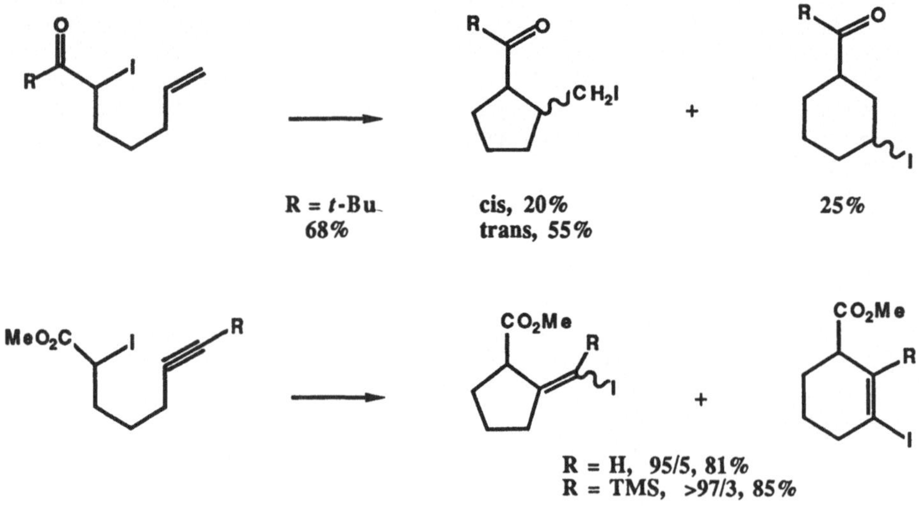
O
R
I
R
O
CH2I
+
R
O
I
R = t-Bu
68%
cis, 20%
trans, 55%
25%
MeO2C
I
R
CO2Me
R
I
+
CO2Me
R
I
R = H, 95/5, 81%
R = TMS, >97/3, 85%

In the atom transfer method, cyclic radicals are not trapped by hydrogen atom transfer, as in the tin hydride method, but by halogen atom transfer. However, even when this additional functional group is not desired in the product, there are cases where the atom transfer method may still be the method of choice. One such example is presented in Scheme 6. The reduction of iodoester **4** with tri-*n*-butyltin hydride at 0.02 M gives only the reduced acetate **5**. None of the cyclic lactone **6** can be detected at this concentration. The problem is that, even at this relatively low tin hydride concentration, the cyclization is too slow to compete with direct reduction.

Scheme 6

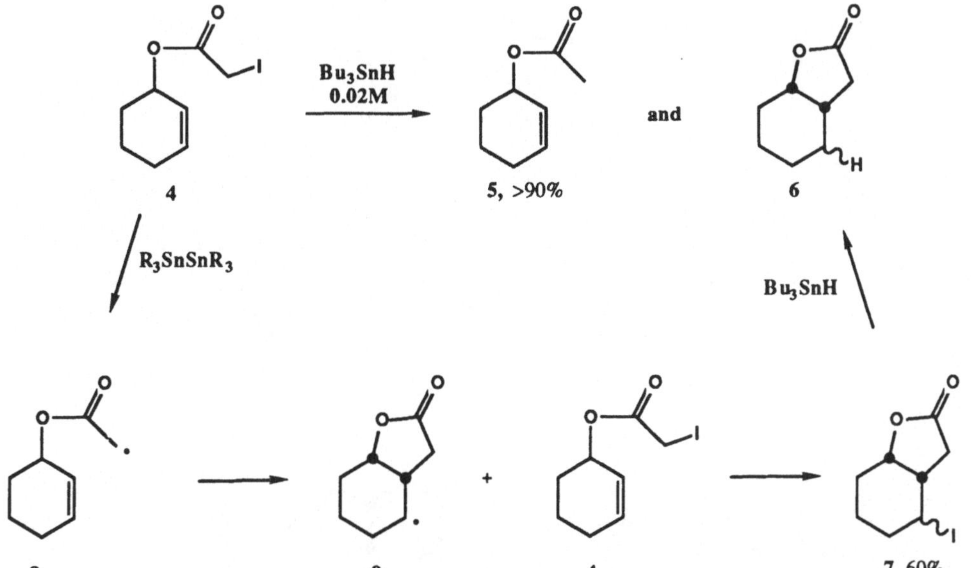

In contrast, isomerization of **4** in the presence of hexabutylditin gives the iodolactone **7** in 60% isolated yield at 0.3 M. If desired, tri-*n*-butyltin hydride can be added directly to the reaction mixture after cyclization (again at 0.3 M). After reinitiation and purification, lactone **6** is isolated, again in about 60% yield. In the atom transfer method, the initially formed radical **8** has a maximum lifetime to cyclize (no tin hydride is present). Following the slow cyclization that forms **9** is the rapid iodine atom transfer reaction with **4**. Thus, the lifetime of radical **8** is long compared to that of radical **9**. In the tin hydride method, most radicals abstract hydrogen at a similar rate and, to a good approximation, all radicals then have similar lifetimes with respect to chain transfer. The ability to establish different lifetimes for radicals is a powerful feature of the atom transfer method, and its value goes beyond the simple conductance of slow cyclizations of resonance-stabilized radicals (see Section 4).

Interesting regioselectivities are observed with iodoketones in which the carbonyl group is endocyclic to the forming ring. With terminal alkene acceptors, exclusive formation of the 6-endo product is observed as shown in Figure 3. As the substitution bias is altered, increasing amounts of the 5-exo product are formed. Kinetic experiments indicate that the product ratios for all of these iodocarbonyl cyclizations are kinetic; reverse cyclization cannot compete with iodine atom transfer. We attribute the endo preference in these systems (first observed by Clive[9]) to a stereoelectronic effect. If one assumes that overlap between the pi-orbital of the radical and the carbonyl is maintained in the cyclization, the geometric constraints of the system make 5-exo approach (with effective orbital overlap) very difficult. The promise of a general method for a substituent-controlled 6-endo closure is attractive for synthetic applications.

Figure 3

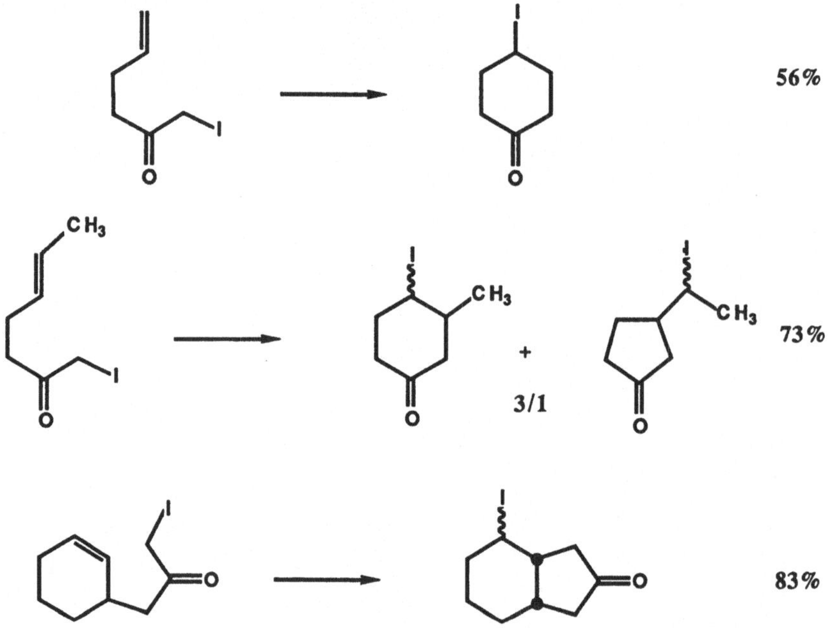

56%

73%

3/1

83%

3.2 α-Iodomalonates

Iodomalonates are among the best compounds discovered so far for the atom transfer method because they are outstanding iodine atom donors and because they add relatively rapidly to electron rich alkenes due to the electrophilic nature of the malonyl radical.[10] Thus, both steps in an atom transfer chain may be better with malonates than with simple esters or ketones. The cyclization of the simple iodomalonate shown in Scheme 7 is readily accomplished by brief irradiation (10 min) in the presence of hexabutylditin.[11] A 90/10 mixture of the 5-exo/6-endo products is isolated in good yield. In synthetically useful transformations, the iodide products can be directly reduced (with tri-n-butyltin hydride) or lactonized (by heating).

Scheme 7

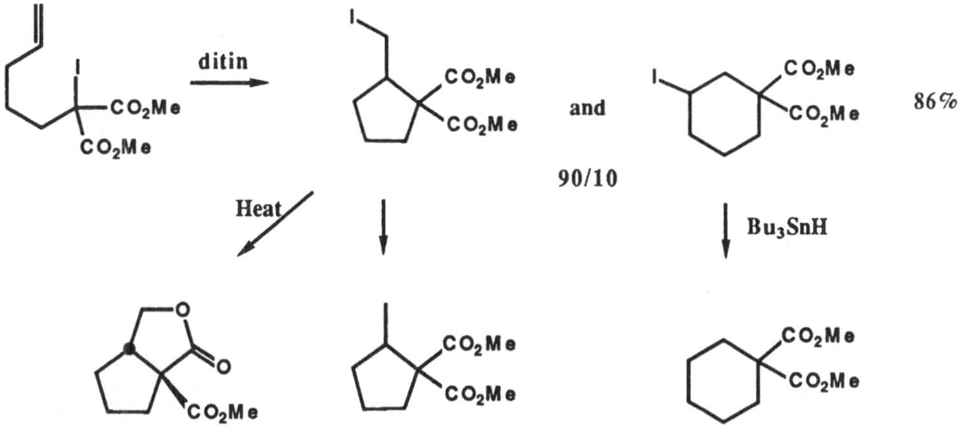

90/10

86%

The use of halomalonates as precursors of malonyl radical provides a strong contrast to the classic work of Julia.[12] As illustrated in Scheme 8, Julia used C–H bonds as radical precursors and partial or complete equilibration of intermediate cyclic radicals was often observed. In the indicated example, only the 6-endo product was formed. However, we generated from the bromomalonate the same cyclic radical as Julia and Maumy but we observed only formation of the kinetic 5-exo product.

Scheme 8

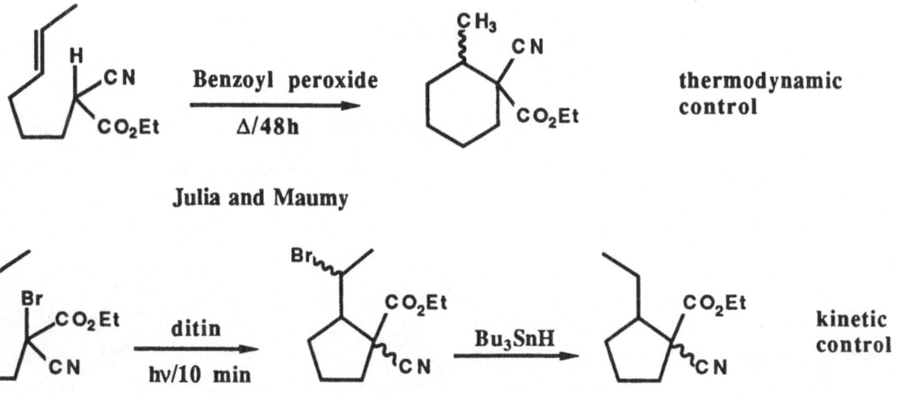

The reason for this dichotomy lies in the difference in atom donor capabilities of the radical precursors. This is shown in Scheme 9. The same radical 10 is generated in both reactions and it closes kinetically to the 5-exo product 11. In the Julia reaction, there is no good trap for 11 and equilibration ensues via reverse cyclization to give the more stable product 12. Ultimately, 12 must abstract a hydrogen from either the starting malonate or the solvent (cyclohexane) to provide 13. In our reaction, the initial cyclic radical 11 abstracts a halogen from the precursor (to give 14) much more rapidly than it can equilibrate to the thermodynamic product. This is a perfect example of how alteration of the atom donor capability of the precursor can control the product ratio.

Scheme 9

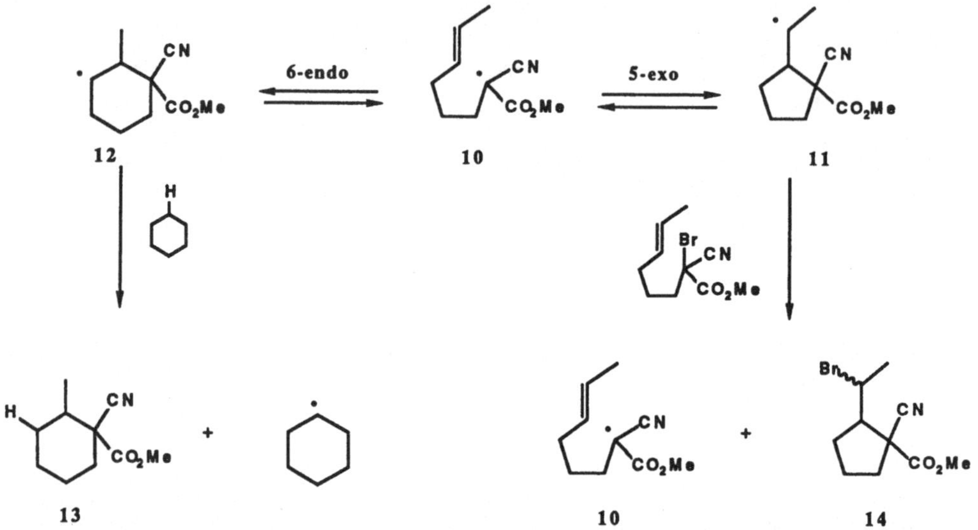

Figure 4 shows representative examples of the cyclization reactions which we have investigated and reveals some interesting substituent effects. In particular, the introduction of an alkyl group on the internal carbon of an alkene promotes kinetic formation of the product of endo addition. It is noteworthy that 6-endo, 6-exo, and 7-endo closures all succeed. These are often difficult reactions to conduct by tin hydride chemistry.

Figure 4

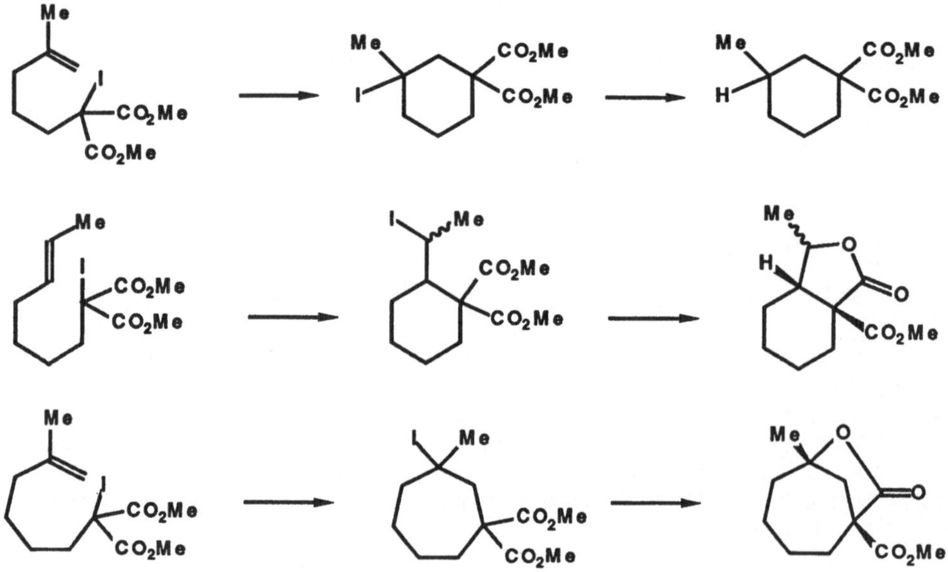

Atom transfer addition reactions of iodomalonates are also possible and provide a mild alternative to the traditional alkylative methods for the functionalization of malonates. The addition of a methyl-substituted iodomalonate to 1-hexene is illustrated in Scheme 10. The intermediate adduct can be isolated, deiodinated, or lactonized. Some examples which illustrate different facets of iodomalonate additions are collected in Figure 5.

Scheme 10

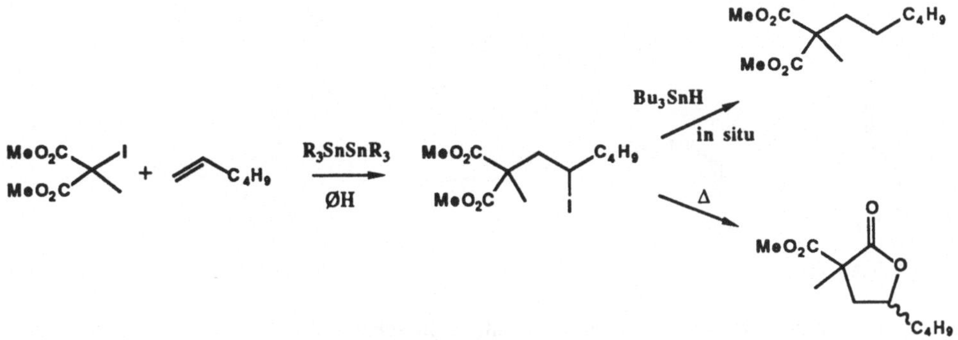

Figure 5

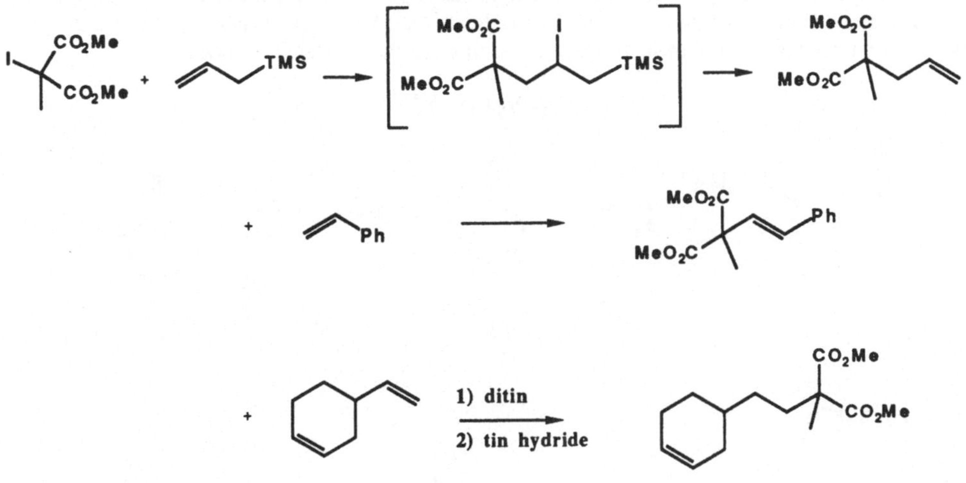

4. ATOM TRANSFER ANNULATION

The ability to assemble a sequence of radical reactions is a major asset for organic synthesis. Certain sequences which are not possible with tin hydride reagents can be conducted by the iodine atom transfer method. We have recently developed several new methods to form rings in a strategy which we term atom transfer annulation. The rings are formed by the sequential execution of a radical addition and cyclization reaction. Two new carbon–carbon bonds are formed.

A typical example of an annulation method based on 4-iodo-1-butyne is outlined in Scheme 11.[13] Irradiation of this iodoalkyne (2.5 equiv) and methyl acrylate (1.0 equiv) in the presence of a catalytic amount of hexabutylditin for 2 h resulted in the formation of an (iodomethylene)cyclopentane in 65% yield. This compound was formed as a 2.5/1 mixture of E/Z stereoisomers and a small amount of the 6-endo product was also isolated. As in the examples given above, the iodide products can be directly reduced with tin hydride; however, attempts to form the same reduced product by treating the iodoalkyne and methyl acrylate directly with tin hydride products were not successful.

Scheme 11

2.5 eq.   **2.5/1   E/Z**

The mechanism of this annulation is outlined in Scheme 12. Well-precedented addition of the butynyl radical **15** to the electron deficient alkene is the first step in the sequence. The so-formed radical **16** is more resonance stabilized than the starting radical and will not abstract iodine from the starting iodide. Instead, it closes predominantly in a 5-exo fashion to give **17**. Rapid iodide atom abstraction by **17** from the starting material gives the product **18** and the initial radical **15**. The rapid transfer of iodine to the final radical is the key step because it transfers the chain before any undesired reactions of **17** (such as addition to the acrylate) can occur.

Scheme 12

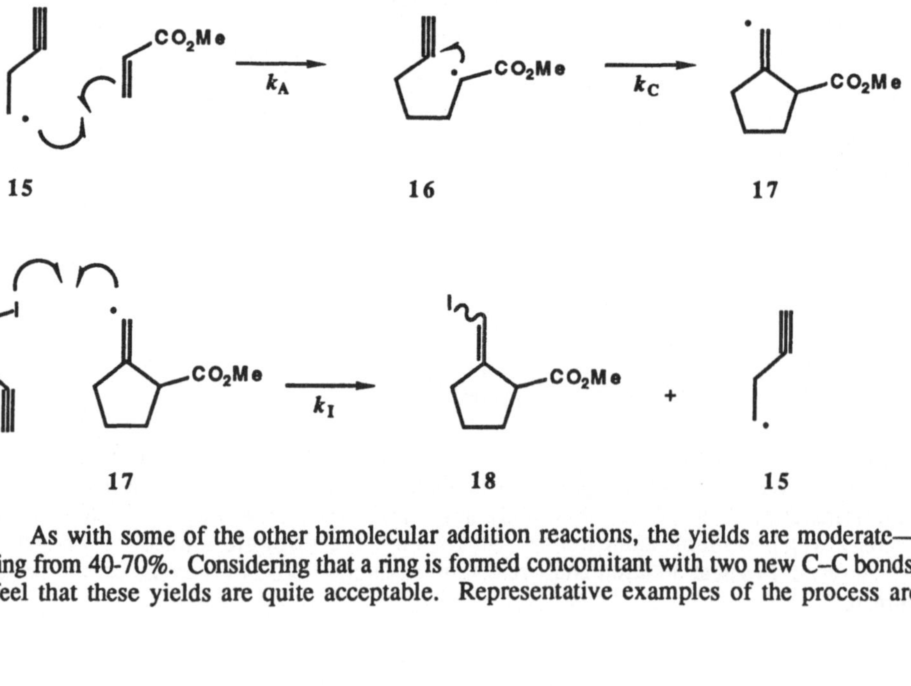

15   **16**   **17**

17   **18**   **15**

As with some of the other bimolecular addition reactions, the yields are moderate—ranging from 40-70%. Considering that a ring is formed concomitant with two new C–C bonds, we feel that these yields are quite acceptable. Representative examples of the process are contained in Figure 6. The indicated yields are those of pure products which were obtained by column chromatography of the crude reaction mixture.

Figure 6

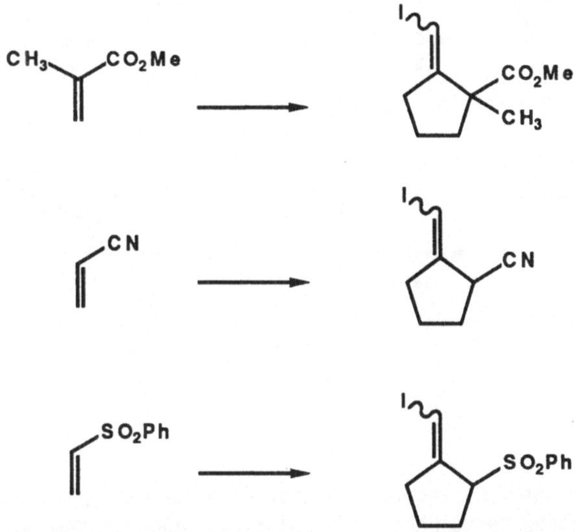

56%
10/1 E/Z

41%
2/1 E/Z

68%
7/1 E/Z

We have also capitalized on the excellent reactivity profile of iodomalonates in atom transfer reactions to design two related annulation procedures with allyl- and propargyl iodomalonates. A generalized reaction for each process, along with several specific products and their respective isolated yields, is outlined in Scheme 13.[14]

Scheme 13

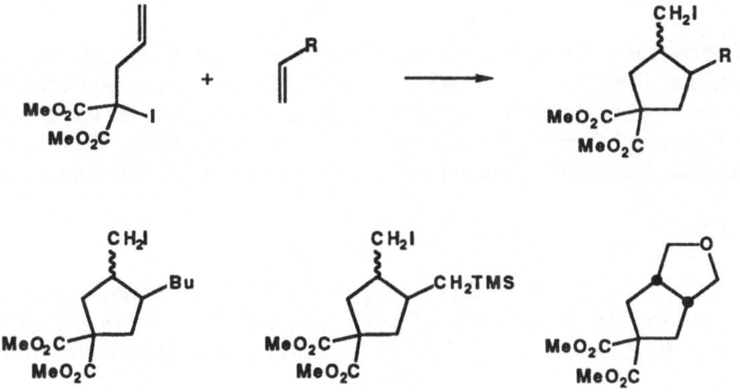

58%, 4/1, c/t

71% 5/1, c/t

40%

50

Scheme 13 (cont.)

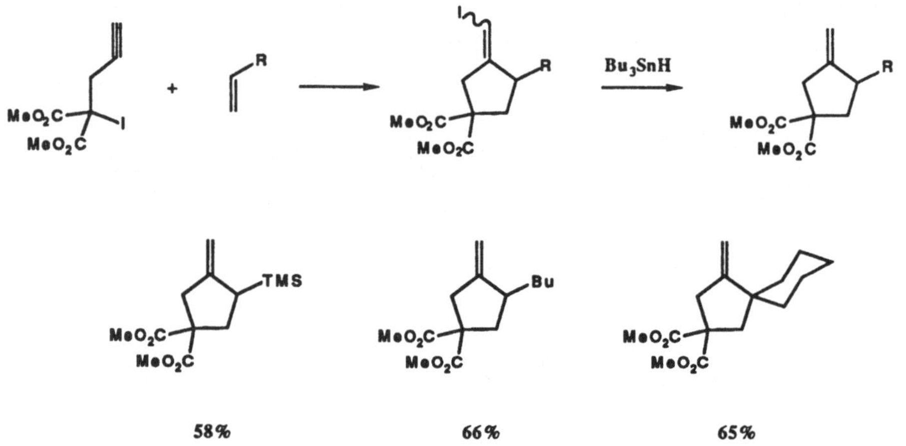

58%   66%   65%

5. CONCLUSION

The application of iodine atom transfer to mediate free radical reactions is a powerful synthetic tool. As with other radical methods, the vast body of knowledge on kinetics of radical cyclization and addition reactions is a great aid in the process of planning viable new reactions. Especially during the last year, some important rate constants for halogen atom transfer have also become available. With many applications still untapped, this continues as an exciting and fruitful area of research from which contributions to both synthetic organic chemistry and fundamental free radical chemistry can be made simultaneously.

ACKNOWLEDGEMENTS

I am especially grateful to the Ph.D. students whose work at the University of Pittsburgh made this lecture possible. They are: Dr. Meng Hsin Chen, Chi-Tai Chang, and Dooseop Kim. I am also grateful to the National Institutes of Health for major funding and for an Career Development Award, to the Sloan and Dreyfus Foundations for Fellowships, and to Eli Lilly, E. M. Merck, Hoffmann-LaRoche, American Cyanamid, and Stuart Pharmaceuticals for unrestricted support.

REFERENCES

1   Reviews: (a) Giese, B. "Radicals in Organic Synthesis: Formation of Carbon-Carbon Bonds", Pergamon Press: Oxford, 1986. (b) Ramaiah, M. *Tetrahedron* 1987, *43*, 3541. (c) Neumann, W. P. *Synthesis* 1987, 665. (d) Curran, D. P. *Synthesis* , in press.

2   Crich, D. *Aldrichimia Acta* 1986, *20*, 35.

3   Curran, D. P.; Chen, M.-H. *Tetrahedron Lett.* 1985, *26*, 4991.

4   Curran, D. P.; Chen, M.-H.; Kim, D. *J. Am. Chem. Soc.* 1986, *108*, 2489.

51

5   Newcomb, M.; Sanchez, R. M.; Kaplan, J. *J. Am. Chem. Soc.* **1987**, *109*, 1195.

6   Curran, D. P.; Kim, D. *Tetrahedron Lett.* **1986**, *27*, 5821.

7   Crimmins, M. T.; Mascarella, S. W. *Tetrahedron Lett.* **1987**, *28*, 5063.

8   Curran, D. P.; Chang, C.-T. *Tetrahedron Lett.* **1987**, *28*, 2477.

9   Clive, D. L. J.; Cheshire, D. R. *J. Chem. Soc., Chem. Commun.* **1987**, 1520.

10 For atom transfer addition reactions of malonates and malononitriles, see: Giese, B.; Horler, H.; Leising, M. *Chem. Ber.* **1986**, *119*, 444. Riemenschneider, K.; Bartels, H. M.; Dornow, R.; Dreschel-Grau, E.; Eichel, W.; Luthe, H.; Matter, Y. M.; Michaelis, W.; Boldt, P. *J. Org. Chem.* **1987**, *52*, 205. Bartels, H. M.; Boldt, P. *Liebigs Ann. Chem.* **1981**, 40.

11 Curran, D. P.; Chang, C. T. submitted for publication.

12 Julia, M. *Acc. Chem. Res.* **1971**, *4*, 386; *Pure Appl. Chem.* **1974**, *40* 553. A nice discussion of the conrtributions from the Julia group is contained in Surzur, J. M. in "Reactive Intermediates", Abramovitch, R. A.; ed. Vol. 2, Chap. 3, Plenum Press, NY, **1982**.

13 Curran, D. P.; Chen, M.-H. *J. Am. Chem. Soc.*, **1987**, *109*, 6558.

14 Ph. D. Thesis of M.-H. Chen, University of Pittsburgh, 1987.

ALKYL IODIDES AS SOURCE OF ALKYL RADICALS, USEFUL FOR SELECTIVE SYNTHESES

E. Vismara, F. Fontana and F. Minisci
Dipartimento di Chimica del Politecnico
Piazza Leonardo da Vinci 32, 20133 Milano
Italy

ABSTRACT - Several methods of using alkyl iodides as sources of alkyl radicals have been developped. The combination of enthalpic factors in generating alkyl radicals and polar factors in the reactivity of the radicals has allowed to develop a variety of selective syntheses. Particularly the alkyl radicals were generated by iodine abstraction from alkyl iodides by aryl radicals from aroylperoxides or diazonium salts and by the methyl radical from t-BuOOH, (t-BuO)$_2$, DMSO, acetone, acetic acid, (MeCOO)$_2$. Nucleophilic and electrophilic radicals of any kind can be easily and selectively generated in this way: the first ones selectively react with electron-deficient substrates (heteroaromatic bases, pyrilium salts, diazonium salts, quinones, electronpoor olefins), the second ones with electron-rich aromatics and olefins.

A new trend of organic synthesis involves selective reactions by carbon-centered radicals[1]. Organic halides have been largely utilized as sources of carbon-centered radicals. Most of these sources involve halogen abstraction (eq.1) in chain processes or electron-transfer reduction (eq.2)

$$R-X + Y. \longrightarrow R. + X-Y \qquad (1)$$

$$R-X + e \longrightarrow R. + X^- \qquad (2)$$

Reaction 2 can be induced under chemical, electrochemical, photochemical or radiolytic conditions and in any case necessarily requires highly reducing media. Also the halogen abstraction (eq.1) was mostly carried out under reducing conditions (particularly by organometallic radicals

F. Minisci (ed.), Free Radicals in Synthesis and Biology, 53–69.
© 1989 by Kluwer Academic Publishers.

54

from the corresponding hydrides, such as $R_3Sn\cdot$, $R_3Si\cdot$ etc.). Reactions (1) and (2) are often very fast,[3] particularly with iodine derivatives, due to the low energies of the C-I bonds (50-60 Kcal/mol), and that makes very selective the radical source[2].

On the other hand the reducing medium is at the same time the main limitation for the synthetic use of the carbon-centered radicals, which are often reduced fast under the reaction conditions. For example high rate constants have been estimated for the reduction of alkyl radicals by tin hydrides[3] or Cr(II) salts[4] (eqs.3, 4)

$$R\cdot + H\text{-}Sn' - \xrightarrow{k} R\text{-}H + \cdot Sn' - \qquad k \sim 10^6 \ M^{-1}s^{-1} \qquad (3)$$

$$R\cdot + Cr^{2+} + H^+ \xrightarrow{k} R\text{-}H + Cr^{3+} \qquad k \sim 10^6\text{-}10^8 \ M^{-1}s^{-1} \ (4)$$

Alkyl iodides, the most reactive among the alkylhalides, have been largely utilized in the synthesis according to the general free-radical chains[5] of Scheme 1

$$R\text{-}I + \cdot MRn \longrightarrow R\cdot + I\text{-}MRn$$
$$R\cdot + HMRn \xrightarrow{k_1} R\text{-}H + \cdot MRn$$
$$R\cdot + S \xrightarrow{k_2} RS\cdot$$
$$RS\cdot + HMRn \longrightarrow RS\text{-}H + \cdot MRn$$

Scheme 1

The synthetic success of this Scheme is strongly affected by the relative values of the rate constants k_1 and k_2. The relative concentrations of the substrate (S) and the hydrogen donor (HMRn) can contribute to improve the selectivity. Thus the most spectacular synthetic applications involve intramolecular reductive alkylations, due to the high values of k_2 in the intramolecular processes.

Another limitation concerns the use of substrates with high electron affinity, such as diazonium salts. In these cases the interaction either of the reagent or of the nucleophilic radical (RnM$\cdot$) with the substrate (generally through an electron-transfer process) competes with the generation of the alkyl radical from the alkyl iodide. Thus, for example, the free-radical diazocoupling process, a reaction of large synthetic interest which we[6] have recently developped, involves the reductive alkylation of diazonium salts (eqs.5 and 6)

$$R\cdot + N\overset{+}{\equiv}N\text{-}Ar \longrightarrow R\text{-}N = \overset{+}{\underset{\bullet}{N}}\text{-}Ar \tag{5}$$

$$R\text{-}N = \overset{+}{\underset{\bullet}{N}}\text{-}Ar + e \longrightarrow R\text{-}N = N\text{-}Ar \tag{6}$$

All the radical sources utilized according to the Scheme 1 could be in principle used for this purpose. Practically, they are not suitable because they react faster with the diazonium salts than with the alkyl iodides.

Obviously radical sources such those in eqs.1 and 2 are not suitable for general processes of large synthetic interest, in which the inter- or intramolecular reactions of the alkyl radicals are followed by an oxidative step according to the general Scheme 2

$$R\text{-}I \longrightarrow R\cdot$$

$$R\cdot \xrightarrow[\text{ox.}]{k_3} \text{Products}$$

$$R\cdot + S \xrightarrow{k_4} RS\cdot$$

$$RS\cdot \xrightarrow[\text{ox.}]{k_5} \text{Products}$$

Scheme 2

Important inter- and intramolecular reactions, such as aromatic substitutions and in general oxidative additions to unsaturated systems[7] (olefins, quinones, imines, oximes etc.) fall within the general Scheme 2. In these cases the competition of the oxidation of the radical R. of the Scheme 2 is much less limiting than that of eqs.(3) and (4) in the reductive alkylation because the rate constants k_3 are generally much lower than k_4 and k_5.

A recent approach[8,5] which utilizes alkyl iodides as sources of alkyl radicals, is based on iodine-transfer chain processes. The key step is the fact that small differences in the strength of the C-I bonds are reflected in large variations of rates and equilibria of the iodine abstraction from alkyl iodides by carbon-centered radicals (eq.7)

$$R\text{-}I + R'\cdot \rightleftharpoons R\cdot + R'\text{-}I \tag{7}$$

Thus inter- and intramolecular syntheses, based on the general Scheme 3, have been developped[8].

$$R-I \longrightarrow R.$$

$$R. + S \longrightarrow RS.$$

$$RS. + R-I \xrightarrow{k_6} RS-I + R.$$

Scheme 3

The energy of the C-I bond higher in RS-I than in R-I and the consequent high rate constants k_6 are the driving force of these chain processes.

We[9] had already developped the same concept related to eq.(7) since 1982 in different areas of synthesis (mainly those falling within the Scheme 2), where the above mentioned processes are not suitable because a reducing medium (Scheme 1) is not compatible and iodine-transfer processes (Scheme 3) are unfavourable from enthalpic point of view.

Our synthetic developments are the result of combined enthalpic and polar factors. The enthalpic factor governs the equilibria of eq.(7), whereas the polar factor governs the selectivity of the reactions of the alkyl radicals. When both factors work in the same direction, that is the radical R. is favoured in the equilibrium of eq.(7) and it is more reactive than the radical R'. with suitable substrates for polar reasons, highly selective syntheses can be achieved. To fulfil the last condition the radical R. must be more nucleophilic than R'. with electron-deficient substrates and more electrophilic with electron-rich substrates.

At first we started[6,9a-c] with the most favourable kinetic conditions by using aryl radicals to abstract iodine from alkyl iodides (eq.8) because the enthalpic factor is the most favourable.

$$R-I + Ar. \xrightarrow{k} R. + Ar-I \qquad k > 10^9 \ M^{-1}s^{-1} \qquad (8)$$

The iodine abstraction is a very fast, practically irreversible process. Considering that the rate constants of most of the possible competitive reactions of the aryl radicals (included the reactions with the most common solvents) are in the range of $10^5-10^7 \ M^{-1}s^{-1}$ it results a very selective source of alkyl radicals.

We have utilized the source of alkyl radicals (8) in three synthetic areas: substitution of heteroaromatic bases[9b,c] and reductive alkylation of diazonium salts[6] by nucleophilic alkyl radicals and substitution of homocyclic aromatics by electrophilic alkyl radicals[9c].

The substitution of protonated heteroaromatic bases by nucleophilic carbon-centered radicals is a general reaction of great synthetic interest[7]; it reproduces most of the numerous aspects of the Friedel-Crafts aromatic substitution, but with opposite reactivity and selectivity. In utilizing the radical source (8) for this substitution the aryl radicals were generated from the two classical sources: aroyl peroxides and diazonium salts. With benzoyl peroxide the reaction stoichiometry is shown by eq.(9)

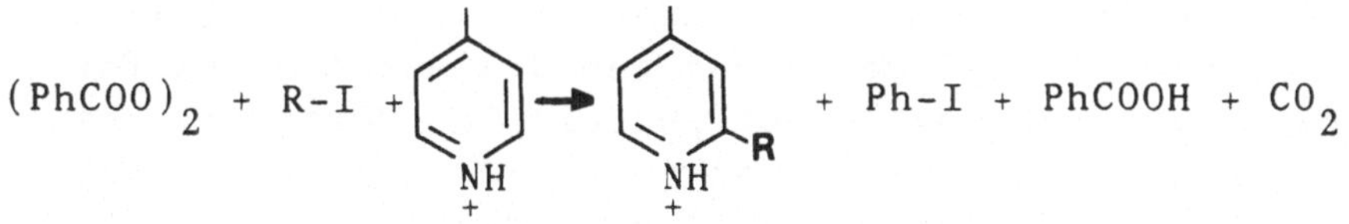

$$(PhCOO)_2 + R\text{-}I + \underset{\underset{+}{NH}}{\bigcirc} \longrightarrow \underset{\underset{+}{NH}}{\bigcirc}\text{-}R + Ph\text{-}I + PhCOOH + CO_2 \qquad (9)$$

A thermal or redox decomposition of the peroxide initiates a free-radical or redox chain according to the mechanism of the Scheme 4.

$$(PhCOO)_2 \longrightarrow 2PhCOO\cdot \longrightarrow 2Ph\cdot + 2CO_2$$

$$(PhCOO)_2 + Fe^{2+} \longrightarrow Fe^{3+} + PhCOO^- + PhCOO\cdot \longrightarrow Ph\cdot + CO_2$$

$$Ph\cdot + R\text{-}I \longrightarrow Ph\text{-}I + R\cdot$$

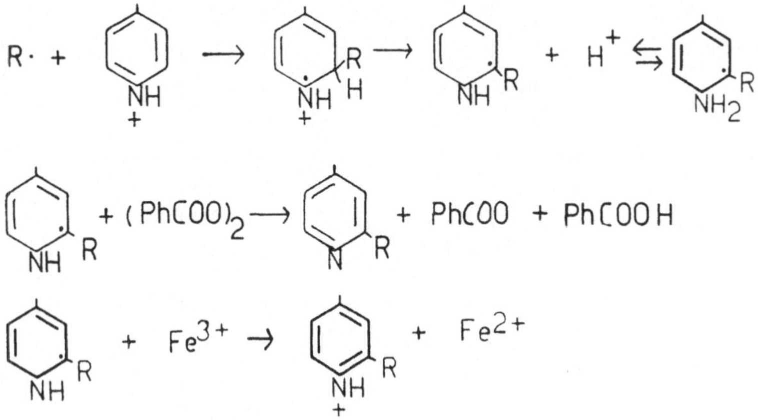

Scheme 4

The stoichiometry with diazonium salts is shown by eq.(10)

$$\text{Pyridine} + R\text{-}I + ArN_2^+ \longrightarrow \text{Pyridine-}R + Ar\text{-}I + N_2 + H^+ \quad (10)$$

The mechanism is similar to that of Scheme 4 with the difference that the aryl radicals are generated according to the eq.(11)

$$ArN_2^+ + Fe^{2+} \longrightarrow Ar\cdot + N_2 + Fe^{3+} \quad (11)$$

Even if both reactions (eqs.9 and 10) give good results in many cases (Table I), there are several structural limitations. The use of acyl peroxides and diazonium salts for large scale preparations is not practical. Moreover, the reaction does not work with t-alkyl iodides and acyl peroxides, which do not lead to t-alkyl radicals because of competitive ionic reaction.

TABLE I - Alkylation according to eq.34

Substrate	RI	Orientation (%)	Yields(%)
4-Cyanopyridine	n-BuI	2(75); 2.6(25)	96
"	i-BuI	2(58); 2.6(42)	98
"	i-Pr	2(66); 2.6(34)	100
Isoquinoline	EtI	1(100)	85
"	$C_6H_{11}I$	1(100)	92
Quinaldine	"	4(100)	88
"	n-BuI	4(100)	93
"	i-BuI	4(100)	98
Lepidine	n-BuI	2(100)	88
"	i-PrI	2(100)	98
"	$C_6H_{11}I$	2(100)	95
"	$EtoCOCH_2CH_2I$	2(100)	93
"	I OH	2(100)	85
Acridine	$C_6H_{11}I$	9(100)	94
Benzothiazole	i-PrI	2(100)	90

A further limitation is due to the reactivity of aroyloxy radicals: even if the decarboxylation is a fast process,
other competitive reactions can be faster. In particular
aroyloxy radicals add fast to electronrich olefins (eq.12)
and aromatics (eq.13) and abstract hydrogen from C-H bond activated by enthalpic and polar effects (alcohols, ethers,
amines) (eq.14)

$$ArCOO^{\cdot} \quad + \quad \rangle = \langle \quad \longrightarrow \quad ArCOO - \overset{|}{\underset{|}{C}} - \overset{|}{\underset{|}{C}} \qquad (12)$$

$$(13)$$

$$ArCOO^{\cdot} \quad + \quad R\text{-}H \quad \longrightarrow \quad ArCOOH \quad + \quad R^{\cdot} \qquad (14)$$

The main limitation with diazonium salt is due to the
high addition rates of the nucleophilic radicals to the diazonium group[6], which compete with the heterocyclic sobsti-
tution and under stoichiometric redox conditions lead to the
"free-radical diazocoupling" (eq.15)

$$RI + 2ArN_2^+ + 2Fe^{2+} \longrightarrow R\text{-}N = N\text{-}Ar + ArI + N_2 + 2Fe^{3+} \quad (15)$$

The mechanistic sequence involves eqs.(11), (8), (5) and (16)

$$R\text{-}N = \overset{+}{\underset{\cdot}{N}}\text{-}Ar + Fe^{2+} \longrightarrow R\text{-}N = N\text{-}Ar + Fe^{3+} \qquad (16)$$

When the alkyl iodide bears electronwithdrawing groups
(COOR, CN, COR, NO_2, SO_2R etc.) in α -position the iodine
abstraction by the aryl radical is still faster than from
unsubstituted alkyl iodides,because the corresponding
radical is stabilized by the substituent, the C-I bond is
weakened and the enthalpic factor governs kinetics and equilibria of the abstraction. However, the resulting radicals
have electrophilic character and do not react with protonated heteroaromatic bases or diazonium salts, whereas they
react fast with electron-rich aromatics or olefins. Thus,
for example, the radical $.CH_2COOH$, generated from $I\text{-}CH_2COOH$
and $(PhCOO)_2$, does not react with protonated heteroaroma-

tic bases, but it easily reacts with benzene according to eq.17

$$\text{(benzene)} + ICH_2COOH + (PhCOO)_2 \longrightarrow \text{(ring with } CH_2COOH) + PhI + PhCOOH + CO_2 \tag{17}$$

The mechanism involves the selective iodine abstraction by the phenyl radical (eq.18), addition to the benzene ring of the electrophilic radical (19) and rearomatization in a chain process (eq.20), which can be made more selective in the presence of redox catalysis (eqs.21, 22)

$$HOOCCH_2I + Ph. \longrightarrow HOOCCH_2. + PhI \tag{18}$$

$$\text{(benzene)} + \cdot CH_2COOH \longrightarrow \text{(cyclohexadienyl radical with } H\ CH_2COOH) \tag{19}$$

$$\text{(cyclohexadienyl radical, } H\ CH_2COOH) + (PhCOO)_2 \longrightarrow \text{(ring with } CH_2COOH) + PhCOOH + PhCOO \cdot \tag{20}$$

$$\text{(cyclohexadienyl radical, } H\ CH_2COOH) + Fe^{3+} \longrightarrow \text{(ring with } CH_2COOH) + Fe^{2+} \tag{21}$$

$$(PhCOO)_2 + Fe^{2+} \longrightarrow PhCOO. + PhCOO^- + Fe^{3+} \tag{22}$$

It is sufficient, however, that the electronwithdrawing group is in β-position and the polar character of the radical is drastically changed and in benzene solution (therefore with benzene in large excess), only the heteroaromatic base is attacked without significant addition to the solvent[9c] (eq.23)

$$+ \text{JCH}_2\text{COOH} + (\text{PhCOO})_2 \longrightarrow \text{(ring with CH}_2\text{CH}_2\text{COOH)} + \text{PhI} + \text{PhCOOH} + \text{CO}_2 \quad (23)$$

(ring, +NH) (ring, +NH, CH$_2$CH$_2$COOH)

The intrinsic limitation, which reduces the range of application of reactions of the type (17), is the fact that the synthetic interest increases with the electron-rich nature of the substrates (aromatics and olefins), due to the electrophilic character of the alkyl radicals involved, but at the same time the competitive reactions (12) and (13) become more important. The use of diazonium salt, in this sense, is less limiting.

All these results and considerations suggested that more general, simple, practical and cheap sources of alkyl radicals from alkyl iodides would have been of undoubted interest.

In pursuing this purpose we[9d] have followed the same cencept related to eq.(8), by using the methyl radical to abstract the iodine atom instead of an aryl radical. The methyl radical is the least stable among the alkyl radicals, the strength of the C-I bonds ranging from 56.5 Kcal/mol for CH_3-I to 52.1 Kcal/mol for t-Bu-I[10]. Since rates and equilibria[10] (Table II) are also in this case governed by the enthalpic factor it was possible to use a source of methyl radical to general alkyl radicals of any kind according to the equilibria of eq.24

$$\text{CH}_3\cdot + \text{RI} \rightleftharpoons \text{CH}_3\text{I} + \text{R}\cdot \quad (24)$$

Thus we have developed six general, simple, practical and cheap sources of alkyl radicals from alkyl iodides, based on the reaction (24). At first we have utilized these radical sources for the selective heteroaromatic substitution; subsequently we have utilized the same sources for a variety of different syntheses.

i) A clean source of the methyl radical is offered by the thermal (eq.25) or redox (eq.26) decomposition of diacetylperoxide

TABLE II - Equilibrium constants[10],
K, for the reaction

$$CH_3 \cdot + RI \rightleftharpoons CH_3I + R \cdot$$

R.	K
Et.	20.1
i-Pr.	468
t-Bu.	1.7×10^4

$$(MeCOO)_2 \longrightarrow 2Me \cdot + 2CO_2 \tag{25}$$

$$(MeCOO)_2 + Fe^{2+} \longrightarrow Me \cdot + MeCOO^- + CO_2 + Fe^{3+} \tag{26}$$

The stoichiometry of the heteroaromatic substitution is shown by eq.(27)

$$> 90\% \tag{27}$$

The mechanism is identical to that described by the Scheme 4; the only difference is the source of the alkyl radical (eq.24). However, the equilibria of eq.(24) (Table II) are less favourable than those of eq.(8) and the fact that they are shifted at right is not in itself a sufficient condition to have a high selectivity because the reaction rates of Me. and R. radicals can be quite different. Thus when the enthalpic factor governs the reactivity, as for the results of Table II, the methyl radical is more reactive than primary, secondary, tertiary and generally α-substituted alkyl radicals; that can counterbalance the unfavourable equilibria. The radical source can become selective when polar effects are important. Now, the addition rates of the alkyl radicals to protonated heteroaromatic bases are strongly affected by the nucleophilic character of the radicals with a transition state similar to a charge-transfer complex[7] (eq.28)

$$(28)$$

When the radical R. is more nucleophilic than the Me. radical (practically all the primary, secondary and tertiary alkyl radicals without electron-withdrawing groups in -position), the faster addition to the heterocyclic ring (eq.28) together with the favourable equilibrium (eq.24) makes the overall reaction highly selective and only the radical R. is involved in the substitution.

The great advantage of using $(MeCOO)_2$ compared with $(PhCOO)_2$ is due to the fact that the radical MeCOO. decarboxylates much faster than ArCOO. (intermolecular reactions of MeCOO. are practically not known) so that the limitations discussed with $(ArCOO)_2$ (eqs.12-14) do not exist with $(MeCOO)_2$. The disadvantage is due to the fact that $(MeCOO)_2$ is not a commercial product because of its dangerousness as pure compound. However, solutions of $(MeCOO)_2$ in MeCOOH can be easily prepared from Ac_2O and Na_2O_2 and these solutions can be utilized with safety for the heteroaromatic substitution.

ii) Hydrogen peroxide reacts with alkyl iodides and protonated heteroaromatic bases in DMSO in the presence of catalytic amount of Fe(II) salt leading to a highly selective alkylation of the heteroaromatic ring (eq.29)

$$(29)$$

90%

A complex, but selective, redox chain is involved. The first interaction involves the well-known redox decomposition of H_2O_2 by Fe(II) salt (eq.30)

$$H_2O_2 + Fe^{2+} \longrightarrow HO\cdot + HO^- + Fe^{3+} \tag{30}$$

The high reactivity and the low selectivity of the HO. radical with a large variety of organic and inorganic compounds is controlled by using DMSO as solvent. The HO. radical reacts fast also with DMSO (eq.31) and the possible, fast, unselective and competitive reactions with other substrates are minimized by the excess of solvent. The radical adduct selectively undergoes β-scission, acting thus as selective source of Me. radical

$$HO\cdot + MeSOMe \longrightarrow Me - \overset{\displaystyle HO \quad O\cdot}{\underset{}{S}} - Me \longrightarrow MeSO_2H \quad + \quad Me^\cdot \tag{31}$$

The further steps of the heteroaromatic substitution are similar to those of i).

In this way also complex substrates, such as iodosugars can be utilized with good selectivity. For example, 6-iodo-1,2,3,4-diisopropyliden-α-galactose reacts with lepidine giving selectively the corresponding C-nucleoside (eq.32)

$$ \tag{32}$$

Since a large variety of iodosugars is available, this new method seems particularly suitable for the synthetic approach to C-nucleosides with heteroaromatic bases, included the purine bases, characterized by great interest for the biological activity (i.e. tiazofurin and selenoazofurin have been shown to be effective against L1210 leucemia and Lewis lung carcinoma in mice[11].)

iii) The redox decompostion of t-BuOOH (eq.33) is another
simple and cheap source of methyl radical

$$Fe^{2+} + t\text{-BuOOH} \longrightarrow Fe^{3+} + OH^- + t\text{-BuO} \cdot \longrightarrow Me \cdot + MeCOMe \quad (33)$$

High temperatures and protic solvents favour the -scis-
sion of t-BuO. radical compared with other competitive
intermolecular reactions. In this way a simple, cheap
and effective source of methyl radical was obtained and
the stoichiometry of the heteroaromatic alkylation is
shown by the eq.34

$$(34)$$

> 90%

iv) Thermal or photochemical decomposition of $(t\text{-BuO})_2$ also
provides t-BuO. and Me. radicals (eq.35)

$$(t\text{-BuO})_2 \longrightarrow 2t\text{-BuO} \cdot \longrightarrow 2Me \cdot + 2MeCOMe \quad (35)$$

A redox catalysis by Fe(II)/Fe(III) is useful also in
this case for a selective rearomatization of the pyri-
dinyl-type radical intermediate (eq.36)

$$(36)$$

The Fe(II) salt is not reoxidized by $(t\text{-BuO})_2$, but by
t-BuO. .

v) The thermal decomposition of H_2O_2 in acetone and catalytic amounts of acids proved to be another simple and
cheap source of methyl radical, useful for the heteroaromatic alkylation (eq.37)

$$\text{(pyridinium)} + \text{R-I} + \text{CH}_3\text{COCH}_3 + \text{H}_2\text{O}_2 \longrightarrow \text{(pyridinium-R)} + \text{CH}_3\text{I} + \text{CH}_3\text{COOH} + \text{H}_2\text{O} \tag{37}$$

The methyl radical is generated by the thermal decomposition of the acetone peroxide (eq.38)

$$\text{Me}_2\overset{\text{OH}}{\underset{\text{C}}{|}}\text{-O-O-}\overset{\text{OH}}{\underset{\text{CMe2}}{|}} \longrightarrow 2\text{Me}_2\overset{\text{OH}}{\underset{\text{C-O.}}{|}} \longrightarrow 2\text{MeCOOH} + 2\text{Me.} \tag{38}$$

In this case the redox catalysis by metal salts is harmful because the decompositon of H_2O_2 takes place and no substitution occurs.

vi) The thermal or redox decomposition of $S_2O_8^{2-}$ in the presence of the acetate of the heteroaromatic base and alkyl iodides leads to the selective alkylation of the heteroxyclic ring (eq.39)

$$\text{(pyridine)} + \text{CH}_3\text{COO}^- + \text{RI} + \text{S}_2\text{O}_8^{2-} \longrightarrow$$
$$\longrightarrow \text{(pyridine-R)} + \text{CH}_3\text{I} + \text{CO}_2 + 2\text{SO}_4^{2-} + \text{H}^+ \tag{39}$$

The overall mechanism is similar to those described in the previous sections. The methyl radical is generated by electron-transfer oxidation of the acetate ion (eq.40)

$$\text{MeCOO}^- + \text{SO}_4^{-\cdot} \longrightarrow \text{Me.} + \text{CO}_2 + \text{SO}_4^{2-} \tag{40}$$

These new methods (i-vi) to generate alkyl radicals from alkyl iodides of any kind are more convenient than those based on the use of aroyl peroxides and diazonium salts. The structural limitations with aroyl peroxides and diazonium salts are eliminated, the reagents are cheaper, the experimental conditions more simple. They can be utilized for other selective syntheses by nucleophilic alkyl radicals and electrondeficient substrates, such as quinones, diazonium salts, pyrilium salts, iminium salts, electrondeficient olefins.

Thus we have utilized all the procedures i-vi for the alkylation of quinones in chain processes. The alkyl radicals arising from eq.24 react fast with quinones (eq.41) and the radical adduct is oxidized by the radical source in a chain process (i.e. eq.42)

$$ \qquad (41) $$

$$ \qquad (42) $$

Polar effects are important in the addition of nucleophilic radicals to quinones; however also the enthalpic factor related to the stability of the intermediate radical adduct, plays a significant role. Thus also the methyl radical reacts fast with quinones and its addition to the quinone ring competes, at some extent, with the iodine abstraction, particularly from primary alkyl iodides, characterized by the less favourable equilibrium (Table II). This competition can be overcome by using an excess of alkyl iodide or by keeping low the stationary concentration of the quinone during the reaction. Polar effects are dominant also in the addition of the nucleophilic alkyl radicals to the diazonium group leading to the selective diazocoupling reaction (formation of alkylaryl azocompounds). This is not a chain process, but stoichiometric amounts of reducing metal salts are required so that only the procedure ii) is particularly suitable[9d]; it allows to work at low temperature with stoichiometric amount of Fe(II) salt and involves the sequence of eqs.(30), (31), (24), (5), (16).

When the alkyl iodide bears an electron-withdrawing group (COOH, COR, CN, NO_2, SO_2R etc.) in the α position, the iodine abstraction by the methyl radical is still more favourable for polar and enthalpic reasons. However, the resulting radical has electrophilic character and does not react with electrondeficient substrates, such as protonated heteroaromatic bases, diazonium salts, pyrilium salts etc. Reversing the polar character of the radical, it is necessary to reverse the polarity of the substrate in order to re-establish the thermodynamic and kinetic conditions for selective syntheses, by using electron-rich substrates (aromatics and olefins). Thus, for example, the use of iodoacetic acid with anisole (eq.43) or naphtalene (eq.44) determines particularly favourable conditions because the radical $.CH_2COOH$ is present in large excess in the equilibrium with the methyl radical (eq.45) and it is more reactive than the methyl radical in the addition to the aromatic ring.

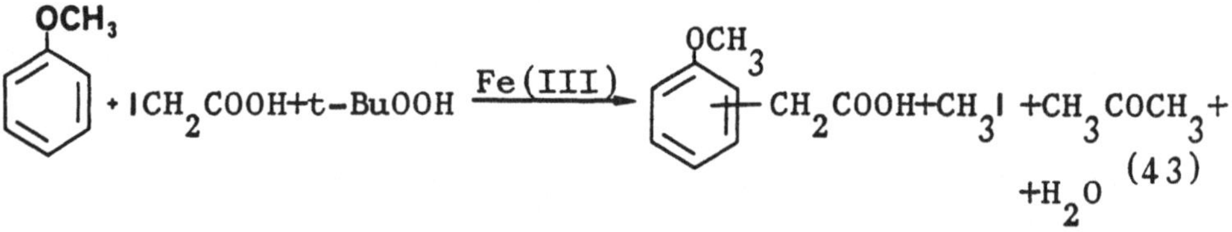

$$
\text{OCH}_3\text{-C}_6\text{H}_5 + ICH_2COOH + t\text{-BuOOH} \xrightarrow{Fe(III)} \text{OCH}_3\text{-C}_6\text{H}_4\text{-}CH_2COOH + CH_3I + CH_3COCH_3 + H_2O \quad (43)
$$

0% 79, m% 5, p% 16

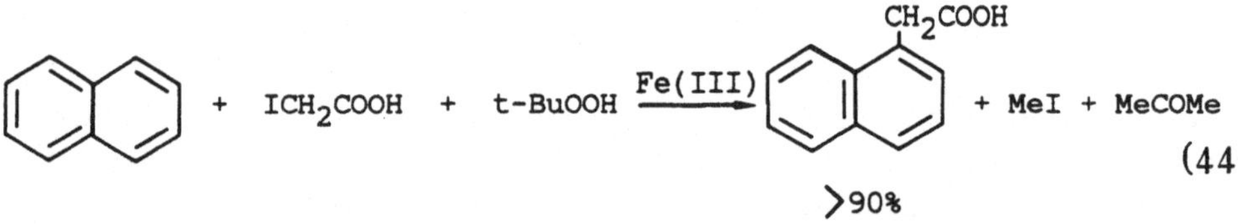

$$
\text{C}_{10}\text{H}_8 + ICH_2COOH + t\text{-BuOOH} \xrightarrow{Fe(III)} \text{C}_{10}\text{H}_7\text{-}CH_2COOH + MeI + MeCOMe \quad (44)
$$

$>90\%$

$$
Me. + ICH_2COOH \rightleftharpoons MeI + .CH_2COOH \quad (45)
$$

REFERENCES

1. B. Giese, "Radicals in Organic Syntheses : Formation of Carbon-Carbon Bonds". Ed. J.E. Boldwin, Pergamon Press, Oxford, 1986.
2. C. Chatgilialoglu, K.U. Ingold and J.C Scaiano, J.Am. Chem.Soc. 1982, 104, 5123.
3. C. Chatgilialoglu, K.U. Ingold and J.C. Scaiano, J.Am. Chem.Soc. 1981, 103, 7739.
4. A. Citterio, F. Minisci and M. Serravalle, J.Chem.Res.(S) 1981, 198.
5. W. P. Neumann, Synthesis, 665, 1987.
6. A. Citterio and F. Minisci, J.Org.Chem. 1982, 47, 1759.
7. F. Minisci, Top.Curr.Chem. 62, 1, 1976; Acc.Chem.Res. 1975, 8, 165; "Fundamental Research in Homogeneous Catalysis" Plenum Publ.Corp., vol.4, 173, 1984; F. Minisci and E. Vismara, "Organic Synthesis: Modern Trends". Ed. O.Chizhov, 229, 1987.
8. See for example, D.P. Curran and M.H. Chen, J.Am.Chem.Soc. 1987, 109, 6558.
9.a)Ref.6; b) F.Minisci, V. Tortelli, E. Vismara and G. Castoldi, Tetrahedron Letters 25, 3897 (1984); c) F. Minisci, E. Vismara, F. Fontana, G. Morini, M. Serravalle and C. Giordano, J.Org.Chem. 1986, 51, 4411; d) F. Fontana, F. Minisci and E. Vismara, Tetrahedron Letters 28, 6373, 1987.
10. J.A. Hawari, J.M. Kanabus-Kamiuska, D.D.M. Waymer and D. Griller, "Substituent Effects in Radical Chemistry", H. G. Vihe Ed., D. Reidel Publishing Co., 91, 1986.
11. R.K. Robins and G.R. Revankar, Med.Res.Rev. 5, 273 (1985).

INTRODUCTION OF A SINGLE ELECTRON-WITHDRAWING CARBON-SUBSTITUENT INTO π-DEFICIENT N-HETEROARENES BY MEANS OF FREE-RADICAL REACTIONS: STRATEGIES AND SYNTHETIC APPLICATIONS

Gottfried Heinisch
Institute of Pharmaceutical Chemistry,
University of Vienna
Währinger Straße 10
A-1090 Wien, Austria

ABSTRACT - π-Deficient N-heteroarenes, having two or more ring carbon-atoms free that can be attacked by a nucleophilic carbon-centered radical, in general yield polysubstituted products in homolytic alkoxy-carbonylation and homolytic aroylation reactions. It is demonstrated that the problem of multiple substitution can be overcome by performing the alkoxycarbonylation reaction in the presence of an appropriate organic solvent. Moreover it was found in pyrazine and pyridazine series that also the homolytic aroylation can be restricted to monosubstitution by employing N-heteroarene carboxylic acids; decarboxylation of the keto acids obtained provides convenient access to diaza-benzophenones. Substitution reactions by carbon-centered radicals thus are shown to be of high preparative value also with regard to the introduction of a single electron-withdrawing substituent into N-heteroarenes.

π-Deficient N-heteroaromatic systems play an important role as subunits of a large variety of valuable drug molecules. Accordingly, methods permitting the regioselective introduction of carbon functional groups into ring carbon atoms of such heteroarenes are of high interest in medicinally orientated organic synthesis. Since electron-poor aromatics in general are not susceptible to conventional electrophilic substitution processes, much attention has been paid over the last years

a) to the directed metalation of N-heteroarenes and reactions of the species thus obtained with electrophiles[1] and
b) to reactions of protonated N-heteroarenes with nucleophilic carbon-centered radicals (*Minisci*-type reactions[2]).

From the point of view of medicinal chemistry the latter methodology appears of particular practical value, since the reactions can be performed under extremely simple experimental conditions in aqueous medium and a wide range of easy-to-generate radicals can be employed. The homolytic alkylation, α-hydroxymethylation, α-alkoxyalkylation, α-N-amido-alkylation and amidation reactions are now well established methods.

71

F. Minisci (ed.), Free Radicals in Synthesis and Biology, 71–79.
© *1989 by Kluwer Academic Publishers.*

Their application in syntheses of substituted pyridines, pyridazines, pyrimidines, pyrazines and fused derivatives thereof has been reviewed extensively[2,3,4].

The radical alkoxycarbonylation, acylation and aroylation, however, until very recently were considered to be of rather limited preparative value, since these reactions in general result in polysubstitution un- less the heteroaromatic substrate has only a single free ring position that can be attacked by a nucleophilic radical. The reactions displayed in Scheme 1 may serve as illustrative examples. The favoured formation of disubstituted products is explained by the markedly increased elec- tron-deficiency in the initially formed monosubstitution products, which thus exhibit enhanced reactivity towards a nucleophilic radical[2].

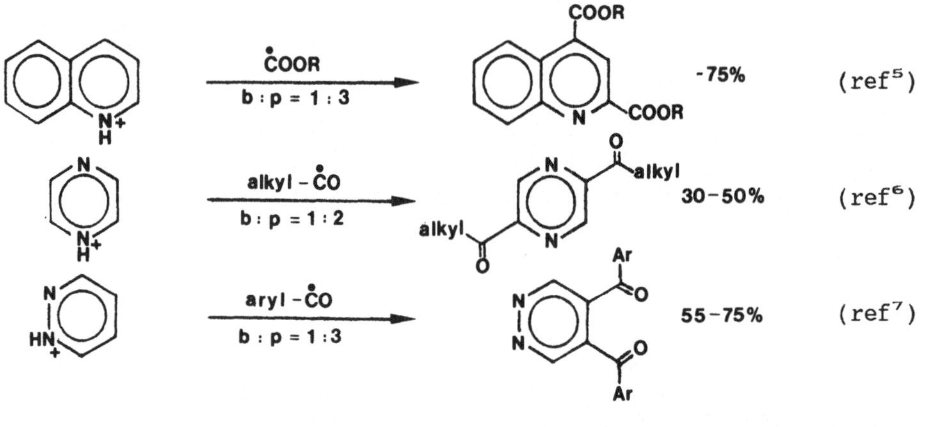

SCHEME 1

With N-heteroaromatics having more than one reactive position free the introduction of a single electron-attracting carbon-substituent so far only could be achieved by using a large excess of the heterocyclic base and consequently at very low conversion rates. The successful syntheses of monoacyl quinoxalines[8] and monoacyl 4-cyano-pyridines[6] in reasonable yields by means of homolytic acylation represent exceptional cases, which arise from the fact that these compounds precipitate from the reaction mixtures.

The problem of multiple substitution in radical alkoxycarbonylation and aroylation reactions now could be overcome by applying a simple two- phase-system technique or by employing heteroaromatic carboxylic acids as protected species, respectively.

Introduction of a single carboxylic ester function into π-deficient N-heteroarenes

When an alkoxycarbonyl group is introduced into an azine or a diazine molecule a decrease of the basicity of the system as well as an increase of its lipophilicity will result. Therefore it should be possible to remove the initially formed monoalkoxycarbonylated product from the acidic aqueous reaction medium simply by carrying out the radical substitution in the presence of an appropriate organic solvent. In preliminary experiments using diethylether, toluene and dichloromethane the latter proved to be most suitable to suppress additional radical attack.

The alteration of product distribution in the homolytic ethoxycarbonylation of 4-pyridinecarbonitrile, caused by addition of varying amounts of dichloromethane to the reaction medium[9], is illustrated in Table I. Even with a large excess of radicals, sufficient for improving the conversion rate up to 100%, multiple substitution is drastically suppressed.

Table I: Product distribution (mol-%) in the homolytic ethoxycarbony-lation of 4-pyridinecarbonitrile (10 mmol)

base:peroxide ratio	ml CH$_2$Cl$_2$ added	conversion rate (%)	products		not identi-fied
3:1 [5]	0	19	14	3	3
1:3	0	76	36	38	2
1:3	30	94	80	6	9
1:3	150	51	42	3	6
1:10	150	100	85	4	11

Similar results are obtained with 3-pyridinecarbonitrile: as shown in Table II, again monosubstitution is clearly favoured, when the reaction is carried out in the presence of an organic layer. This educt of course affords a mixture of cyanopyridinecarboxylic acid esters due to three reactive positions free.

This simple two-phase-system procedure also provides convenient access to a so far unexplored class of compounds, namely "mixed esters" derived from N-heteroaromatic dicarboxylic acids.

Similar to alkyl cyanopyridinecarboxylates, which hitherto were accessible only by cumbersome multistep-syntheses, such compounds may be
anticipated to represent versatile synthons for the construction of
differently disubstituted pyridines, particularly when there is an
extreme difference in the reactivity of the two ester functions like
with t-butyl methyl pyridinedicarboxylate. In an analoguous manner also
in the pyridazine series "mixed esters" derived from dicarboxylic acids
are now available in satisfactory yields.

With C-alkylsubstituted N-heteroarenes somewhat lower yields are
observed due to expectedly reduced conversion rates caused by desacti-
vation of the π-deficient system by the alkyl group. However, with 4-
picoline or 4-alkylpyridazines again substitution could be restricted to
the attack of a single ethoxycarbonyl radical upon addition of dichloromethane to the reaction medium. This approach also enabled us to prepare
1,2-diazines bearing a masked aldehyde function as well as a single
ester group.

Table II: Homolytic alkoxycarbonylation of heteroaromatic
          substrates in a two-phase system

educts	products (% yield of isolated, analytically pure material)			remarks
	33%	22%	21%	separable by mplc[10]
	83%			ref[10]
	81%			ref[10]
	30%	25%		separable by mplc[10]
	90%			ref[11]
	74%			ref[10]
	75%			ref[11]

Table II. continued

educts	products (% yield of isolated, analytically pure material)	remarks
	53%	ref[12]
	R=H 39% R=CH₂Ph 70% R=Ph 50%	ref[13] ref[13] ref[11]
	32%	ref[11]
	89%	ref[12]
	54%	ref[11]
	65%	ref[13]
	67%	ref[13]
	31%	ref[16]

Finally, alkoxycarbonylation in the presence of an appropriate amount of dichloromethane permits the preparation of 2-pyrazine- and 4-pyrimidinecarboxylic acid esters in a single step from the parent heteroaromatics. In the 1,2-diazine series this procedure provides convenient access to 4,5-dicarboxylates, when pyridazine or 3-alkyl-derivatives thereof are employed as the educts as shown in Table II.

It has to be stressed that in accordance with previous findings in homolytic substitution reactions with pyridazines[4], the 1,2-diazine system is attacked by alkoxycarbonyl radicals preferentially at the ß-carbon atoms C-4 and C-5, which is in accordance with LUMO-energies calculated for the protonated 1,2-diazine system[14]. With this hetero-cycle the synthetic utility of the radical alkoxycarbonylation under standard conditions is restricted not only by additional substitution at C-3 and C-6 but also by the formation of substantial amounts of N-ethoxycarbonylated products. According to recent investigations[15], this exceptional radical attack at the nitrogen-atom in *Minisci*-type reactions with heteroarenes most probably occurs as outlined in Scheme 2.

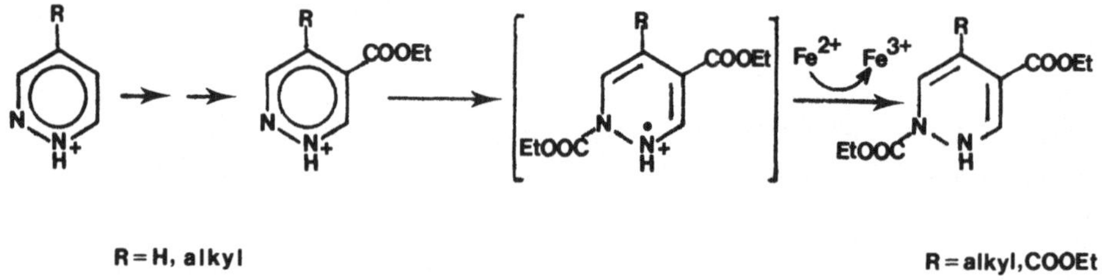

R = H, alkyl   R = alkyl, COOEt

SCHEME 2

By applying the above discussed two-phase-system technique the initially formed 4,5-disubstituted pyridazine derivatives, however, can be protected from further radical attack at the α-carbon atoms as well as at the nitrogen-atom. Some selected examples given in Scheme 3 show that compounds of this type represent versatile key intermediates in syntheses of so far not accessible diaza-analogs of pharmaceutically interesting bi- and tricyclic systems[16,17].

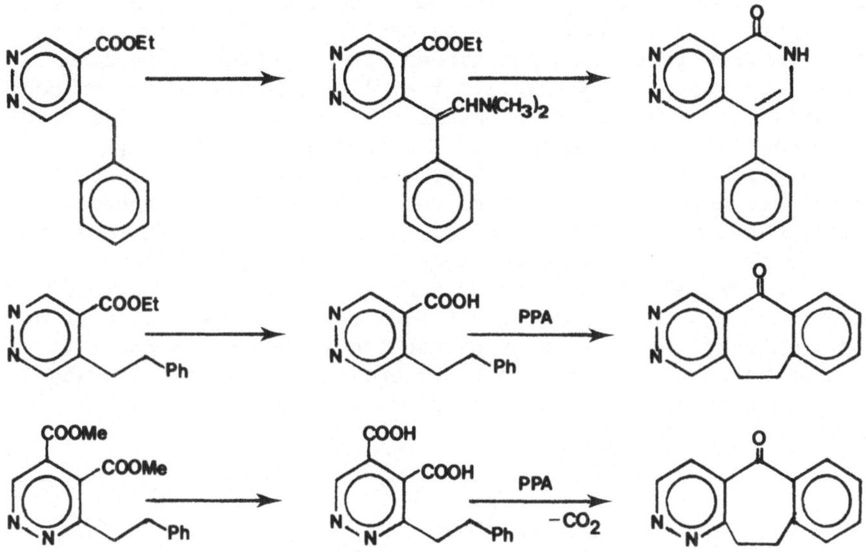

SCHEME 3

Introduction of a single aroyl group into π-deficient N-heteroarenes

In view of the widespread occurence of the diarylmethyl moiety in biologically active compounds the benzophenones are important starting materials in drug syntheses. For the preparation of monoaroyl-N-hetero-arenes (= aza-benzophenones) the two-phase strategy to suppress multiple substitution is not applicable. The pronounced lipophilicity of aromatic carbaldehydes, serving as precursors for aroyl radicals, requires the performance of these reactions in the presence of acetic acid and thus in a homogeneous medium. In reactions of protonated pyrazine employing acyl radicals generated from lower alkylaldehydes it is possible to avoid the addition of the co-solvent, as recently shown by Williams and co-workers, who succeeded in the preparation of monoacylpyrazines in reasonable yields[18] by carrying out the reactions in a heterogeneous medium (water/alkylaldehyde)*. Attempts to obtain pyrazine-derived diaza-benzophenones under these conditions, however, resulted in yields < 10%[18].

----

* It should be emphasized that this strategy for restricting acylation to the introduction of a single substituent, which resembles closely to our two-phase-system procedure, was developed independently by the above mentioned authors short time after our first preliminary note on this subject[9].

Considering that carboxylic acids derived from π-deficient N-heteroarenes in general can be decarboxylated easily (particularly when an
additional electron-attracting substituent is attached to the ring
system) a simple two-step synthesis for aryl pyrazinylketones now could
be developed[19]. As illustrated in Scheme 4, 2-pyrazinecarboxylic acid
affords reasonable yields (30-40%) of C-5 aroylated products (R= 4-CH$_3$O,
4-F, 3-Cl, 4-Cl) upon treatment with aromatic carbaldehydes/t-BuOOH/
FeSO$_4$ in the presence of acetic acid. Due to significantly enhanced
lipophilicity of these products, their isolation in turn can be accomplished simply by addition of water. Upon heating in a Kugelrohr-
apparatus decarboxylation takes place smoothly to give the desired
benzophenone-analogs in satisfactory yields (60-90%). This concept of
employing a carboxylic function as protecting group in the homolytic
aroylation of heteroarenes, having two reactive positions free, was
found to be of high synthetic utility also in pyridazine series[20,21]
(see Scheme 4).

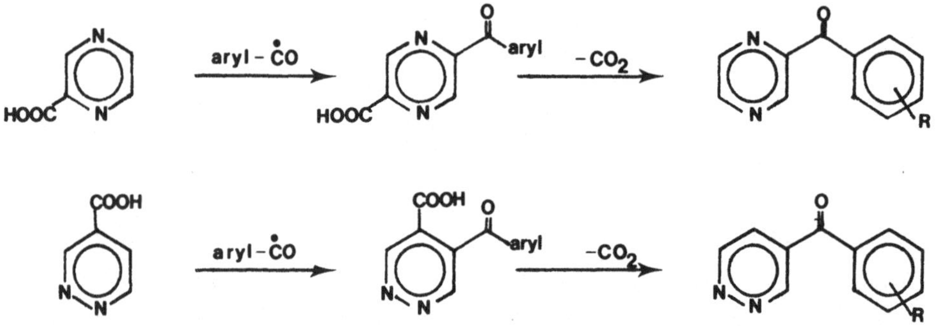

SCHEME 4

Conclusion

The discussed results clearly demonstrate the high preparative value
of homolytic alkoxycarbonylation reactions also with π-deficient N-
heteroaromatic substrates having more than one reactive position free.
Performance of these reactions in the presence of a suitable organic
layer prevents the initially formed monosubstitution product from
further radical attack. The problem of multiple substitution in homolytic aroylation reactions can be overcome by using N-heteroarene-
carboxylic acids as starting materials, as shown with pyrazine and pyridazine. In this case the carboxylic function serves as a protecting
group, which easily can be removed after an aroyl moiety has been introduced into the heteroaromatic system.

Acknowledgement

The author gratefully acknowledges support of these investigations by the "Fonds zur Förderung der wissenschaftlichen Forschung" (Projekt No.P6260). The author also wants to express his gratitude to all co-workers mentioned in the references, in particular to Dr.G.Lötsch, for enthusiastic co-operation.

REFERENCES

1. M. Mallet and G. Quéguiner, *Tetrahedron*, **38**, 3035 (1982)
    F. Marsais and G. Quéguiner, *Tetrahedron*, **39**, 2009 (1983);
    F. Marsais, A. Cronnier, F. Trecourt and G. Quéguiner,
    J.Org.Chem., **52**, 1133 (1987).
2. F. Minisci, *Synthesis*, **1973**, 1; F. Minisci, *Top.Curr.Chem.*, **62**, 1
    (1976); F. Minisci in *"Substituent Effects in Radical Chemistry"*
    (ed.by H.Viehe et al.), D.Reidel Publishing Co., Dordrecht, **1986**,
    p. 391.
3. T. Sakamoto and H. Yamanaka, *Heterocycles*, **15**, 583 (1981).
4. G. Heinisch, *Heterocycles*, **26**, 481 (1987).
5. R. Bernardi, T. Caronna, R. Galli, F. Minisci and M.Perchinunno,
    Tetrahedron Lett., **1973**, 645.
6. T. Caronna, G. Fronza, F. Minisci and O.Porta, *J.Chem.Soc. Perkin
    Trans II*, **1972**, 2035.
7. M. Braun, G. Hanel and G. Heinisch, *Monatsh.Chem.*, **109**, 63 (1978).
8. G. Gardini and F. Minisci, *J.Chem.Soc.[C]*, **1970**, 929.
9. G. Heinisch and G. Lötsch, *Angew.Chem.Int.Ed.Engl.*, **24**, 692 (1985).
10. G. Heinisch and G. Lötsch, *Heterocycles*, **26**, 731 (1987).
11. G. Heinisch and co-workers, unpublished results.
12. G. Heinisch and G. Lötsch, *Tetrahedron*, **42**, 5973 (1986).
13. G. Heinisch and G. Lötsch, *Tetrahedron*, **41**, 1199 (1985).
14. G. Quéguiner and A. Turck, personal communication.
15. M. Gebauer, G. Heinisch and G. Lötsch, *Tetrahedron*, in the press.
16. W. Dostal, G. Heinisch and G. Lötsch, *Monatsh.Chem.*, in the press.
17. P. Boamah, N. Haider, G. Heinisch and J. Moshuber, *J.Heterocycl.
    Chem.*, in the press.
18. Y. Houminer, E. Southwick and D.Williams, *J.Heterocycl.
    Chem.*, **23**, 497 (1986).
19. G. Heinisch and G. Lötsch, *Synthesis*, **1988**, 119.
20. G. Heinisch and I. Kirchner, *Monatsh.Chem.*, **110**, 365 (1979).
21. G. Heinisch, I. Kirchner, I. Kurzmann, G. Lötsch and
    R. Waglechner, *Arch.Pharm.(Weinheim)*, **316**, 508 (1983).

FREE RADICAL SYNTHESIS OF PYRIDYLDERIVATIVES: INFLUENCE OF THE SOLVENT

Rosanna Bernardi, Tullio Caronna*,
Dianella Coggiola and Sergio Morrocchi
Dipartimento di Chimica, Politecnico di Milano
Piazza Leonardo da Vinci 32
20133 Milano, ITALY

ABSTRACT. The photochemically induced reactions of the azaheteroaromatic bases are strongly affected by the medium. For example, changing the acidity of the solution results in a change of mechanism of the reaction from that of hydrogen abstraction to the electron-transfer. 2,4-Dicyanopyridine (1) is a substrate that works only by electron-transfer, but the two charged species, radical anion and radical cation, created by photochemical excitation give rise to very different reactions depending on the polarity of the solvent. In non polar medium, if (1) is photochemically excited in the presence of an alkene, substitution of a cyano group with an allyl group occurs, resulting in the formation of 2-cyano-4-allyl pyridine and 4-allyl-2-cyano pyridine. In polar solvents, however, there is the formation of a new ring between position 5 of the heterocyclic ring and the carbon atom of the cyano group in position 4 , resulting in a cyclic imino derivative. An explanation of the two different paths of the reaction are given in terms of the stabilization of the intermediate ionic couple.

The heteroaromatic bases are an important class of substances because of their uses and properties in chemistry and in biology. There are a number of different ways to introduce the substituents on these bases, however, the photochemical induced reactions offer wide possibilities in the synthesis of derivatives that are difficult to obtain by other means. For example, a synthetically useful reaction is one in which allyl substituents are introduced by a photochemically initiated substitution reaction of cyano substituted heterocycles. (1) (Figure 1). The reaction has been demonstrated to be of general applicability concerning the heterocyclic derivative, the only limitation being that the cyano group must be located in position α or γ to the heterocyclic nitrogen. The mechanism that is operating is the direct allyl hydrogen abstraction from the excited base.
A similar result, though with a different mechanism, is obtained in HCl acidified solution, in which both allyl derivatives and alkyl chlorinated derivatives are formed. (2) (Figure 2).

81

F. Minisci (ed.), Free Radicals in Synthesis and Biology, 81–88.
© *1989 by Kluwer Academic Publishers.*

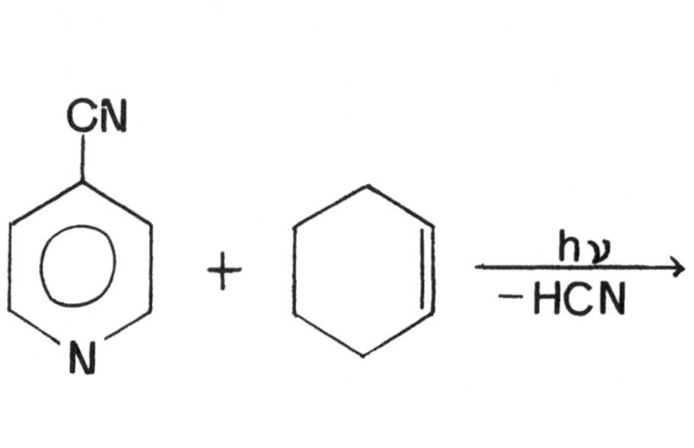

Figure 1: Photochemical induced allyl substitution in neutral medium

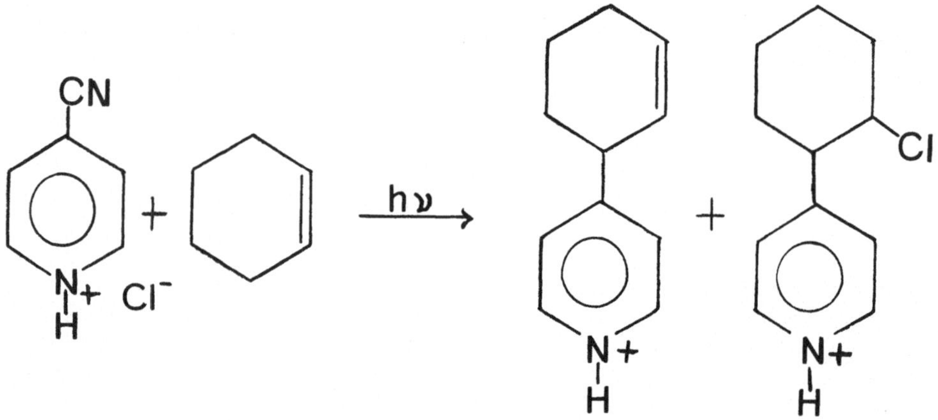

Figure 2: Photochemical induced substitution in HCl acidified medium

The ratio between the two kinds of products depends on the degree of substitution of the alkene and on the acididy of the medium. In this case, the mechanism that is operating involves a photoinduced electron transfer. (3)

Photoinduced electron transfer is a mechanism that is also operating as the first step in the reactions involving 2,4-dicyanopyridine, the difference being that, in this case, the introduction of an additional cyano group allows the electron transfer to occur in neutral medium. (4) After the electron is transferred from the donor to the excited base, the two charged species, the radical cation and the radical anion, may react together or move apart, each give rise then to their own series of reactions. Which path is followed, depends on the ability of the solvent to stabilize the ionic couple in a cage.

MATERIALS AND METHODS.
All reactions were carried out in a Rayonet RPR-100 photochemical
reactor equipped with low pressure mercury lamps using quartz vessels.
All reagents are commercially available. The preparative reactions were
run dissolving 2 mmole of (1) in 50 ml of the appropriate solvent,
deareated by bubbling nitrogen for 20 minutes, adding 10 mmole of
alkene and irradiating overnight.
The competitive reactions in acetonitrile were run as reported above,
the only difference being that 5 mmole of each of the two selected
alkenes were added and the irradiation time was 1 hour. The resulting
mixture was then analyzed by GC. The parallel reactions in benzene were
run as the preparative procedure, but the four resulting solutions for
the four used alkenes were irradiated contemporaneally in a merry-go-
round for 5 hours. The resulting mixture was then analyzed for by GC.

RESULTS.
Four different solvents were used to test the effect of solvent polarity on the path of the photochemical reaction. In a non polar solvent,
such as benzene or cyclohexane, the reactions proceed with the substitution of the cyano group analogous to the reaction we reported above.
(Figure 3)

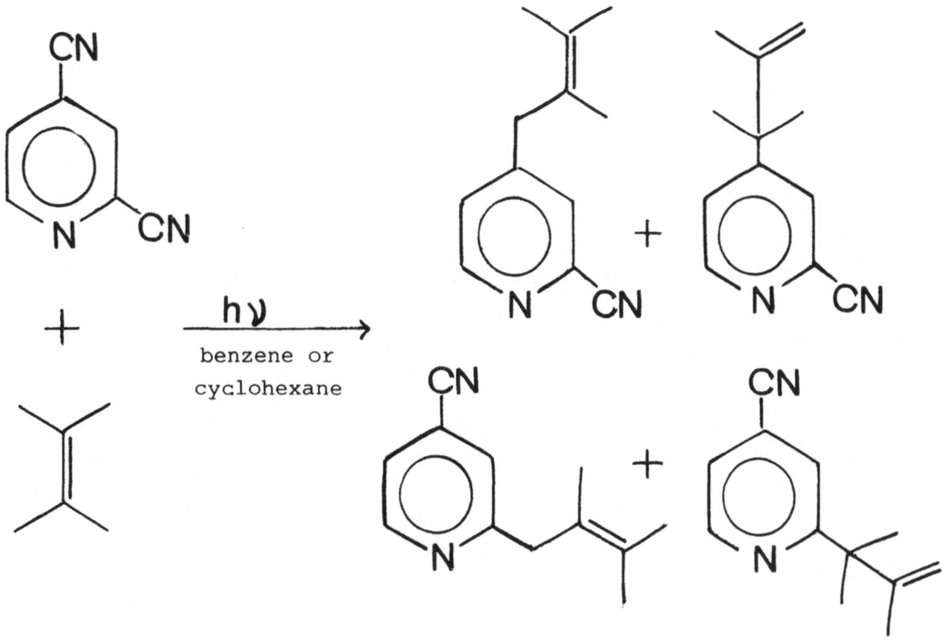

Figure 3: Photochemical induced substitution of 2,4-dicyanopyridine and
alkene in non polar solvent.

The alkenes used, the relative ratio of the 2 and 4 isomers, the total
yields, and the relative ratios of reactivity determined in parallel
experiments are reported in Table I.

Table 1

Alkene	4 CN-2 Allyl Pyridine	2 CN-4 Allyl Pyridine	Total[*] Yield	Relative[°] Reactivity

Reaction of 2,4-dicyanopyridine with alkenes in non polar medium
Alkenes, isomers distribution, yields and relative reactivities

Alkene	4 CN-2 Allyl Pyridine	2 CN-4 Allyl Pyridine	Total[*] Yield	Relative[°] Reactivity
(structure)	36.4	63.6	65	13
(structure)	39.8	60.2	47	8.2
(structure)	43.7	54.3	40	6.3
(structure)	54.7	45.3	13	1

[*] 2 mmole of 2,4-dicyanopyridine, 10 mmole of alkene in 20mL of benzene or cyclohexane irradiated at 254 nm for 15h
[°] 2 mmole of 2,4-dicyanopyridine, 10 mmole of alkene in 20mL of benzene or cyclohexane and irradiated in a merry-go-round for 5h

Table II

Reaction of 2,4-dicyanopyridine with alkenes in polar medium
Alkenes, isomers distribution, yields and relative reactivities

Alkene	Regioisomer Relative Ratio		Total* Yield	Relative° Reactivity
			100	10.7
	78.8	21.2	100	3.4
	41.7	58.3	100	2.5
	36.4	63.6	64	1

*2 mmole of 2,4-dicyanopyridine, 10 mmole of alkene in 20mL of acetone or acetonitrile irradiated at 254 nm for 15h
°2 mmole of 2,4-dicyanopyridine, 5 mmole each of the selected alkenes in 20mL of acetone or acetonitrile and irradiated for 1h

The course of the reaction is completely different using acetonitrile or acetone as solvent: a cyclic derivative was obtained which is not observed in the non polar solvents. (Figure 4)

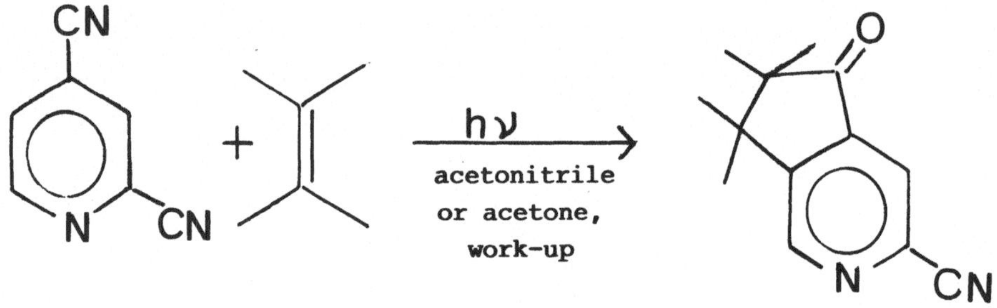

Figure 4: Photochemical induced substitution of 2,4-dicyanopyridine and alkene in polar solvent.

The ketone forms as a result of the work-up procedure in which the corresponding imino derivative is hydrolyzed. It is possible, however, to obtain direct evidence for the involvement of this intermediate by reduction with sodium borohydride to the corresponding amine.
In Table II are reported the alkenes, the total yields, the relative ratio of the regioisomers, where possible, and the relative reactivity determined in competitive reactions.

DISCUSSION.
It seems reasonable to assume that the first step of these reactions is an electron transfer from the alkene to the excited base.
For the reactions run in non polar solvents, one observes from the data in Table I that the relative reactivities are inversely proportional to the ionization potential of the alkenes, that is to say, that as the electron is more easily transferred from the alkene to the heterocycle, the yields of products are increased.
Owing to the fact that the reaction proceeds via substitution of the cyano group, an explanation may be that there is a transfer of a proton from the radical cation to radical anion and the resulting radicals may then dimerize and undergo HCN elimination yielding the final products. Furthermore, a calculation of the spin density of 2,4-dicyanopyridinyl radical indicates that both carbon atoms attached to the cyano groups have a higher spin density than the rest of the positions, and this explains the regioselectivity of the reaction. (4)
In the case of the polar solvents, evidently the cyano group and position 5 on the pyridinium radical anion undergo attack by the radical cation. Again, the trend of the reactivities follows the ionization potential for the linear alkenes, methylcyclohexene being an exception probably for entropic reasons. The costruction of a new ring "freezes" the carbon atoms of the radical cation so that the decrease in the degrees of freedom is less for methylcyclohexene and the reactivity is higher.
(Figure 5)

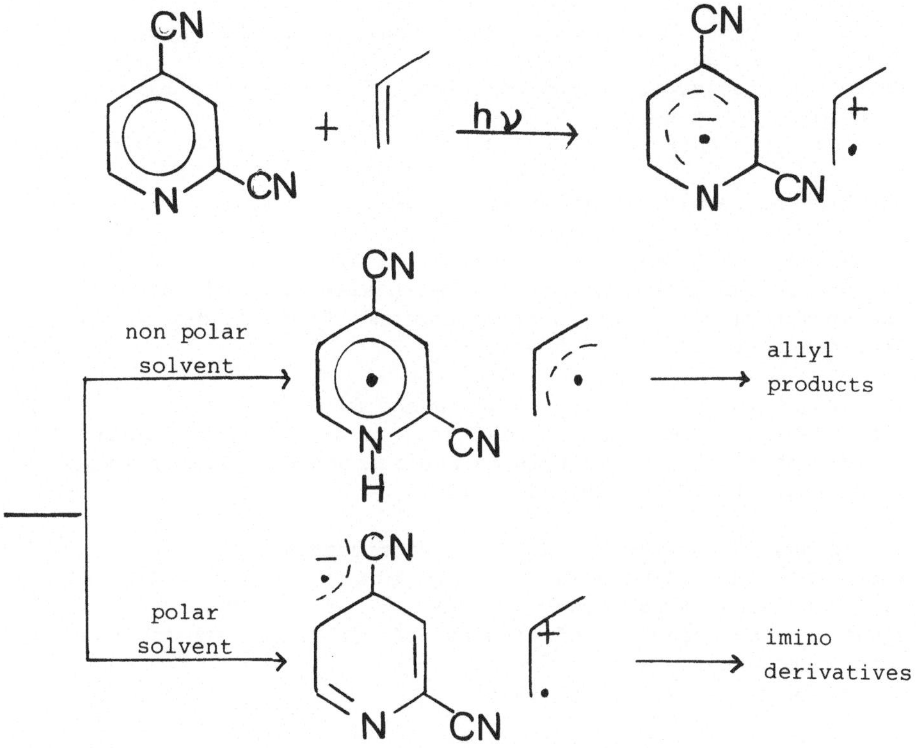

Figure 5: Proposed mechanism for the reaction of 2,4-dicyanopyridine in non polar and polar solvent.

In our opinion, the only difference lies in the fact that, in polar solvent, the two parts of the intermediate ionic couple are farther apart and this allows for a greater negative charge density on both position 5 and the cyano group. In non polar solvent, however the couple is tighter and the proton is more easely transferred to the radical anion.

ACKNOWLEDGMENT

We thank the M.P.I. for financial support.

88

REFERENCES

(1) R. Bernardi, T. Caronna, S. Morrocchi, P. Traldi and
    B.M. Vittimberga .
    Photoinitiated substitution reactions of heterocyclic bases.
    Reactions with alkenes.
    J. Chem. Soc. Perkin I 1607 (1981)

(2) Unpublished results from our laboratory

(3) T. Caronna, A. Clerici, D. Coggiola and S. Morrocchi
    Photoreactions of 4-Cyanopyridine with alkenols; Influence of
    the medium on the reaction mechanism and photoproducts formation.
    Tetrahedron Lett. **22,** 2115 (1981)

    T. Caronna, A. Clerici, D. Coggiola and S. Morrocchi
    Photochemical substitution reactions of 4-Pyridinecarboni-
    trile with aliphatic alcohols in neutral and acidified medium.
    J. Heterocyclic Chem. **18,** 1421 (1981)

(4) T. Caronna, S. Morrocchi and B. M. Vittimberga
    Importance of acidity on an energetically unfavorable
    electron-transfer reaction--An extention of the Rhem-Weller
    equation. Photoreaction of triplet 2,4-Pyridinedicarbonitrile
    with 2-Propanol.
    J. Am. Chem. Soc. **108,** 2205, (1986)

Photoinitiated Radical Reactions of 1-(4'-Pyridyl)-4-Pyridone and its N-Protio and N-Methyl Derivatives. Dimerization and Polymerization Reactions

Bruno M. Vittimberga, John Cosgrove and Howard R. Heyman, Dèpartment of Chemistry, University of Rhode Island, Kingston, RI 02881, USA.

ABSTRACT. 1-(4'-Pyridyl)-4-pyridone (1) was prepared from 4-pyridone and acetic anhydride by the method of Arndt and Kalischek[1]. When 1 is irradiated at 300 nm in deaerated solutions of 1° or 2° alcohols in the cavity of an ESR spectrometer, an intense blue radical forms immediately. Emission and quenching studies indicate that this radical forms through an $S(\pi,\pi^*)$ state of 1. Further studies revealed that the blue radical is identical with the alkylviologen cation radical. The presence of water in these reaction mixtures results in the formation of a dimer product determined to be a salt of 11,12-di(4-pyridyl)-11,12-diazatricyclo[5.3.11^{2,6}] dodecane-4,9-dione.[2,3] Viologen forms by a very rapid 4,4' coupling-elimination reaction of the radical form of 1, while the dimer results from a head-to-tail dimerization of two molecules, very likely with an aziridine structure, solvated by water molecules. Cyclic voltammetry shows that 1 undergoes irreversible reduction at −1.2V and immediately converts to viologen after one cycle with the related reversible cyclic voltammagram. Irradiation of 1 with chloropyridine in 2-propanol yields a polypyridine that has different physical properties than the polypyridine obtained from the thermal polymerization of chloropyridine alone. A precipitate forms which was shown to be a radical which when dissolved in water undergoes a series of spontaneous color changes.

INTRODUCTION

As a part of a continuing study on the photoiniated radical forming reactions of six membered N-heterocycles, we prepared and studied 1-(4'-pyridyl)-4-pyridone(1) and its protio and N-methyl derivatives. When 1 is irradiated as the N-protio or N-methyl salt in deaerated solutions of 1° or 2° alcohols, an intense blue radical is obtained that has an ESR spectrum similar to that of methylviologen cation radical (2). We became interested in determining the structure of this blue radical with the hope that 1 might be a precursor to 2 in the photochemical reaction with alcohol

In the past, 4,4'-bipyridinium salts have been studied extensively, especially with regard to the ease with which they undergo one electron reduction.[4,5] 1,1'-Dimethyl-4,4'-bipyridinium dichloride (methylviologen) is probably the most studied member of this family of compounds because of its widespread use as a herbicide, electron

89

F. Minisci (ed.), Free Radicals in Synthesis and Biology, 89–95.
© 1989 by Kluwer Academic Publishers.

transfer indicator, and electron carrier in solar energy conversion systems.[6,7] We were also interested in other reactions of the radical of 1 both in relation to its dimerization and as an initiator of polymerization with chloropyridine.

$$\underline{1} \qquad\qquad\qquad\qquad \underline{2}$$

MATERIALS AND METHODS

All solvents used in this study were analytical grade. The 4-pyridone was technical grade (90%) obtained from the Aldrich Chemical Co. All ESR spectra were taken on either a Varian E104 or a Bruker ER 220D spectrometer. The radical of 1 was generated in situ by irradiation of solutions of the base in the cavity of an ESR spectrometer. ESR spectra were taken either with a flat, quartz flow-through cell using solutions that had been deaerated by nitrogen bubbling for a minimum of twenty minutes before irradiating or with pyrex tubes which were 5 inches long and 2 mm i.d. and subjected to three freeze-vacuum-thaw cycles before irradiating. Irradiations were carried out with a 1000-W high pressure mercury lamp which had the same orientation with respect to the sample for each run and focused on the sample.

RESULTS

When a deaerated alcoholic solution of 1 is irradiated at 300 nm, the solution turns dark blue immediately. When this irradiation is performed with the sample in the cavity of an ESR spectrometer, a complex spectrum is obtained which is very similar to that of methylviologen cation radical. Illumination studies indicate that radical formation occurs from an $S(\pi,\pi^*)$ state of 1. The radical obtained showed strong absorption in the UV-visible spectrum at 385, 395, and 605 nm (ε_m = 11,200, 4:1 2-propanol/H_2O) and was similar to the spectrum of 2.

Final confirmation of the structure of this radical was obtained by an ENDOR study which showed that the hyperfine splitting constants of the radical obtained from 1 are the same as those of the ESR spectrum of 2.

The quantum yield for radical formation of 2 was measured and found to be a function of the concentration of 1. The quantum yield peaked at a concentration of 1 equal to 0.01M(Φ=0.054). No isotope effect was observed with CD_3OD/H_2O indicating that the radical forming process most likely involves electron transfer rather than hydrogen atom abstraction. The radical, 2, is also formed chemically by treating 1 with powdered zinc or by electron transfer from thermally generated diphenyl ketyl.

Cyclic voltammagrams of alkylated 1 were very revealing. The N-methyl derivative of 1-(4'-pyridyl)-4-pyridone shows an irreversible voltammagram with a redox potential at –1.2V. Interestingly, after one cycle the voltammagram reverts to that of a viologen-type compound with the related reversible cyclic voltammagram.[8]

When irradiations of 1 are performed with samples containing water, an interesting dimer forms as a product of the reaction. Irradiation of either 1 in hydrochloric acid acidified 4:1 2-propanol-water or the N-methyl derivative of 1 in unacidified 4:1 2-propanol-water in pyrex results in a complex reaction leading to the formation of 4-pyridone hydrochloride (45% yd) as the major product. Photohydration of the pyridine ring also occurs yielding an amorphous solid and an ammonium salt.[9] During the irradiation, crystals began to form in the reaction flask after approximately 3 hours. Irradiation was then continued for a total of 48 hours. This crystalline compound (3) which was collected by filtration does not form in neutral solution, nor does it form in the presence of oxygen. (Eq. (1)). It is soluble in water, slightly soluble in methanol, and insoluble in essentially all other common organic solvents. Infrared analysis (KBr) showed 3 to be an amine salt (3200–2800 cm^{-1}) containing a carbonyl group (1630 cm^{-1}).

(1)

1·2HCl

3

The ^{13}C NMR spectrum of 3 in D_2O shows six different types of carbon atoms with peaks in the off-resonance spectrum at δ190(s), 159(s), 142(d), 111(d), 58(d), and 44(t). The proton NMR spectrum shows peaks at 8.5(d,d), 7.6(d,d), 5.2(d), 3.2(d,d), and 2.8(d). The melting point was greater than 300°C and the elemental analysis corresponded to a hydrated dihydrochloride salt of a dimer of 1. The yield based on the formula $C_{20}H_{22}N_4O_2^{2+} \cdot 2Cl^- 2H_2O$ was 5 percent.

Treatment of 3 with sodium hydroxide converts it to the free base as shown by infrared spectroscopy and elemental analysis. This product is insoluble in essentially all common solvents. Based on the data available to us at that time, we were not able, to assign an unequivocal structure to 3. Fortunately, this compound forms monoclinic crystals from aqueous ethanol,[3] so it was subjected to single crystal X-ray diffraction analysis.[3] The structure, 3, that was provided by this method is supported by all spectral and analytical data. A similar structure also forms with the N-methyl derivative of 1.

Irradiation of 1 with 4-chloropyridine in 2-propanol yielded a polypyridine unlike the polypyridine obtained from 4-chloropyridine alone. If both 1 and 4-chloropyridine are present in solution, irradiation at 250 to 350 nm causes an immediate color change. If irradiation is continued for a longer period of time a dark precipitate will form in the flask. This precipitate can only be separated from solution by centrifugation and is only partly soluble in water, methanol, or ethanol. When this material is dissolved in one of these solvents it exhibits an interesting series of color changes. The solution is initially green and over the course of about an hour passes through the following color sequence: green, blue-green, violet, red-violet, rose red. This color change will occur in the dark but will not occur in the absence of oxygen. Treatment of an aqueous solution of the dark precipitate with base results in an immediate loss of color and an absorption appears at 340 nm in the UV spectrum which then disappears. This absorption at 340 nm has been assigned to a trimer.[10,11]

DISCUSSION

The aim of this work is to examine the photoinitiated radical reaction of salts of 1-(4'-pyridyl)-4-pyridone that evidence indicates is initiated by an electron transfer rather than a hydrogen atom abstraction. This electron may be supplied by a 1° or 2° alcohol in a photochemical reaction, or by diphenyl ketyl or zinc in a chemical reaction. In the photochemical reaction an electron is most likely transferred to 1 in its $S_1(\pi,\pi^*)$ state and becomes delocalized over the π system involving both rings. The spin density in the 4 position of ring B results in rapid coupling with a similar radical to form an unstable dimer. Perhaps, due to the bulkiness of the pyridone groups, elimination of pyridone occurs rapidly to form a viologen structure (Eq. (2)). Photoreduction to the blue cation radical, 2, then follows.

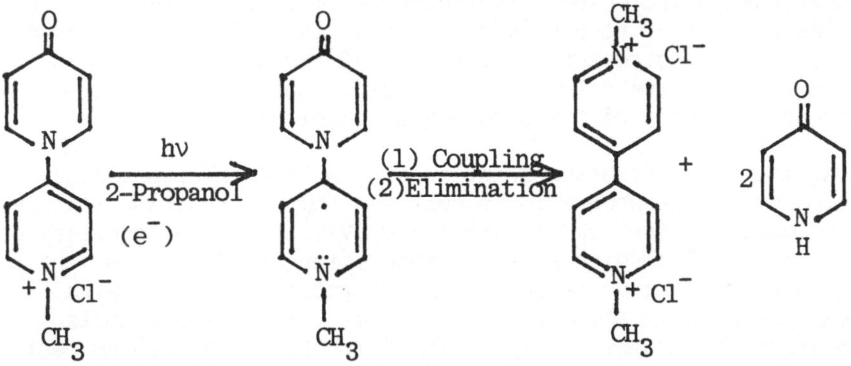

$$\text{(2)}$$

The $S(\pi,\pi^*)$ state of 1 was the most logical choice as the photoreactive state since the emission spectrum, which in ethanol shows fluorescence at 325 nm and a weak phosphorescence at 400 nm, changes very little as the polarity of the solvent is changed, and radical formation occurs in the presence of piperylene.

The mechanism that we would like to propose for the formation of 3 is shown in Scheme 1.

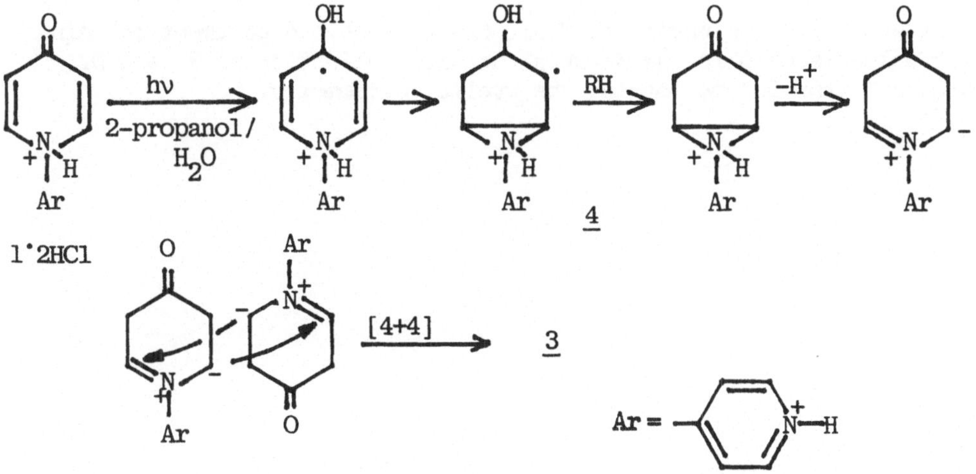

Scheme 1. Proposed mechanism for the photoinitiated dimerization of 1·2HCl in 2-propanol.

94

In an attempt to trap the proposed aziridine intermediate, (4), dimethylacetylene dicarboxylate was added to the reaction medium. We were not able to isolate products involving the trapping agent, though the formation of 3 was inhibited under these conditions. Noteworthy is the fact that the formation of 3 is regiospecific. The orientation of two molecules of 1 prior to cycloaddition most likely is determined by the intermolecular electrostatic attraction of the carbonyl group on one molecule and the positive nitrogen on another resulting in loose attachments at both ends of the bimolecular complex.

Regarding the polymerization reaction of 1 with 4-chloropyridine, this reaction goes best when acid is omitted from the reaction mixture. When this polymer is dissolved in water the UV/Vis spectrum consists of one peak at about 635 nm with a strong absorption at 287 nm. As the solution ages there is a decrease in the absorption at 635 nm with a simultaneous increase in absorption at 525 nm. If the solution is allowed to stand for a week, or is treated with base, the 525 nm peak disappears and a new absorption appears at 255 nm along with the peak at 287 nm. An ESR spectrum of the polymer, taken as soon as it is dissolved in water, gives a one line signal centered at 3351 gauss with a g value of 2.0024. As the color of the solution changes the one line signal decreases in intensity and becomes broader. At the completion of this color change there is no ESR signal detectable.

ACKNOWLEDGEMENTS

We would like to thank Dr. Clair Cheer of this department for his very valuable help in the X-ray structure elucidation of 3, and Dr. Richard Durand for performing the cyclic voltammetry.

REFERENCES

1. F. Arndt and A. Kalischek, Ber., **63B**, 587(1930).

2. For the photoinitiated substitution reactions of $\underline{1}$ with benzophenone in acidified 2-propanol-water. Please see J. P. Cosgrove and B. M. Vittimberga, J. Heterocyclic Chem., (1984) **21**, 1277.

3. For a discussion of the X-ray analysis please see C. J. Cheer, J. P. Cosgrove and B. M. Vittimberga, Acta Cryst. (1984) **C40**, 1474.

4. L. A. Summers, "The Bipyridinium Herbicides," Academic Press: New York, NY, 1980; p. 28, 69.

5. A. S. Hopkins, A. Ledwith, M. F. Stan, Chem. Comm. (1970) 494, and references therein.

6. A. A. Akavein, D. I. Linscott, Residue Rev., (1968) **23**, 97.

7. M. Gretzel and J. Kiwi, Nature, (1979) **281**, 657.

8. Cyclic voltammetry was done on a Princeton Applied Research PAR 273 Potentiostat/Galvanostat utilizing a two compartment cell with a saturated calomel reference electrode removed from the working compartment. Platinum electrodes were used for both the counter and working electrodes.

9. Ammonium chloride was not isolated, but alkylamines were isolated with N-alkylated $\underline{1}$.

10. Part of this work was presented as the 10th IUPAC Symposium on Photochemistry, Interlaken, Switzerland, July, 1984.

11. Elemental analysis was performed by Micro Analysis, Inc., Wilmington, DE.

RADICAL REACTIONS WITH ORGANOCOBALT COMPLEXES

B. Giese[*], A. Ghosez, T. Göbel, J. Hartung, O. Hüter, A. Koch, K. Kroder, R. Springer
Institut für Organische Chemie
Technische Hochschule Darmstadt
Petersenstr. 22, D-6100 Darmstadt, Germany

ABSTRACT. Organocobalt complexes are used as radical precursors in biomimetic syntheses. Mechanistic studies demonstrate that carbon cobalt bonds can be either cleaved homolytically or heterolytically. In addition to these reactions, reductive eliminations occur as product forming steps.

Phosphoenol pyruvate reacts as C-3 unit with aldehyde functions of carbohydrates in enzymatic aldol reactions[1].

$$CH_3 \overset{O}{\underset{\|}{C}} CO_2H$$

$$RCHO + H_2C = C \underset{CO_2^-}{\overset{O\,\textcircled{P}}{<}} \xrightarrow{\text{Enzyme}} RCH \overset{OH}{\underset{\|}{+}} CH_2 \overset{O}{\underset{\|}{C}} CO_2H$$

Erythrose	$\longrightarrow$	Shikimic Acid
Arabinose	$\longrightarrow$	KDO
Aminomannose	$\longrightarrow$	Neuraminic Acid

97

F. Minisci (ed.), Free Radicals in Synthesis and Biology, 97–106.
© 1989 by Kluwer Academic Publishers.

Because respective in vitro reactions fail in the absence of the enzymes, we have started biomimetic syntheses using radical methodology. For these reactions one has to find methods by which aldehyde functions of sugars are converted into radicals. Suitable radical precursors could be generated from aldehydes with trimethylsilyl phenyl selenide[2] or from acetals with HBr.

As biomimetic synthons of phosphoenol pyruvate we have used methylenemalonic ester[3] or an estersubstituted allylstannane[4].

Application of these methods to the carbohydrate arabinose yields derivatives of KDO.

Because the diastereoselectivity of the radical CC bond forming step is rather low and the generation of the free aldehyde from sugars are multistep syntheses, the reactions of cyclic carbohydrate radicals were studied. From easily accessible glycosylbromides the radical CC bond formation occurs with high diastereoselectivity[3,4]).

But the alkenes used in these syntheses differ from phosphoenol pyruvate because they are C-4 instead of C-3 units. A better biomimetic synthon is ethoxyacrylonitrile. This alkene reacts with glycosyl-Co complexes and gives substitution products[5].

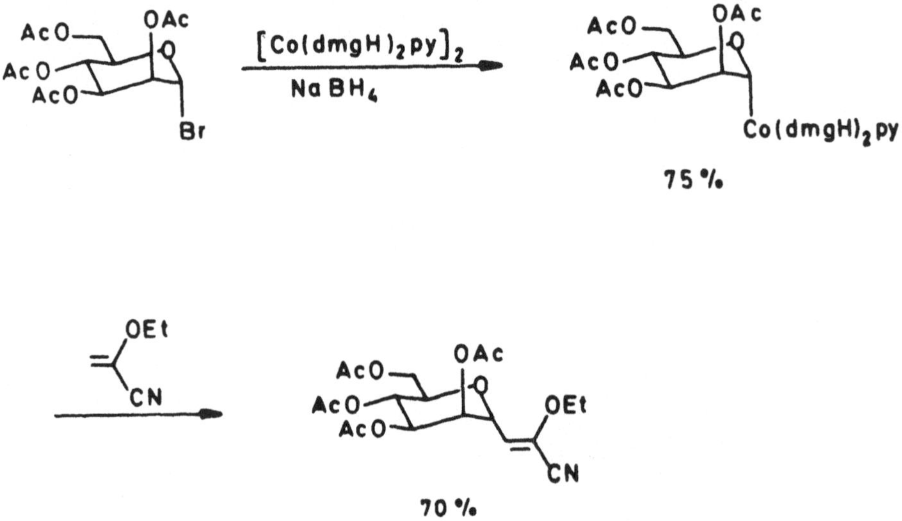

75 %

70 %

The sugar-Co complexes can be synthesized from glycosylbromides and Co[I]-complexes. Photolysis gives carbohydrate radicals that react with several radical traps[5].

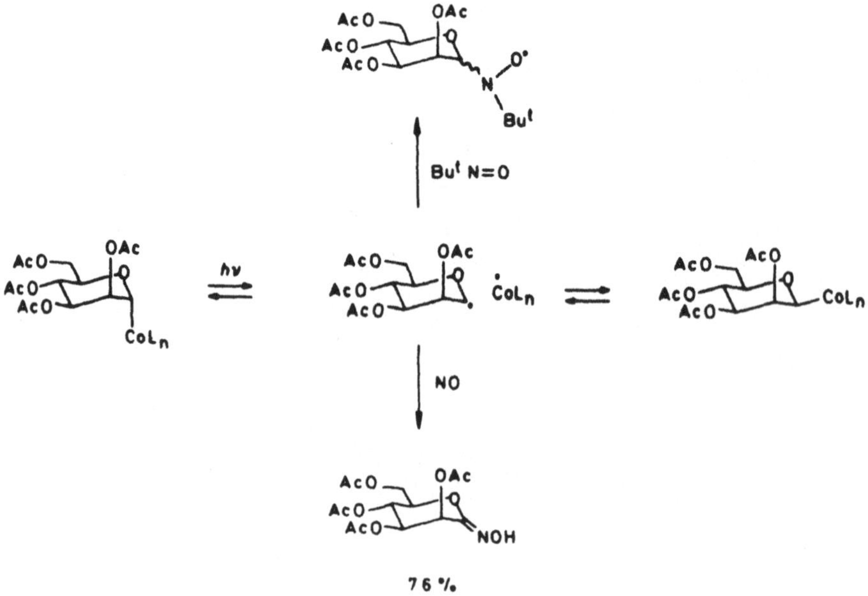

76 %

With alkenes these radicals lead after combination with the Co^{II}-complex to new Co-organo compounds. Depending upon the substituent Y the products are formed either via solvolysis or reductive elimination.

$$RCH_2-\underset{\underset{D}{|}}{C}HCN$$

$$\uparrow CH_3OD$$

$$\underset{\underset{CoL_n}{|}}{R} \rightleftharpoons \underset{\underset{CoL_n}{|}}{R\cdot} \xrightarrow{H_2C=CHY} RCH_2-\underset{\underset{CoL_n}{|}}{\overset{\cdot}{C}}HY \underset{h\nu}{\rightleftharpoons} RCH_2-\underset{\underset{CoL_n}{|}}{C}HY$$

$$\xleftarrow{h\nu}$$

$$\downarrow -HCoL_n$$

$$RCH=CHPh$$

Thus, glycosyl-Co complexes react with different alkenes either to addition or substitution products.

The question whether these reactions occur via "free" radicals or not, could be answered by selectivity experiments.

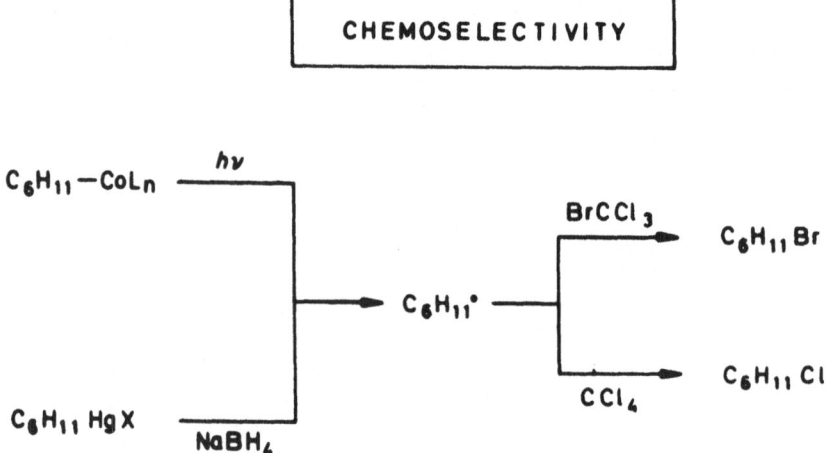

Studies on their stereoselectivity show that glycosyl and cyclohexyl radicals generated from Co-complexes behave like "free" radicals. But experiments with primary alkyl-Co complexes and the $BrCCl_3/CCl_4$ competition system[6] show surprising results.

The plot of the activation enthalpies against steric E_S parameters gives a straight line for radicals[7]. Whereas the cyclohexyl-Co complex fits into this correlation the n-hexyl-Co complex has a completely different (higher) chemoselectivity.

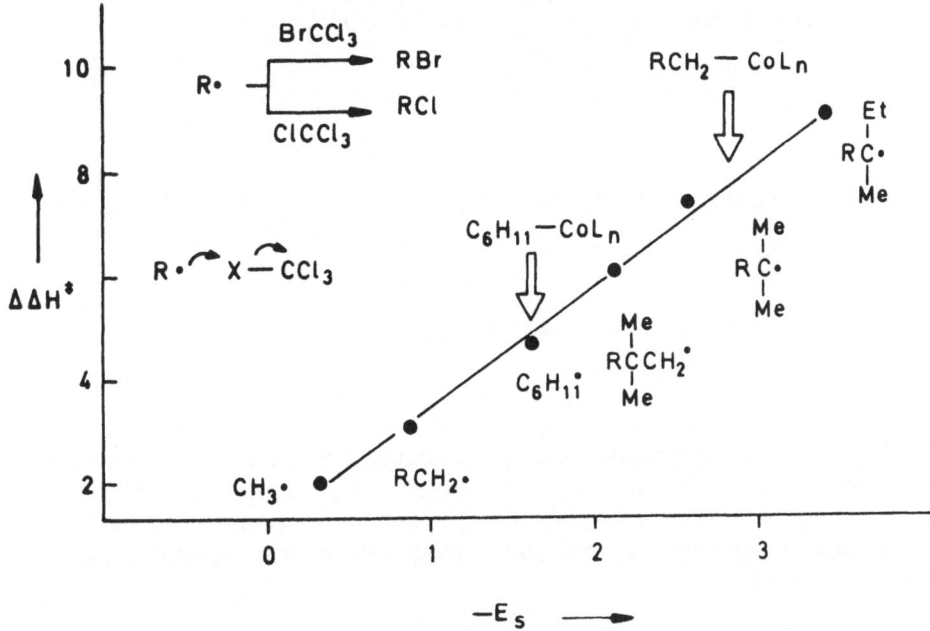

This different behaviour of primary systems can also be observed with the hexenyl-Co complex that cyclizes very slowly under irradiation. But in the presence of BrCCl3 the hexenyl-Co complex reacts very rapidly. Acyclic products are formed even if the BrCCl3 concentration is so low that cyclization should occur if free radicals are the intermediates. The absence of cyclization products demonstrates that the excited Co-complex can react with XCCl3 before the Co-C bond is cleaved completely. But this seems to be the case only in Co-complexes with primary alkyl groups that form strong Co-C bonds.

These reactions with cobaloximes are similar to vitamin B_{12} catalyzed CC-bond forming syntheses, first observed by Scheffold[8].

$$RBr \ + \ H_2C=CHY \ \xrightarrow[\substack{Zn \ / \ NH_4Cl \\ DMF}]{Vit. \ B_{12}} \ R-CH_2-CH_2Y$$

$$RBr \ \xrightarrow{^{I}CoL_n} \ R-CoL_n \ \xrightarrow{H_2C=CHY} \ \underset{\underset{CoL_n}{|}}{RCH_2CHY} \ \xrightarrow{^{III}CoL_n} \ RCH_2CH_2Y$$

$$^{III}CoL_n \ \xrightarrow{Zn} \ ^{I}CoL_n$$

In these B_{12} catalyzed syntheses reduction products are formed. Two questions concerning these reactions had not been answered: 1) Are "free" radicals involved in the first step, and 2) what is the source for the hydrogen in the second step? Reactivity and stereoselectivity experiments prove that at least with cyclohexyl systems the first step occurs via free radicals.

FREE RADICAL ADDITION

a· Reactivity of Alkenes

$$R\cdot \ + \ H_2C=CHY \ \longrightarrow \ R-CH_2-\overset{\bullet}{C}HY$$

b· Stereoselectivity of Radicals

But the second step is an ionic reaction, as deuteration experiments show.

$$\boxed{\text{IONIC} \quad \text{CLEAVAGE}}$$

a. Trapping

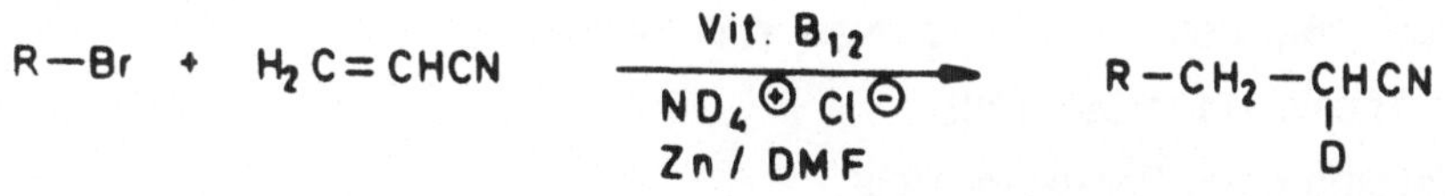

$$R{-}Br \;+\; H_2C{=}CHCN \quad \xrightarrow[\begin{array}{c} ND_4^{\oplus}\,Cl^{\ominus} \\ Zn\,/\,DMF \end{array}]{\text{Vit. }B_{12}} \quad \underset{\overset{|}{D}}{R{-}CH_2{-}CHCN}$$

b. Stereochemistry

$$R{-}Br \;+\; HC{\equiv}CHPh \quad \xrightarrow[Zn]{\text{Vit. }B_{12}} \quad \underset{H}{\overset{R}{>}}C{=}C\underset{H}{\overset{Ph}{<}} \;+\; \left(\underset{H}{\overset{R}{>}}C{=}C\underset{Ph}{\overset{H}{<}} \right)$$

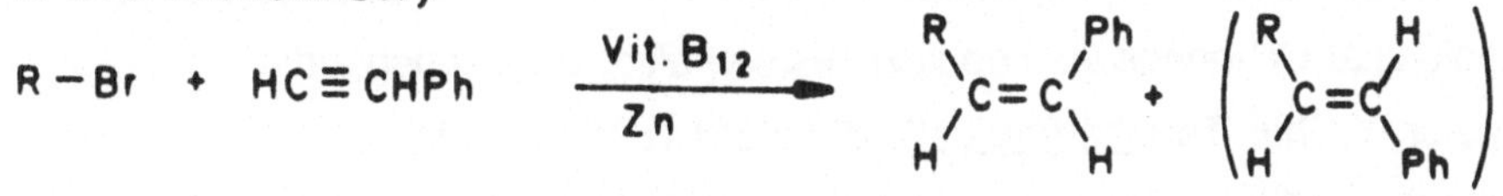

This is in accord with stereochemical studies which show that the formation of products cannot occur via hydrogen abstraction steps.

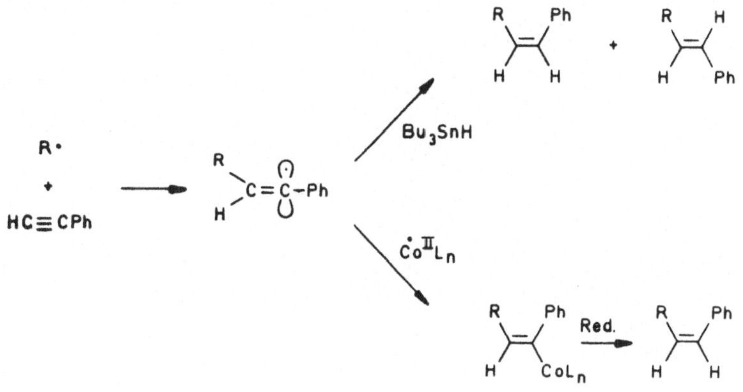

ACKNOWLEDGEMENT: This work was supported by VW-Stiftung and the Deutsche Forschungsgemeinschaft.

REFERENCES

1) A.L.Lehninger, "Biochemistry", Worth, New York 1977.

2) R.Rupaner, Thesis, Darmstadt 1988.

3) R.Muhn, Diploma work, Darmstadt 1988.

4) T.Linker, Diploma work, Darmstadt 1988.

5) B.Giese, A.Ghosez, T.Göbel, Chem. Ber. i. press. For the formation of intermolecular CC-bonds via cobaloximes see also: B.P.Branchaud, M.S.Meier, Y.Choi, Tetrahedron Lett. 29 (1988) 167; V.F.Patel, G.Pattenden, J.Chem.Soc.Chem.Commun. 1987, 871.

6) B.Giese, K.Keller, Chem.Ber. 112 (1979) 1743; see also: K.Herwig, P.Lorenz, C.Rüchardt, Chem.Ber. 108 (1975) 1421.

7) B.Giese, Angew.Chem. 89 (1977) 162.

8) R.Scheffold, Chimia 39 (1985) 203.

NEW WAYS OF GENERATING STANNYL RADICALS FOR ORGANIC SYNTHESES

K. Baines[1], R. Dicke, W. P. Neumann[*], and K. Vorspohl[2]
Lehrstuhl für Organische Chemie I, University of Dortmund
Otto-Hahn-Strasse 6
D-4600 Dortmund 50
Federal Republic of Germany

ABSTRACT. Stannyl radicals, powerful reagents in free radical reactions including cyclizations and tandem reactions, commonly are generated from organotin hydrides. But, sometimes drawbacks are given by the very strong hydrogen donating power of the latter. We report now independent sources for $R_3Sn^{\cdot}$: a) thermolysis of distannanes with bulky substituents, b) thermolysis of bisstannyl pinakols via stannyl ketyl radicals, and c) convenient photolysis of benzyl tin compounds including 9-stannyl-9,10-dihydro anthracene showing absorption up to 400 nm under preparative conditions. Examples of advantageous applications of these sources for $R_3Sn^{\cdot}$ with or without combination with H donors of appropriate strength are given and discussed.

1. INTRODUCTION

Stannyl radicals, mostly of the type $R_3Sn^{\cdot}$, and here again mostly $Bu_3Sn^{\cdot}$, are used for organic syntheses via free radicals, at present in a rapidly growing scale. Besides mechanistically simple dehalogenations of organic chlorides, bromides, and iodides, often additions to alkenes, dienes, and alkynes are used. A special advantage is given, when the intermediate carbon radicals can undergo intra- or intermolecular additions, cycloadditions, or rearrangements before the final stabilization occurs, very often by H donation. Mostly organotin hydrides R_3Sn-H, especially Bu_3Sn-H, are used both for generation of stannyl radicals and as H donors. The main progress is given for complicated organic target molecules, in general biomolecules. Multistep regio-, stereo-, and enantioselective syntheses could be successfully carried out in one-pot procedures with high yields, often by tricky handling of the kinetics. A recent and comprehensive review is available[3].

This recent boom has a surprising aspect, too, because the basic organotin chemistry leading to it has been investigated thoroughly 20-25 years ago (for a critical report, see ref.[4]), e. g.:

F. Minisci (ed.), Free Radicals in Synthesis and Biology, 107–114.

First Use of Radicals $R_3Sn^{\cdot}$

$$R'-Hal \;+\; R_3Sn^{\cdot} \;\longrightarrow\; R_3Sn-Hal \;+\; R'^{\cdot}$$

H.G. Kuivila et. al. 1962

$$\underset{}{{\supset}C=C{\subset}} \;+\; R_3Sn^{\cdot} \;\longrightarrow\; \underset{R_3Sn}{{\supset}C-\dot{C}{\subset}}$$

W.P. Neumann et. al. 1961

Mostly used: $Me_3Sn^{\cdot}$, $Et_3Sn^{\cdot}$, $Bu_3Sn^{\cdot}$, $Ph_3Sn^{\cdot}$

2. THE PROBLEM

Besides the great and still increasing number of successful syntheses, several fields became visible, where the use of stannyl radicals is essential, but the strong radical scavenger R_3Sn-H gives rise to un-desired side reactions (byproducts, loss of yield, more complicated workup and purification steps) or cannot be tolerated at all. An example for the latter case is the elegant alkyl iodide radical chain reaction (D. P. Curran et al.)[5] where the stannyl radicals are gene-rated from hexaalkyl distannanes:

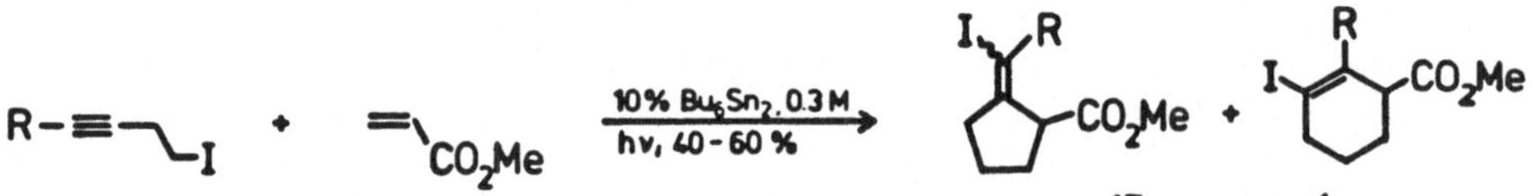

R−≡−\\−I + =\CO₂Me →[10% Bu₆Sn₂, 0.3 M / hv, 40-60 %]

D. P. Curran et al. 1987

15 : 1

E/Z = 3:1

The photolytic splitting of hexaalkyl distannanes seems obvious and is widely preferred in the case of spectroscopic work. But, for prepara-tive use we have found considerable drawbacks, especially for the butyl derivative. The short waved absorption of the latter needs the applica-tion of rather hard UV-irradiation being absorbed also by many of the organic reaction partners present in the mixture. This may cause diffe-rent problems. Another drawback is the splitting of C-Sn bonds besides the desired Sn-Sn splitting ($D_{Sn-Sn} \geq D_{C-Sn}$) which leads to undesired side and consecutive reactions, and products as well:

Photoreactions of $R_3Sn-SnR_3$

$$R_3Sn-SnR_3 \xrightarrow{h\nu} [R_3Sn-SnR_3]^{T*} \xrightarrow{k_a} 2\,R_3Sn^{\cdot}$$

$$\downarrow k_b \qquad \xrightarrow{-2R^{\cdot}} \quad \xrightarrow{+2R'Hal} 2R_3SnHal$$

$R=Me: k_a > k_b$ $R_3Sn-SnR_2^{\cdot} + R^{\cdot} \longrightarrow \frac{2}{n}(R_2Sn)_n + R-R$ etc.

$R=Et: k_a \approx k_b$ $+R_6Sn_2 \Big| S_H2$ $R=Et, Bu \ldots\ldots$

$$R_4Sn + \frac{1}{n}(R_2Sn)_n \, . \quad R=Me$$

Another drawback for the application of tin hydrides has been found, when the hydrogen abstraction is - in spite of all kinetic tricks like big excess of one partner, extreme dilution or titrating slowly - too fast, and prevents (fully or partly) desired intermediate reactions of the carbon radical. Two examples may illustrate this[6,7]:

$$\xrightarrow[C_6H_6. \; \Delta]{Bu_3SnH, \; AIBN}$$

M.D.Bachi et al. 1987

0.02 M	20 %	50 %
0.003 M	50 %	5 %

2-10 exs.

$$\xrightarrow[85-98\%]{Bu_3SnH, \; AIBN}$$

Ch. Walling et al. 1972

0.355 M	54	:	46
0.016 M	92	:	8

Substantial amounts of the undesirable (in these cases: open-chained) byproduct have to be tolerated, and a 2-10 fold bromide excess was used.

Therefore, a need for other sources of stannyl radicals is given, not connected with the presence of organotin hydrides. This is a challenge for organotin research and chemists to look for suited and capable sources amongst the known reactions, or to find new ones.

3. PREVENTION OF THE PROBLEM BY USE OF OTHER SOURCES FOR STANNYL RADICALS

The organotin chemistry has to be checked for factual or potential sources of stannyl radicals not being organotin hydrides.

3.1. Use of Hexaorgano Distannanes with Bulky Substituents

Regarding the dissociation enthalpy D of the Sn-Sn bond, 66 kcal/mol, a lowering down to 8.5 kcal/mol can be effected by introducing bulky substituents into the hexaorgano distannanes, and therefore a lowering of the temperature of spontaneous and reversible dissociation into stannyl radicals. This avoids irradiation and allows the use of distannanes by simply warming them up. In the most striking example,[8] even at room temperature the esr signal of $R_3Sn^\cdot$ is detectable[8] (additional examples are given in ref.[8]):

$$R_3Sn - SnR_3 \underset{}{\overset{T}{\rightleftharpoons}} 2\,R_3Sn\cdot$$

R =	T	D_{Sn-Sn} kcal/mol	Raman ν_{Sn-Sn} cm^{-1}	
⬡–	> 230°C (dec.)	⩾ 60	138 (splitting)	
Me–⬡–Me, Me	⩾ 180°C	49 ±2	102	
Et–⬡–Et, Et	⩾ 100°C	27 ±2	92	10 G → , g = 2.0075, Lw ~ 5G
iPr–⬡–iPr, iPr	⩾ 20°C	8.5 ±1	–	

These stannyl radicals exhibit a typical, sufficient or even high reactivity towards non-bulky partners. The desired action, e. g. halogen abstraction from an organic halide, can be completed by addition of a H donor of adjusted donor strength, like cumene or THF.

Thus, the hexakis (triethylphenyl) distannane mentioned above shows the ESR signal of the ketyl at 100°C. At 140°C (o-xylene) it reacts completely with 6-bromohexene-1 (which gave unsatisfying results with R_3SnH, see above[7]), and yields exclusively the desired ring closure product, methyl cyclopentane, after 3.5 hrs. No hexene-1 could be detected (GLC). In this procedure, no longer excess bromohexene was necessary.

3.2 Bisstannyl Pinakols as Thermal Sources for Stannyl Radicals

Bisstannyl benzpinakol, as a good example, is stable at room temperature and can easily be stored. It is prepared conveniently, as we have found, from benzophenone and the distannane. This works best with R = Me. There are other ways for the preparation, too, e. g. the reaction of the corresponding pinakol with R_3Sn-NEt$_2$, giving the bistannyl pinakol and 2 mol HNEt$_2$. When warming its solution, it dissociates into stannyl ketyl radicals[2] reversibly. But, at 60°C or above, it is split into benzophenone and stannyl radicals which give the typical reactions[9,10]:

$$Bu_3Sn-SnBu_3 + 2\ Ph_2C=O \xrightarrow{h\nu} \begin{array}{c} Ph_2C-CPh_2 \\ |\quad\ | \\ Bu_3SnO\ \ OSnBu_3 \end{array}$$

$$Ph_2C=O + Bu_3Sn^{\bullet} \xleftarrow{\ \Delta\ } 2\ Ph_2\overset{\bullet}{C}-OSnBu_3$$

Additions to C=C and C≡C:

$$+R-Hal \xrightarrow{-Bu_3Sn^{\bullet}} R^{\bullet}$$

$$+H_2C=CHR \rightarrow Bu_3SnCH_2-\overset{\bullet}{C}HR \longrightarrow Products$$

$$HC\equiv CR \xrightarrow{THF\ or\ Cumene} Bu_3SnCH=CHR$$

$$R^{\bullet} \xrightarrow{R'-H} R-H + R'^{\bullet}$$

$$Products$$

$$Bu_3SnCH_2-\overset{\bullet}{C}HR \xrightarrow{THF\ etc.} Bu_3SnCH_2-CH_2R$$

Other partners give smooth reactions, however, even at room temperature. Iodine gives spontaneously and quantitatively R_3SnI, boiling CCl_4 gives R_3SnCl and products of the CCl_3 radical, whereas $BrCCl_3$ reacts exothermally. N-Stannyl succinimide is formed from NBS exothermally. 1,2-Dibromo alkanes yield, at 80°C, the corresponding alkenes (in benzene), 1,2-dibromo alkenes the alkynes. Benzoyl peroxide is split exothermally forming the benzoic acid stannyl ester (besides a little carbon dioxide and other products). Additions can also be effected: quinone yields the hydroquinone bisstannyl ether in benzene, azobenzene the N,N'-bisstannyl hydrazobenzene[9]).

It is likely, that in many cases the partners attack the bistannyl pinakol directly. The R_3Sn residue in the latter behaves, therefore, as a stabilized, easily available stannyl radical.

This could be a versatile and promising source of stannyl radicals for syntheses. It can be combined with H donors of appropriate strength like cumene or THF as to be seen above. Following our first communication, very recently stannyl radicals of this origin have been applied for additions to C=N groups, too[11].

3.3 Photochemical Generation of Stannyl Radicals from Precursors with Longwaved Absorption

In earlier investigations, we found a considerably low dissociation enthalpy of the R_3Sn-benzyl (and the R_3Sn-allyl) bond when we irradiated at the absorption wavelength of the benzyl moiety by means of a monochromator. The stannyl substituent is easily split off from the molecule as a stannyl radical.

This prompted us to irradiate the benzyl trimethyltin or tributyltin compound with 1-hexyne in THF[2,10] and we isolated 70% of the hydrostannation product of the alkyne:

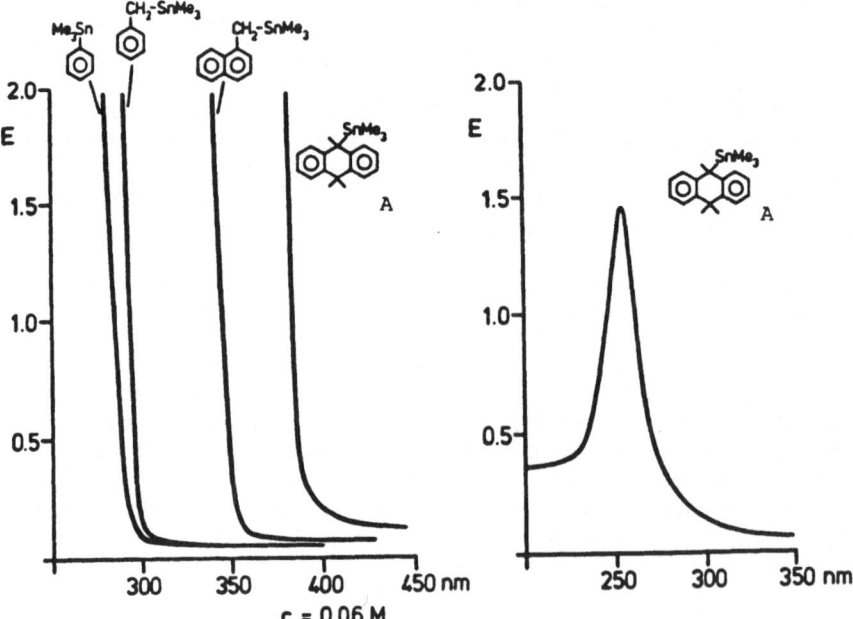

The longer-waved and more intense light absorption of α-trialkylstannyl naphthalene gave rise to a comparable yield of adduct in only one fourth of the irradiation time under the same conditions, see above.

9-Trialkylstannyl-9,10-dihydro anthracene A exhibits an even longer-waved absorption of 255 nm and, in addition, a remarkable tailing up to 400 nm, see below. Under the same conditions, with 1-hexyne an equivalent yield is obtained after only one third of the irradiation time with the naphthalene derivative mentioned above.

The foregoing diagram shows the absorptions of the dihydro anthracene derivative A mentioned (λ_{max} = 255 nm), and those of all the stannyl compounds dealt with in this section in a more concentrated solution, 0.06 M, what is more realistic for preparative purposes. As can be seen, for A a soft irradiation at 350-400 nm is completely sufficient for splitting it. This allows the presence by far of most of the organic educts for free radical syntheses to be considered here.

Additions to acrylonitrile and the dehalogenation of n-octyl bromide may provide further information. 9,10-dihydro anthracene works not better, as H donor, than THF does. Again, the stannyl dihydro-anthracene A gives the more satisfying results:

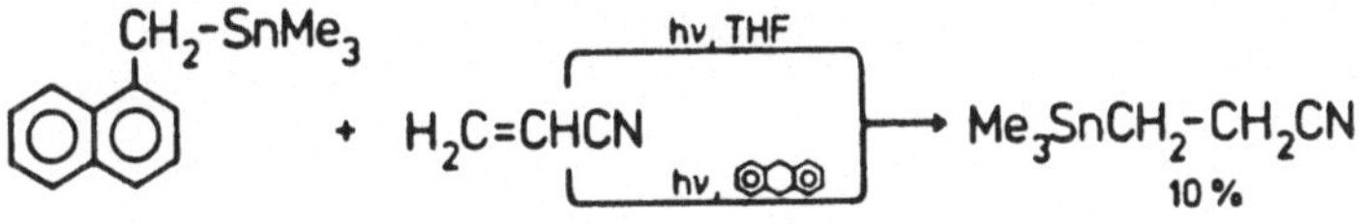

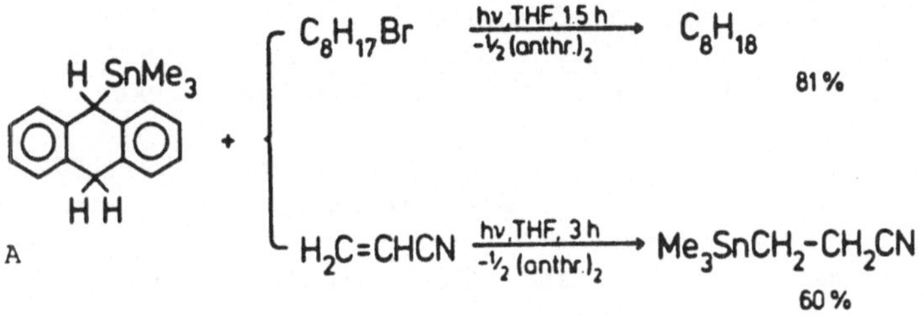

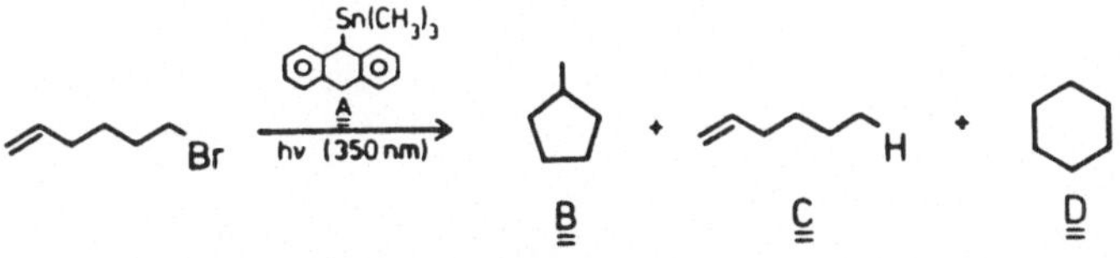

[A]	[RBr]	solvent	yield [%] B	C	D	ratio B/C
0.10	0.150	benzene	44	0	0	100/0
0.05	0.075	THF/Oc	88	<1	<2	>99/1
0.11	0.081	THF	73	<1	<2	>99/1

This enabled us to try to resolve some of the problems described in Section 2. So we took the 6-bromohexene-1 which gave, with R_3Sn-H, considerable amounts of the undesired hexene-1 C besides the desired ring closure product, methyl cyclopentane B[7]. With A in benzene, however, the yield is not exciting, but there is only the target molecule B, and no C or D. A yield of 90% could be achieved, with a ratio B/C better than 99/1. This is encouraging for further application of this source, moreover, since no excess bromohexene had to be used.

ACKNOWLEDGEMENTS

We are grateful to the Government of Canada for the NSERC Postdoctoral Fellowship for one of us (K.B.) and to the Fonds der Chemischen Industrie for support.

REFERENCES

1) Postdoctoral Fellow 1987/88 from the University of Toronto, Canada
2) K. Vorspohl, Dr. rer. nat. Thesis, University of Dortmund 1986
3) W.P. Neumann, Synthesis 1987, 665, with 190 references.
4) W.P. Neumann, The Organic Chemistry of Tin, J. Wiley (1970). Further references are cited there.
5) D.P. Curran, M.H. Chen, J. Am. Chem. Soc. 109, 6558 (1987)
6) M.D. Bachi, A. DeMesmaeker, N. Stevenart-DeMesmaeker, Tetrahedron Lett. 28, 2637 (1987)
7) C. Walling, A. Cioffari, J. Am. Chem. Soc. 94, 6059 (1972)
8) A.F. El-Farargy, M. Lehnig, W.P. Neumann, Chem. Ber. 115, 2783 (1982). Further references are given there.
9) H. Hillgärtner, W.P. Neumann, B. Schroeder, Liebigs Ann. Chem. 1975, 586
10) R. Dicke, W.P. Neumann, K. Vorspohl, 5th ICOCC, Padova 1986, Abstracts P. 41
11) D.J. Hart, F.L. Seely, J. Am. Chem. Soc. 110, 1631 (1988)

SILANES AS NEW REDUCING AGENTS IN ORGANIC SYNTHESIS

C. Chatgilialoglu
I.Co.C.E.A.
Consiglio Nazionale delle Ricerche
40064 Ozzano Emilia (Bologna)
Italy

ABSTRACT. The rationalization of thermodynamic and kinetic data of silane/silyl radical systems allows the characterization of a class of silanes as new reducing agents. In fact, tris(trimethylsilyl)silane and trialkylmercaptosilanes would appear to offer an attractive alternative to tributyltin and tributylgermanium hydrides respectively in radical chain reactions.

INTRODUCTION

The majority of radical reactions of interest to synthetic chemists are chain processes in which radicals are generated by some initiation process, and which then undergo a series of propagation steps generating fresh radicals and finally disappear, usually by mutual coupling or disproportionation. For a general reduction reaction:

$$RZ + MH \longrightarrow RH + MZ \tag{1}$$

where RZ is the group to be reduced and MH is the organometallic hydride, the following radical displacements may be involved as propagation steps:

$$RZ + M^{\bullet} \longrightarrow (R\dot{Z}M) \longrightarrow R^{\bullet} + ZM \tag{2}$$

$$R^{\bullet} + MH \longrightarrow RH + M^{\bullet} \tag{3}$$

where $(R\dot{Z}M)$ is a reactive intermediate or a transition state.

Probably the best known and most useful free radical reaction is the reduction of alkyl halides by tributyltin hydride(1). This reaction involves a two-step chain process:

$$R^{\bullet} + Bu_3SnH \longrightarrow RH + Bu_3Sn^{\bullet} \tag{4}$$

$$Bu_3Sn^{\bullet} + RX \longrightarrow Bu_3SnX + R^{\bullet} \tag{5}$$

However, the formation of tributyltin halides often makes work-up and

115

F. Minisci (ed.), Free Radicals in Synthesis and Biology, 115–123.
© *1989 by Kluwer Academic Publishers.*

product isolation difficult. Moreover, the tin compounds themselves are toxic and create a disposal problem.

Trialkylsilyl radicals are more reactive in halogen atom abstractions than $Bu_3Sn\cdot$, but silanes are rather poor H-atom donors towards alkyl radicals and therefore do not support chain reactions except at elevated temperatures. As the kinetic and thermodynamic factors that control propagation steps 2 and 3 are fundamental ones for the invention of new free radical reducing agents, the purpose of this article is to report some recent data concerning silyl radicals and rationalize them from the point of view of the reducing ability of silanes.

BOND DISSOCIATION ENERGIES OF SILANES

The factors which control carbon-hydrogen bond dissociation energies (BDE), eq. 6 and 7, are reasonably well understood. Bond dissociation energies are lowered when conjugated radicals are formed, when the radical center has an adjacent heteroatom, or when the bond dissociation

$$R-H \longrightarrow R^\cdot + H^\cdot \tag{6}$$

$$BDE(R-H) = \Delta H_f(R^\cdot) + \Delta H_f(H^\cdot) - \Delta H_f(R-H) \tag{7}$$

relieves steric compression in R-H(2). From the values collected in table I it can be seen that the factors influencing the strength of carbon-hydrogen bonds are essentially unimportant in the silicon congeners,

TABLE I. Bond Dissociation Energies for Si-H and Analogous C-H Bonds (in kcal/mol)

CH_3-H	105.1	SiH_3-H	90.3
$MeCH_2-H$	98.2	$MeSiH_2-H$	89.6
Me_2CH-H	95.1	Me_2SiH-H	89.4
Me_3C-H	93.2	Me_3Si-H	90.3
$PhCH_2-H$	88.0	$PhSiH_2-H$	88.2
F_3C-H	106.7	F_3Si-H	100.1
Cl_3C-H	95.8	Cl_3Si-H	91.3
		H_3SiSiH_2-H	86.3
Me_3SiCH_2-H	99.2	$Me_3SiSiMe_2-H$	85.3

namely, (i) <u>the absence of a weakening effect of substituent alkyl groups</u> <u>on Si-H bonds</u>. The C-H bond weakening by substituent methyl groups in the simple alkanes has been interpreted either by the methyl group inductive effect or by the hyperconjugation in the corresponding radicals. Walsh has rationalized the behaviour of silanes by the lower electronegativity of silicon versus carbon (1.8 and 2.6 on the Pauling scale), which provides a poorer acceptor of electron density from the substituent alkyl groups(3). (ii) <u>the ineffectiveness of delocalization by phenyl</u> <u>substitution</u>. Si-H bond weakening by phenyl substitution is only ca. 1.5 kcal/mol, in contrast to the 10 kcal/mol observed for phenyl substituted C-H bonds. As pointed out by Walsh the small size of sila-benzyl stabilization is not unexpected in view of the weakness of bonding in sila olefins(3). Although bond dissociation energies for cumulative phenyl substitution are unknown, judging from the reactivity of these silanes towards <u>tert</u>-butoxyl radicals (cf. table II) the ineffectiveness of delocalization by further phenyl substitution is evident(4). (iii) <u>the profound effect produced by an adjacent heteroatom</u>. The most interesting observation is that the difference in BDE between trifluoro- and trichloro-substituted silane and methane are similar,

TABLE II. Rate Constants for the Reaction of <u>tert</u>-Butoxyl Radicals with Some Silanes at ca. 300 K.

Subsrate	$\underline{k},^{\S}$ $M^{-1}s^{-1}$	Substrate	$\underline{k},^{\S}$ $M^{-1}s^{-1}$
$n\text{-}C_5H_{11}SiH_3$	3.0×10^6	$PhSiH_3$	2.5×10^6
		Ph_2SiH_2	6.5×10^6
Et_3SiH	4.6×10^6	Ph_3SiH	7.7×10^6

§Corrected rate constants; values represent the attack at single Si-H bond.

i.e., 9 and 11 kcal/mol. However, while $BDE(F_3C\text{-}H)$ is similar to $BDE(CH_3\text{-}H)$ with $BDE(Cl_3C\text{-}H)$ being ca. 10 kcal/mol weaker, the $BDE(F_3Si\text{-}H)$ is ca. 10 kcal/mol stronger than $BDE(SiH_3\text{-}H)$ or $BDE(Cl_3Si\text{-}H)$. One way to rationalize this striking observation is to suppose that the electron-withdrawing properties of F and Cl are opposed to stabilization associated with the interaction involving the p-type non bonding electrons on the halogen and the single electron on the central atom; while in the halocarbons the latter effect seems to be more pronounced, in silanes the general electron-withdrawing effect is more likely to predominate. Although electron spin resonance data largely support these conclusions, this kind of argument is still a gross oversimplification. With this in mind one would expect the Me_3Si substituent to weaken the Si-H bond strength. In fact the Si-H bond in disilanes is ca. 4-5 kcal/mol weaker than in monosilanes. Recently we found that tris(trimethylsilyl)silane, $(Me_3Si)_3SiH$, has a bond dissociation energy of 79.0 kcal/mol indicating that cumulative Me_3Si substitution produces a greater weakening(5).

It is clear that the electronegative character of the substituent plays an important role in the strength of the Si-H bond. However, simple electronic effects do not completely account for this behaviour since Si-H bond dissociation energies do not correlate with the group electronegativities of the ligands (see Figure 1). One way of rationalizing this behaviour is to separate the trends of second and third row substituents. By doing this one can see that a correlation exists between the sum of second row substituent group electronegativities and bond dissociation energies of silanes (Figure 1). The extra stabilization of the third row substituted silyl radicals may arise from a bonding interaction between the substituent d-orbitals and the semi-occupied molecular orbital (SOMO) on the central silicon atom as well as by hyperconjugation in the case of the tris(trimethylsilyl)silyl radical. Electron spin resonance data support this interpretation(7). However, it is possible that steric compression in $(Me_3Si)_3SiH$ is relieved on radical formation thus making the Si-H bond quite weak.

Based on the above analysis, we were led to study the behaviour of the trialkylmercaptosilanes, $(RS)_3SiH$. Preliminary work on trimethyl-mercaptosilane as a reducing agent(8), suggests a bond dissociation energy

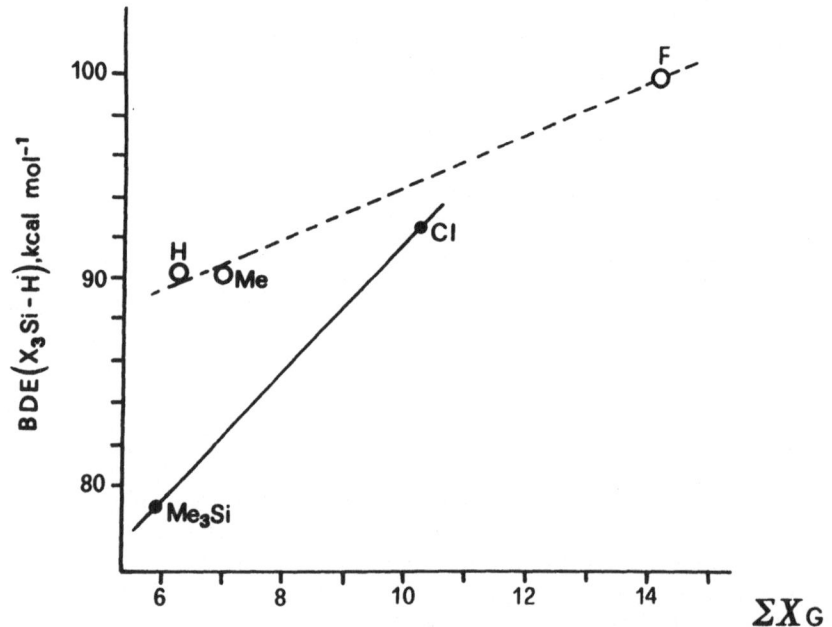

Figure 1. Bond dissociation energies of silanes, BDE(X_3Si-H), vs. ΣX_G substituent group electronegativities (taken from ref. 6).

similar to trialkylgermane, i.e, 82-84 kcal/mol, in excellent agreement
with expectations. In fact, in figure 1 the $(MeS)_3SiH$ lies close to the
line drawn between Cl_3SiH and $(Me_3Si)_3SiH$. However, due to the scarsity
of points in figure 1, these kind of arguments are still somewhat
speculative. Work is now in progress to remedy this.

SILANES AS REDUCING AGENTS

Alkyl halides are reduced to hydrocarbons by organosilanes in good yields
(50-90%) when catalysed by Lewis acids, although secondary reactions can
often predominate(9), viz.,

$$RX + Et_3SiH \xrightarrow{AlCl_3} RH + Et_3SiX \qquad (8)$$

However, this methodology is further limited to (i) alkyl halides that
donot undergo extensive structural rearrangement with Lewis acids,
(ii) phenyl-substituted alkyl halides for which a Friedel-Crafts alkylation process does not predominate and (iii) functionalized alkyl halides
possessing neither a nitro nor a cyano substituent.

On the other hand, trialkylsilanes are poor reducing agents in a
free radical chain process(10). That is, although trialkylsilyl radicals
are very reactive in halogen atom abstraction their corresponding silanes
are rather poor H-atom donors towards alkyl radicals and therefore not
do support chain reactions except at elevated temperatures. As we have
already mentioned above the silicon-hydrogen bonds can be dramatically
weakened by multiple substitution of third-row groups at the Si-H function.
In fact the silicon-hydrogen bond in $(Me_3Si)_3SiH$ is 79 kcal/mol and is
11 kcal/mol less than that of Et_3SiH. This result suggests that $(Me_3Si)_3SiH$
might be a good hydrogen donor and that the compound would be capable
of sustained radical chain reduction of organic halides. This expectation
turned out to be correct and we report some of our results in table III(11).

The reduction of alkyl halides by tris(trimethylsilyl)silane can be
achieved under a variety of conditions. This reduction reaction involves
a two-step free radical chain process, viz.,

$$R^\cdot + (Me_3Si)_3SiH \longrightarrow RH + (Me_3Si)_3Si^\cdot \qquad (9)$$

$$RX + (Me_3Si)_3Si^\cdot \longrightarrow R^\cdot + (Me_3Si)_3SiX \qquad (10)$$

as indicated by the fact that the reaction is catalysed by thermal or
photochemical sources of free radicals and retarded by common inibitors(11).
The data in table III show that the reductions were very efficient for
alkyl iodides, bromides and chlorides with the following reactivity
order: $RI > RBr > RCl$.

The absolute rate constants for the reactions of tris(trimethylsilyl)silyl radicals with a large nubmer of organic halides have been
measured in solution by using laser flash photolysis for compounds having
rate constants $> 10^6$ $M^{-1}s^{-1}$ and competitive experiments for compounds
having rate constants $< 10^6$ $M^{-1}s^{-1}$ (12). The reactivities vary over a

TABLE III. Reduction of Halides by Tris(trimethylsilyl)silane

Halides	Reaction Time (h)	GC yield RH%
$C_{18}H_{37}Cl$	5.0	93
—Cl	2.5	95
$C_{16}H_{33}Br$	0.5	100
$C_{18}H_{37}I$	0.1	100

wide range at ca. 300 K from ca. $10^5 M^{-1}s^{-1}$ for chlorides to ca. $10^9 M^{-1}s^{-1}$ for iodides. The trends in reactivity are those which would be expected for an individual radical; that is, (i) for a particular R group the rate constants decrease along the series X = I > Br > Cl, (ii) for a particular X the rate constants decrease along the series benzyl > tert-alkyl > sec-alkyl > primary alkyl > phenyl. For most halides the rate constants for halogen abstraction by tris(trimethylsilyl)silyl radicals are about an order of magnitude slower than the corresponding reactions of triethylsilyl radicals. On the other hand, although for bromides and iodides tris(trimethylsilyl)silyl and tributyltin radicals have essentially similar reactivities, with chlorides the tris(trimethylsilyl)-silyl radicals react an order of magnitude faster than the corresponding reactions of $Bu_3Sn^·$ radicals. These results suggest that tris(trimethyl-silyl)silane is a more efficient reducing agent than tributyltin hydride for organic chlorides.

The rate constant for hydrogen abstraction from $(Me_3Si)_3SiH$ by primary alkyl radicals has been measured using the 5-hexenyl "radical clock" and led to $k=6x10^6 M^{-1}s^{-1}$ which is ca. 4 times slower than the corresponding reaction with tributyltin hydride(11). For tin hydride it has been shown that the rate constant for the primary, secondary and tertiary alkyl radicals are essentially the same at room temperature(13). This often convenient property however, has not yet been confirmed for tris(trimethylsilyl)silane(14).

It is worth noticing that tributyltin hydride has been frequently employed in systems where the initially formed organic radical must first undergo a unimolecular or intermolecular reaction with the formation of a carbon-carbon bond if the desired product is to be obtained (cf. figure 2, where M=Sn)(15). In certain cases the latter process is slow relative to its reduction by tin hydride and a slightly less active hydrogen donor that can nevertheless fulfil the other requirements of these chain processes could provide a very useful alternative to the usual tin hydride

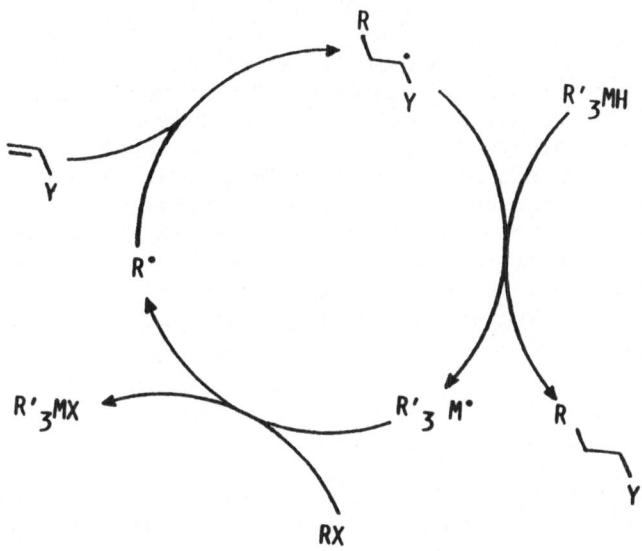

Figure 2. Radical addition of alkyl halides to alkenes in the presence of group 14 organometallic hydrides.

method. We suggest that tris(trimethylsilyl)silane has the necessary chemical properties to fulfil these requirements. That is, $(Me_3Si)_3Si\cdot$ abstracts halogen atoms with rates similar or even faster to those of the $Bu_3Sn\cdot$radical while $(Me_3Si)_3SiH$ is only ca. 1/4 as good as Bu_3SnH as an H-atom donor to primary alkyl radicals.

One drawback of the above mentioned methods is that the product, once formed, must be separated from a full 1 equiv. of organometallic halide as group 14 organometallic hydride, R_3MH, is employed in stoichiometric amounts. For bromides and iodides it has already been shown that is often possible to work with catalytic amounts of tin hydride by reducing the tributyltin halide formed with $NaBH_4$ to regenerate the tin hydride(1,15,16). A method for catalytic dehalogenation via tris(trimethylsilyl)silane was also developed(17): the organic halide is treated with excess of sodium borohydride and a catalytic amount of $(Me_3Si)_3SiH$ or its corresponding halide, the reaction being initiated photochemically. The catalytic cycle is given in eq. 11 and 12.

$$(Me_3Si)_3SiX + NaBH_4 \longrightarrow (Me_3Si)_3SiH + NaX + BH_3 \quad (11)$$

$$(Me_3Si)_3SiH + RX \longrightarrow (Me_3Si)_3SiX + RH \quad (12)$$

The reaction is carried out in relatively polar solvents like ethanol or ethyleneglycoldimethyl ether, in the absence of air, and preferably at room temperature. Yiels are good to excellent; for example 1-bromohexadecane has been reduced to the corresponding hydrocarbon in 98% yield.

In recent years tributylgermanium hydride has also been used as a viable alternative to tin hydrides, particularly for radicals which undergo relative slow rearrangements, since it is a much less active hydrogen donor towards alkyl radicals. Furthermore, Ingold and coworkers recently reported detailed kinetic studies of hydrogen abstraction from Bu₃GeH by alkyl radicals as well as of halogen abstraction from organic halides by the corresponding germyl radicals(18). Based on the rationalization of the bond dissociation energies of silanes (see supra) we were led to study the behaviour of the trialkylmercaptosilanes. Preliminary work suggests that trimethyl- and triisopropylmercaptosilanes function as free radical reducing agents which rivals tributylgermanium hydride(8). Work on the kinetics and synthetic scope of these reactions is in progress.

CONCLUSIONS

Contrary to what is generally believed, the reactivity of silanes containing silicon-hydrogen bonds towards free radicals may vary over a wide range. Thus, although the importance of stabilization in the alkyl radicals is essentially unimportant in the silicon congeners, the "(d-3p)π-type bonding" and/or hyperconjugation together with the electronegative character of the substituents can play a very important role.

Tris(trimethylsilyl)silane and trialkylmercaptosilanes appear to offer an alternative to tributyltin and tributylgermanium hydrides respectively in radical chain reduction reactions of organic halides. Furthermore, these economically available silanes are probably far more acceptable reducing agents than tin and germanium compounds from an ecological and toxicological point of view.

ACKNOWLEDGEMENT

I thank all my previous and present coworkers whose names are listed in the references. Particularly, I wish to thank Dr. D. Griller for his active partecipation on this project and NATO for a travel grant which made this collaboration possible.

REFERENCES

(1) For a recent review see: B. Giese, Angew. Chem. Int. Ed. Engl., 24, 553 (1985).
(2) D.F. McMillen, and D.M. Golden, Annu. Rev. Phys. Chem., 33, 493 (1982).
(3) R. Walsh, Acc. Chem. Res., 14, 246 (1981).
(4) C. Chatgilialoglu, K.U. Ingold, J. Lusztyk, A.S. Nazran, and J.C. Scaiano, Organometallics, 2, 1332 (1983).
(5) J.M. Kanabus-Kaminska, J.A. Hawari, D. Griller, and C. Chatgilialoglu, J. Am. Chem. Soc., 109, 5267 (1987).

(6) J. Murray, J. Am. Chem. Soc., 106, 5842 (1984) and 107, 7271 (1985).

(7) C. Chatgilialoglu, and S. Rossini, Bull. Soc. Chim. Fr., in press.

(8) C. Chatgilialoglu, and G. Seconi, unpublished results.

(9) M.P. Doyle, C.C. McOsker, and C.T. West, J. Org. Chem., 41, 1393 (1976).

(10) J. Lusztyk, B. Maillard, and K.U. Ingold, J. Org. Chem., 51, 2457 (1986) and references therein.

(11) C. Chatgilialoglu, D. Griller, and M. Lesage, J. Org. Chem., in press.

(12) C. Chatgilialoglu, D. Griller, and M. Lesage, J. Org. Chem., Submitted for publication.

(13) C. Chatgilialoglu, K.U. Ingold, and J.C. Scaiano, J. Am. Chem. Soc., 103, 7739 (1981).

(14) C. Chatgilialoglu, D. Griller, and M. Lesage, studies are currently in progress.

(15) B. Giese, Radicals in Organic Synthesis: Formation of Carbon-Carbon Bonds, Pergamon Press: New York, 1986.

(16) E.J. Corey, and J. Williams Suggs, J. Org. Chem., 40, 2554 (1975).

(17) C. Chatgilialoglu, D. Griller, and M. Lesage, Italian Patent: 48150A87 (1987).

(18) J. Lusztyk, B. Maillard, S. Deycard, D.A. Lindsay, and K.U. Ingold, J. Org. Chem., 52, 3509 (1987); K.U. Ingold, J. Lusztyk, and J.C. Scaiano, J. Am. Chem. Soc., 106, 343 (1984).

ACYL, IMIDOYL, AND OTHER HETERO-SUBSTITUTED FREE RADICALS AS INTERMEDIATES IN THE SYNTHESIS OF CYCLIC ORGANIC COMPOUNDS

M.D. BACHI, E. BOSCH, D. DENENMARK AND D. GIRSH
Department of Organic Chemistry
The Weizmann Institute of Science
Rehovot 76100
Israel

ABSTRACT. The tributylstannane-induced cyclization of various chloro, thio, and selenoderivatives of carbonic and carboxylic acids were studied. It was found that the intramolecular addition of acyl radicals, or synthetic equivalents thereof, to double bonds follows regioselectively the exo-mode in the formation of both five and six membered rings. Alkyl-γ-lactones and thionolactones, as well as six-membered cyclic oxo-compounds were obtained in high yields. The current understanding of the scope and limitations of these reactions are discussed.

The high effectiveness of the intramolecular addition of carbon centered free-radicals to carbon-carbon multiple bonds in the construction of cyclic systems has recently been widely acknowledged.[1-3] A convenient method for the site specific generation of carbon-centered free-radicals involves the chemospecific homolysis of a C-Halogen, C-S or C-Se bond by trialkyltin radicals derived from the corresponding tin hydrides.[4] The first reports on the application of this method for the synthesis of multifunctional complex organic compounds involved the intramolecular addition of alkyl,[5] vinyl[6], and aryl[7] radicals. Although these early reports were subsequently followed by scores of others[1-4] the analogous cyclization of acyl radicals has rarely been employed for synthetic purposes.

The first report on the intramolecular addition of acyl radicals to double bonds, dated 1964, is related to the cyclization of campholenic aldehyde to camphor by diacetylperoxide.[8] It was followed by three additional reports on peroxide-induced cyclization of unsaturated aldehydes,[9,10] and by two contrasting reports on the reductive cyclization of acyl chlorides induced by tributyltin hydride (TBTH) and azoisobutyronitrile (AIBN).[11,12] Thus, 5-hexenoyl chloride was reported to undergo endo-cyclization in one report[11] and exo-cyclization in another.[12] An esr study indicated that at low temperatures citronelloyl radical does not undergo ring closure.[13]

We reported some preliminary results in 1986 concerning the synthesis of α-alkylidene-γ-lactones by the intramolecular addition of alkoxycarbonyl free-radicals to acetylenes.[14] This method, which is described in Scheme I, is of wide applicability as it allows the conversion of homopropargylic alcohols to the corresponding α-alkylidene-γ-lactones in almost quantitative yield. In conjunction with a new desilylation procedure, it provides a new approach to α-methylene-γ-lactones.[15] In the present communication, we describe some recent results

125

F. Minisci (ed.), Free Radicals in Synthesis and Biology, 125–134.
© *1989 by Kluwer Academic Publishers.*

Scheme I

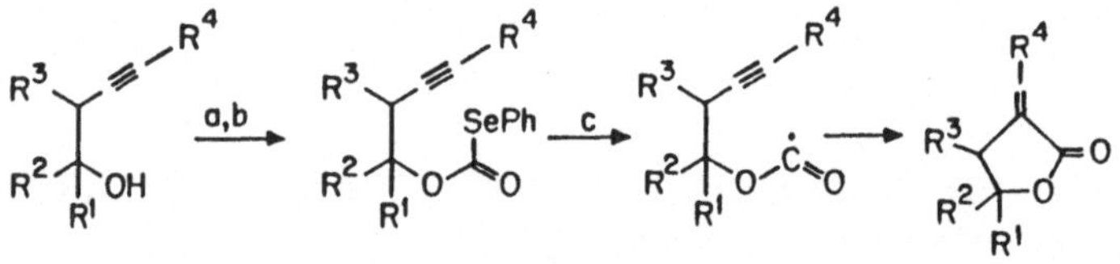

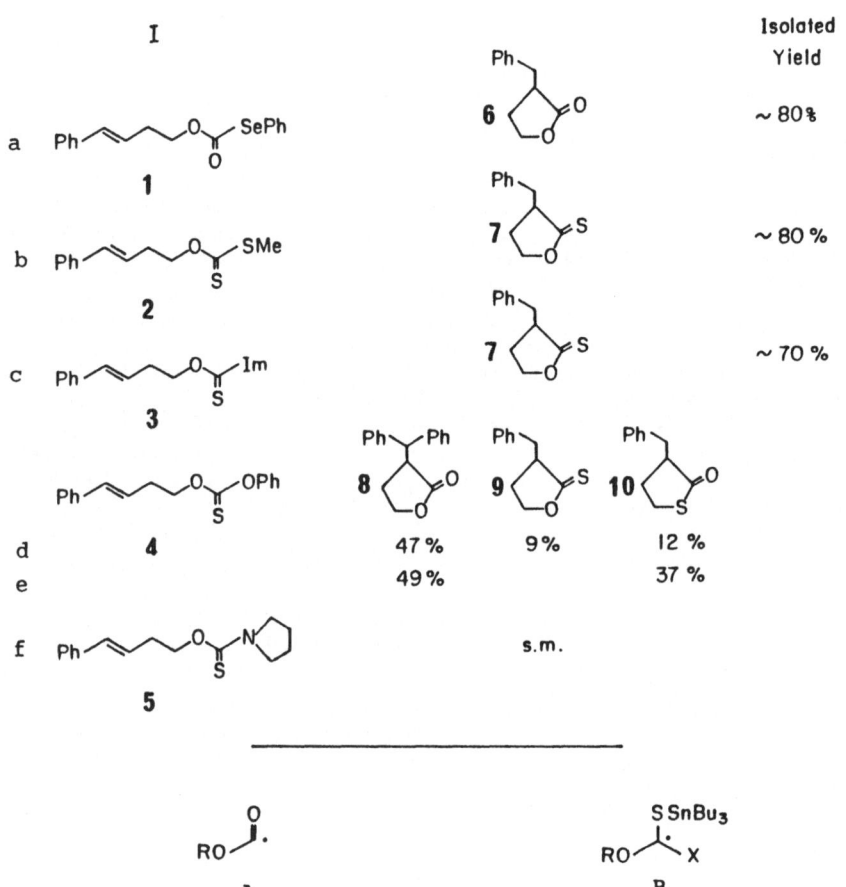

(a) $COCl_2$; (b) PhSeH, Pyridine; (c) Bu_3SnH, AIBN

contributing to the assessment of the scope and limitation of free-radical carbolactonization and of other reactions involving the intramolecular addition of acyl radicals and synthetic equivalents thereof, to multiple bonds.

In a series of experiments presented in Scheme II, carbonic acid derivatives of

Scheme II

$I (0.02 M) + 1.1 eq$ $Bu_3SnH + 0.1 eq$ AIBN

phenylhomoallyl alcohol were treated with TBTH (1.1 eq) and AIBN (0.1 eq) in boiling benzene (or toluene, entry e). The phenylselenocarbonate **1** cyclized to the lactone **6** in excellent yield through the intermediacy of an alkoxycarbonyl σ radical of type **A**. The dithiocarbonate **2** and the thioimidazolide **3** were converted, in good yield, into the thionolactone **7**, through the intermediacy of a tris-hetero-substituted p-radical of type **B** as shown, in the case of the dithiocarbonate **2**, in Scheme III. The free-radical ring closure of the phenylthionocarbonate **4** resulted in a mixture of the diphenylmethyl-lactone **8**, the thionolactone **9** and the thiolactone **10**, which are presumably obtained through the reaction mechanism proposed in Scheme IV. The failure of the thiocarbamate **5** to undergo homolytic cyclization[16] is attributed to excessive stability of the free-radical of type **B** (X=N$\leq$). Selenocarbonates thus seem to be the best of the tested substrates for the TBTH/AIBN induced carbolactonization. Some reactions of selencarbonates with TBTH/AIBN are summarized in Scheme V. Except for one special case (entry g) all reactions performed under dilute conditions (0.02 M in benzene at 80°C) afforded the corresponding lactones in high yields. Products deriving from direct hydrogen atom transfer to the alkoxycarbonyl radical were only observed in experiments performed with a high concentration of TBTH

Scheme III

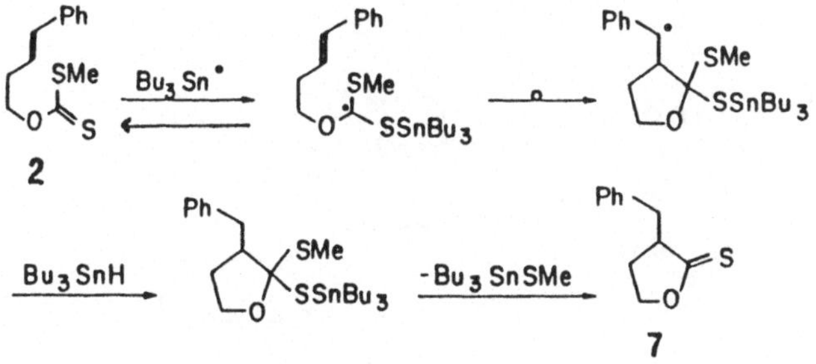

Scheme IV

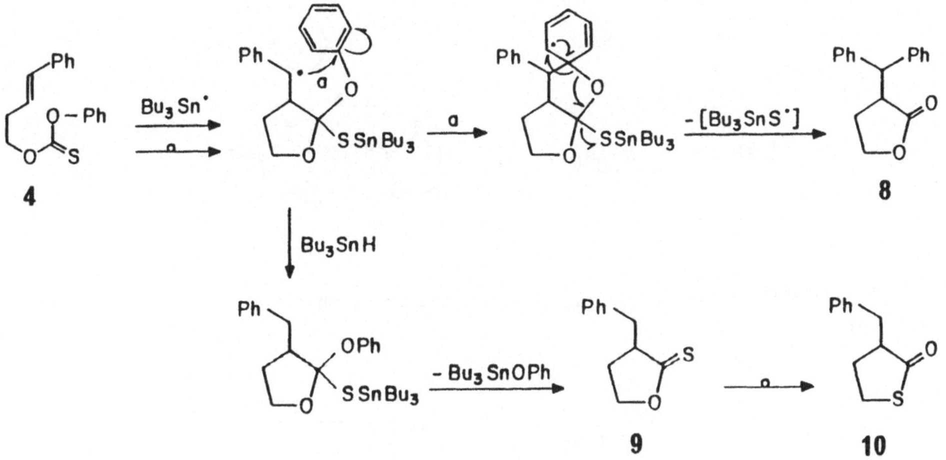

(entries c and e). A decarboxylation product was obtained only from a benzylselenocarbonate (entry g) in which an especially stabilized radical is obtained after decarboxylation of the corresponding intermediate of type **A**.[17] These results indicate that the intramolecular addition of alkoxycarbonyl radicals to double bonds at position 5 follows the exo-mode exclusively. The rate of cyclization under standard conditions (0.02 M) is sufficiently high to compete with hydrogen transfer and with decarboxyation processes.[18] The double bond does not need any special activation and can carry an additional substituent at position 5 (entry d). No side reactions were observed even in the case of the more sensitive selenocarbonate deriving from a tertiary alcohol (entry h).

Scheme V

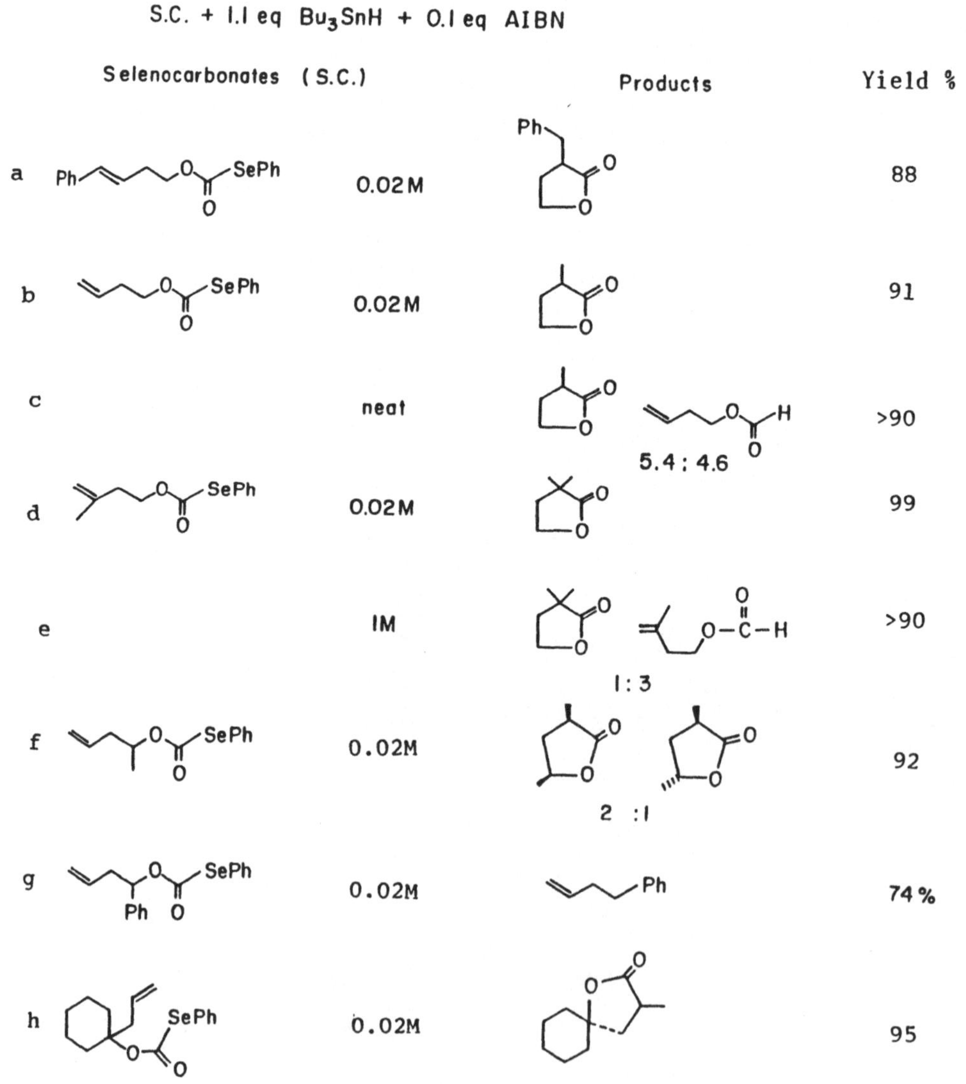

$$S.C. + 1.1 \ eq \ Bu_3SnH + 0.1 \ eq \ AIBN$$

Selenocarbonates (S.C.)		Products	Yield %
a	0.02 M		88
b	0.02 M		91
c	neat	5.4 : 4.6	>90
d	0.02 M		99
e	1M	1 : 3	>90
f	0.02 M	2 : 1	92
g	0.02 M		74 %
h	0.02 M		95

In order to evaluate the suitability of acyl radical annelation for the preparation of 6-membered cyclic ketones, we repeated the reaction of citronelloyl chloride **11** with TBTH/AIBN,[11] and compared it to the same reaction with the selenoester of citronelloic acid **12** (Scheme VI). Both compounds underwent free-radical cyclization affording ($\sim 85\%$) of a mixture (~ 1.5:1) of menthone and isomenthone along with citronellal ($\sim 10\%$) when the reactions were performed in boiling toluene (0.02 M solution) with TBTH (1.1eq) and AIBN (0.1eq). With both citronelloyl chloride and phenylselencitronelloate, high dilution techniques (e.g addition of TBTH and AIBN in boiling toluene over a period of 90 min) totally suppressed the formation of citronellal. However, while in the case of citronellyl chloride the yield of cyclic products was reduced to 45%, it was increased to 98% (1.6:1 menthone/isomenthone) in the case of selenoester **12**. The superiority of selenoester over acyl chloride in homolytic cyclizations induced by TBTH/AIBN seems to derive mainly from the higher stability of the selenoester to polar side-reactions.[19]

Scheme VI

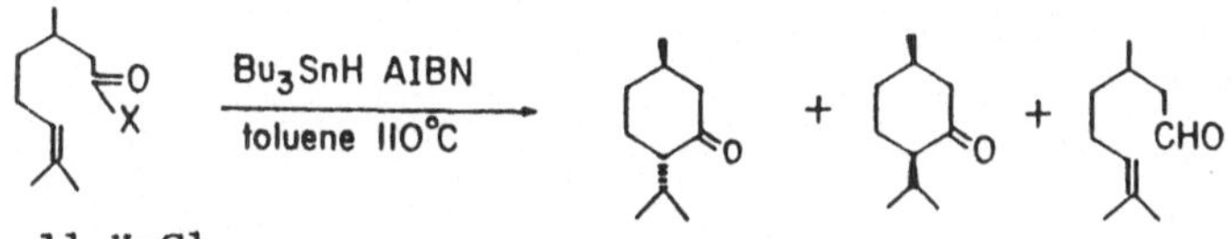

11, X=Cl
12, X=SePh

Scheme VII

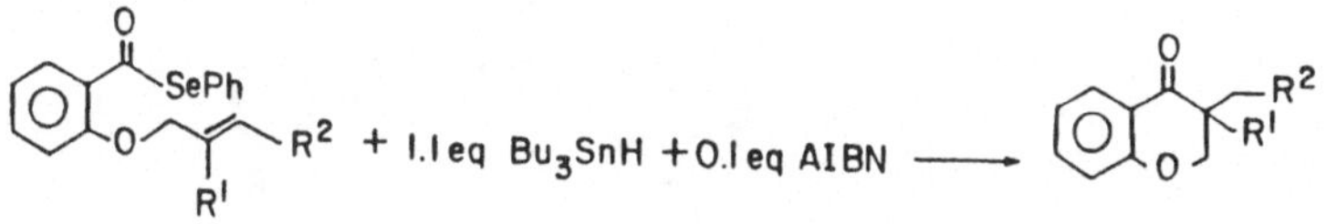

I

	I	Product	(isolated yields)
a	R^1=H; R^2=Me	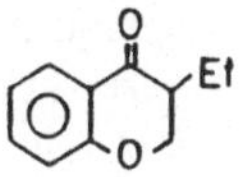	(87%)
b	R^1=R^2=H	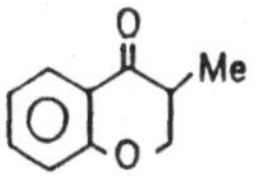	(90%)
c	R^1=Me; R^2=H	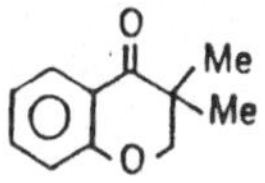	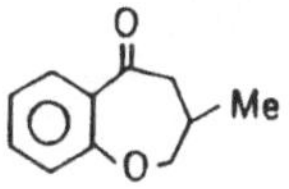 (79%) ... (8%)

Scheme VII describes cyclizations of some selenoesters of salicyclic acid derivatives induced by TBTH/AIBN. The reactions follow exclusively (entry a,b), or predominantly (entry c) the 6-*exo* addition mode. No products deriving from either inter- or intramolecular hydrogen transfer to the intermediate acyl radicals were observed. The selectivity of the 6-*exo-trig* cyclization of acyl radicals is thus higher than that of alkyl radicals, which are more prone to intramolecular hydrogen atom abstraction through a 6-membered ring transition state,[22] and is similar, or even superior, to that of vinyl radicals.[6] Treatment of the selenoester **13** with TBTH/AIBN under the standard conditions afforded a mixture of the chromanone **14** and the chromanone **15** (Scheme VIII). While **14** is the product of the regular reductive cyclization of the acyl radical **16**, the chromanone **15** is a secondary product resulting from reaction of **14** with additional tributylstannyl radicals to give radicals **17**, which may lose benzyl radicals through β-elimination.

Scheme VIII

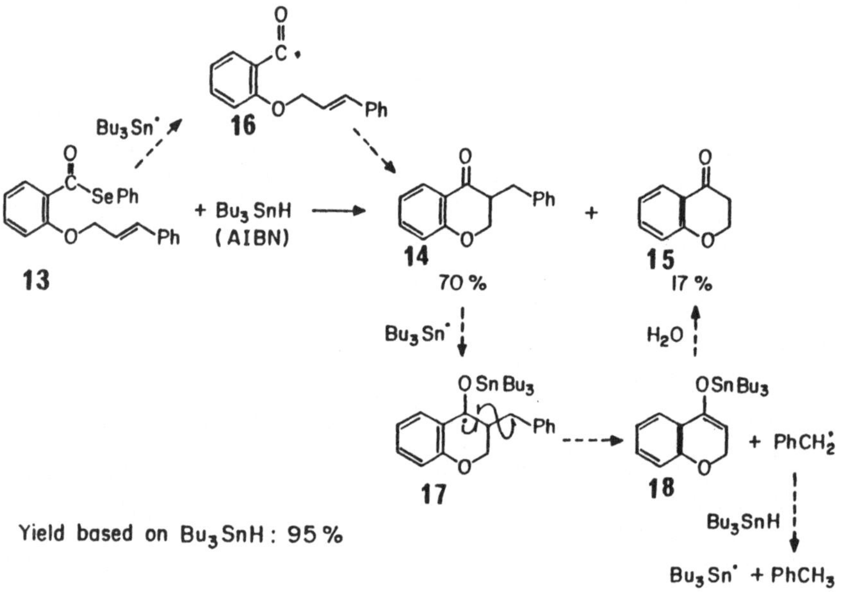

Yield based on Bu₃SnH : 95 %

In an alternative approach, phenylselenoimides **19** were cyclized by TBTH/AIBN and the resulting imines **20** were hydrolyzed to the oxo-compounds **21**. The reaction is general (Scheme IX) but the yields of products are highly dependent on the character of the substituent R, which in certain cases alter the overall reaction path. For example, treatment of the phenylselenoimine **22** with

Scheme IX

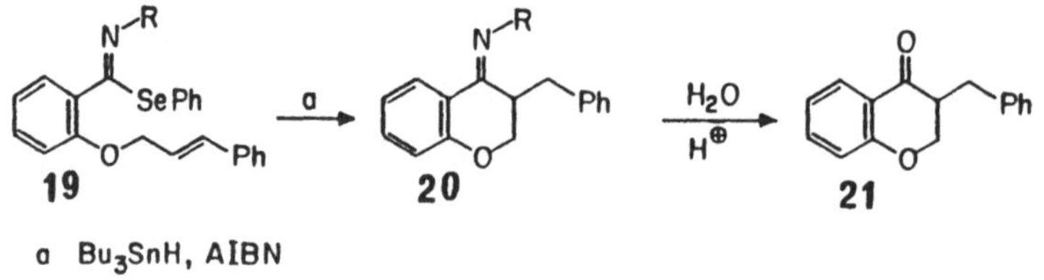

a   Bu₃SnH, AIBN

TBTH/AIBN under the standard conditions afforded, after workup, a 1:1 mixture of the chromanone **25** and the nitrile **26**. As shown in Scheme X, the former derives from cyclization of the imidoyl radical **23**, while the nitrile arises from the same intermediate through the β-elimination of a benzyl radical.

Scheme X

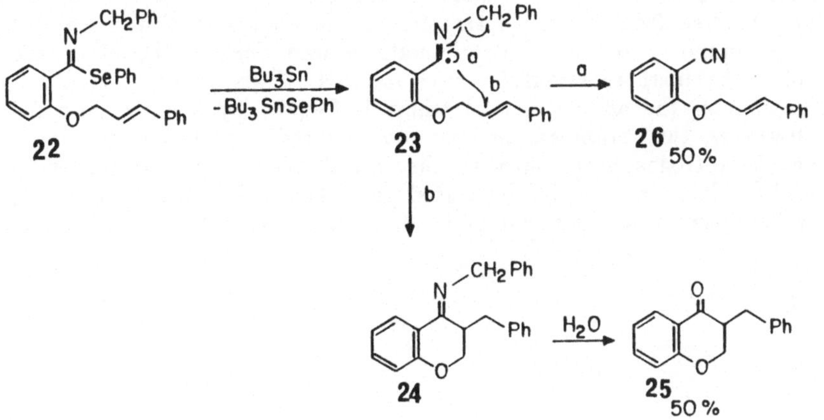

Scheme XI

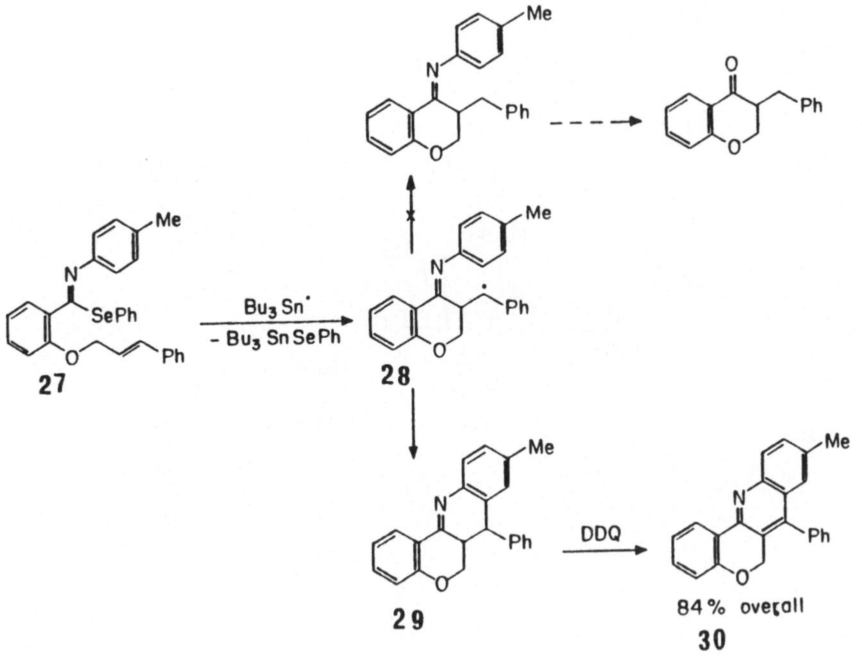

In another case, Scheme XI, the intramolecular addition of the imidoyl radical, generated from the selenoimidate **27** under standard conditions, was followed by an intramolecular homolytic aromatic substitution of the intermediate radical **28** to give the hydroaromatic compound **29**. Oxidation of **29** afforded the aromatic compound **30**. This accidental observation opens a new approach to the synthesis of polycyclic heterocyclic compounds through tandem homolytic addition and substitution.[23] In all the experiments, the imidoyl radicals added to the double bond exclusively through the *exo* mode in the same fashion as the acyl radicals. No products derived from hydrogen transfer to the imidoyl radical were obtained.

In close similarity to the thiocarbamate **5** (see Scheme II), other thioamides were found to be inert to tributylstannyl radicals. This property is attributed to the stabilization of the adduct radical which is flanked by a sulfur and a nitrogen atom. However, this stabilization can be reduced by decreasing the electron density on the nitrogen atom through the introduction of an electron-withdrawing substituent.[24] This practice is illustrated in Scheme XII, which shows the free-radical cyclizations of a thiosuccinimide derivative **31** and a saccarine derivative **32**.

Scheme XII

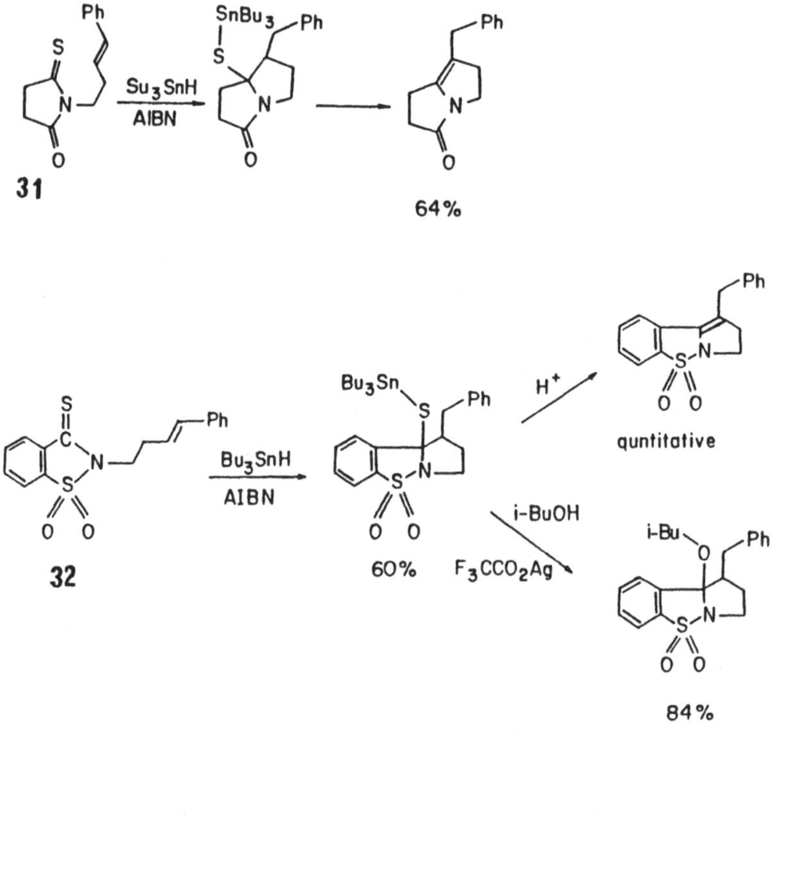

REFERENCES

1. Hart, D.J. *Science,* **1984,** *223,* 883.

2. Giese, B. "Radicals in Organic Synthesis: Formation of Carbon-Carbon Bonds", Pergamon Press, Oxford, **1986.**

3. Ramaiah, M. *Tetrahedron,* **1987,** *43,* 3541.

4. Neumann, W.P. *Synthesis,* **1987,** 665.

5. See for example: (a) Buchi, G.; Wuest, H. *J. Org. Chem.* **1979,** *44,* 546. (b) Bachi, M.D.; Hoornaert, C. *Tetrahedron Lett.* **1981,** *22,* 2689; 2693. (c) Bachi, M.D.; Hoornaert, C. *Tetrahedron Lett.* **1982,** *23,* 2025. (d) Hart, D.J.: Tsai, Y.-M. *J. Am. Chem. Soc.* **1982,** *104,* 1430.

6. Stork, G.; Baine, N.H. *J. Am. Chem. Soc.* **1982,** *104,* 2321.

7. Ueno, Y.; Chino, M.; Okawara, M. *Tetrahedron Lett.* **1982,** *23,* 2575.

8. Dulou, R.; Chretien-Bessiere, Y.; Desalbres, H. *C.R. Acad. Sc. Ser. C,* **1964,** *258,* 603.

9. (a) Montheard, J.-P. *C.R. Acad. Sc. Ser. C,* **1965,** *260,* 577. (b) Chatzopoulos, M.; Montheard, J.-P. *C.R. Acad. Sc. Ser. C,* **1975,** *280,* 29.

10. Kampmeier, J.A.; Harris, S.H.; Wedegaertner, D.K. *J. Org. Chem.* **1980,** *45,* 315.

11. Cekovic, Z. *Tetrahedron Lett.* **1972,** 749

12. Walsh, E.J. Jr.; Messinger, J.M.II; Grudoski, D.A. Allchin, C.A. *Tetrahedron Lett.* **1980,** *21,* 4409.

13. Davies, A.G.; Sutcliffe, R. *J. Chem. Soc., Chem. Comm.* **1979,** 473.

14. Bachi, M.D.; Bosch, E. *Tetrahedron Lett.* **1986,** *27,* 641.

15. Bachi, M.D.; Bosch, E. *Tetrahedron Lett.* **1988,** *29,* 0000.

16. The stability of a thiocarbamate to tributylstannane was previously observed: Barton, D.H.R.; McCombie, S.W. *J. Chem. Soc., Perkin I* **1975,** 1574.

17. For a similar decarboxylation, see: Beak, P.; Moje, S.W. *J. Org. Chem.* **1974,** *39,* 1320.

18. For example, the rate constant for decarboxylation of t-butyloxycarbonyl radical is $7.9 \times 10^5 \text{sec}^{-1}$ at 80°C, see: Griller, D.; Roberts, B.P. *J. Chem. Soc., Perkin Trans I.* **1972,** 747.

19. While the present work was in progress, the tributylstannane-induced cyclization of a seleno ester (32%),[20] and the photolytically-induced cyclization of an acyl xanthate (70%)[21] were reported.

134

20. Crich, D.; Fort, S.M. *Tetrahedron Lett.* **1987,** *28,* 2895.

21. Delduc, P.; Tailhan, C.; Zard, S.Z. *J. Chem. Soc., Chem. Comm.* **1988,** 308.

22. Leonard, W.R.; Livinghouse, T. *Tetrahedron Lett.* **1985,** *26,* 6431.

23. A different approach to the synthesis of heterocyclic compounds through the intermediacy of imidoyl radicals was recently described: Leardini, R.; Tundo, A.; Zanardi, G.; Pedulli, G.F. *Synthesis,* **1985,** 107. (b) Leardini, R.; Nanni, D.; Pedulli, G.F.; Tundo, A.; Zanardi, G.; *J. Chem. Soc., Perkin Trans I* **1986,** 1591.

24. Padwa, A.; Nimmesgern, H.; Wong, G.S.K. *J. Org. Chem.* **1985,** *50,* 5620.

SOME STUDIES ON CARBONYL RADICAL CYCLIZATIONS AND ON THE SYNTHESIS OF 2-DEOXY-β-D-GLYCOSIDES BY STEREOSELECTIVE RADICAL REACTIONS

David Crich*, Simon M. Fortt and Timothy J. Ritchie
Department of Chemistry, University College London,
20 Gordon Street, London WC1H OAJ, UK.

ABSTRACT. The effect of substituents on the mode (*exo*- or *endo*-) and efficiency of 6,7-olefinic carbonyl radical cyclizations has been studied. A new method for the synthesis of 2-deoxy-β-D-glycosides, in which the key step is a stereoselective radical decarboxylation, has been developed.

1. CARBONYL RADICAL CYCLIZATIONS

Since its discovery some 20 years ago the vitamin D_3 metabolite 1α,25-dihydroxyvitamin D_3 (1) has challenged the imagination of many synthetic chemists worldwide and various approaches have been devised for its total synthesis[1] as well as for its partial synthesis[2] from vitamin D_3 . We have been interested in synthesizing the ketones (2) or (3), by carbonyl radical cyclizations, with a view to their further elaboration into 1α,25-dihydroxyvitamin D_3 by adaptations of the Lythgoe method or the more recent Solladié route[3].

TBS = Bu^tMe_2Si

In the first instance we were attracted to what appeared to be a very direct entry into α-methylene-cyclohexanones, as (2), involving the cyclization of 6,7-acetylenic carbonyl radicals. However, and in stark contrast to the elegant work of Bachi[4] on the formation of α-methylene-γ-lactones by the cyclization of 3,4-acetylenic alkoxycarbonyl radicals, we were unable to obtain[5] any cyclohexanones

135

F. Minisci (ed.), Free Radicals in Synthesis and Biology, 135–143.
© *1989 by Kluwer Academic Publishers.*

from such reactions. Thus we turned our attention to the cyclization of 6,7-olefinic carbonyl radicals, an area in which there was some slight literature precedent[6], and have been mainly concerned with a systematic study of the effect of substituents on both the mode (*exo*- or *endo*-) and the efficiency of such cyclizations. In general carbonyl radicals were generated by the action of tributyltin hydride and a catalytic amount of azoisobutyronitrile (AIBN) on the corresponding selenol esters, according to Graf[7], at 80 °C in benzene and occasionally photolytically at room temperature in order to minimize decarbonylation.

The selenol ester (4) was prepared from the corresponding acid (5) by treatment of the derived acyl chloride with sodium phenylselenide in ethanol. The acid (5) was prepared from monomethyl glutaryl chloride by treatment with bis(trimethylsilyl)acetylene according[8] to Nicolaou followed by protection of the resultant ketone as its 1,3-dioxolane, saponification and concomitant protodesilylation and finally hydrogenation over Lindlaar's catalyst. On reaction with tin hydride and AIBN (4) gave[9], unexpectedly, the cycloheptanone (7) and the aldehyde (6) in 32 and 55% yields respectively. A further selenol ester (8) was prepared from diethyl malonate by a sequence of reactions involving alkylation with 3-chloropropional diethylketal, condensation with bis(trimethylsilyl)acetylene and titanium tetrachloride, Krapcho deethoxycarbonylation, partial hydrogenation, saponification and reaction of the derived acyl chloride with sodium phenylselenide. Like (4), (8) gave[9] unexpectedly the cycloheptanone (10) and the aldehyde (9) in 27 and 35% isolated yields respectively with only a trace of the expected cyclohexanone being formed. Clearly the cyclizations of the carbonyl radicals derived from both (4) and (8) are inefficient and lead preferentially to cycloheptanones. In a further experiment the selenocarbonate (11) was allowed to react with tin hydride and AIBN under the standard conditions when the major isolated product was the formate (12). That the poor cyclization yields from (11) and presumably (4) and (8) are not due to competing Barton type δ-hydrogen abstraction was demonstrated[9] by the reaction of (11) with tributyltin deuteride giving the deuterioformate (13) as major product in which deuterium was incorporated into the aldehydic function and not into the allylic position.

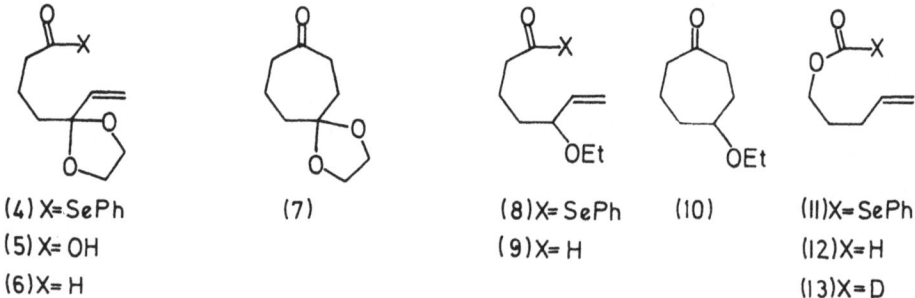

(4) X=SePh   (7)   (8) X=SePh   (10)   (11) X=SePh
(5) X=OH   (9) X=H   (12) X=H
(6) X=H   (13) X=D

The preferential formation of the cycloheptanones (7) and (10) may be the result of direct *endo*-mode cyclization favoured by the conformation of the carbonyl radical intermediate or alternatively the result of kinetic ring closure in the *exo*-mode followed by rapid rearrangement to the thermodynamically more stable *overall endo*-mode product as has been demonstrated[10] for vinyl radical cyclizations. The corresponding acyl migration in 3-ketoalkyl radicals has been observed[11] on several occasions. Unfortunately all our efforts so far aimed at differentiating between the two possibilities have proved fruitless.

The selenol esters (14) and (15) were prepared from ethyl 3-ketohept-6-enoate, itself prepared by alkylation of the ethyl acetoacetate dianion with allyl bromide, by either ketalization or reduction and silylation followed in both cases by saponification, formation of the acid chlorides and their reaction with sodium phenylselenide. Under the standard conditions (14) gave[9] with tin hydride an inseparable mixture of cyclohexanone (16) and cycloheptanone (17) in 86% combined yield. However the ratio of (16):(17) was now 6:1 in favour of the cyclohexanone. Similarly selenol ester (15) gave the cyclohexanones (18) and the cycloheptanone (19) in 54 and 17% yields respectively. Evidently removal of the ketal or ether functionality from the 5- position, as in (4) and (8) to the 3- position, as in (14) and (15) leads to greatly improved cyclization yields and reverses the *overall* mode of cyclization from *endo*-(cycloheptanones) to *exo*-(cyclohexanones).

(14)   (15)   (16)

(17)   (18)   (19)

We next sought to prepare a substrate carrying oxygen functionality at both the 3- and 5- positions as in our target molecules (2) and (3). This was achieved by alkylation of the ethyl acetoacetate dianion with acrolein followed by silylation and ketalization to give the ethyl ester corresponding to (20) which was converted to (20) in the usual manner. Treatment of this selenol ester with tin hydride and AIBN under the standard conditions gave the cyclohexanones (21) and the cycloheptanone (22) in 72 and 24% isolated yields respectively. Thus the regiodirecting effect of oxygen functionality at the 3-position overcomes that at the 5-position and allows the preparation of a 2-methyl-3,5-dioxygenated cyclohexanone related to the target molecules (2) and (3).

(20)   (21)   (22)

Finally we turned our attention to the possible introduction of the α-methylene group as in (2). We considered that this would be best achieved[12] by *syn*-elimination of a sulphoxide formed by oxidation of a molecule such as (3). This requires the preparation of 7-phenylthiohept-6-enoic acid selenol esters and the cyclization of a carbonyl radical onto a vinyl sulphide moiety. To the best of our knowledge there are no known examples of carbonyl, or even simple alkyl radical cyclizations, onto vinyl sulphides but several examples of alkyl[13] and one of the closely related vinyl[14] radical onto enol ethers have been published recently. Regarding the mode, *exo*- or *endo*-, of cyclization onto the vinyl sulphide it was predictable that the terminal phenylthio moiety would reduce the rate of direct *endo*-cyclization by sterically hindering attack at the terminal position and also stabilize the radical formed on *exo*-cyclization and so retard any rearrangements thereby favouring cyclohexanone formation regardless of the mechanism for the formation of cycloheptanones (7), (10), (17) and (19) above. A model compound (23) was readily prepared by addition of thiophenol to the acetylenic acid (5)(CH=CH$_2$ = C≡CH) prepared *en route* to (5) followed by selenol ester formation according to Grieco[15] with tributylphosphine and *N*-phenylselenophthalimide. As predicted treatment of (23) with tin hydride and AIBN under the standard conditions led to clean cyclization to the cyclohexanone (24) which was isolated in 72% yield together with a minor amount of the aldehyde resulting from straight reduction.

(23)   (24)   (25)

(26)   (27)

Having completed our model studies we are currently endeavouring to synthesize (3) and (2) by quenching the ethyl acetoacetate dianion with β-phenylthioacrolein[16] followed by stereoselective reduction of the so formed hydroxyketone (25), according to Davis[17], to the *anti*-diol (26), which after protection should be readily transformable into

the key selenol ester (27). On the basis of the above model of studies
we fully expect (27) to yield (3) cleanly and in high yield on
treatment with tin hydride and AIBN and hope to be able to report on
this in the near future.

2. SYNTHESIS OF 2-DEOXY-β-D-GLYCOSIDES BY STEREOSELECTIVE RADICAL
REACTIONS

The 2-deoxy-β-D-glycoside moiety occurs in a wide range of natural
products of potential pharmaceutical importance. The antitumor
antibiotic Olivomycin A[18] (28) containing no less than 3 such residues
serves as an excellent example. Classical glycosidic coupling methods
lead, in the absence of any anchimeric assistance from the 2-position
as in the 2-deoxy series, predominantly to α-glycosides. To the best
of our knowledge only one route[19] has so far been developed for the
preparation of 2-deoxy-β-D-glycosides leaving an almost virgin field
for the synthetic chemist's imagination to play upon. The approach
outlined below is highly novel in so much as it relies upon a
stereoselective radical reaction for the introduction of the correct
configuration at the anomeric centre.

(28)

It is now well known[20] that the tetraacetylglucosyl radical is
trapped from the α-face leading to the formation of tetraacetyl-α-
C-glucosides with good stereoselectivity. This stereoselectivity has
been explained[21] in terms of the boat conformation adopted by this
particular glucosyl radical. The method outlined below is based on
the stereoselectivity *predicted* for the quenching of, not simple
glycosyl radicals, but 1-alkoxyglycosyl radicals which we considered
might be generated from a suitable mixed orthoester (29) with tin
hydride as illustrated in Scheme 1.
Examination of the literature revealed work in which it had been
demonstrated[22] that *t*-butoxyl radicals abstract hydrogen significantly
more rapidly from *cis*-2-methoxy-4-methyltetrahydropyran than from the
trans-isomer indicating a kinetic preference for axial hydrogen
abstraction in such molecules. More importantly it was also
demonstrated[23] that the same radical was generated in each case, that

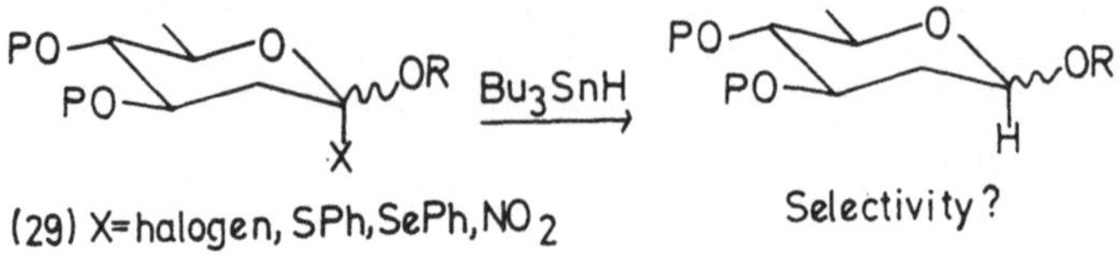

(29) X=halogen, SPh, SePh, NO$_2$

Selectivity?

Scheme 1

the radical centre was pyramidal and most importantly that the single
electron was in the axial position. Thus we attempted to synthesize
mixed orthoesters as in (29) in order to be able to react them with
tin hydride and convert Ingold's spectrocopic results[23] into a
practical and novel stereoselective O-glycoside synthesis.
Unfortunately all our attempts at the preparation of such mixed orthoesters proved fruitless, although some interesting compounds were
prepared[24] along the way. Eventually we turned our attention to the
possibility of alkoxyglycosyl radical generation by the
decarboxylation of 3-deoxyulosonic acid glycosides using Barton's
recently discovered[25] O-acyl thiohydroxamate chemistry.
3-Deoxyulosonic acids, typified by N-acetylneuraminic acid, and their
O-glycosides, occur widely in nature and so at least in this modified
approach we were secure in the knowledge that the key intermediates
would be stable and isolable. We required therefore a rapid,
efficient and general synthesis of 3-deoxyulosonic acids and their
glycosides and after some experimentation came up with[26] the route
outlined in Scheme 2 which allows the formation of gram quantities of
the key glycosyl donor (30) from the glycal in just 2 days.

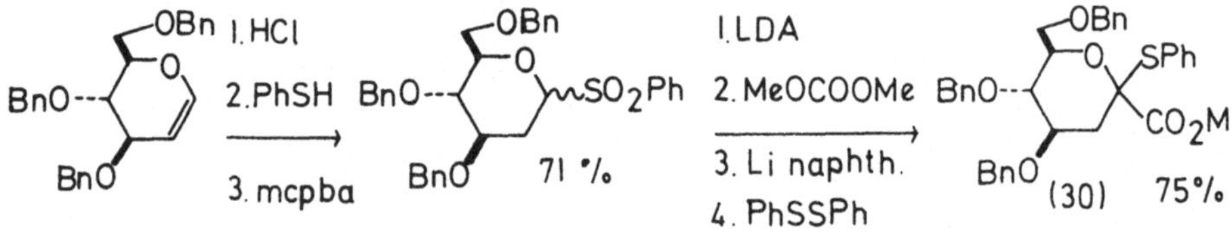

Scheme 2

In a test case (30) was coupled with methanol by means of either
N-bromosuccinimide or mercuric acetate, followed by saponification to
give the ulosonic acid glycoside (31). This glycoside (31) was
reacted with dicyclohexylcarbodi-imide and N-hydroxypyridine-2-thione
to give the bright yellow O-acyl thiohydroxamate which was not
isolated but photolysed (W) in the presence of the *relatively*
odourless, cheap commercial t-dodecyl mercaptan to give a 40% isolated
yield of the 2-deoxyglycosides (32) and (33) in which, as expected,
the β:α ratio was excellent (10:1). (Scheme 3).

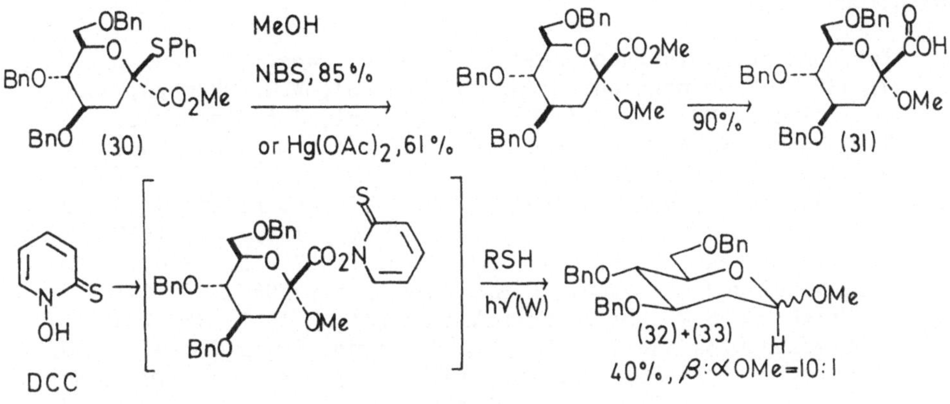

Scheme 3

Subsequently the model glycosyl donor (30) was coupled with
p-cresol using mercuric chloride in dichloromethane. Saponification
gave the ulosonic acid glycoside which was decarboxylated by means
of its derived O-acyl thiohydroxamate to give the β-tolyl-2-
deoxyglucoside (34) with apparently no trace of the α-anomer
(Scheme 4).

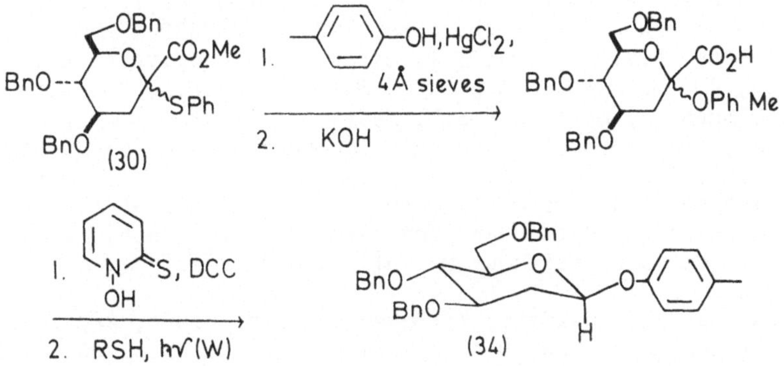

Scheme 4

Having adequately demonstrated the viability of the principle of
our 2-deoxy-β-D-glycoside synthesis, and given the ready availability
of a range of glycals as starting materials, we are now working
towards its application in the olivomycin A and related fields.

ACKNOWLEDGEMENTS. We thank the SERC for financial assistance and the
Nuffield Foundation for an equipment grant.

142

REFERENCES

1. For some key references see: B. Lythgoe, *Chem. Soc. Revs.*, 1980, *9*, 449; P.E. Georghiou, *ibid*, 1977, *6*, 83; W.H. Okamura, *Acc. Chem. Res.*, 1983, *16*, 81; E.G. Baggiolini, J.A. Iacobelli, B.M. Hennessy, A.D. Batcho, J.F. Sereno and M.R. Uskokovic, *J. Org. Chem.*, 1986, *51*, 3098.
2. D.R. Andrews, D.H.R. Barton, K.P. Cheng, J-P. Finet, R. Hesse, G. Johnson and M.M. Pechet, *J. Org. Chem.*, 1986, *51*, 1635.
3. G. Solladié and J. Hutt, *J. Org. Chem.*, 1987, *52*, 3560.
4. M.D. Bachi and E. Bosch, *Tetrahedron Lett.*, 1986, *27*, 641.
5. D. Crich and S.M. Fortt, *Tetrahedron Lett.*, 1987, *28*, 2895.
6. M. Chatzopoulos and J.P. Montheard, *Compt. Rendu C.* 1975, *280*, 29; Z. Cekovic, *Tetrahedron Lett.*, 1972, 749; also see: D.H.R. Barton and D. Crich, *J. Chem. Soc.*, *Perkin Trans. 1*, 1986, 1603; D.J. Coveney, V.F. Patel and G. Pattenden, *Tetrahedron Lett.*, 1987, *28*, 5949, P. Delduc, C. Tailhan and S.Z. Zard, *J. Chem. Soc.*, *Chem. Commun.*, 1988, 308.
7. J. Pfenninger, G. Heuberger and W. Graf, *Helv. Chim. Acta*, 1980, *63*, 2328.
8. K.C. Nicolaou, R.E. Zipkin, R.E. Dolle and D.D. Harris, *J. Am. Chem. Soc.*, 1984, *106*, 3548.
9. D. Crich and S.M. Fortt, *Tetrahedron Lett.*, 1988, *29*, 0000.
10. A.L.J. Beckwith and D.M. O'Shea, *Tetrahedron Lett.*, 1986, *27*, 4525; G. Stork and R. Mook, *ibid*, 1986, *27*, 4529.
11. H. Riemann, A.S. Capomaggi, T. Strauss, E.P. Olivetto and D.H.R. Barton, *J. Am. Chem. Soc.*, 1961, *83*, 4481; A.L.J. Beckwith, D.M. O'Shea, S. Gerba and S.W. Westwood, *J. Chem. Soc.*, *Chem. Commun.*, 1987, 666; P. Dowd and S.C. Choi, *J. Am. Chem. Soc.*, 1987, *109*, 3493.
12. D. Crich and L.B.L. Lim, *J. Chem. Res. S.*, 1987, 353.
13. M. Ladlow and G. Pattenden, *Tetrahedron Lett.*, 1985, *26*, 4413; J. Ardisson, J.P. Ferezou, M. Julia and A. Pancrazi, *ibid*, 1987, *28*, 2001.
14. H. Urabe and I. Kuwajima, *Tetrahedron Lett.*, 1986, *27*, 1355.
15. P.A. Grieco, J.Y. Jaw, D.A. Claremon and K.C. Nicolaou, *J. Org. Chem.*, 1981, *46*, 1215.
16. M. Iwao and T. Kuraishi, *Tetrahedron Lett.*, 1985, *26*, 6213.
17. S. Anwar and A.P. Davis, *J. Chem. Soc.*, *Chem. Commun.*, 1986, 831.
18. W.A. Remers, *The Chemistry of Antitumor Antibiotics*, Wiley Interscience, New York, 1979, vol. 1.
19. J. Thiem and M. Gerken, *J. Org. Chem.*, 1985, *50*, 954 and references therein; also see: K.C. Nicolaou, T. Ladduwahetty, J.L. Randall and A. Chucholowski, *J. Am. Chem. Soc.*, 1986, *108*, 2466.
20. B. Giese and J. Dupuis, *Angew. Chem. Int. Ed. Engl.*, 1983, *22*, 622; R.M. Adlington, J.E. Baldwin, A. Basak, and R.P. Kozyrod, *J. Chem. Soc.*, *Chem. Commun.*, 1983, 944; G.E. Keck, E.J. Enholm and D.F. Kachensky, *Tetrahedron Lett.*, 1984, *25*, 1867.
21. H.G. Korth, R. Sustmann, B. Giese and J. Dupuis, *J. Chem. Soc.*,

Perkin Trans II, 1986, 1453.

22. K. Haydey and R.D. McKelvey, *J. Org. Chem.*, 1976, *41*, 2222.
23. V. Malatesta, R.D. McKelvey, B.W. Babcock and K.U. Ingold, *J. Org. Chem.*, 1979, *44*, 1872.
24. D. Crich and T.J. Ritchie, *Tetrahedron*, 1988, *45*, 2319.
25. D.H.R. Barton, D. Crich and W.B. Motherwell, *J. Chem. Soc.*, *Chem. Commun.*, 1983, 939, idem, *Tetrahedron*, 1985, *41*, 3901; for a recent review see: D. Crich, *Aldrichimica Acta*, 1987, *20*, 35.
26. D. Crich and T.J. Ritchie, *J. Chem. Soc.*, *Chem. Commun.*, 1988, 0000.

STEREOSELECTIVE FORMATION OF LINEAR AZA-TRIQUINANES BY THREE CONSECUTIVE RADICAL RING CLOSURES

Lucien Stella *[a], Doug Boate[a] and Eric Guittet[b]
a - Laboratoire de Chimie Organique B, Université
    d'Aix-Marseille 3, F - 13397 Marseille Cedex 13, France
b - Institut de Chimie des Substances Naturelles, CNRS,
    F - 91198 Gif sur Yvette, France

ABSTRACT. Treatment of the trienyl N-chloroamine $\underline{5}$ with titanium trichloride affords regio- and stereoselectively, as the major product (nearly half of the overall yield) the 8-aza-4-$\underline{anti}$-chloromethyl-$\underline{cis}$, $\underline{syn}$, $\underline{cis}$ tricyclo $(6,3,0,0^{2,6})$ undecane $\underline{\textbf{10d}}$ by serial homolytic cyclisations.

Introduction

Intramolecular additions of carbon or heteroatom-centered radicals to carbon-carbon double or triple bonds are a topic of intense current interest in synthetic organic chemistry (1,2). In this context, we have shown that homolytic cyclizations of many N-chloroalkenylamines can lead to a variety of β-functionalized, substituted, fused or bridged aza-heterocycles in high yields, in a regio and, possibly, stereoselective manner (3).

In an extension of the titanium trichloride mediated N-chloroamine radical cyclization studies (4), we examined the triple homolytic cyclization of the N-chloro derivative of the N-allyl-$\underline{E}$-4,7-octadienylamine $\underline{\textbf{4}}$. This suitable amine $\underline{\textbf{4}}$ was easily prepared from readily available starting materials as shown in scheme 1.

The key step was the sodium-sand mediated opening of the cyclic β-chloro-ether $\underline{1}$ which afforded the alcohol $\underline{2}$ as its $\underline{\textbf{E}}$ isomer, an assignment which was confirmed by its NMR and IR spectra. The amine $\underline{\textbf{4}}$ can be converted essentially quantitatively into its N-chloroderivative $\underline{5}$ upon treatment with sodium hypochlorite. The N-chloroamine $\underline{5}$, which has three suitably disposed olefinic moieties has been shown (3) to undergo intramolecular radical cyclization upon treatment with aqueous titanium trichloride, to afford in 75% overall yield, a mixture of six isomeric bicycles ($\underline{\textbf{9a-b}}$) and tricycles ($\underline{\textbf{10a-d}}$) as indicated by GC-MS analysis (scheme 2).

F. Minisci (ed.), Free Radicals in Synthesis and Biology, 145–153.

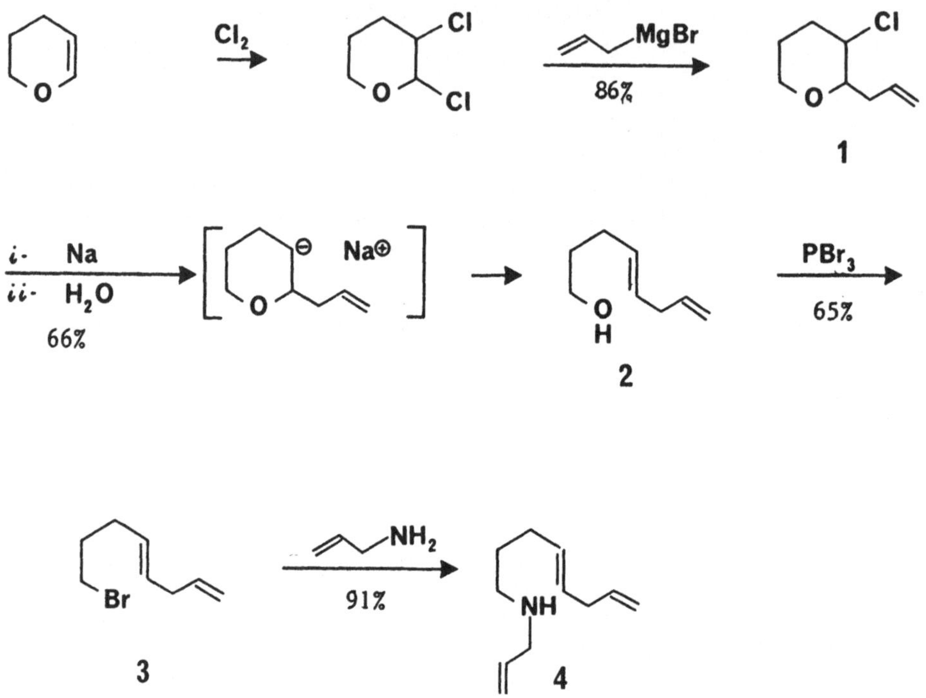

Scheme 1

Aside from the monocyclic product which has never been detected, two bicyclic (**9**) and four tricyclic (**10**) products are expected. This prediction is based upon the relative stereochemistry of the resultant methylene radical and the allylic side chain in the carbon-centered radicals **8** after the second radical cyclisation (ie 7 → 8). If the methylene radical and the allylic substituent are in a syn relationship, as in **8a** and **8d**, the carbon-centered radical will undergo an easy third cyclization to afford the corresponding tricyclic product (ie either **10a** or **10b** from **8a**, or either **10c** or **10d** from **8d**). However, if the methylene radical and the allylic substituent are in an anti relationship as in **8b** and **8c**, the third cyclization is disfavored with the result that both **8b** and **8c**, after chlorine atom transfer with **5** lead to the respective bicyclic products **9a** and **9b**.

The four minor components (present in 11, 6, 6 and 4% relative G.C. yield) obtained in the reaction of **5** with titanium trichloride were not assigned structures as they could not be isolated (3). A bicyclic and a tricyclic amine were isolated (27 and 46% relative G.C. yield respectively) but the stereochemistry was not assigned. The aim of this study was to fully assign the absolute stereochemistry of each of the isolated isomers.

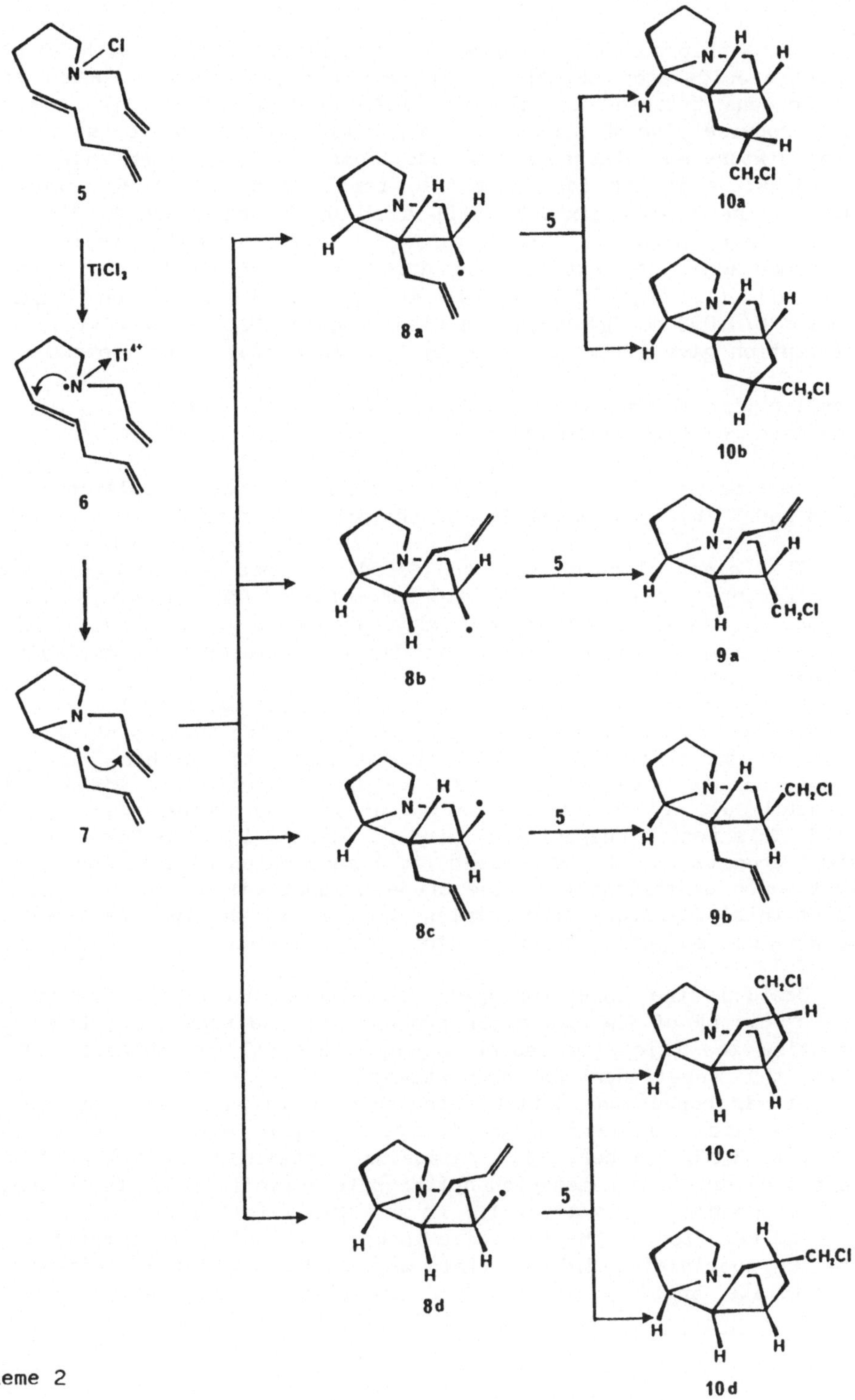

Scheme 2

148

Results

The N-chloroamine **5** (2.1 g, 10.5 mmol) was diluted with 50 ml of an acetic acid-water solution (1:1) and the mixture was cooled to ca 0°C. Aqueous titanium trichloride (15%) (1.8 ml, 0.15 ǝq.) was then added dropwise over 40 min. Once the purple-violet colour persisted, the cooled mixture was stirred for an additional 1h and, after confirmation of a negative result for iodometric test, 8N potassium hydroxide was added to the cooled mixture until a pH of 9 was reached. The basic mixture was then filtered through celite and extracted with dichloromethane. The extract was dried and concentrated to afford a crude oil (1.95 g, 93% yield) which was distilled on Kugelrohr (100-105°C/0.003 mm Hg) to give a clear liquid (820 mg, 42% yield). The distillation gave a mixture shown by G.C. to contain four isomers :

retention time (min)   : 6.14   6.69   7.33   7.76
normalized relative yield (%) : 27   5   10   58

It is noteworthy that the two major isomers previously isolated (3) (R_T = 6.14 and 7.76) account for 85% of the product mixture.

The clear liquid was subjected to medium pressure liquid chromatography (silica ; 85% dichloromethane/10% diisopropylamine/5% methanol ; fractions vizualized with phosphomolybdic acid - ethanol spray) to afford the following two fractions which were suitable for examination by NMR spectroscopy.

Fraction A, containing a 5.6/1 mixture of two bicyclic isomers ; G.C. R_T = 6.14 and 6.69 and fraction B containing an essentially pure tricyclic isomer ; G.C. R_T = 7.76. Attempts to further purify the product mixture obtained from MPLC by preparative G.C. (5% OV-17 on chromosorb W), either isothermally or with a temperature gradient failed. This result suggest that product decomposition is occuring under thermal conditions. The poor yield (ie : mass balance) obtained from the bulb-to-bulb distillation of the crude product mixture would appear to support this. Therefore, the samples isolated as the two fractions from MPLC were not subjected to any further purification.

Besides the spectroscopic techniques used to assign the stereochemistry of the two major products of the reaction, attempts to derivatise the tricyclic isomer through chemical manipulation of the chloromethyl substituent has been attempted.

It was hoped that suitable crystals could have been obtained for X-ray analysis. Treatment of the tricyclic isomer with potassium iodide and then with either (i) potassium phtalimide, (ii) potassium succinimide or (iii) p-methylphenylthiolate anion in attempts to prepare the corresponding derivatives **11**, **12** and **13** all failed.

The reaction of the tricyclic isomer with potassium phtalimide and potassium succinimide did not yield any products and the reaction with the thiolate anion gave only the disulfide and unchanged starting material.

11   R = [phthalimido group]

12   R = [succinimido group]

13   R = —S—[benzene ring]—Me

In another attempt to obtain suitable crystals for X-ray analysis, the picrate salt derivative of the tricyclic isomer was prepared. The long thin needles obtained (m.p. 170–172°C dec.) proved to be unsuitable for X-ray analysis.

If the major tricyclic product has cis-syn-cis stereochemistry, as in **10d**, then it is not surprising that the above nucleophiles do not undergo the desired substitution as the tricyclic nucleus would hinder back side substitution.

The absolute stereochemical assignments were based on spectroscopic data listed below for each isolated isomer.

Major product of the reaction : tricyclic with GC R_T = 7.76. The chemical shifts in brackets refer to the nor-chloromethyl substituted tricyclic amine **14** (5).

10   **14**

a) Carbon Chemical Assignment (CDCl$_3$ – 50,309 MHz)

C_1	73.0	(73.1)
C_2	49.6	(50.5)
C_3	24.8	(26.0)
C_4	41.0	(24.9)
C_5	31.2	(31.3)
C_6	43.3	(44.0)
C_7	53.5	(53.6)
C_9	59.8	(60.5)
C_{10}	35.6	(32.3)
C_{11}	36.1	(32.7)
C_{12}	48.5	–

b) <u>Proton Chemical Assignment</u> (CDCl$_3$ – 400 MHz)

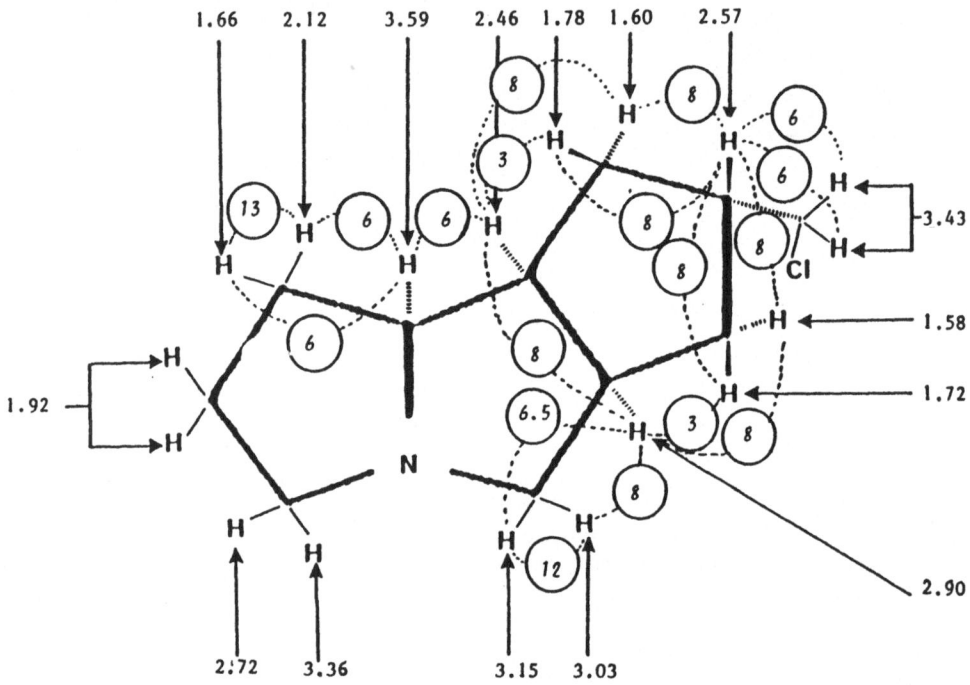

c) <u>Infrared Spectrum</u> : 2940, 2920, 2830, 2800, 1460, 1440, 1285, 1095, 785 cm-1

d) <u>GCMS</u> : A GCMS was run on a sample of the major tricyclic product and confirmed that the sample was essentially pure. Mass Spec. data : m/e 199 (21,0%, M$^+$), 198 (16.4%, M-1), 170 (16.0%), 164 (36,3%, M-35), 136 (20.6%), 122 (15.5%), 83 (100%).

Second major product of the reaction : Inseparable mixture (5.6/1) of two bicyclic isomers <u>9</u>

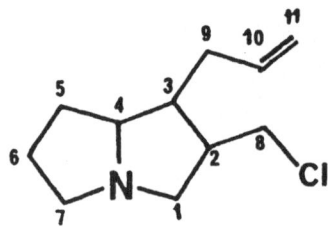

9

a) <u>Carbon Chemical Assignment</u> ($CDCl_3$ - 50,309 MHz)

	Major isomer GC R_T 6.14	Minor isomer GC R_T 6.69
C_1	54.7	55.9
C_2	47.8*	44.2*
C_3	47.0*	44.1*
C_4	70.9	68.7
C_5	**	**
C_6	**	**
C_7	58.1	58.5
C_8	44.7	45.9
C_9	**	**
C_{10}	135,6	136.2
C_{11}	117.1	116.5

* may be reversed
** the three upfield resonances
   at 36.3, 31.7 and 25.8 ppm
   have not been assigned

* may be reversed
** the three upfield resonances
   at 33.9, 29.7 and 26.5 ppm
   have not been assigned.

b) <u>Proton Chemical Assignment</u>

Major (5.6)

Minor (1)

H_{10}	5.76 (ddt, 1H) (18, 10 and 7 Hz)	5.78 (ddt, 1H) (18, 10, 7 Hz)
H_{11}	5.12 (d, 1H, 18Hz) 5.05 (d, 1H, 10 Hz)	5.08 (d, 1H, 18 Hz) 5.02 (d, 1H, 10 Hz)
H_8	3.73 (dd, $\underline{ABX}$, 1H) 3.59 (dd, $\underline{ABX}$, 1H)	3.71 (dd, $\underline{ABX}$, 1H, 11 et 3.5 Hz) 3,50 (dd, $\underline{ABX}$, 1H, 11 et 6.5 Hz)
H_4 + H_7	3.53 (m, 2H)	3.26 (m, 2H)
H_7	3.18 (ddtt, 1H,	2.94 (ddd, 1H,
H_2	2.76 (dt, 1H	2.59 (dt, 1H,
H_9	2.58 (q, 1H 2.50 (dt, 2H	2.43 (q, 2.32 (, 2H

b) <u>Infrared Spectrum</u> : 3080, 2960, 2910, 2865, 2800, 1640, 1440, 995, 910, 725 cm^{-1}

Discussion

First of all, we can discuss the absolute stereochemistry of the three-ring system in the tricyclic isolated product by comparing our ^{13}C NMR results with those reported for the <u>cis-syn-cis</u> analogous system **14** (5). The bridgehead carbon, α to nitrogen (C_1) resonates at 73.0 ppm and is in excellent agreement with the value (73.1 ppm) reported for the same carbon of the analogous system **14**. This deshielding of the bridgehead carbon <u>alpha</u> to nitrogen, in this type of system, is consistent with other chemical shifts reported (6) for <u>N</u>-bridgehead <u>cis</u> ring functions. The essentially identical carbon resonances for the bridgehead carbons (C_2 and C_6) imply that our tricyclic molecule **10** have the same cis-syn-cis stereochemistry as **14**. Consequently, either **10c** or **10d** can be proposed on this basis. In addition, no Bohlmann bands (7), attributed to α-hydrogens oriented trans antiperiplanar to the nitrogen lone pair, could be assigned in the infrared spectra in the region 2600-2800 cm-1.

The calculated (MM2 or AMBER parameters) stable structure which fits best with our ^{1}H NMR data appears to be the cis-syn-cis **10d** with an α (or anti) chloromethyl side chain. Moreover, the boat like conformer of the chloromethyl substituted cyclopentane ring seems to be preferred because of steric interactions between the side chain and the C_2 and C_6 hydrogen atoms.

Since the major cyclized product from a 1-substituted hex-5-enyl system is the cis-isomer (7), any subsequent ring closure affords a cis-fused bicyclic system. The formation of **15** by reaction of cyanopropyl radicals with diallylamines provides a good example (8)

Similarly, the major product **17** from the ring closure of the dienyl radical **16** contains reactive centers suitably disposed for a second ring closure to afford exo- and endo- **18**, kendo/kexo = 1.4 (9).

16   **17**   **18**

Cyclisation of the radical **19** is relatively slow (9) and occurs mainly in the endo-mode (**21**) presumably because of the strain engendered in formation of the trans-(3,3,0) bicyclooctane system by exo- ring-closure (**20**).

19   **20** (1/2.2)   **21**

Since the length of C-N bond is less than that of C-C and the bond angle C-N-C is less than that of C-C-C, the minimum C-1-C-5 distance in the radical 22 is less than it is in hex-5-enyl, while the C-1-C-6 distance is greater. Consequently radicals containing a N atom in the chain (e.g. 22) should show enhanced rates of exo-ring closure (10).

22

This, and the fact that the reaction in presence of $TiCl_3$ is not a rapid redox chain process (by ligand transfer reaction from metallic salt to carbon radical) but a chain reaction (by chlorine transfer from N-chloroamine 5), well explain why bicyclic compound resulting from either 8a or 8d are not obtained preferentially. On the contrary, we do not observe cyclisation of either 8b or 8c but both 9a and 9b respectively.

Aknowledgement

We thank C. Fontaine for the high field (400 MHz) nuclear magnetic resonance study.

References

(1) For recent reviews, see : (a) Beckwith, A.L.J. ; Ingold, K.U. In Rearrangements in Ground and Excited States ; de Mayo, P., Ed. ; Academic : New York, 1980, pp 162-283. (b) Surzur, J-M. In Reactive Intermediates ; Abramovitch, R.A., Ed. ; Plenum : New York, 1981 ; vol. 2, Chapter 3. (c) Beckwith, A.L.J., Tetrahedron, 1981, 37, 3073 and references therein. (d) Griller, D. ; Ingold, K.U. Acc. Chem. Res. 1980, 13, 317.

(2) (a) Hart, D.J. Science, 1984, 223, 883. (b) Stork, G. ; Mook, Jr. R. J. Am. Chem. Soc. 1987, 109, 2829. (c) Curran, D.P. ; Chen, M.-H. J. Am. Chem. Soc. 1987, 109, 6558. (d) Clive, D.L.J. ; Boivin, T.L.B. ; Angoh, A.G. J. Org. Chem. 1987, 52, 4943. (e) Padwa, A. ; Nimmergern, H. ; Wong, G.S.K., J. Org. Chem., 1985, 50, 5620.

(3) Stella, L. Angew. Chem. Int. Ed. Engl. 1983, 22, 337.

(4) (a) Stella, L., Raynier, B. ; Surzur, J.M. Tetrahedron, 1981, 37, 2843. (b) Surzur, J.M. ; Stella, L. Tetrahedron Letters, 1974, 2191.

(5) Chastanet, J. ; Roussi, G. Heterocycles, 1985, 23, 653.

(6) Jones, T.H. ; Blum, M.S. ; Fales, H.M. ; Thompson, C.R. J. Org. Chem. 1980, 45, 4778.

(7) Beckwith, A.L.J. ; Easton, C.J. ; Serelis, A.K. J. Chem. Soc., Chem. Commun. 1980, 482.

(8) Hawthorne, D.G. ; Johns, S.R. ; Willing, R.I. Aust. J. Chem. 1976, 29, 315.

(9) Beckwith, A.L.J. ; Philipon, G. ; Serelis, A.K. Tetrahedron letters, 1981, 22, 2811.

(10) a) Beckwith, A.L.J. ; Hawthorne, D.G. ; Solomon, D.H. Aust. J. Chem. 1976, 29, 995 ; b) Corfield, G.C. Chem. Soc. Rev. 1972, 1, 523 ; c) Solomon, D.H., J. Macromol. Sci. 1975, A9, 97.

SYNTHETIC APPLICATIONS OF SUBSTITUTION AND ADDITION REACTIONS PROMOTED BY CERIUM(IV) AMMONIUM NITRATE

Enrico Baciocchi[*] and Renzo Ruzziconi
Department of Chemistry, University of Rome "La Sapienza",
P.le A. Moro 5, 00185 Roma, Italy
Department of Chemistry, University of Perugia, Via Elce
di Sotto 10, 06100 Perugia, Italy

ABSTRACT. Cerium(IV) salts, and expecially cerium(IV) ammonium nitrate, (CAN), are very efficient in promoting a highly selective side-chain functionalization of methylbenzenes. With CAN the process leads to aromatic aldehydes and benzyl derivatives, depending on the reaction conditions. In di- and polymethylbenzenes selective functionalization of the methyl group with electron donating substituents at the ortho and (or) para position is possible.

Reactivity can significantly be increased by adding Br⁻, at the expense, however, of selectivity, which is significantly reduced since the attacking species is now Br·. Both selectivity and reactivity are instead very high in the photochemical reaction of CAN with methylbenzenes in MeCN leading to benzyl nitrates; the reacting species is NO_3·, formed by photolysis of CAN. The reaction can be extended to cycloalkanes and moreover, if carried out in the presence of dioxygen, made catalytic in CAN.

CAN in AcOH is also a very useful reactant for ring functionalization (formation of phenyl acetates or quinones) of aromatic compounds. In methylaromatics nuclear attack can be the favored process in the presence of electron donating substituents meta to the methyl group. With alkenylaromatic compounds CAN leads to 1,2-dinitrate adducts in good yield. Under irradiation the reaction becomes much faster and can also be extended to unactivated alkenes.

Reaction of CAN with carbonyl compounds forms electrophilic α-keto carbon radicals which can promote homolytic aromatic substitutions or add to alkenes to give addition products where the formation of a new carbon carbon bond is accompanied by the introduction of the nitrate group, which can undergo further functionalization. Among the homolytic aromatic substitutions, malonylation is very useful expecially with heteroaromatic compounds. The addition reactions to alkenes (particularly, conjugated dienes, styrenes, vinyl acetates) allow the synthesis of a variety of compounds (vinylcyclopropanes, 1,4-disubstituted trans-2-butenes, 1,4-dicarbonyl compounds, diydrofurans and furans) to be carried out.

Ce(IV) is a bona fide one electron oxidant since it can only exchange one electron in redox processes to give Ce(III) which is its unique reduced state. The Ce(IV)/Ce(III) couple is characterized by pos-

155

F. Minisci (ed.), Free Radicals in Synthesis and Biology, 155–185.
© 1989 by Kluwer Academic Publishers.

itive and relatively high E°value (between 1 and 1.7V depending upon the
ligands and reaction conditions) so that Ce(IV) salts, and particularly
the commercially available cerium(IV) ammonium nitrate (heretofore
indicated as CAN), have found many interesting applications in organic
synthesis. [1,2]

During the last few years our group has intensively investigated the
mechanistic aspects of CAN induced reactions of alkyl[3,6] and alkenylaromatic compounds[7,8]. Even though the main aim of this study has been
that of obtaining mechanistic information, it has also provided us with a
number of results which can find some significative exploitation in organic
synthesis, expecially in sectors, such as those of oxidative substitution and
addition reactions, little explored in the past. The review and the discussion of these results is the subject of this paper.

As many other one electron oxidants, Ce(IV) either can directly
interact with the substrate (direct reactions) or it can promote the
formation of a species which then acts as the effective reagent (indirect
reactions). These two types of reactions will be considered separately in the
following discussion.

A. SUBSTITUTION REACTIONS OF ALKYLAROMATIC COMPOUNDS AND ALKANES

1.DIRECT REACTIONS

1.1 Side-chain substitutions of alkylaromatic compounds

The side-chain functionalization of alkylaromatic compounds is a very
important process both from the theoretical and practical point of view.
CAN is a very effective reagent in this respect, being able to convert
methylbenzenes to benzyl derivatives[3,6,9,10] or benzaldehydes[11,12] with
good efficiency and selectivity (eqs 1 and 2).

$$ArCH_3 \xrightarrow{\text{CAN, AcOH}} ArCH_2ONO_2 \quad + \quad ArCH_2OAc \qquad (1)$$

$$ArCH_3 \xrightarrow{\text{CAN, AcOH 50\%}} ArCHO \qquad (2)$$

Particularly effective is the reaction described in eq. 1 where yields
around 90% are obtained with methylbenzenes more electron rich than toluene
in a 25-80 °C temperature range. Since benzyl nitrate can easily be
converted into benzyl acetate either under the reaction conditions or by
reflux of the reaction mixture in the presence of AcO⁻, this procedure is
one of the best methods to prepare benzyl acetates directly from

methylbenzenes. Less useful appears the formation of aldehydes (eq. 2) since in several cases it is nearly impossible to avoid substantial formation of benzyl nitrates as side products[13].

The intermolecular selectivity of the process is very high as shown by the values of relative reactivity for a number of methylbenzenes: mesitylene, 0.0048; pseudocumene, 0.006, isodurene, 0.22; durene, 1; pentamethylbenzene, 10; p-methoxytoluene, 85; hexamethylbenzene, 330[5,6]. Clearly, the reaction rate is strongly increased by the presence of electron donating groups. Moreover, it can also be noted that the overall reactivity is larger when electron donating groups, i.e. OCH_3 or CH_3, are ortho or para to the methyl group. Owing to this high intermolecular selectivity the raction can easily be controlled to give monosubstituted compounds.

Still more significant can be the intramolecular selectivity. Thus, in the reaction of isodurene with CAN in AcOH, 88% of the reaction product is 2,4,6-trimethylbenzyl acetate[9]. Selectivity is strongly influenced by the presence of substituents and in a series of 5-Z-1,2,3-trimethylbenzenes, the reaction takes place nearly exclusively at the 2-position when Z is a + R group such as Br, OCH_3, $OCOCH_3$[3]. With these substrates CAN is significantly more selective than other one electron oxidants. For example the free radical bromination of 5-bromo-1,2,3-trimethylbenzene forms only 62% of the product coming from attack at the methyl group in the 2-position. Still smaller selectivity is observed with $Co(OAc)_3$.[3] With electron withdrawing (-R,-I) substituents, selectivity of CAN-induced reactions is however much smaller. Thus, when Z is $CO_2C_2H_5$ there is only a slight preference for attack at the 2-position.

The peculiar selectivity characteristics of CAN-promoted side-chain oxidations of methylbenzenes are certainly related to the reaction mechanism, which has been established, on the basis of kinetic studies, to be that described in eqs 3-5.[5]

An electron transfer process occurs in the first step of the reaction to give a radical cation which is then deprotonated to form a benzyl radical. The last step, eq. 5, is an oxidative ligand transfer from CAN to the radical to form the benzyl nitrate.

$$ArCH_3 + Ce^{(IV)}ONO_2 \longrightarrow ArCH_3^{+\cdot} + Ce^{(III)}ONO_2 \qquad (3)$$

$$ArCH_3^{+\cdot} \longrightarrow ArCH_2\cdot + H^+ \qquad (4)$$

$$ArCH_2\cdot + Ce^{(IV)}ONO_2 \longrightarrow ArCH_2ONO_2 + Ce^{(III)} \qquad (5)$$

Since there is evidence that in AcOH the equilibrium shown in eq. 6 is established[14] it is also possible that the ligand transfer involves an acetic acid molecule, thus directly leading to benzyl acetates in addition

$$Ce(NO_3)_6^= + 2AcOH \longrightarrow Ce(NO_3)_4(CH_3CO_2H)_2 + 2NO_3^- \qquad (6)$$

to benzyl nitrates.

The strong sensitivity of the process to the electronic effect of the substituent can be ascribed to the substantial development of positive charge occurring in the transition state of the slow step, which generally is that where the radical cation is formed.

The observed intramolecular selectivity is determined in the step where the radical cation is deprotonated. For example, from 5-Z-1,2,3-trimethylbenzene radical cation 1, two benzyl radicals are formed: 2 and 3 (eq. 7). Since no significant difference in the stability of 2 and 3

$$(7)$$

can reasonably be expected, the strong preference for deprotonation at the 2-methyl group of 1 has probably to be explained on the basis of the spin density distribution in the aromatic radical cation. Generally, when a + R substituent is present, the largest spin density is at the position para to the substituent itself, [15-17] which may make the para methyl group the more reactive one.

Further examples of this high intramolecular selectivity have recently been found in the naphthalene series. Thus, 1,2-dimethylnaphthalene reacts with CAN in 50% aqueous AcOH to form 78% of 2-methyl-1-naphthaldehyde and only 5% of 1-methyl-2-naphthaldehyde.[18]

When the solvent is changed from acetic acid to methanol the CAN-promoted reaction of methylbenzenes leads to the formation of benzyl methyl ethers (eq. 8), presumably by a mechanism similar to that suggested for the reactions in acetic acid.[19] The scope of the process is however more limited, in view of the fact that the stability of CAN in MeOH is less than in AcOH. Thus, only substrates as electron rich as or more than durene can react faster than the solvent.

An interesting observation is that in MeOH further fast reaction of the formed benzyl methyl ether is possible leading to benzaldehyde derivatives. This possibility has been exploited for the synthesis of p-methoxybenzaldehyde which can be obtained, in more than 90% yield, from p-methoxytoluene.[20] Another example[20] which also illustrates the high intramolecular selectivity of the CAN-promoted reactions is in eq. 8

86 %

$$(8)$$

DMSO too has found some use as the solvent for the CAN-promoted side-chain substitutions of methylbenzenes.[21] With respect to AcOH, yields are, however, lower since the stability of CAN in DMSO is less than in AcOH.

DMSO offers the advantage of allowing the first formed benzyl nitrate to rapidly react with other nucleophiles such as H_2O, pyridine, alkaline thiocyanate, cyanide and azide, in the same reaction medium, at room temperature. In this way a variety of functionalized benzyl derivatives can be obtained in an one pot process. As in MeOH the reaction is limited to substrates as electron rich as or more than durene.

1.2. Nuclear functionalization of aromatic compounds

CAN can react with alkylaromatic compounds to give ring attack in addition to the already discussed side-chain substitution. The nuclear reaction is of course the only one observed with not alkylated aromatic derivatives. For example, anisole reacts with CAN in AcOH to form a mixture of 55% p-acetoxyanisole and 45% o-acetoxyanisole, with an overall selectivity of 90%.[22] Since the acetoxy derivatives have a significant tendency to undergo further oxidation, the best yields are obtained when an excess of substrate is used. Thus, naphthalene reacts with CAN to give acetoxynaphthalene (α/β ratio = 49) in nearly quantitative yield (with respect to CAN used) when the naphthalene: CAN molar ratio is 4: 1.[23]

The nuclear acetoxylation can also compete with side-chain substitution in some alkylaromatic compounds, the extent of this competition depending on the substrate structure. Generally, the extent of nuclear substitution is low or negligible if the alkyl group has a + R group in the para or ortho position. Accordingly, in the series of polymethylbenzenes, nuclear acetoxylation (ca 50%) is observed only in the reaction with mesitylene, a substrate where the methyl groups are meta to one another.[6,24]

Since the nuclear acetoxylation, likewise the side-chain substitution, occurs via the formation of an intermediate radical cation (eqs 9-11) the competition between the nuclear and the side-chain reaction is

$$ArH \quad + \quad Ce^{IV} \quad \longrightarrow \quad ArH^{+}\cdot \quad + \quad Ce^{III} \qquad (9)$$

$$ArH^{+}\cdot \quad + \quad AcOH \quad \longrightarrow \quad Ar\overset{OAc}{\underset{H}{\cdot}} \quad + \quad H^{+} \qquad (10)$$

$$Ar\overset{OAc}{\underset{H}{\cdot}} \quad + \quad Ce^{IV} \quad \longrightarrow \quad ArOAc \quad + \quad Ce^{III} \quad + \quad H^{+} \qquad (11)$$

determined from the two possible reaction modes of the alkylaromatic radical cation with the base B⁻ (scheme 1), that is attack at the CH_3 group

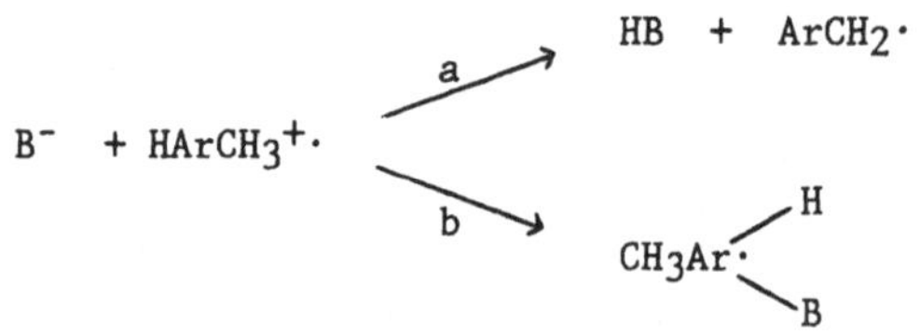

Scheme 1

(path a) or at the ring carbon (path b). B⁻ can be a ligand of the metal or the conjugate base of the solvent. Like the intramolecular selectivity between non equivalent methyl groups, also the competition between pathway a and pathway b depends on the spin density distribution in the radical cation. When a + R group is para at the methyl group spin density is much higher at the methyl protons than at the ring protons.[17,25] The preferred reaction of the radical cation is therefore that involving side-chain deprotonation. Conversely, if the + R group is in the meta position with respect to the methyl group, spin density at the ring protons becomes of comparable magnitude than that at the methyl protons and the nuclear reaction has more chance to compete.

Thus, whereas p-methoxytoluene reacts with CAN in acetic acid exclusively to give p-methoxybenzyl acetate, m-methoxytoluene reacts to form the ring substituted acetoxy derivatives without any evidence for side chain substitution.[24]

The tendency of CAN to give nuclear substitution product is higher than that observed with other one electron oxidants. For example, with $Co(OAc)_3$ in AcOH meta-methoxytoluene forms only the side-chain acetoxylated compound.[24] Moreover, 2-methylnaphthalene reacts with CAN in AcOH to give 1-acetoxy-2-methylnaphthalene nearly exclusively. Side-chain substitution accounts only for the 2% of the reaction product. With cobaltic acetate 76% of the reaction product is 2-acetoxymethyl-naphthalene.[24]

The difference between CAN and Co(OAc)$_3$ cannot be ascribed to the different nature of the ligands as exclusive nuclear acetoxylation of m-methoxytoluene is also observed with Ce(OCOCH$_3$)$_4$.[26] Problably, the phenomenon is due to the operation of a different reaction mechanism: electron transfer with CAN, hydrogen atom transfer with Co(OAc)$_3$.

Recently, however, it has been reported that, albeit in low yield (22%), m-methoxybenzaldehyde is obtained in the reaction of cerium(IV) trifluoroacetate with m-methoxytoluene in aqueous CF$_3$CO$_2$H.[27] It is possible that in a strongly acidic medium the ring attack becomes reversible, at least to some extent, as shown in Scheme 2,

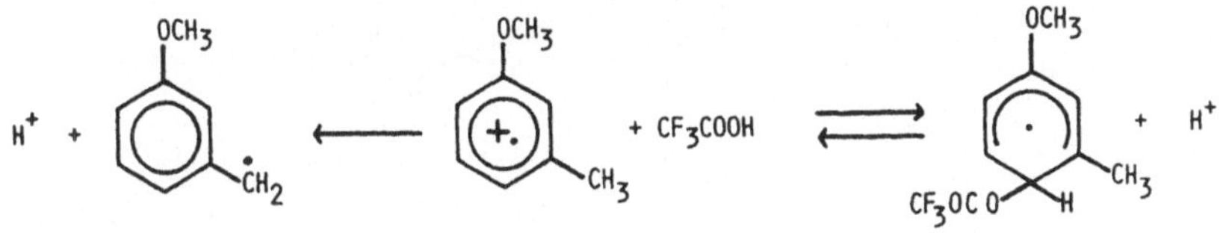

Scheme 2

thus favoring the irreversible side-chain deprotonation reaction of the radical cation.

In several respects, the tendency for nuclear acetoxylation of the CAN/AcOH system resembles that exhibited by anodic oxidation, which is not surprising since both reactions take place via the formation of an intermediate radical cation. There is, however, a peculiar difference between the two processes. In the electrochemical reaction nuclear acetoxylation is only possible in the presence of AcO$^-$.[28] This is not required for the CAN promoted reaction; in fact, addition of AcO$^-$ to a CAN solution drastically reduces the oxidizing behavior of the system, since the replacement of NO$_3^-$ ligands by AcO$^-$ ligands, forms a species with a lower oxidation potential (the E° value for the Ce(IV)/Ce(III) couple in AcOH is 1,16V (vs NHE) in the presence of AcO$^-$ and 1,34V (vs NHE) in the presence of NO$_3^-$[29]).

In aqueous media the Ce(IV)-promoted nuclear reaction of aromatic compounds leads to quinone derivatives. This reaction is synthetically useful in the naphthalene series and in this case the reagent of choice is ceric ammonium solfate (CAS). Thus, in dilute sulfuric acid, naphthalene is converted by CAS to 1,4-naphthoquinone in a yield of 90%[30].

Interestingly, the same reaction can be performed in a two-phase system using ammonium persulfate in the presence of catalytic amounts of CAS.[31]

Another interesting reaction is the oxidative demethylation of 1,4-dimethoxyarenes to give p-quinones promoted by CAN in MeCN/H$_2$O. The yields of this reaction are significantly increased by the presence of catalytic amounts of pyridine and pyrazine carboxylic acids, having the carboxylic groups in the α-positions.[32] An example is in eq. 12 where the catalyst is 2,4,6-pyridinetricarboxylic acid (PTA).

$$(12)$$

The actual role of the pyridinecarboxylic acids is not clear, even though it seems likely that they intervene by complexing cerium(IV).

2.INDIRECT REACTIONS

2.1. Side-chain functionalization of alkylaromatic compounds

Bromination Reactions. The reactions we have examined so far have a serious drawback in being practically limited to alkylaromatics more electron rich than toluene. With alkylaromatics substituted by electron with-drawing groups, relatively high temperatures (80°C or more) are requested, with consequent considerable decomposition of the oxidant.

The side-chain functionalization of electron-poor methylbenzenes by CAN can, however, be performed if the reaction is carried out in the presence of Br⁻ions.[33] An efficient side-chain bromination takes place, as shown by the results in Table 1.

On the basis of these results CAN/Br⁻ in AcOH can represent a valid alternative to N-bromosuccinimide for the side-chain bromination of electron-poor toluene derivatives. Moreover, since at the end of the reaction bromides can quantitatively be converted into acetates in the same reaction medium, the one-pot side-chain acetoxylation of the same substrates is possible as well.

Table 1. Yields of Side-chain Brominated Products in the CAN-Promoted
Oxidation of R-Substituted Toluenes in AcOH-KBr at 80 °C[a]

R	t/h	$ArCH_2Br$, yield, %[b]
$3-CO_2H$	2.5	50
$4-NO_2$	2	50
$2,4-Cl_2$	2	80
$2,6$ Cl_2	2	77
$4-Br$	1.5	77
H	1	80
$4-Me$	8[c]	50

a) Substituted toluene (4.1 mmol); CAN (8,2 mmol), KBr (4.1 mmol) in
   16 ml AcOH. b) Isolated product, against the starting material. c) At
   room temperature.

The effective reacting species is very probably the bromine atom
generated by reaction of Ce(IV) with Br⁻. A possible reaction scheme is
reported in the eqs. 13-16.

$$Ce^{IV}ONO_2 \;+\; Br^- \longrightarrow Ce^{IV}Br \;+\; ONO_2^- \qquad (13)$$

$$Ce^{IV}Br \longrightarrow Ce^{III} \;+\; Br\cdot \qquad (14)$$

$$Br\cdot \;+\; ArCH_3 \longrightarrow ArCH_2\cdot \;+\; HBr \qquad (15)$$

$$ArCH_2\cdot \;+\; Ce^{IV}Br \longrightarrow Ce^{III} \;+\; ArCH_2Br \qquad (16)$$

A first reaction of CAN with Br⁻ to give a new species,
indicated in the scheme as Ce(IV)Br, where Br⁻ has replaced some of NO_3^-
ligands, has to be postulated to account for the observation that only
benzyl bromides are obtained in the process.

It is, however, unlikely that this new species be the effective reagent,
as we have found that the intramolecular selectivity of side-chain bromination
of 4-substituted-o-xylenes (eq. 17, Z = t-Bu,Cl) does not significantly
change when CAN/Br⁻ is replaced by NBS, $S_2O_8^=$/Br⁻ or Co(OAc)₃/Br⁻.[34]

$$\text{(17)}$$

Table 2. Intramolecular Selectivity ($\underset{\sim}{4}/\underset{\sim}{5}$ Molar Ratio) in Some Side-chain Brominations of 4-Z-o-Xylenes (eq. 17)

Brominating System	T/°C	$\underset{\sim}{4}/\underset{\sim}{5}$ Molar Ratio	
		Z = t-Bu	Z = Cl
CAN/Br⁻/AcOH	60	1.8	2.6
Co(OAc)$_3$/Br⁻/AcOH	60	1.75	2.7
Na$_2$S$_2$O$_8$/Br⁻/AcOH	115	1.6	3.1
NBS/CCl$_4$/ABN	60	1.55	3.0

These results (Table 2) suggest that the same reacting species, very probably the bromine atom, is operating under all the reaction conditions.

Nitrooxylation Reactions. Recent work in the Ridd's Group has shown that the side-chain functionalization of methylbenzenes by CAN in MeCN to give benzyl nitrate is a little useful reaction, since substantial ring nitration also occurs.[35] We have, however, discovered that a much faster and efficient reaction takes place at room temperature when the reaction mixture is irradiated by a high pressure mercury lamp.[36] This photochemical process represents a very useful method for the one pot side-chain nitrooxylation of methylbenzenes, which can also be applied to electron-poor toluenes and cycloalkanes.[37] Some results are collected in Tables 3 and 4.

Table 3. Photochemical Nitrooxylation of R-Substituted Toluenes Promoted by CAN in MeCN at Room Temperature[a].

R	t/h	$ArCH_2ONO_2$(yield, %)[b]
2,4Cl$_2$	2.5	47
4-Br	2	70
H	1	75[c]
4-OAc	2.5	80
4-CH$_3$	0.5	96
3,5-(CH$_3$)$_2$	1	85
3-CH$_3$	1.5	67

a) CAN: substrate molar ratio 2:1; 150W high pressure mercury lamp, pyrex filter; b) Based on the starting material; c) No product is formed after 3h in the dark.

Flash photolysis experiments have shown than when CAN is irradiated in MeCN a transient forms which has the spectroscopic characteristics of the nitrate radical.[38] Since the rate of decay of this transient is strongly enhanced by the addition of an alkylaromatic compound or a cycloalkane, it seems reasonable to conclude that in photochemical reactions the effective reacting species is the nitrate radical generated in the light promoted redox process shown in eq. 18.

Table 4. Photochemical Reactions of Some Cycloalkanes Promoted by CAN in MeCN at Room Temperature

Substrate	Products (yield)[b]
Adamantane	1-adamantyl nitrate (35%), N-(1-adamantyl)acetamide(56%) 2-adamantyl nitrate (5%), N-(2-adamantyl)acetamide(4%)
Norbornane	2-exo-norbornyl nitrate(80%)[c], 2-endo-norbornyl nitrate (20%)[c]
Cyclohexane	Cyclohexyl nitrate (30%)[d]

a) CAN: substrate molar ratio 1:1; 125W high pressure mercury lamp, pyrex filter. Reaction time 2-5 h; b) With respect to CAN; c) Determined after reduction to alcohols; d) 52% of cyclohexane has been recovered unchanged.

$$Ce^{IV}ONO_2 \longrightarrow Ce^{III} + ONO_2 \cdot \qquad (18)$$

A detailed mechanistic investigation[38,39] has later allowed us to propose a mechanistic dichotomy for the reaction of this radical with alkylaromatics. With substrate more electron rich than toluene the photochemical side-chain mitrooxylation takes place by an electron transfer mechanism (eqs. 19-21) whereas with substrates less electron rich than toluene and with cycloalkanes a H-atom transfer mechanism is the most likely one.[37,39]

$$ArCH_3 + NO_3 \cdot \longrightarrow ArCH_3^{+} \cdot + NO_3^{-} \qquad (19)$$

$$ArCH_3^{+} \cdot \longrightarrow ArCH_2 \cdot + H^{+} \qquad (20)$$

$$AeCH_2 \cdot + Ce^{IV}ONO_2 \longrightarrow ArCH_2ONO_2 + Ce^{III} \qquad (21)$$

<u>Autoxidation Reactions</u>. All the processes discussed so far are stoichiometric reactions which require two moles of CAN to convert one mole of substrate into products.

It has been, though, found that when the photochemical CAN-promoted reaction of alkylaromatics are run in the presence of oxygen, an autoxidation process takes place where CAN is used in catalytic amount.[40] This reaction becomes particularly efficient in the presence of HNO_3. For example, the photochemical autoxidation of p-xylene at room temperature with 2% CAN leads to p-tolualdehyde in 27% yield. The reaction mechanism reported in Scheme 3 has been suggested

$$Ce^{IV}ONO_2 \xrightarrow{\ h\nu\ } Ce^{III} + NO_3 \cdot$$

$$NO_3 \cdot + ArCH_3 \longrightarrow ArCH_2 \cdot + HNO_3$$

$$ArCH_2 \cdot + O_2 \longrightarrow ArCH_2O_2 \cdot \longrightarrow Products$$

$$ArCH_2O_2 \cdot + Ce^{III} \longrightarrow ArCH_2O_2Ce^{IV}$$

$$ArCH_2O_2Ce^{IV} + HNO_3 \longrightarrow ArCH_2O_2H + Ce^{IV}ONO_2$$
$$\downarrow$$
$$Products$$

Scheme 3

Once that the benzylic radical is formed its reaction with O_2 gives the benzylperoxy radical which can be converted into products or, under the action of light, promote Ce(III) to Ce(IV) oxidation as already suggested by Sheldon and Kochi to rationalize the Ce(IV)-catalyzed photochemical decarboxylation of pivalic acid in the presence of O_2.[41] The role of HNO_3 might be that of converting $ArCH_2OOCe^{IV}$ into $Ce^{IV}ONO_2$ and the hydroperoxide. Nitrate ligands are thus reintroduced into the Ce(IV) coordination sphere.

An important contribution of a direct hydrogen atom abstraction by peroxy radicals (eq. 22) to the oxidation chain is made unlikely by the following observations: (i) the photochemical autoxidation of p-xylene in

$$ ArCH_2O_2\cdot \quad + \quad ArCH_3 \quad \longrightarrow \quad ArCH_2O_2H \quad + \quad Ar\overset{\cdot}{C}H_2 \qquad (22) $$

the presence of AIBN produces only 7% of p-tolualdehyde; (ii) some competitive experiments have shown that the substrate selectivity of the autoxidation reaction is very similar to that observed in the stoichiometric photochemical nitrooxylation; (iii) toluene is more reactive than cumene and methyl group is more reactive than isopropyl group in p-cymene, whereas the reverse is expected in both cases if reaction (22) would play a significant role.

CAN-catalyzed photochemical autoxidation takes place also with adamantane, norbornane and cyclohexane.[37] The results are particularly good with adamantane since, by using 2% of CAN, it is possible to convert this substrate in oxidation products (1- and 2-adamantanol and 2-adamantanone) with an overall yield of 95%. The selectivity in 1-adamantanol is as high as 87%. The reaction is, however, much less efficient with norbornane (20% yield of 2-norborneol; exo/endo ratio = 1) and with cyclohexane (10% yield of cyclohexanol).

2.2 Nuclear substitutions of aromatic compounds

It is well known that electrophilic carbon radicals like $\overset{\cdot}{C}H_2X$ and $\overset{\cdot}{C}HXZ$ (X and Z electron withdrawing groups, i.e. COR, NO_2, CO_2R, CN) are generated by reactions of CH_3X and CH_2XZ, respectively, with metal ions (M^{n+}), which are one electron oxidants (eq. 23). These radicals can then react with aromatic compounds or alkenes to give

$$ CH_3X \quad + \quad M^{n+} \quad \longrightarrow \quad \cdot CH_2X \quad + \quad M^{(n-1)+} \qquad (23) $$

substitution and addition products according to the sequences in eqs. 24 and 25 (Y is a ligand of the metal or the conjugate base of the solvent).[42]

$$\cdot CH_2X \ + \ ArH \ \longrightarrow \ (HArCH_2X)\cdot \ \xrightarrow{\ M^{n+}\ } \ ArCH_2X \ + \ H^+ \qquad (24)$$

$$\cdot CH_2X \ + \ \overset{}{\diagup}\!\!\diagdown_R \ \longrightarrow \ X\!\!\diagup\!\!\underset{O}{\diagdown}\!\!\diagup\!\!\diagdown_R^{\ \cdot} \ \xrightarrow{\ M^{n+},\,Y^-\ } \ X\!\!\diagup\!\!\underset{O}{\diagdown}\!\!\diagup\!\!\diagdown_R^{\ Y} \qquad (25)$$

The success of this reaction rests on the relatively difficult oxidation of the electrophilic carbon radical by the metal ion. Thus, the radical has sufficient time to react with the aromatic substrate or the alkene.

The oxidative homolytic aromatic substitution has been studied by several workers, with particular regard to the reactions promoted by $\cdot CH_2COCH_3$,[43,45] $\cdot CH_2NO_2$[46,47] and $\cdot CH_2CO_2H$[44,45] generated via oxidation o. acetone, nitromethane and acetic acid, respectively, with Mn(III) or Ce(IV) salts.

In our laboratory this process has recently been exploited to perform the aromatic homolytic malonylation.[48] This reaction occurs with fair to good yields under very mild conditions and in very short time when CAN is reacted in MeOH with dimethyl malonate in the presence of aromatic compounds as electron rich as or more than benzene. The mildness of the conditions used is very remarkable expecially when it is considered that the reaction of benzene with CAN and acetone in AcOH to give propiophenone requires reflux temperature and gives only 22% of the product.[44] Some of our results are reported in Table 5.

Yields are generally larger when electron donating substituents are present in the benzene ring, as expected for an electrophilic free radical such as $\cdot CH(CO_2Me)_2$. With monosubstituted benzenes the reaction appears not very valuable from the synthetic point of view since the three isomeric malonyl derivatives are formed. However, with naphthalene only the α-isomer is obtained and formation of only one isomer is similarly observed in several reactions of heteroaromatic compounds.[49] In view of the already mentioned mildness of the reaction conditions, this procedure recommends itself for the malonylation of the often acid-sensitive π-electron excessive heteroaromatic compounds.

Another notation concerns the good yield observed in the reaction with mesitylene, which suggests that the process is not very sensitive to steric effects. This is in line with the general belief that free radical reactions are characterized by reagent-like transition states.

Table 5. Formation of Dimethyl Arylmalonates from the Reaction of ArH with Dimethyl Malonate Promoted by CAN in MeOH

ArH $ArCH(CO_2Me)_2$, yield%[a]

Benzene   53
Toluene   59 (mixture of isomers)[b]
Anisole   87 (mixture of isomers)[c]
Chlorobenzene   33 (mixture of isomers)[d]
Mesitylene   66
Naphthalene   50 (α-isomer)
Thiophene   50 (α-isomer)
Furan   56 (α-isomer)
2-Acetylfuran   39 (α-isomer)

---

a) with respect to CAN; b) 50.8%, ortho; 21.4%, meta; 27.8%, para; c) 82.3%, ortho; 1.7%, meta; 16.0%, para; d) 50%, ortho; 50%, meta and para.

In spite of this, however, the reaction exhibits a quite significant intermolecular selectivity as we have found that the overall reactivity of anisole is 7.3 times larger than that of toluene.

Before concluding this section two additional CAN-promoted nuclear substitutions of the indirect type can be mentioned.

In the presence of alkali, metal iodides, or iodine CAN can promote the nuclear iodination of polymethylbenzenes, polymethoxybenzenes and naphthalenes, with high conversion and good selectivity.[50] More recently, it has also been observed that arylphosphonate can be prepared in good yields in an one-step synthesis when arenes are reacted with tri- or diethylphosphites in the presence of CAN.[51]

B. ADDITION TO ALKENES

1. DIRECT REACTIONS

1.1 Formation of dinitrate adducts

In MeCN CAN reacts with alkenylaromatic compounds, in the temperature range of 25-40° C, to give dinitrate adducts (eq. 26) in high yields.[7,8]

$$\text{ArCH} = \text{CHR} \xrightarrow{\text{CAN, MeCN}} \text{ArCHONO}_2\text{CHONO}_2\text{R} \qquad (26)$$

A kinetic study[7] has shown that for substrates, more electron rich than styrene an electron transfer mechanism is probable, eqs 27-29. A radical cation is first formed which should remain closely associated with the reduced Ce(III) species. In the second step the radical cation is converted into a radical which finally reacts with CAN by a ligand transfer mechanism to form the dinitrate adduct.

$$\text{ArCH} = \text{CH}_2 + \text{Ce}^{IV}\text{ONO}_2 \longrightarrow (\text{ArCHCH}_2)^{+\cdot}\ \text{Ce}^{III}\text{ONO}_2 \qquad (27)$$

$$(\text{ArCHCH}_2)^{+\cdot}\ \text{Ce}^{III}\text{ONO}_2 \longrightarrow \text{Ce}^{III} + \text{Ar}\overset{\cdot}{\text{C}}\text{HCH}_2\text{ONO}_2 \qquad (28)$$

$$\text{Ar}\overset{\cdot}{\text{C}}\text{HCH}_2\text{ONO}_2 + \text{Ce}^{IV}\text{ONO}_2 \longrightarrow \text{ArCHONO}_2\text{CH}_2\text{ONO}_2 + \text{Ce}^{III} \qquad (29)$$

For substrates less electron rich than styrene a free radical addition, involving the direct transfer of the nitrate group to the alkene (eq. 30), appears to be the most likely mechanism.

$$\text{ArCH} = \text{CH}_2 + \text{Ce}^{IV}\text{ONO}_2 \longrightarrow \text{Ar}\overset{\cdot}{\text{C}}\text{HCH}_2\text{ONO}_2 + \text{Ce}^{III} \qquad (30)$$

Interestingly, addition reaction is the only process observed also with α- and β-methylstyrene. Apparently, allylic substitution has little or no chance to compete with the addition process.

The stereochemistry of the addition is predominantly anti (90%) with cyclic alkenylaromatics such as acenaphthene or benzofuran, which is presumably due to a steric effect. With acyclic olefins, such as trans- and cis- β-methylstyrene or trans- and cis-stilbene a stereoconvergent process is observed, the same mixture of diastereoisomeric products being formed from both the geometric isomers, in line with the suggested mechanisms involving the formation of species (like radical ions and radicals) which can quite easily equilibrate. Moreover the electron transfer step must have some degree of reversibility since cis-stilbene is partially converted into trans-stilbene under the reaction conditions.[8]

2. INDIRECT REACTIONS

2.1 Formation of dinitrate adducts

The nitrooxylation reaction described in the preceding section becomes much faster when it is carried out under irradiation by a high pressure mercury lamp.[52] As shown by the results reported in Table 6 the reaction can also be extended to unactivated olefins such as cyclohexene and 1-octene. Also in this case the active species is $NO_3 \cdot$ generated, as described before, in the photolysis of CAN. As in the reaction with alkylaromatic compounds, two mechanisms can be envisaged.

Table 6. Photochemical Nitrooxylation of Alkenes with CAN in CH_3CN at Room Temperature[a].

Substrate	$ArCHONO_2CH_2ONO_2$, yield,%[b]
styrene	86
3-chlorostyrene	61
4-chlorostyrene	73
4-methylstyrene	70
3-trifluoromethylstyrene	75
2,4-dichlorostyrene	95
α-methylstyrene	96
trans-β -methylstyrene	87[c]
trans-β -methylstyrene	-[d]
cyclohexene	60
1-octene	91[e]

----

[a]Substrate and CAN, 4×10^{-2}M (125W, high pressure mercury lamp, reaction time 5-20 min). [b]Yield with respect to CAN used, considering a CAN: alkene 2:1 stoichiometry. [c]The threo/erythro dinitrate ratio is ca. 0.5. [d]Thermal reaction. The reaction time (5 min) is the same as in the corresponding photochemical process. [e]Determined after reduction of the dinitrates to the corresponding diols by $LiAlH_4$.

First, an electron transfer mechanism (eqs. 31-33) is possible, where the first step is the formation of a radical cation. It follows the reaction of the radical cation with NO_3^- to give a α-nitrate radical, which eventually undergoes a ligand transfer reaction with CAN to give the addition product. The second mechanistic possibility is that of a direct

$$NO_3 \cdot \ + \ ArCH=CH_2 \ \longrightarrow \ NO_3^- \ + \ ArCHCH_2^{\cdot +} \qquad (31)$$

$$ArCHCH_2^{\cdot +} \ + \ NO_3^- \ \longrightarrow \ ArCHCH_2ONO_2 \qquad (32)$$

$$ArCHCH_2ONO_2 \ + \ Ce^{IV}ONO_2 \ \longrightarrow \ ArCHONO_2CH_2ONO_2 \ + \ Ce^{III} \qquad (33)$$

attack of $NO_3 \cdot$ to the alkene to give the β-nitrate radical in only one step. On the basis of a kinetic study (flash fotolysis technique)[52] the electron transfer mechanism seems favored, at least for alkenylaromatics.

2.2 Addition reactions of carbonyl compounds

As already discussed, the electrophilic radicals $\dot{C}H_2X$ and $\dot{C}HXY$ (X and Y, electron withdrawing groups) generated by reaction of CH_3X and CH_2XY, respectively, with CAN or other one electron oxidants can be involved in addition reactions to alkenes (eqs 23 and 25).

In this process a new carbon-carbon bond is formed and simultaneously the group Y (ligand of the metal or conjugate base of the solvent) is introduced, which can undergo further reaction, thus widening the synthetic scope of this oxidative addition.

Most of studies in this field have so far concerned addition reactions of carbonyl compounds promoted by $Mn(OAc)_3$ and several important synthetic applications have recently reported in the literature.[53]

It has been considered of interest to study the corresponding CAN promoted oxidative addition to alkenes in some detail for the following reasons: First, CAN, as already shown in other reactions, often exhibits quite peculiar properties with respect to other one electron oxidants. Second, owing to the great ability of CAN to promote oxidation of radicals by a ligand transfer mechanism the step described in eq. 25 should lead to nitrate adducts ($Y = ONO_2$, $M^{n+} = Ce^{+4}$) which could be useful intermediates in synthesis in view of the possible replacement or conversion of the nitrate group.

A first observation was, however, that in the CAN-induced reaction of dimethyl malonate with simple alkenes the formation of the nitrate addition product is accompanied by the formation of a substituted alkene, as shown in eq. 34.

$$(34)$$

35 %   35 %

Probably the intermediate radical, in addition to ligand transfer, can also undergo oxidation by CAN to a carbocation which gives the alkene by proton loss. For this reason, we became mainly concerned with reactions of unsaturated systems, such as conjugated dienes, alkenylaromatics, enol derivatives, where this complication is not possible. Moreover, these systems are more electron rich than simple alkenes and this makes the reaction easier in view of the electrophilic character of the radical.

The addition of some ketones to 1,3-butadiene,[54] as shown in eq. 35 for the case of acetone, leads to a mixture approximately 1:1 of 1,2- and 1,4-adduct, with an overall yield of ca 80%. The 1,4-adduct has a trans

$$(35)$$

geometry. Interestingly, with the unsymmetric methyl ethyl ketone the addition involves the substituted α-carbon exclusively. In this respect CAN exhibits a regioselectivity completely different from that found in the reactions with Mn(OAc)$_3$ where attack to the alkene preferentially occurs from the less substituted side of ketone. Attempts to exploit this reaction by conversion of the nitrate group into other functional groups have so far met with limited success, even if a number of interesting derivatives was obtained (scheme 4).[55]

Apparently, only the product coming from the 1,4-nitrate adduct is recovered. This because the 1,2-nitrate adduct turned out to be very unstable, undergoing extensive decomposition under the reaction conditions.

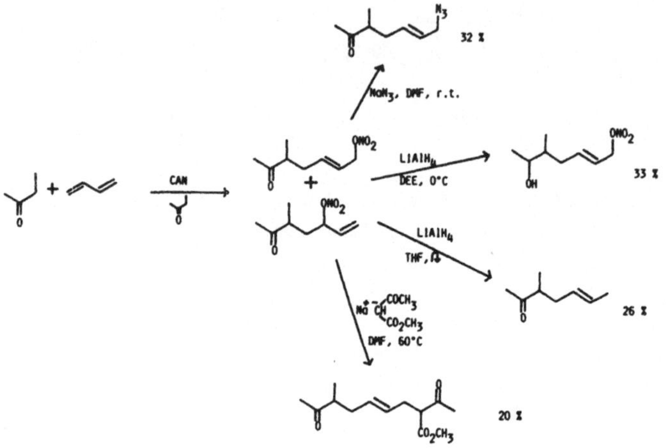

* Yield of isolated product with respect to CAN used.

Scheme 4

A better situation arises when a dialkyl malonate is used as the carbonyl compound. As shown in Scheme 5, in this case too an 1:1 mixture

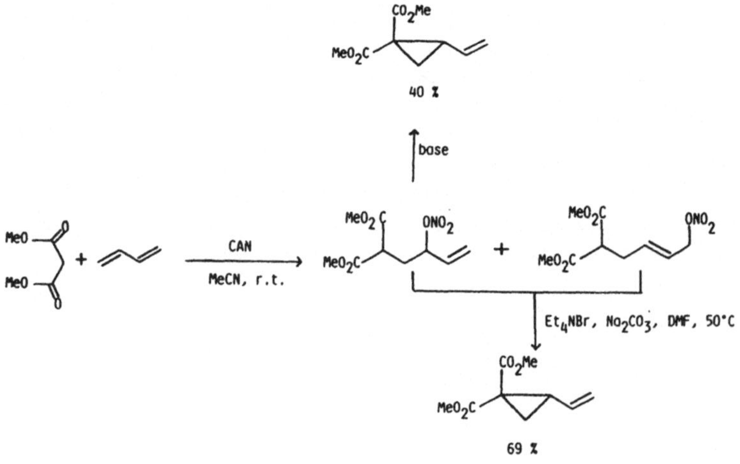

Scheme 5

of 1,2 and 1,4-adduct (overall yield, 80%) is obtained. However, treatment of this mixture with a base (i.e. MeO⁻, CN⁻, Na₂CO₃) converts the 1,2-adduct into a vinylcyclopropane, through the intermediacy of the carbanion 6, whereas the 1,4-nitrate adduct remains unchanged (bottom part of

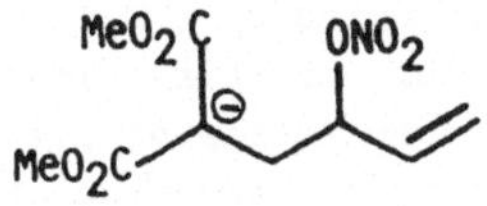

6

Scheme 5). Interestingly, it was found that also most of the 1,4-adduct can be converted into cyclopropane when the mixture is treated with Br⁻ in DMF before reaction with the base (upper part of Scheme 5). It is suggested that Br⁻ converts the 1,4-nitrate adduct into the corresponding bromide adduct which can afford a carbanion capable to undergo an S_N2 reaction leading to the vinylcyclopropane. This process might be much more difficult with the nitrate adduct since NO_3^- is a significantly worse leaving group than Br⁻. When styrene derivatives are used in the place of 1,3-butadiene the CAN-promoted reaction with dimethyl malonate leads only to the 1,2-nitrate adduct, which is then converted into the phenylcyclopropane derivative in 70% yield.

Ring closure is of course no longer possible if an alkylmalonate replaces malonate as the carbonyl compound. In this case it is therefore possible to carry out substitution reactions of the nitrate group as already shown for 2-butanone. An example is shown in Scheme 6 (malonate: diene molar ratio 2:1).

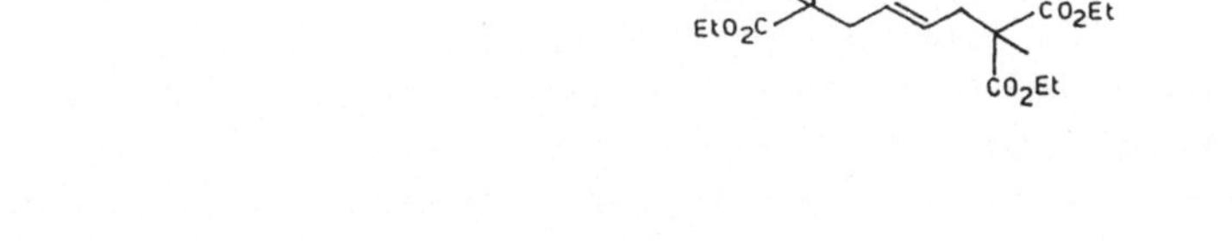

Scheme 6

where the sodium salt of the alkylmalonate itself is used as the nuclophile. A symmetrically substituted trans-4-octene is obtained since only the 1,4 adduct undergoes the replacement reaction of the nitrate group.

Interestingly, if the first oxidative process is carried out with a two-fold excess of substrate, the trans-4-octene can be obtained by simple addition of NaH to the reaction mixture.

Another useful application of this oxidative addition process is represented by a very efficient synthesis of 1,4-dicarbonyl compounds which can be accomplished when ketones are made to react with vinyl or isopropenyl acetate in the presence of CAN.[56]

In this case, the nitrate adduct (eq. 36, R and R^I = alkyl, R^{II} = alkyl,H) forms which can undergo the

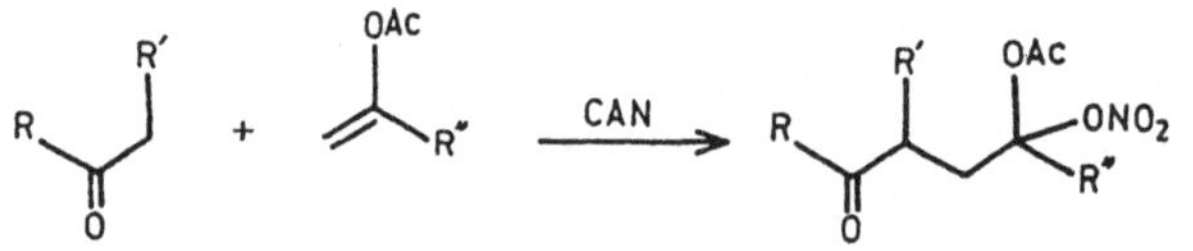

$$(36)$$

reactions shown in Scheme 7

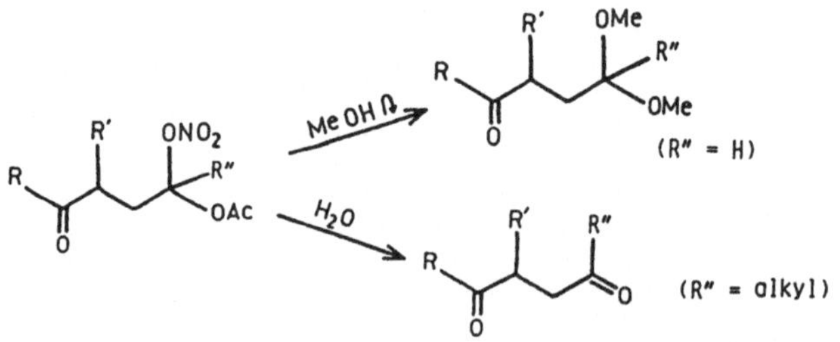

Scheme 7

When R^{II} = alkyl (isopropenyl acetate) 1,4-diketones are formed, whereas when R^{II} = H (vinyl acetate) the dimethyl acetals of a 4-ketoaldehyde is obtained. These reactions are carried out in methanol at room temperature and most of the yields are over 70%, as shown by the results reported in Table 7. Particularly valuable is the reaction directly leading to the partially protected ketoaldehyde which can allow us to immediately perform further reactions at the carbonyl group.

It is interesting to know that the corresponding reactions

with Mn(OAc)$_3$,[57] exhibit yields which are significantly lower than those observed in the CAN-promoted reactions.

Table 7. The CAN-promoted Reactions of Carbonyl Compounds with Vinyl and Isopropenyl Acetate.

Carbonyl compound	Product, yield (%)[a]	
	Reaction with  ⟋OAc	Reaction with 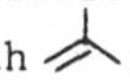 ⟍OAc

---

(structures and yields)

	 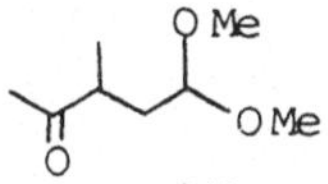 70	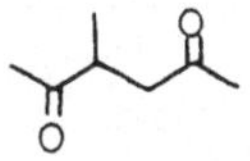 78
	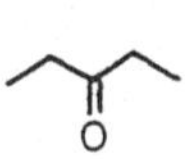 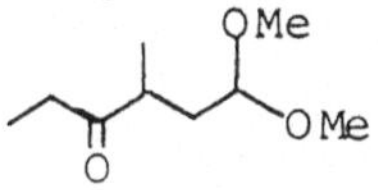 73	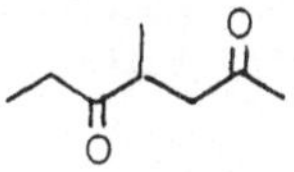 74
	 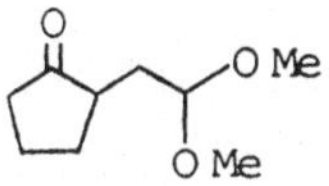 77	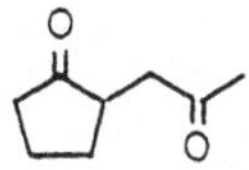 70
	 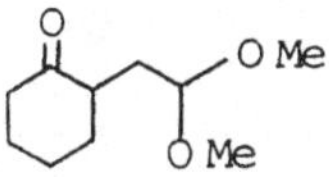 69	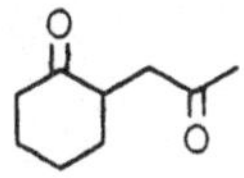 65
	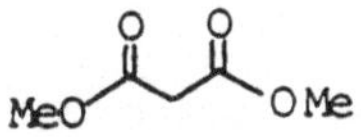 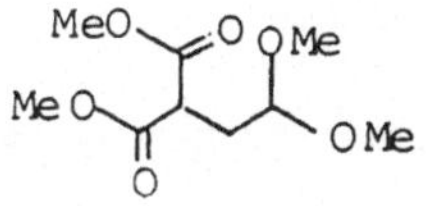 75	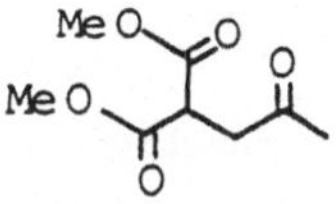 82

(a)Yield of isolated product calculated with respect to CAN.

Attempts to extend this reaction to 2-alkyl substituted vinyl acetates failed when a ketone was used as the carbonyl reagent. However the reaction was successful with 1,3-dicarbonyl compounds. Thus, from dimethyl malonate and 1-heptenyl acetate 70% yield of 7 was obtained. No such reaction occurs with Mn(OAc)$_3$.[53a]

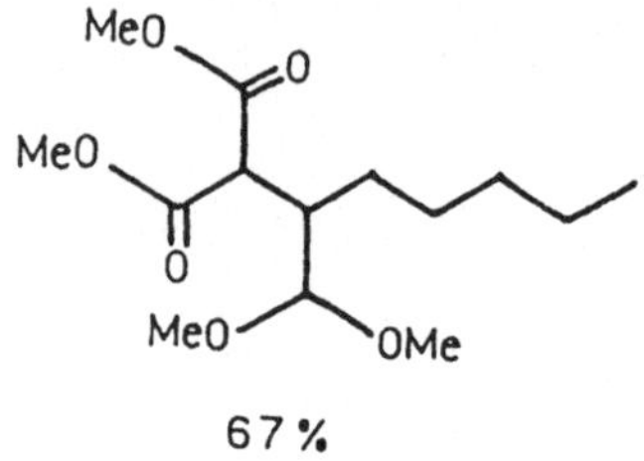

67 %

7

A completely different type of product is obtained when β-diketones or β-ketoester are the carbonyl reagents of these oxidative additions.[54,58] In this case it has been observed that the 1,2 nitrate adduct cyclizes by an intramolecular S_N2 displacement of the nitrate group promoted by the carbonyl group, to give a dihydrofuran derivative. An example is shown in Scheme 8.

The same reaction can also be carried out with Mn(OAc)$_3$ but yields are significantly lower than with CAN

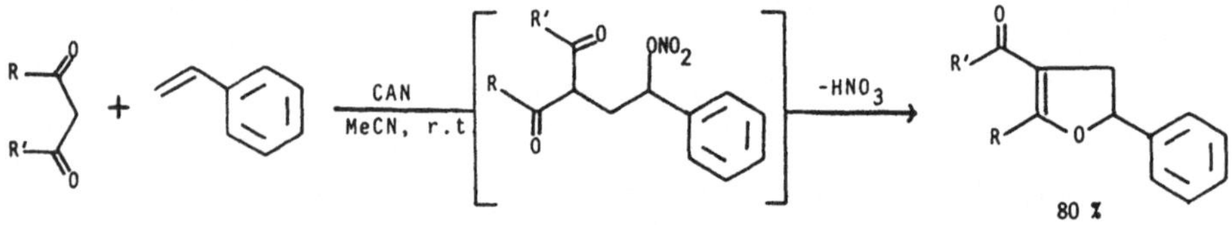

Scheme 8

This reaction can be exploited for the synthesis of furans by

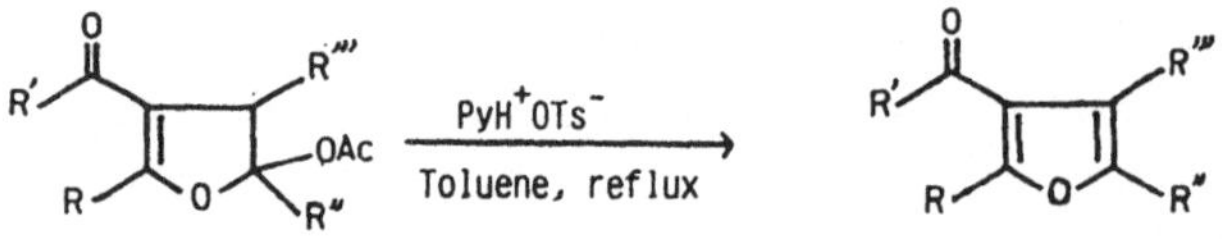

(37)

using enol acetates as the olefinic substrate.[59] In this case a 2-acetoxy substituted dihydrofuran obtains, which loses AcOH affording the furan derivative (eq. 37, R^{II} = H, Me; R^I = Me, alkoxy; R = alkyl). As shown by the results in Table 8 the yields, with respect to the oxidizing agent, are over 50% in the reaction with vinyl acetate, which is quite satisfactory in view of the simple procedure used.

Table 8. Synthesis of 3-Acyl- and 3-Carboalkoxyfurans by CAN Promoted Oxidative Addition of 1,3-Dicarbonyl Compounds to Vinylic Acetates.

1,3-Dicarbonyl compound	Vinylic acetate	Product	Yield,%*

1. — COMe product — 56

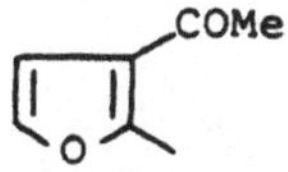

2. — COMe product — 35

3. — CO₂Me product — 52

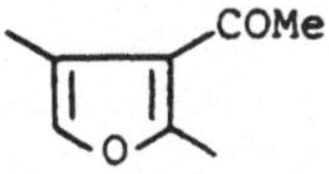

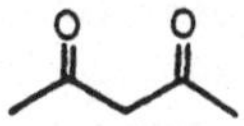

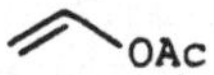

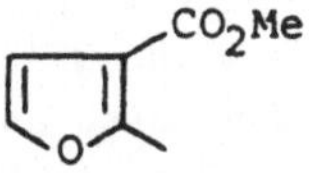

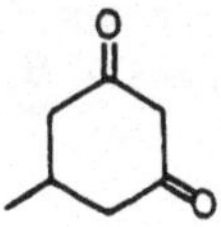

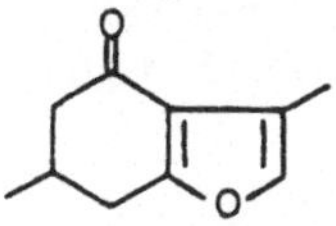

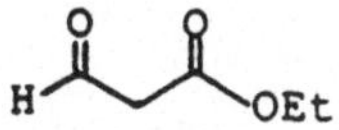

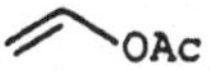

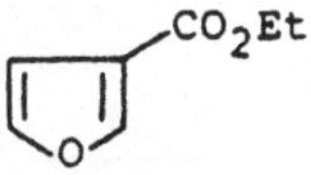

4. — product — 30

5. — CO₂Et product — 9

6. — COMe product — 61

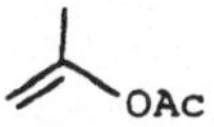

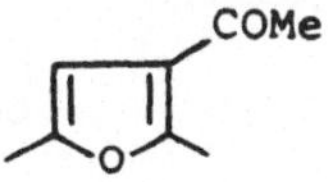

---

* Yield of isolated product calculated with respect to CAN

Lower yields are however observed when the alkene is propenyl acetate. Probably, the presence of a methyl group at the 2-position of vinyl acetate slows down the rate of the addition process by a steric effect. Nevertheless, the reaction remains valuable as shown by the finding that the naturally occurring furanomonoterpene evodone (entry 4 of Table 8) can be prepared with a yield comparable with that (overall) of other multistep syntheses.[60]

An attempt has also been carried out to use ethyl formylacetate as the 1,3-dicarbonyl compound, which would allow us to synthetize unsubstituted 3-carbetoxyfuran. However, only a very modest yield (9%) of purified compound has been obtained, which is probably due to the high tendency of the starting material to undergo autocondensation processes under the reaction conditions.

An interesting observation, which certainly has a bearing with respect to the mechanism of these reactions, has been that the rate of reduction of CAN significantly increases when the alkene is added to the carbonyl compound.[54] This shows that the rate of oxidative addition is faster than the oxidation of the carbonyl compound by CAN. Clearly, such an observation is inconsistent with the hypothesis of an irreversible formation of an α-ketoradical, which then reacts with the alkene in a fast step. In this case the rate of CAN reduction would be unaffected by the added alkene.

It is worth noting that similar results have been reported for corresponding processes involving Mn(OAc)$_3$, conjugated dienes and 1,3 - dicarbonyl compounds.[61] In that case it was suggested that the carbonyl compound and diene react with one another, being both coordinated to Mn^{III}.

It is doubtful that this hypothesis also holds for the Ce^{IV}-induced addition in view of the little or no tendency of lanthanides to form coordination complexes with π-bonding ligands.[62] A suggestion compatible with the rate effects discussed above, is that the attacking species is a Cerium^{III}-coordinated free radical, which is in equilibrium with the CAN-carbonyl complex, possibly through the intermediacy of the species **8**, as indicated in Scheme 9.

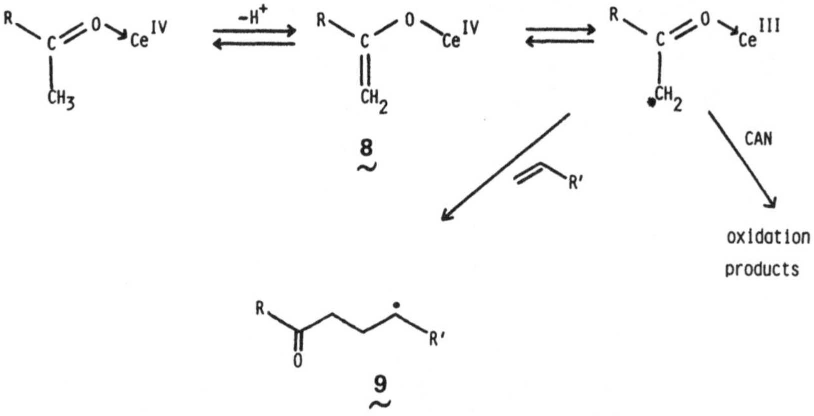

Scheme 9

This complexed radical can either react with the appropriate alkene to give the radical **9** or undergo further reaction with CAN to form the carbonyl compound oxidation products. If the latter reaction is slower than the former, the observed effect of the added alkene on the reduction rate of CAN is accounted for.

Complexes of Ce^{IV} with oxygen compounds are well known,[62,63] and more significantly, it has been shown that a complex between CAN and acetone is the intermediate in the oxidation of ketones by this salt.[64] The mechanism reported in Scheme 9 might hold for the corresponding $Mn(OAc)_3$-promoted reactions and interestingly equilibria somewhat similar to the ones reported in Scheme 9 have also been proposed for this oxidant.[65]

To get further mechanistic insight, the relative reactivities of substituted styrenes have been measured and plotted against the σ values of the substituents.[66]

Good correlation have been obtained from which the ϱ values reported in Table 9 have been calculated

---

Table 9. Values of the Reaction Constant ϱ for the Oxidative Addition of $RCOCH_2COR^I$ to Substituted Styrenes Promoted by CAN in MeOH and CH_3CN

R	R^I	ϱ	
		MeOH	MeCN
Me	Me	-1.75	-1.17
Me	OMe	-1.77	-0.51
OMe	OMe	-1.80	-

(a) at 20°C. Hammett σ values as the substituent constants.

A first observation is that the reaction is significantly more selective in MeOH than in MeCN. Tentatively, this difference, particularly significant in the reaction with methyl acetoacetate, may be attributed to a lower stability in the latter solvent of the cerium coordinated radical. This because MeCN may coordinate cerium ions better than MeOH, thus decreasing the capability of these ions to stabilize the α-keto radical.

In MeCN selectivity is higher with acetylacetone than with methyl acetoacetate, a result which may be explained with the higher stability of $MeCO\overset{\cdot}{C}HCOMe$ with respect to $MeCO\overset{\cdot}{C}HCO_2Me$. Accordingly, an acetyl group should be more effective than a carbomethoxy group in stabilizing a carbon

182

radical.

In MeOH, however, the selectivity is nearly unsensitive to the structure of the carbon radical. In line with the previous suggestion, it may be proposed that in MeOH the stabilization of the free radical due to cerium coordination is enough large as to make the system little sensitive to the nature of substituents at the carbonyl groups. Certainly, further detailed mechanistic study of the process is necessary in order to reach reliable conclusions.

C. CONCLUDING REMARKS

Summing up, CAN appears to be a very useful reactant in particular for the functionalization, either nuclear or in the side-chain, of alkylaromatics and for the oxidative addition of carbonyl compounds to alkenes. In the former process very remarkable is the intramolecular selectivity, which can make CAN the reagent of choice for selective functionalizations. In the oxidative additions of great value is certainly the possibility to carry out the reactions under very mild conditions.

Nearly all the reactions we have examined require CAN in stoichiometric amount and this sometimes can represent a drawback in the use of this oxidant, in view of its high molecular weight and the relatively high cost. However, the electrochemical regeneration of CAN is a quite easy process[20] and this can help in overcoming the above problem.

D. ACKNOWLEDGMENTS.

The authors wish to thank the Italian National Council of Research (CNR) and the Ministry of the "Pubblica Istruzione" for the financial support which has made possible the work described in this paper.

REFERENCES

1. T.L. Ho, Synthesis, 347 (1973).
2. H.B. Kagan, J.L. Namy, Tetrahedron, 42, 6573 (1986).
3. E. Baciocchi, A. Dalla Cort, L. Eberson, L. Mandolini, C. Rol, J.Org.Chem., 51, 4544 (1986).
4. E. Baciocchi, R. Ruzziconi, J.Chem.Soc.Chem.Commun 445 (1984).
5. E. Baciocchi, C. Rol, L. Mandolini, J.Am.Chem.Soc. 102, 7597 (1980).
6. E. Baciocchi, C. Rol, L. Mandolini, Tetrahedron Lett., 3343 (1976).
7. E. Baciocchi, C. Rol, G.V. Sebastiani, A. Zampini, J.Chem.Soc. Chem.Commun., 1045 (1982).
8. E. Baciocchi, C. Rol, G.V. Sebastiani, Gazz.Chim.Ital., in press.
9. E. Baciocchi, L. Eberson, C. Rol, J.Org.Chem., 47, 5106 (1982).
10. E. Baciocchi, L. Mand.lini, C. Rol, J.Org.Chem., 42, 3682 (1977).
11. W. Trahanovsky, L. Brewster Young, J.Org.Chem., 31, 2033 (1965).
12. L. Syper, Tetrahedron Lett. 4493 (1966).
13. L.A. Dust, E.W. Gill, J.Chem.Soc.(C), 1630 (1970).
14. T.W. Martin, A Henshall, J.M.Burk, R.W. Glass, unpublished work quoted by T.W. Martin, J.M. Burk, A.Henshall, J.Am.Chem.Soc., 88, 1097 (1966)
15. R.M. Dessau, S.Shih, E.I. Heiba, J.Am.Chem.Soc., 92, 412 (1970).
16. W. T. Dixon, D. Murphy, J.Chem.Soc.Perkin Trans.II, 1823 (1976).
17. D.N. Ramakrishna Rao, M.C.R. Symons, J.Chem.Soc.Perkin Trans.II, 991 (1985).
18. L.K. Sydnes, S.H. Hansen, I.C.Burkow, L.J. Saethre, Tetrahedron, 41, 5205 (1985).
19. A. Dalla Cort, A. La Barbera, L. Mandolini, J.Chem.Res.(S), 44 (1983).
20. S.Torii, H. Tanaka, T. Inokuchi, S. Nakane, M. Kada, N. Sailo, T. Sirakawa, J.Org.Chem., 47, 1647 (1982).
21. A. Dalla Cort, L. Mandolini, Gazz.Chim.Ital., 114, 283 (1984).
22. E. Baciocchi, S. Mei, C. Rol, L. Mandolini, J.Org.Chem., 43, 2919 (1978).
23. E. Baciocchi, C. Rol, G.V. Sebastiani, unpublished results.
24. E. Baciocchi, C. Rol, G.V. Sebastiani, Gazz.Chim.Ital., 112, 513 (1982).
25. M.C.R. Symons, L.Harris, J.Chem.Res.(S), 268 1982.
26. E. Baciocchi, C. Rol, G.V.Sebastiani, J.Chem.Res.(S), 232 (1983).
27. M. Marocco, G.Brilmyer, J.Org.Chem., 48, 1487 (1983).
28. L.Eberson, J.Am.Chem.Soc., 89, 4669 (1967).
29. R. Palombari, C. Rol, G.V. Sebastiani, Gazz.Chim.Ital., 116, 87 (1986).
30. M. Periasamy, V.Vivekananda Bhatt, Synthesis, 330 (1977).
31. J.Skarzewski, Tetrahedron, 40, 4997 (1984).
32. L.Syper, K.Kloc, J.Mlochowski, Tetrahedron, 36, 123 (1980).
33. E. Baciocchi, C. Rol, G.V. Sebastiani, B. Serena, J.Chem.Res.(S), 24 (1984).
34. E. Baciocchi, M. Crescenzi, to be published.
35. S. Dinkturk, J.H.Ridd, J.Chem.Soc.Perkin Trans.2, 961, 965 (1982).
36. E. Baciocchi, C. Rol, G.V.Sebastiani, B. Serena, Tetrahedron Lett., 1945 (1984).

184

37. E. Baciocchi, T.Del Giacco, C. Rol, G.V. Sebastiani, Tetrahedron Lett., 1941 (1987).
38. E. Baciocchi, T. Del Giacco,, C. Rol, G.V. Sebastiani, Tetrahedron Lett., 541 (1985).
39. E. Baciocchi, T. Del Giacco, S. Murgia, G.V. Sebastiani, J.Chem.Soc.Chem.Commun. , 1246 (1987).
40. E. Baciocchi, T. Del Giacco, C. Rol, G.V. Sebastiani, Tetrahedron Lett. 3353 (1985).
41. R.A. Sheldon, J.K.Kochi, J.Am.Chem.Soc., 90, 668 (1968).
42. For pioneering work in the field see (a) A.I. Heiba, R.M. Dessau, J.Am.Chem.Soc. 93, 524 (1971); (b) E.I. Heiba, R.M. Dessau, J.Am.Chem.Soc., 93, 995 (1971); (c) E.I. Heiba, R.M. Dessau, W.J.Koehl, Jr. J.Am.Chem.Soc., 91, 138 (1969).
43. M.E. Kurz, Tsu-yn R.Chen, J.Org.Chem., 43, 279 (1978).
44. M.E. Kurz, V. Baru, P. Nhi Nguyen, J.Org.Chem., 49, 1603 (1984).
45. C. Gardrat, Bull.Soc.Chim.Belg. 93, 397 (1984).
46. M.E. Kurz, P. Ngoviwatachai, T. Tantrarant, J.Org.Chem., 46, 4668 (1981).
47. M.E. Kurz, P. Ngoviwatachai, J.Org.Chem., 46, 4672 (1981).
48. E. Baciocchi, D. Dell'Aira, R. Ruzziconi, Tetrahedron Lett., 2763 (1986).
49. E. Baciocchi, R. Ruzziconi, unpublished work.
50. T. Sugiyama, Bull.Chem.Soc.Jpn. 54, 2847 (1981).
51. H. Kottmann, J. Skarzewski, F.Effenberger, Synthesis, 797 (1987).
52. E. Baciocchi, T. Del Giacco, S.M. Murgia, G.V. Sebastiani, Tetrahedron, submitted for pubblication.
53. (a) E.J. Corey, A.K. Ghosh, Tetrahedron Lett., 175 (1987); (b) R. Mohan, S.A. Kates, M.A. Dombroski, B.B. Snider, Tetrahedron Lett., 845 (1987); (c) B.B. Snider, R. Mohan, S.A. Kates, Tetrahedron Lett., 841 (1987); (d) W.E. Fristad, J.R. Peterson, A.B. Ernst, J.Org.Chem., 50, 3143 (1985); (e) W. Fristad, S. Hershberger, J.Org.Chem. 50, 1026 (1985); (f) E.J.Corey, A. W. Gross, Tetrahedron Lett., 4291 (1985); (g) E.J. Corey, M. Kang, J.Am.Chem.Soc., 106, 5384 (1984).
54. E. Baciocchi, R. Ruzziconi, J.Org.Chem. 51, 1645 (1986).
55. E. Baciocchi, R. Ruzziconi, Gazz.Chim.Ital., 116, 671 (1986).
56. E. Baciocchi, G. Civitarese, R. Ruzziconi, Tetrahedron Lett., 5357 (1987).
57. R.M. Dessau, E.I. Heiba, J.Org.Chem., 39, 3457 (1974).
58. E. Baciocchi, R. Ruzziconi, unpublished work.
59. E. Baciocchi, R. Ruzziconi, Synthetic Commun., in press.
60. M. Miyashita, T. Kumazowa, A. Yoshikoshi, J.Chem.Soc.Chem.Commun., 362 (1978).
61. M. G. Vinogradov, M.S. Pogosyan, A.Ya. Shteinshneider, G.I. Nikishin, Izv. Ak.Nauk. SSSR,Ser.Khim 2077 (1981).
62. F.A. Cotton, G.Wilkinson "Advanced Inorganic Chemistry", Interscience, New York, 1966 pp 1061-1064.
63. M. Mendelsohn, E.M. Arnett, H. Freiser, J.Phys.Chem., 64, 600 (1960).
64. S. Venkatakrishnan, M. Santappa, Z.Phys.Chem. (Munich), 16, 73 (1958).

65. J.K. Kochi in "Free Radicals"; J.K. Kochi Ed.; Wiley-Interscience: New York, 1973, Vol. I, Chapter 11, p. 656.
66. E. Baciocchi, R. Ruzziconi, unpublished work.

USE OF HOMOLYTIC REDOX PROCESSES IN ORGANIC SYNTHESES

A. Citterio
Dipartimento di Chimica del Politecnico
P.za L. da Vinci 32
20133 Milano, Italy

R. Santi
Donegani Institute, Montedison
Via Fauser, 4
28100 Novara, Italy

ABSTRACT. The design of new radical stoichiometric redox reactions useful in the synthesis of organic compounds is discussed taking examples of inter and intramolecular oxidative alkylation of aromatics and reductive alkylation of olefins promoted by metal salts. The analysis is focoused mainly on the reactivity of ∝-carbonylalkyl radicals generated by Mn(III)acetate oxidation of carbonyl compounds or by addition of nucleophilic carbon free radicals to α-enones. Efficient propagations and terminations results in good yield and selectivity and increase also the efficiency of the initiation step.

In recent years much attention has been paid to basic and applied research on organic free radical chemistry. The increase of knowledge on structure-reactivity and reactivity-selectivity relationship and thermodynamic parameters was helpful in the design many new radical reactions showing high chemo-, stereo- and regioselectivity which are useful for the synthesis of speciality chemicals, even the more structurally complicated ones[1] . The investigation has been mainly concentrated on radical chain processes, identifying the structural requirements for fast and selective propagation steps (i.e. addition to π-systems, fragmentations, atom abstractions, etc.). In these processes the generation of carbon-centered radicals at a specific site of the organic substrate is intrinsically determined in the propagation sequence of the chain, whereas the initiation step, which occurs by thermal, photochemical and redox activation of an extraneous radical initiator, and the termination

F. Minisci (ed.), Free Radicals in Synthesis and Biology, 187–212.
© 1989 by Kluwer Academic Publishers.

188

step contribute only marginally to the reaction products.[2]
The recent attention placed on electron-trasfer processes[2] has emphasized the importance and utility of redox-chain processes in organic synthesis (i.e. $S_{NR}1$, photochemical, electrochemical and metal catalyzed oxidations and reductions) and renewed the interest for thermal stoichiometric processes.

This paper has as its aim the identification of some metodologies of preparative value in the field of oxidative and reductive homolytic stoichiometric processes, which involve addition steps of carbon radicals to aromatic and olefins.

Oxidative and reductive stoichiometric processes (fig. 1) are in principle conceptually simple; they imply a two electron oxidation or reduction of an organic substrate (RH or R_{ox}, respectively) by a metal species considered to have a monoelectronic change of oxidation number.

Fig. 1.

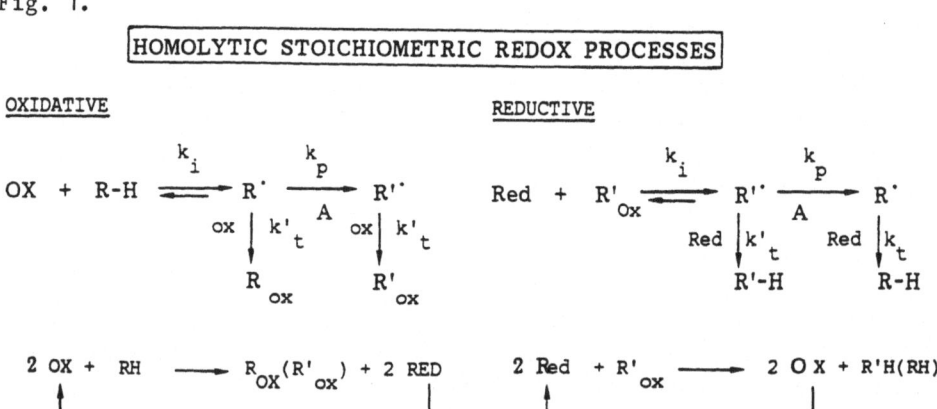

The first electron is trasferred in the initiation step to give radicals and the second in the termination step without regeneration of an active metal species. The propagation step can be completely absent or can involve insertion or elimination of a structural fragment, (A), giving new radical species which are selectively oxidized or reduced, and preventing the propagation cycle of a radical chain or catalytic process. Therefore a stoichiometric amounts of metals must be used and the process is autoinhibited. From a synthetic point of view, the onus of the formation of metal byproducts can be abated combining the chemical reaction with an electrochemical regeneration of the oxidized or reduced metal species in an indirect electrosynthetic process[3].

The overall selectivity of these redox reactions depends primarily on the initiation and termination steps, the propagation steps contribute also, when present. Therefore, stoichiometric redox processes, in order to have synthetic interest, must present fast and selective oxido-reductive termination reactions and efficient and selective initiation. In addition, the basic principle involved in free radical chain processes that the propagation step must be fast and usually exothermic in order to compete with fast radical dimerization/disproportionation reactions, must also be followed.

We will take in consideration in this paper the synthetic possibilities offered by an appropriate choice of efficient propagations by addition to aromatics and olefins and efficient terminations by oxidation or reductions of the radical intermediates using two examples of typical radical redox generation processes : 1) the oxidation of carbonyl compounds by Mn(III) acetate and 2) the reduction of hydroperoxides by Ti(III) and Fe(II) salts. The study takes advantage of the strong influence that the substituents in position α to the radical centre exert on the redox properties of carbon-centered radicals[2a,4] (fig. 2). Radicals α-substituted with electron-withdrawing groups have a high electron affinity, which increases when the number of electron-withdrawing groups increased; therefore they are at the same time good oxidants in electron-transfer processes and good electrophiles in bond making/breaking processes. On the other hand, radicals substituted with electron-releasing substituents present low ionization potentials and are strongly reducing and nucleophilic species. The nucleophilic and electrophilic properties of radicals are related to the oxido-reductive stabilizing interaction in the transition state of a bond making/breaking radical process and to the SOMO-LUMO or SOMO-HOMO frontier orbital interaction (fig. 3).

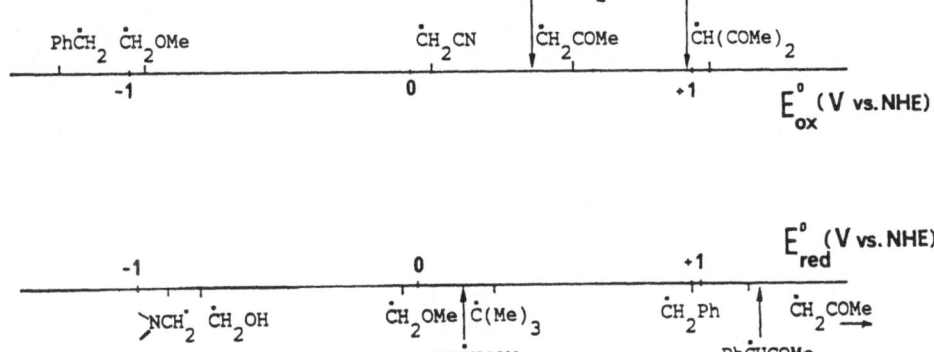

Fig. 2 Oxidation and reduction potentials of some representative carbon centered free radicals[2a,4].

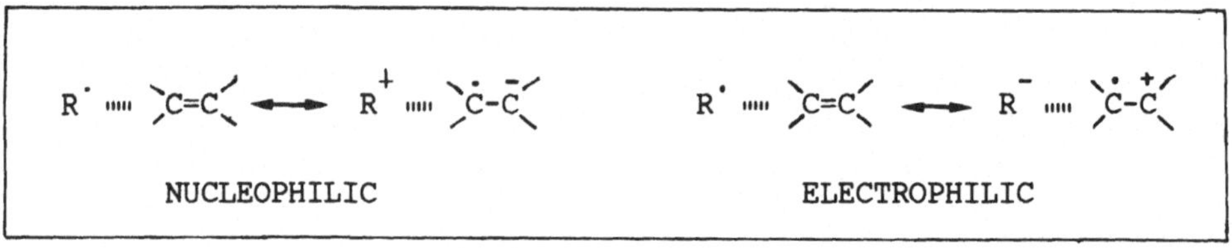

$$R^{\cdot} \cdots \; \diagdown C=C \diagup \longleftrightarrow R^{+} \cdots \; \diagdown \overset{\cdot}{C}-\overset{-}{C} \diagup \qquad R^{\cdot} \cdots \; \diagdown C=C \diagup \longleftrightarrow R^{-} \cdots \; \diagdown \overset{\cdot}{C}-\overset{+}{C} \diagup$$

NUCLEOPHILIC   ELECTROPHILIC

Fig. 3 Transition state determining the nucleophilic and electrophilic behaviour of radicals in addition to π-systems.

These properties can be usefully exploited in obtaining products of "umpolung" of the ionic reactivity of an organic substrate[1b,5].

1. OXIDATION OF CARBONYL COMPOUNDS BY Mn(III)ACETATE.

Mn(III)acetate is a well known oxidant of a variety of electron-rich organic substrates[6]. In particalur, it promotes the oxidative deprotonation of nitroalkanes, ketones, acids, etc., generating carbon-free radicals α-substituted with electron-withdrawing groups(eq. 1)[6]. The more extensive use of this initiation process was done in combination with addition processes to olefins (eq. 4) to give a large variety of intersting polyfunctionalized products[7]. Only a few reports are known on the oxidation of radicals 2 to α-acetoxy or α-halo derivatives (eq. 3)[8] and of addition to aromatics (eq. 5)[9], whereas hydrogen abstraction from C-H bonds (eq. 6) was seldom observed.

$$\text{Mn(III)} \; + \; \underset{R}{-\overset{|}{C}-H} \; \longrightarrow \text{Mn(II)} \; + \; H^{+} \; + \; \underset{R}{-\overset{|}{C}{}^{\cdot}} \tag{1}$$

$$\underline{1} \qquad\qquad\qquad\qquad \underline{2}$$

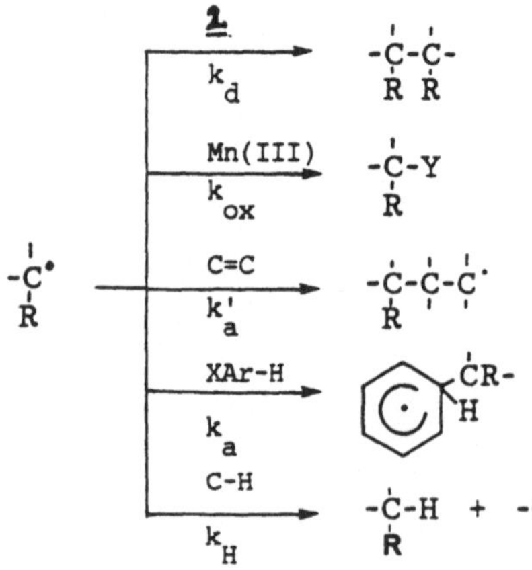

$$\underline{\mathbf{2}} \xrightarrow{\;k_d\;} \underset{R\;\;R}{-\overset{|}{C}-\overset{|}{C}-} \tag{2}$$

$$\xrightarrow[k_{ox}]{\text{Mn(III)}} \underset{R}{-\overset{|}{C}-Y} \tag{3}$$

$$\underset{R}{-\overset{|}{C}{}^{\cdot}} \xrightarrow[k'_a]{C=C} \underset{R}{-\overset{|}{C}-\overset{|}{C}-\overset{|}{C}{}^{\cdot}} \tag{4}$$

$$\xrightarrow[k_a]{XAr-H} \; \text{aromatic radical} \tag{5}$$

$$\xrightarrow[k_H]{C-H} \underset{R}{-\overset{|}{C}-H} \; + \; -\overset{|}{C}{}^{\cdot} \tag{6}$$

We focused our attention on the homolytic aromatic substitution process (eq. 7).

$$X\text{-}ArH + 2 \ MnL_3 + R\text{-}\underset{\underset{R''}{|}}{\overset{\overset{R'}{|}}{C}}\text{-}H \longrightarrow X\text{-}Ar\text{-}\underset{\underset{R''}{|}}{\overset{\overset{R'}{|}}{C}}\text{-}R + 2 \ MnL_2 + 2 \ HL \qquad (7)$$

Previous work in this field concerned simple substrates 1 (acetone[9c], nitromethane[9d] and acetic acid[9a,b]) generally used, as the aromatic substrate, in large molar excess on Mn(III)acetate. These reactions present a low molar productivity and a limited positional selectivity. We attemted to improve the potentiality of the process by using as substrates alkanes 1,1-disubstituted with electron-withdrawing groups (1, R' = or ≠ R'' = COOalkyl, COOH, CN, NO_2, SO_2R, PO(OEt)_2, etc.) and electron-rich aromatic, in order to increase the initiation efficiency and the addition rate of α-carbonylalkyl radicals to aromatics owing to the higher acidity of the carbonyl compound and the electrophilic behaviour of the radical.

1.1. Two electron oxidation of dialkyl malonates.

We initially investigated the oxidation of malonic esters with anhydrous or dihydrated Mn(III)acetate in acetic acid in the absence of any aromatic substrate. The oxidation to the corresponding α-acetoxy (and to a minor extent to the α-hydroxy) derivative is selectively observed only with dialkyl malonates substituted with strongly electron-releasing groups (i.e. OH or OR) in the α or conjugated with the α-position. α-Aryl and α-alkyl derivatives, conversely, afforded mainly products of dimerization (eq. 9) or disproportionation (eq. 10)(when the substituent was a primary or a secondary alkyl group) or oligomerization (eq. 11)(when the para position of the aromatic ring was unsubstituted (Tab. 1)[10].

$$R\text{-}CH(COOR')_2 + 2 \ Mn(OAc)_3 \longrightarrow R\text{-}\underset{OAc}{\overset{|}{C}}(COOR')_2 + 2 \ Mn(OAc)_2 + AcOH \qquad (8)$$

(R = OH, OAlkyl, 2- or 4-OR-aryl)

$$2 \ R\text{-}CX_2^\bullet \quad \underset{}{\overset{}{\left\{ \begin{array}{l} \longrightarrow X_2RC\text{-}CRX_2 \hfill (9) \\ \overset{R=R\dot{C}H}{\underset{R^\bullet}{\longrightarrow}} RCHX_2 + CR'R'' = CX_2 \overset{R\dot{C}X_2}{\longrightarrow} CRX_2\text{-}CR'R''\text{-}CX_2^\bullet \longrightarrow \hfill (10) \\ \end{array} \right.}}$$

(X = COOR)

$$\qquad (11)$$

The oxidation process is related to a contemporary increase of the reducing properties and decrease of the oxidant properties of malonyl radicals
owing to the presence of the electron-releasing α-substituent. The moderate oxidant Mn(III)acetate (E° = 1.04 V vs. NHE in acetic acid-water
9:1)[11] can therefore oxidize efficiently the radical by ligand or electron-transfer process. No oxidation of the aromatic nucleous takes place.
The α-substituents exerts a substantial influence also on the rate of
Mn(III) decay, the α-OH or OR and aryl groups increasing and the α-alkyl
groups decreasing the oxidation rate. Significantly, different reactivity and   product distribution are observed with diethyl α-2-, 3- and 4-
methoxyphenylmalonates. To exclude the involvement of a s ncronous two-
electron oxidation, the same reactions were carried out also under O_2 at
atmospheric pressure. In all cases the α-acetoxylation product was formed
in trace amounts while the α-hydroxy derivative was on the contrary obtained in high yield (eq. 12). The process becomes essentially a catalyzed autoxidation which involves trapping of malonyl radicals by oxygen
at a rate higher than the oxidation by Mn(III). The regeneration of
Mn(III) under these conditions is quite efficient and with dialkyl malonates this results in high yield of the corresponding α-ketomalonates
with high Mn(III) turnover number (eq. 13)[12].

Tab. 1 - Oxidation of α-substituted diethyl malonates by anhydrous
    Mn(III)acetate in acetic acid.

$$R-\underset{X}{C}(COOEt)_2$$

R	X=OAc	X=OH	Dimers	Oligomers	Rel. rate[e]
H	3(30)[a]	3 (80)[*]	40	3[b]	1
Me	15	4 (89)[*]	24	34[b]	0.8
i-Pr	38	- (78)[*]	20	11[b]	0.2
Ph	10	2 (80)[*]	2	80[c]	6.3
4-Cl-Ph	4	-	90	-	6.6
3-OMe-Ph	8	3 (n.d.)	16	50[c]	7.3
4-OMe-Ph	88	4 (80)[*]	-	-	16
2-OMe-Ph	71	22 (76)[*]	-	-	8.7
1-[4OMe-Naph]	93	- (81)[*]	-	-	17.1
OH	85[a]	- (88)[a*]	-	-	4.2
OMe	90[a]	- (86)[a*]	-	-	3.1

a) Diethyl ketomalonate ; b)Unsymmetrical dimers or telomers
c) Products of benzylic C-para C coupling ; e) Competitive exp.
*) Reactions carried out under oxygen at 100°C

$$R\text{-}\underset{H}{\overset{}{C}}X_2 \;+\; 1/2\; O_2 \;\xrightarrow[\text{Cu(II)}/100°C]{\text{Mn(OAc)}_3}\; 2\; R\text{-}\underset{OH}{\overset{X}{C}}\text{-}X \qquad (12)$$

$$70\text{-}90\%$$

$$CH_2X_2 \;+\; O_2 \;\xrightarrow[110°C]{\text{Mn(OAc)}_3 \;\; 1\%}\; CX_2{=}O \;+\; H_2O \qquad (13)$$

$$90\text{-}95\%$$

$$2\; CH_2X_2 \;+\; O_2 \;\xrightarrow[\text{AcONa, Ac}_2O]{\text{Mn(OAc)}_3/100°C}\; X_2C{=}CX_2 \;+\; 2\; H_2O \qquad (14)$$

$$85\text{-}90\%$$

$$[X = COOMe,\; COOEt,\; COOBu^n]$$

In the presence of acetic anhydride and sodium acetate, the oxidation of dialkyl malonates with oxygen at 100°C, in the presence of Mn(III), results in oxidative dehydrodimerization to tetraalkyl ethylenetetra-carboxylate in 90% yield (eq. 14)[13]. Moreover, the oxidation of α-alkyl α-aryl and unsubstituted dialkyl malonates by Mn(III)acetate in the presence of stoichiometric amounts of alkaline chlorides, bromides, thiocyanates and azides afford the corresponding α-halo or pseudohalo derivative[10] (eq. 15) in high yield. The reaction is general for carbonyl compounds. Conditions to obtain high yield of α-azido and α-thiocyano-α-arylmalonates were specifically investigated and found generally to require the use of Mn(III) in excess owing to the contempo-rary formation of nitrogen and thiocyanogen, respectively.

$$RCH(COOR')_2 \;+\; 2\; Mn(OAc)_3 \;+\; NaX \;\xrightarrow{AcOH}\; R\text{-}CX(COOR')_2 \;+\; 2\; Mn(OAc)_2 \qquad (15)$$

$$X = Cl^-,\; Br^-,\; SCN^-,\; N_3^- \qquad\qquad R = Alkyl,\; Aryl\,(70\text{-}90\%)$$

$$R\overset{\cdot}{C}(COOR')_2 \;+\; MnL_{n-1}^{\prime\prime\prime}X \;\xrightarrow{k_{ox}}\; RCX(COOR')_2 \;+\; Mn^{\prime\prime}L_{n-1}$$

$$(k_{ox} > 10^6\; M^{-1}s^{-1},\; X = SCN,\; N_3\; ;\; k_{ox} \geqslant 10^8\; M^{-1}s^{-1},\; X = Br)$$

The specificity of the process recalls the fast ligand transfer oxidation of carbon radicals by Copper(II) halides and pseudohalides[14] and is cha-racterized by high rate constants. The peculiar behaviour of the ortho and para alkoxy substituted α-arylmalonates can also be observed in this case, since the primary substitution product formed at low temperatures undergoes the substitution of benzylic Br, Cl groups with acetate ion at higher temperatures and internal rearrangement to isothiocyanate in the case of benzylic thiocyanate. Stabilization of carbocations by methoxy and hydroxy groups is certainly important in these substitutions[15] as in the oxidation process, despite the presence of electron-withdrawing alkoxycarbonyl group[16]. In the presence of halide ions Mn(III) becomes a more efficient oxidant and, for instance, esters of arylacetic acids, which are unreactive towards Mn(III)acetate at 70°C, are efficiently α-fuctionalized in high yield[17]. The process is presumed to occur through benzylic hydrogen abstraction by halide radicals. Remarkable is, however, the lower reactivity of the primary oxidation product to further oxidation, which allows for good selectivity (80-88%) at relatively high conversions (70%).

1.2. Intermolecular aromatic substitution by Mn(III) oxidation of alkanes 1,1-disubstituted with electron-withdrawing groups.[18]

The aromatic substitution process was investigated using diethyl methyl-malonate as reference substrate. Table 2 summarizes typical yield and distribution of substitution products obtained with Mn(III) in acetic acid. The alkylation process is observed with aromatics having an ionization potential in the range 7.7 - 8.5 eV. Electron-rich aromatics (i.e. an-

Tab. 2 - Mn(III)acetate induced oxidative substitution of aromatics by diethyl methylmalonate (AcOH, 80°C, N_2, [NaOAc] = 2 M)

$$ArH + CH_3CH(COOEt)_2 + 2\ Mn(OAc)_3 \longrightarrow Ar-\underset{\underset{CH_3}{|}}{C}(COOEt)_2$$

ArH	I.P.	time(h)	Conv.(%)	Yield(%)[b]	Isomer distribution
Antracene	7.43	1	88	- (-)	-
2-OMe-Naph.	7.82	4	61	52 (85)	1(83), 8(13), al.(4)
1-OMe-Naph.	7.70	4	51	43 (84)	4(>95)
Naphthalene	8.09	4	20	19 (93)	1(91), 2(9)
Phenantrene	8.1	8	15	10 (88)	42/8/10/33/7[a]
Anisole (10)	8.39	4	10	15 (80)	2(60), 4(40)
1,4-diOMe-Ph		4	16	12 (92)	2(100)
1,2-diOMe-Ph		4	20	20 (89)	4(96), 3(4)
Benzene (10)	9.25	8	n.d.	<1	-

a) Five isomers (glc ratio) b)based on Mn(III) (based on converted aromatic

thracene) afford only nuclear acetoxylation via an aromatic radical cation
(eq.18)[19]. The process is competitive with the aromatic alkylation with
1- and 2-methoxynaphthalenes.

$$Ar-H \quad + \quad 2 \; Mn(OAc)_3 \quad \longrightarrow \quad Ar-OAc \quad + \; 2 \; Mn(OAc)_2 + AcOH \quad (18)$$

Aromatic having lower electron density are not directly oxidized, but
the yields of the alkylation process (eq. 17) remain moderate and were
found to decrease with the increase of the I.P. of the aromatic compound.
The yield based on converted aromatic are generally high and the oxidation
of the malonate affords mainly dimerization and oligomerization products.
The substitution is quite selective, approaching to typical ionic electrophilic processes. For 2-methoxynaphthalene and phenantrene a lower
selectivity is observed. The result is more related to the steric hind-
rance in the approaching of the bulky tertiary radical to the hindered
position of higher electron density and to a higher reversibility of the
addition than to a real unselectivity of the reaction

Fig 4.Positional selectivity of α-carbonylalkyl radicals in the substitution
of methoxynaphthalenes and anisole.

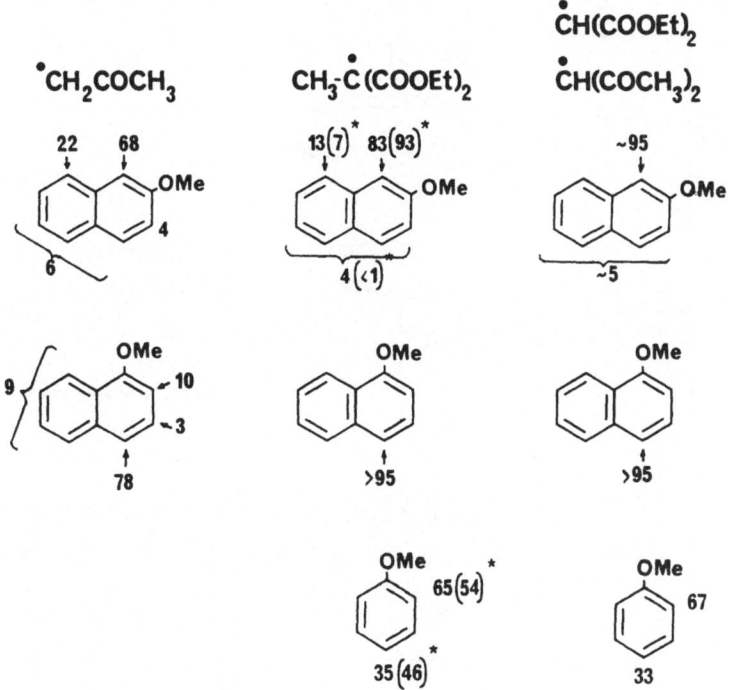

* in the presence of NaOAc (2 M)

This is the consequence of the reversibility of the addition of malonyl radicals to the aromatic ring and of the increased discriminating abili ty of the oxidant towards intermediate cyclohexadienyl radicals[20], owing to the decrease of the redox potential of Mn(III) in the presence of acetate ions[11]. The effect of α-substituents on the alkylation process of 1-methoxynaphthalene is also significant. α-Arylmalonates are oxidized with formation of dimerization and oligomerization or oxidation products without aromatic substitution. The malonylation process is conversely observed with α-alkylmalonates in yield based on Mn(III) progressively decreasing when the steric demand of the alkyl group increase (Me (43%), Et(30%), n-Bu(23%), i-Pr(3%)). The sequence is related to a less effi- cient radical generation since the aromatic acetoxylation product is formed in increasingly higher yield (5, 13, 15 and 32 %, respectively). With diethyl malonate the product of substitution in position 4 of 1- methoxynaphthalene and further benzylic acetoxylation was selectively obtained in high yield (eq. 19). This behaviour is generally observed with other alkoxy substituted aromatics and is the result of the combined high reactivity of ortho and para positions to the alkoxy group of the aromatic towards addition of electrophilic malonyl radicals and selec- tive fast α-acetoxylation of the resulting primary addition product.

The high selectivity for the addition to the position 1 of 2-methoxy-
naphthalene (eq. 20) further supports the hypothesis that steric effects
are responsible for the lower selectivity observed with the same substra
te and diethyl methylmalonate. Dimethyl malonate, on the contrary, presents a more complex product distribution (eq. 21); the addition occurs
again selectively in position 1 but the further oxidation involves also
the ester and the alkoxy group.

Dialkyl malonates does not give isolable complexes with Mn(III), however
the increase of the substitution over direct aromatic acetoxylation in
the presence of acetate ions and the influence of the alkyl substituents
suggest that the initiation step involves α-deprotonation and Mn(III)
complexes. Therefore, we studied the behaviour of easily isolable and
relatively thermally stable Mn(III)acetylacetonate and Mn(III)acetoace-
tate neutral complexes in the presence of electron-rich aromatic compounds.
Mn(AcAc)$_3$ is a well known radical initiator of polymerization of vinyl
monomers in bulk or in solution[21]. Both compounds are found to give the
substitution process on 1 and 2-methoxynaphthalenes selectively at position 4 and 1, respectively, in yield close to those observed in the
malonylation (eq. 22). The main difference is the lower amounts of the
α-acetoxyarylmalonate obtained; however, the primary substitution products exist completely in enolic form and the corresponding Mn(II) complexes are quite stable. Moreover, by using appropriate concentration of
alkaline halide or pseudohalide, these complexes afford good yield of
products of aromatic and benzylic substitution (i.e. eq. 23).

$$R = H \ (41\%); \ OAc \ (5\%)(conv.52\%) \tag{22}$$

$$(40\%) \ (conv. \ 43\%) \tag{23}$$

If we compare the positional selectivity in the aromatic substitution
by acetonyl and α-diacetylmethyl radicals of 1- and 2-methoxynaphthalenes
(fig. 4), the first appear less selective in accord with their lower

electrophilic properties and lower reversibility in the addition to the aromatic ring.

Emiesters of malonic acid or malonic acid itself can also be used in these aromatic functionalization processes. The oxidationoccurs at lower temperature (20-40°C) and involves the α-hydrogens as suggested for reactions with olefins. Emiesters of α-arylmalonic acid are in fact isolated in low yield and are considered the precursors of the products observed : the esters of α-arylalkanoic acid (by thermal decomposition) and α-acetoxy-α-arylalkanoic acid (by oxidative decarboxylation) (eq. 24 and 25, respectively). The first process is generally preferred and this allows to obtain a new synthesis of some antiinflammatory agents α-arylpropionic acids. Thiaprofenic acid ($\underline{3}$) for instance can be obtained in 51% yield with high positional selectivity (93%).

$$Ar-CH(CH_3)COOEt \qquad (24)$$

$$CH_3CH-COOEt \xrightarrow[Mn(III)]{ArH} Ar-\overset{CH_3}{\underset{COOH}{\overset{|}{C}}}-COOEt$$

$$\underset{OAc}{Ar\overset{|}{C}(CH_3)COOEt} \qquad (25)$$

$$51\% \text{ (80\% select.) + 2\% of 3-isomer}$$

Oxidative decarboxylation is more commonly found with malonic acid. Aromatic carboxylic acids can be so obtained in moderate yield but with good selectivity (i.e. eq. 27, where 8 equivalent of Mn(III) was used).

$$55\% \qquad (27)$$

1.3. Intramolecular homolytic aromatic substitution promoted by Mn(III).

An intramolecular process which involves the formation of 5 or 6 membered ring is entropically favoured over the intramolecular analog and an increase of the rate constant of 100 at room temperature is common.[22] Intramolecular aromatic substitution to give cyclic compounds can therefore be expected to be more efficient than the analogous intermolecular processes. Substrates of general formula $\underline{4}$ (R_1 = R_2 = COOEt) with different substituents on the aromatic ring and different chain were investigated in order to determine the synthetic potentiality of eq. 28.

$$\text{(structure)} \underset{\underline{4}}{\overset{}{\underset{CHR_1R_2}{\bigcirc}}}^A + 2\,Mn(OAc)_3 \longrightarrow \text{(structure)}^A\underset{R_1\ R_2}{C} + 2\,Mn(OAc)_2 + 2\,AcOH \quad (28)$$

The effect of the chain length of the residue A is summarized in table 3, as concerns the yield of cyclization products and the initial and competitive rates of oxidation (in acetic acid at 80°C). The rates were found to depend slightly on the structure of the substrate, the discrimination being more pronounced in competitive experiments. The cyclization process is very selective for compounds $\underline{4c}$, $\underline{4e}$ and $\underline{4f}$, but compete with the dimerization of the substrate at the malonic position with compounds $\underline{4b}$ and $\underline{4g}$, and is negligible for compounds $\underline{4a}$ and $\underline{4d}$.

Tab. 3 - Rate constant and cyclization yield for some α-phenylalkyl and α-ω-phenoxyalkylmalonate (X = COOEt).

	4a	4b	4c	4d
Cycl.(yield %)	0	38	91	< 1
$k_{80°C}$ (rel. rate)	(0.11)	1.0×10^{-3} (0.6)	1.8×10^{-3} (1)	3.1×10^{-3} (6)

	4e	4f	4g
Cycl. (yield %)	86	93	19
$k_{80°C}$ (rel. rate)	2.1×10^{-3} (-)	4.6×10^{-3} (4.1)	8.2×10^{-4} (-)

The results indicate a general trend in the cyclization process with the $Ar_26 > Ar_25 > Ar_27 > Ar_24$ and ω-phenoxyalkyl derivatives more reacti-ve than the corresponding phenylalkyl derivatives. The influence of substituents present in the aromatic ring or in the chain on the yield of some Ar_25 and Ar_26 cyclizations is summarized in tables 4 and 5, res-pectively.

Tab. **4** - Mn(III)acetate induced Ar_25 cyclization of diethyl 2-arylethyl-malonates (0.2 M) [AcOH, 70°C, 7h, /Mn(III)/= 0.4 M]

Y—⟨ring⟩—CH₂CH(COOEt)₂

| Y | Conv.(%) | Products (Yield %) | | Isomer distr. |
		Symm.dimer	Cycliz. Pr.	(%)
4-NO₂	80	30	2	-
4-NHCOMe	88	18	32	-
H	90	20	38	-
3-OMe	95	-	90	2(52), 6(48)[b]
3,4-diOMe	70	-	42(67)[a]	2(8), 6(92)[b]
3,4-(CH)₄-	98	-	93	1(96), 3(4)[b]

a)Product of cyclization and benzylic acetoxylation (25% yield)
b) Position of attack in the starting aromatic

The positive effect of the increase of the electron density of the aro-matic ring on the yield of the intramolecular substitution parallel and expand the potentiality of the intermolecular substitution. Here, in fact, also simple phenyl derivatives can undergo the addition of malo-nyl radicals and, almost in the case of Ar_26 cyclization, electron poor aromatic are sufficiently reactive. The effect of substituents is more discriminant with α-arylethyl than α-arylpropylmalonates. The higher reactivity towards Ar_26 than Ar_25 cyclization is also evident from the high yield of tetrahydronaphthalene derivative obtained in the oxida-tion of compound **8**. Significantly, the nuclear acetoxylation was never observed, also with the more electron rich aromatic derivatives investigated (i.e. diethyl α-2-naphthoxyethylmalonate). The presence of p-alkoxy substituents on the aromatic ring favours the further oxida-tion of the cyclic product to the corresponding benzylic acetoxylation product (i.e. eq. 29)[18,23]. The involvement in this process of 4 equi-

Tab. 5 - Mn(III)acetate induced $Ar_2 6$ cyclization of diethyl arylalkylmalonates
(AcOH, 70°C, 4h (except for the nitro derivative (12h), /Mn(III)/=0.4 M

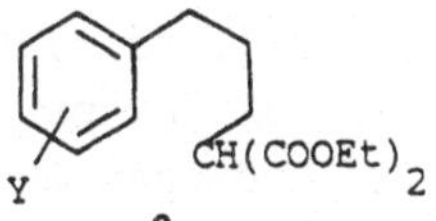

6

Y	Yield(%)
H	85
4-OMe	75
3,4-diOMe	$46^{a}(88)^{b}$
4-NHCOMe	88
4-NO$_2$	60
3-OMe	$86^{c}(95)^{b}$

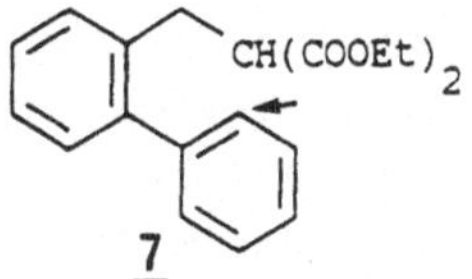

7
92%

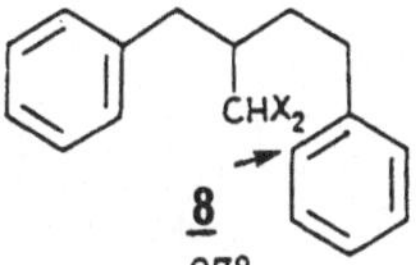

8
87%

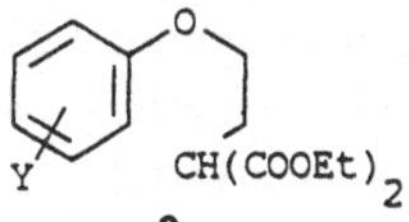

9

Y	Yield(%)
H	90
2,3-(CH)$\frac{}{4}$	91^{d}
3,4-(CH)$\frac{}{4}$	90^{e}

a) o/p ratio =0.09, b) Benzylic acetoxylation product, c)o/p ratio = 1.22
d) Addition to position 2; e) Addition to position 1

valents of Mn(III) explains the fast decay of Mn(III) observed with these substrates.

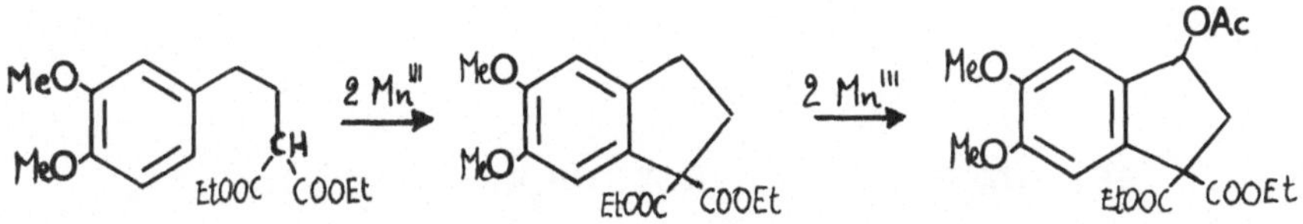

(29)

The clean reactions obtained with compounds 6 and 9, allows to obtain accurate kinetic measurements. The reactions with these substrates are first order in Mn(III) and aromatic substrate until 85% conversion; no retarding effect of Mn(II) can be observed, whereas sodium acetate increases slightly the rate and inhibition by oxygen is observed. Under N_2, electron releasing substituents were found to increase the oxidation rate (4-NO$_2$ < 4-NHCOCH$_3$ < H < 4-OMe < 3-OMe) and an activation enthalpy of 26 Kcal/mol and an activation entropy of 2 u.e. can be determined for compound 9 (Y = 3,4-(CH)$_4$-).
Products of ipso substitution ($Ar_{1,n}$ mode of ring closure) were observed only in trace amounts, suggesting a higher reversibility for steric reasons of the endo($k_{1,n}/k_{-1,n}$) than the exo ($k_{2,n}/k_{-2,n}$) addition to the aromatic double bond.
The yield of the $Ar_2 7$ mode of ring closure can also be increased using

more electron rich aromatic substrates starting from 1d (1%) to 1g (19%)to
diethyl α-3-[2-naphthoxy]propyl malonate (61%).
Mn(III) oxidation of compounds 1a and 1d does not result in the cyclization reaction; the first gives dimers (mainly at the malonic position
and in minor amounts mixed benzylic-malonic dimer) (eq. 31), the latter
affords the benzylic acetoxylation product (eq. 30). The different
result is related to the different rate of 1,5 benzylic hydrogen abstraction, due to the different energy of the two bonds involved, primary for
1a and secondary for 1d (78 and 74 Kcal/mol, respectively)[24]. The energy
of a tertiary malonic C-H bond can be estimated from this result to be
approximately 75-76 Kcal/mol.

(30)

83%

(31)

RCX$_2$-CX$_2$R   RCX$_2$-CH$_2$Ar

85%   3%

Surprisingly, all these cyclization reactions cannot be observed using
some typical sources of carbon free radicals (i.e. dibenzoyl peroxide
or percarbonates) under the conditions found useful for the cyclization on olefinic double bonds[25]. However, sources of nucleophilic carbon
free radicals (i.e. α-oxyhydroperoxides of carbonyl compounds and dialkyl
peroxides) can be usefully applied. The results confirm the strong polar
effects in hydrogen abstraction reaction from arylalkyl malonates[26].
The intramolecular Ar$_2$5 and Ar$_2$6 substitutions are general processes
compatible with the presence of alkyl, aryl, alkoxy and amido substituents on the alkyl chain and on the aromatic ring; however, they fail
with alkylthio and amino substituents.
Replacement of the two alkoxycarbonyl groups by other electron-withdrawing substituents (carbonyl, cyano, nitro, sulphonyl, etc.) results in

efficient cyclization to six membered ring products. The oxidation of these substrates occurs at lower temperature and several labile substituent resist to these mild conditions. Substituted β-ketoesters are the substrates which behave more similarly to dialkyl malonates. In a study on the oxidation of substituted alkyl 5-aryl-3-oxopentanoates, Mn(III) acetate was found to give the corresponding four-electron oxidation product 11 efficiently when elctron-releasing groups on the meta position of the aromatic ring were present, and complex reaction mixtures in the presence of less electron-releasing groups (eq. 32, table 6). The more oxidant Ce(IV) ammonium nitrate (CAN) can be used in these cases in acetic acid:water or in methanol at 10-30°C, affording efficiently the compound 11 (X = OMe, ONO$_2$, OH, depending from the experimental conditions used). All compounds 11 can be rearomatized to the corresponding 2-hydroxynaphthoic acid 12 by refluxing a benzene solution in the presence of silica gel (Tab 6). The method was used for the synthesis of the aromatic chromophore of the antitumor antibiotic neocarzinostatin 12 (Y = 5-methyl, 7-methoxy). The same product was previously synthetized in 11 steps from 2-methoxy-6-methylbenzanilide in 35% overall yield [27]. Similar reactions carried out on substituted

Tab.6 - Synthesis of 2-hydroxy-1-naphthoic acid derivatives by oxidative cyclization of 5-aryl-3-oxo-pentanoic acid derivatives.

$$(32)$$

Metal	Solv.	Y	R$_1$	R$_2$	R	B Yield% (X)	C (Yield %)
Mn(III)	AcOH	H	H	H	Et	30 (OAc)	24
CAN	AcOH-H$_2$O	H	H	H	Et	3(OAc);41(OH) 30(ONO$_2$)	62
Mn(III)	AcOH	3-OMe	Me	H	Me	62(OAc)	69
CAN	AcOH-H$_2$O	"	"	"	"	45(OH)	41
CAN	MeOH	2,5-Me	H	H	Et	18(OMe),20(ONO$_2$)	39
Mn(III)	AcOH	2-Me,4-OMe	H	H	Me	59(OAc)	64
Mn(III)	AcOH	2-OMe,4-Me	H	H	Me	n.d.	55
CAN	MeOH	4-F	Me	H	Me	36(OMe),41(ONO$_2$)	42

acetanilides 13 resulted after hydrolisis and aromatization in the synthesis of quinoline derivatives 15 in 30_60% yield (eq. 33).

$$(33)$$

A further example of intramolecular aromatic substitution of carbonyl derivatives is the one-step synthesis of 3-acylbenzofuranes 18 by four electron oxidation of β-aryloxyethyl carbonyl compounds 16 (eq. 34, tab. 7). The low yield with R_1 = Alkyl and Y = H, Halogen, aryl, can be increased in the presence of electron-releasing methoxy groups. Under optimum conditions the benzofurane 18 (Y = H, R_2 = Et, R_1 = 4-OMe-Ph) was obtained in 49% yield. This compound can be converted in high yield by known procedures to the uricosoric agent benzbromarone.[28]

Tab. 7 - Synthesis of 3-acylbenzofuranes by Mn(III)acetate oxidation of 2-aryloxyethylketones.

$$(33)$$

Y	R_1	R_2	Conv.(%)	Yield (%)
H	Me	H	100	24
H	Me	H	50	17
H	Ph	H	90	42
H	4-OMe-Ph	Et	83	49*
3-OMe	Me	H	100	59
4-Cl	Ph	H	86	30

2. REDUCTIVE ALKYLATION OF α-ENONES BY HYDROPEROXIDES.

Carbon-centered radicals can be obtained easily from hydroperoxides by several different routes; the more useful are the inter and intramolecular hydrogen abstractions, inter and intramolecular addition to olefins and β-fragmentation of the intermediate alkoxy radical (scheme 1).

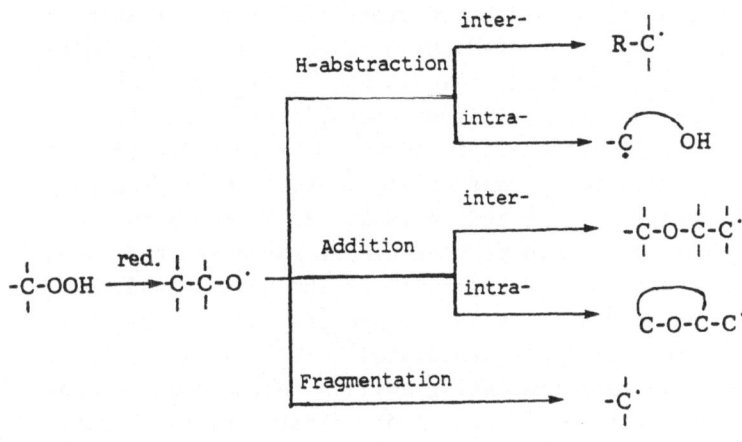

Scheme 1

All these processes are relatively fast[28] to compete efficiently with
the reduction of the alkoxy radical by reducing species to give the
corresponding alcohol under appropriate concentrations of the reductant
and the transfer agent. When the carbon radicals produced are nucleophilic, these can be trapped by olefins substituted with electron-
withdrawing groups[29] generating electrophilic carbon radicals which
can be reduced by a metal salt present in acid medium to give the corresponding hydroderivative (scheme 2). The overall process is a
reductive alkylation of the olefine[30]. The process requires for an

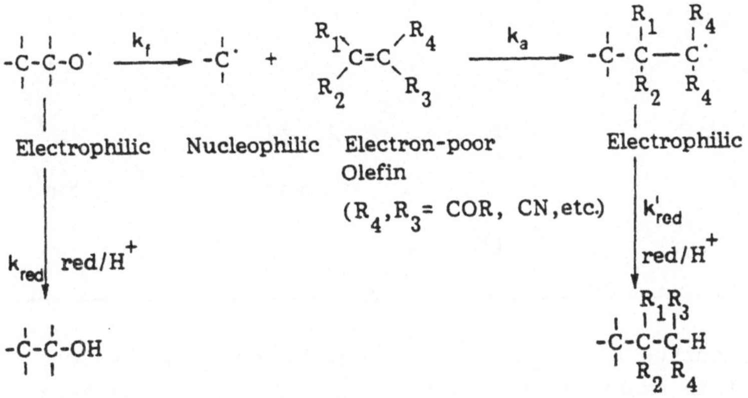

$$-\overset{|}{\underset{|}{C}}-\overset{|}{\underset{|}{C}}-O^{\cdot} \xrightarrow{k_f} -\overset{|}{\underset{|}{C}}^{\cdot} + \underset{R_2}{\overset{R_1}{\diagdown}}C=C\underset{R_3}{\overset{R_4}{\diagup}} \xrightarrow{k_a} -\overset{|}{\underset{|}{C}}-\overset{R_1}{\underset{R_2}{\overset{|}{C}}}-\overset{R_4}{\underset{R_4}{\overset{|}{C}}}{}^{\cdot}$$

Electrophilic   Nucleophilic   Electron-poor   Electrophilic

Olefin

$(R_4, R_3 = COR, CN, etc.)$

k_{red} | red/H$^+$ k'_{red} | red/H$^+$

$$-\overset{|}{\underset{|}{C}}-\overset{|}{\underset{|}{C}}-OH \qquad\qquad\qquad -\overset{|}{\underset{|}{C}}-\overset{R_1}{\underset{R_2}{\overset{|}{C}}}-\overset{R_3}{\underset{R_4}{\overset{|}{C}}}-H$$

Scheme 2

useful synthetic application, propagation steps faster than reduction
of the alkoxy radical ($k_a > k_{red}$) and reduction rates different for differently substituted carbon-centered radicals ($k'_{red} > k_{red}$). From
studies of reductive alkylation of vinyl monomers, it was possible to
estimate the absolute rate constants for reduction of substituted carbon
radicals by aqueous Ti(III), Fe(II) and Cr(II) ions, from polymerization
rate constants[31]. The data are reported in Table 18. Radicals having
electron-withdrawing α-substituents present higher rates and the rates
parallel to a some extent the redox potential of the radical and the
metal ion. The reduction by Cr(II) is a fast process, not very discimi-
nat, and results in organometallic intermediates[32]. Ti(III) and Fe(II)
show, on the contrary, high discriminating ability in the reduction of
carbon radicals and show rates of reduction of α-dicarbonyl and α-carbo-
nyl radicals synthetically useful.

Tab. 8- Rate constants for the reduction of carbon free-radicals by inorganic
ions. (H_2O - AcOH, 20°C).

Radical	k (mol $l^{-1}s^{-1}$)		
	Ti(III)	Fe(II)	Cr(II)
$-\dot{C}H-CHO$	$3.6 \ 10^5$	$7.1 \ 10^4$	$6.0 \ 10^7$
$-\dot{C}H-COMe$	$5.9 \ 10^4$	$0.9 \ 10^3$	$3.4 \ 10^7$
$-\dot{C}H-COOH$	$3.3 \ 10^3$	-	-
$-\dot{C}H-CN$	$6.5 \ 10^2$	-	-
$-\dot{C}H-CONH_2$	$3.2 \ 10^2$	-	-
$-\dot{C}H-COOMe$	$3.3 \ 10^2$	-	-
$-\dot{C}H-Ph$	10	-	-
$-\dot{C}(COOMe)_2$	$6 \ 10^6$	$1.1 \ 10^5$	$>10^7$
$-\dot{C}(COMe)_2$	$4 \ 10^6$	$4.0 \ 10^5$	$>10^7$
$-\dot{C}(CN)_2$	$4 \ 10^4$	-	-

Examples of reductive α-alkylation of ethers by methylvinylketone
(eq. 34) have been reported[33]; they requires the use of a large excess
of the ether to favour the intermolecular hydrogen abstraction step
and substrates symmetric or having quite different C-H bonds. More

selective carbon radical sources need in the synthesis of structurally complex compounds. Some possibilities are offered by the intramolecular 1,5-H abstraction in saturated hydroperoxides (i.e. eq. 35), cyclization of unsaturated hydroperoxides (eq. 37) and β-fragmentation of α-alkoxy-hydroperoxides of carbonyl compounds (eq. 37)[34].

$$Bu^tOOH + 2\ Ti^{3+} + \underset{O}{\overset{X}{\bigcirc}} + RCH=CHCOR' \overset{H^+}{\longrightarrow} \underset{O}{\overset{X}{\bigcirc}}\underset{CHRCH_2COR'}{} + 2\ Ti^{4+} + Bu^tOH \quad (34)$$

X = Bond, R=H, R'=Me (43%)

X = CH$_2$, R=H, R'= Me (51%)

X = O , R=H, R'=Me (58%)

$$R\underset{H}{-}CH(CH_2)_3OOH + 2\ Ti^{3+} + CH_2=CHCOCH_3 + H^+ \longrightarrow RCH(CH_2)_3OH \underset{CH_2CH_2COCH_3}{} + 2\ Ti^{4+} \quad (35)$$

R=H (18%) R=OMe (51%)
R=Et (40%)

$$+ 2\ Ti^{3+} + CH_2=CHCOCH_3 + H^+ \longrightarrow \quad + 2\ Ti^{4+} \quad (36)$$

CH$_2$CH$_2$COCH$_3$

(52%)

$$\underset{(CH_2)_n}{\overset{MeO\ OOH}{\bigcirc}} + CH_2=CHCOCH_3 \overset{Ti^{3+}/H^+}{\longrightarrow} MeOOC-(CH_2)_{n+2}COCH_3 \quad (37)$$

n=5 (50%)
n=6 (41%)

Several different electron-poor olefins can be used in the processes and Fe(II) can replace Ti(III) when the olefine is not easily polymerized. As observed previously, the possibility to have intramolecular processes increases the propagation steps by addition to π-systems and determines better yield and selectivity by using strictly stoichiometric amounts of reagents. An analysis of the problem of the intramolecular version of the reductive alkylation allows us to hypothesize as optimum substrates the tertiary allyl hydroperoxides. These compound can be considered in fact masked α-enones, since the intermediate alkoxy radical can fragment the

β- C-C bond giving carbon radicals   (scheme 3). Addition of the resulting to the α-enone and subsequent reduction of the α-carbonylalkyl radical by metal ions result in products of reductive rearrangement of the allyl hydroperoxide.

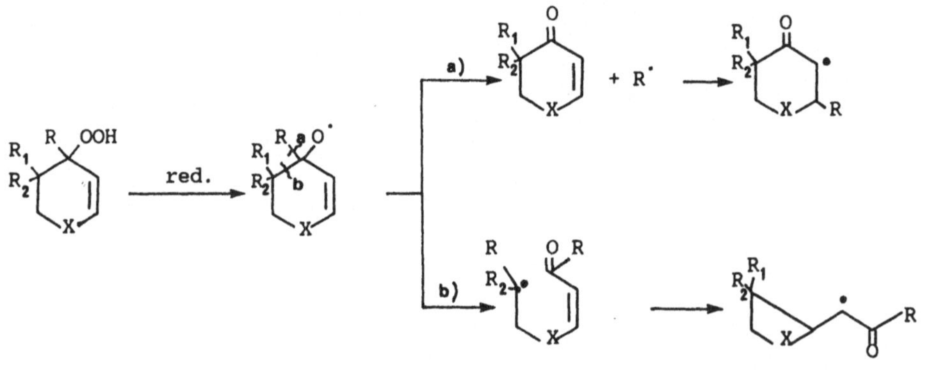

Scheme 3

The selectivity and the yield of the process are governed by the rate of fragmentation of the alkoxy radical intermediate, a reaction which is strongly influenced by steric and stereoelectronic effects. Tertiary alkyl radical can in fact be efficiently obtained and C-C bonds which are coaxial with the radical centre are generally more easily made and broken[35]. The angles α and β are determinant for the fragmentation of intermediate radical 19.

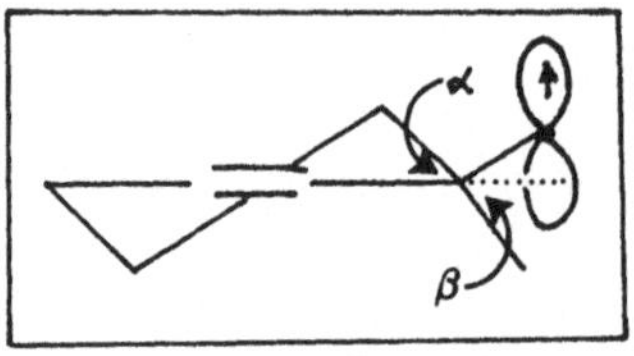

19

Some typical examples of these reductive rearrangements are reactions 38, 39, 40 and 41. Examples of application in the steroid field have been reported[36] (i.e. eq. 42 and 43).

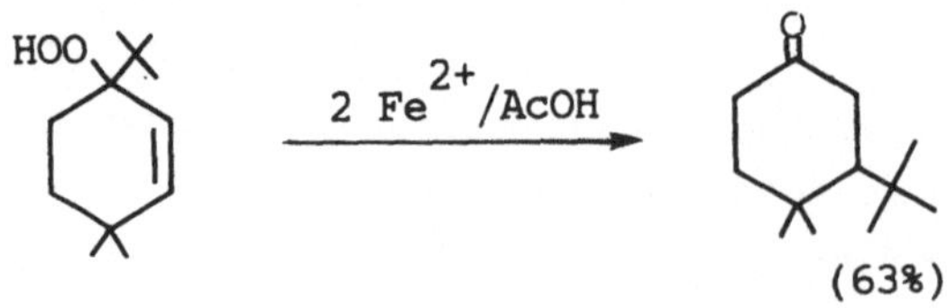

(38)

(63%)   TiCl$_3$/DMF-H$_2$O (58%)

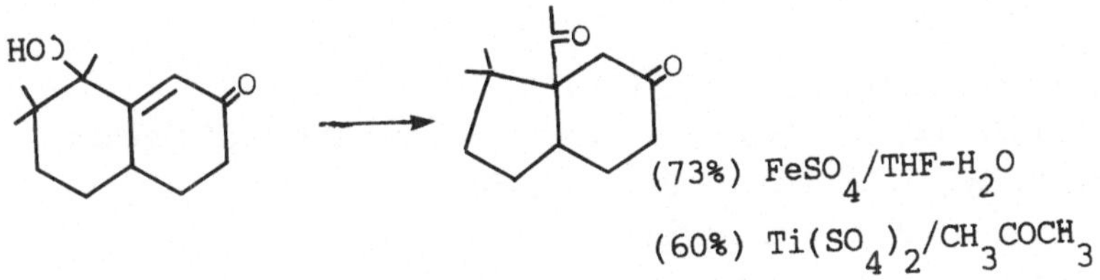
HO
(73%) FeSO4/THF-H2O
(60%) Ti(SO4)2/CH3COCH3

(39)

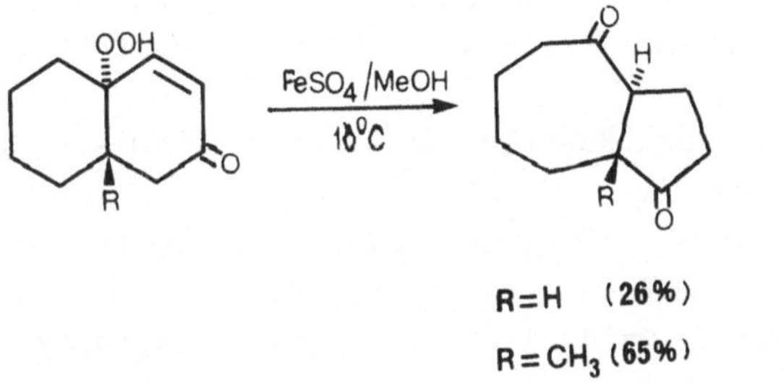
OOH
FeSO4/MeOH
10°C
R
H
R=H (26%)
R=CH3 (65%)

(40)

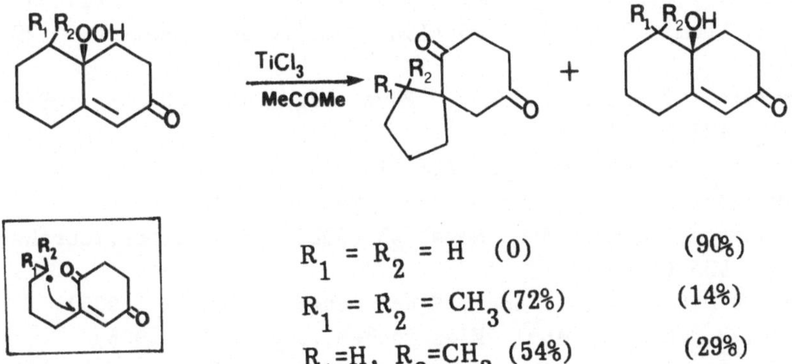
R1 R2 OOH
TiCl3
MeCOMe
R1 R2
+
R1 R2 OH
R1 = R2 = H (0) (90%)
R1 = R2 = CH3 (72%) (14%)
R1=H, R2=CH3 (54%) (29%)

(41)

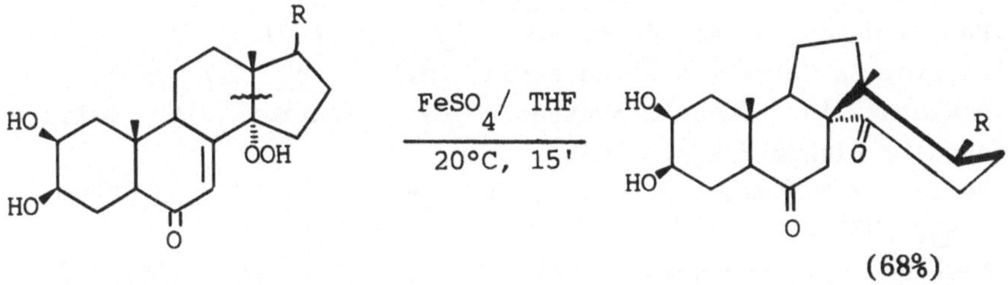
R
HO
HO
OOH
FeSO4/ THF
20°C, 15'
(68%)

(42)

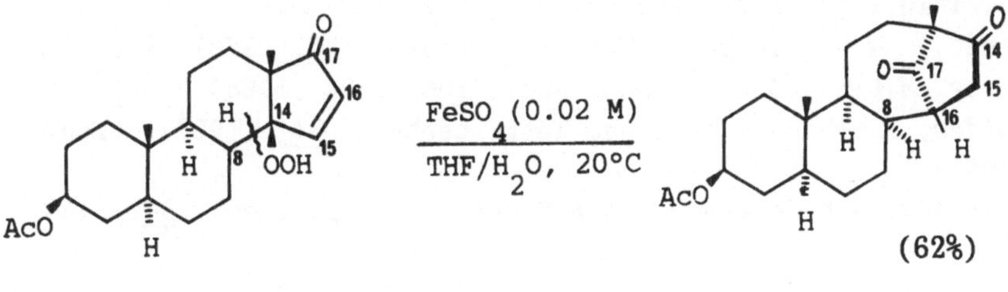
AcO
OOH
FeSO4(0.02 M)
THF/H2O, 20°C
(62%)

(43)

REFERENCES

1 a - M.Ramaiah, Tetrahedron, 43, 3541 (1987)

  b - B.Giese, "Radicals in organic synthesis: formation of Carbon-Carbon bonds", Pergamon Press, Oxford 1986

  c - D.J.Hart, Science, 223, 883 (1984)

  d - F.Minisci, A.Citterio, Advances in Free Radicals Chemistry, (G.H. Williams Ed.), Vol. 6, p. 65, Heyden, London (1980)

  e - D.J.Devies, M.J.Parrot, Free Radicals in Organic Synthesis, Springer, New YorK (1978)

2 a - L.Eberson, "Electron Transfer Reaction in Organic Chemistry", Spinger Verlag, 1987

  b - W.J.Mijs, C.R.H.I. De Jonge, "Organic synthesis by oxidation with metal compounds", Plenum Press, New York, 1986

  c - J.K.Kochi, "Metal-Catalyzed Oxidation of Organic Compounds", Academic Press, New York (1981)

  d - N.G.Connelly, V.E.Geiger, Adv. Organometal. Chem.,23, 1 (1984)

  e - M.Chanon, Chem. Rew., 83, 425 (1983); Bull. Soc. Chem. Fr., 209 (1985)

  f - O.Hanmerich, V.D.Parker, Adv. Phys. Org. Chem.,20. 55 (1984); ibid. 19, 131 (1983)

  g - M.A.Fox, Adv. Photochem., 13, 238 (1986)

  h - G.J.Kavarnos, N.J.Turro, Chem. Rev., 86, 401 (1986)

3 a - R.E.W.Jansson, Chem. Eng. News, 43 (1984), M.M.Baizer,Tetrahedron, 40, 935 (1984)

  b - H.Wendt, H.Schneider, J. Appl. Electrochem., 16, 401 (1986)

  c - G.Kreysa, H.Medin, J. Appl. Electrochem., 16, 757 ((1986)

4 a - D.D.M.Wagner, J.J.Dannenberg, D.Griller, Chem. Phys. Lett., 131, 189 (1986) and references

  b - R.G.Pearson, J. Am. Chem. Soc., 108, 6109 (1986)

  c - P.S.Rao, E.Hayon, J. Am. Chem. Soc., 96, 1287 (1974)

5 a - A.Citterio, La Chimica e L'Industria. (Milan). 63 417 (1981)

6 a - W.J.DeKlein, in "Organic Synthesis by oxidation with metal compounds", Plenum Press, (1986)

7 a - E.I.Meiba, R.M.Dessau, J. Org. Chem., 39, 3457 (1974); J. Chem. Soc., 96, 7977 (1974)

  b - I.N.Nishino, K.H.Kurosawa, Bull. Chem. Soc. Jpn., 56, 3527 (1983), ibid., 58, 217 (1985)

  c - A.B.Ernst, W.E.Fristad, Tetr. Letters 26, 3761 (1984); E.J.Corey, M.Kang, J. Am. Chem. Soc., 106, 5384 (1984)

  d - B.B.Snider, R.Mohan, S.A.Kates, Tetr. Letters, 841 (1987). ibid, 845 (1987)

  e - W.E.Fristad, J.R.Peterson, A.B.Ernst. J. Org. Chem., 50, 3143 (1985) f - F.Z.Yang, H.K.Trost, W.E.Fristad, Tetr. Letters, 1492 (1987)

8 a - G.J.Williams, N.R.Munter, Can. J. Chem., 54, 3830 (1976);
N.K.Dunlaj, M.R.Sabol, D.S.Watt, Tetr. Letters, 5839 (1984)

9 a - H.L.Finkbeiner, J.B.Bush Jr.. Discuss. Far. Soc., 46, 150 (1968)

b - R.E. van der Ploeg, R.W. de Korte, E.C.Koojman, J. Catal., 10, 52 (1968)

c - M.G.Vinogradov,G.I.Nikishin, Isv, Akad Nauk. SSSR, Ser.Khim., 1674 (1972); ibid. 12, 272 (1976); Zh. Org. Chim.,1, 947 (1975)

d - M.E.Kurz, R.T.Y.Chen, J. Org. Chem., 43, 239, (1978)

e - M.E.Kurz, V.Baru, P.N.Nguyen, J. Org. Chem., 49, 1603 (1984)

10 A.Citterio, R.Santi, C.Finzi, S.Strologo, J. Chem. Res., (S), 120 (M) 1301 (1988)

11 R.H.G. Khrishnam, R.V.Venkat, B.Sethuram, R.T.Naraneeth, J. Electroanal. Chem., 133, 317, (1982)

12 Eur. Pat., 167053 (1987)

13 It. Pat. 19097/A (1987)

14 J.K.Kochi, Organometallic Mechanisms and Catalysis, Academic Press, N.Y. (1978)

15 H.Tonellato, O.Rossetto, A.Fava, J. Org. Chem., 34, 4032 (1969)

16 Carbenium ions ⍺-substituited by electron withdrawing groups are stable species in strongly acid medium (P.G.Gassman, Tidwell, Acc. Chem. Res., 16, 279 (1983); X. Creary, Acc.Chem.Res. 18, 3 (1985)

17 H.Yonemura, H.Nishino, K.Kurosawa, Bull.Chem.Soc.Jpn, 60,809 (1987)

18 A.Citterio, R.Santi, D.Fancelli, A.Pagani, S.Bonsignore, Gazz.Chim. Ital., 118, 1855 (1988)
A.Citterio, R.Santi, T.Fiorani, S.Strologo, J.Org.Chem., in press

19 P.J.Andrulis, M.J.S.Dewar, J.Am.Chem.Soc., 88, 5483 (1966)

20 S.Steenken, N.V.Raghavan, J.Phys.Chem., 24 3101 (1979),
A.Citterio, F.Minisci, V.Franchi, J.Org.Chem., 45, 4572 (1980)

21 Gmelin Handbuch der Anorganischen Chemie, Springer Verlag, Berlin (1980), Sy.N. 56, Section 1, 190, Section 2, p. 85

22 K.Ingold, in "Free Radicals", J.K.Kochi Ed., Wiley (1973)

23 L.Eberson, J.Am.Chem.Soc., 89, 4669 (1967)
J.M.Davidson, C.Triggs, J.Chem.Soc. A, 1331 (1968)
E.Heiba, R.M.Dessau, W.j.Koehl, J.Am.Chem.Soc., 91 138 (1969)
J.R.Gilmore, J.M.Mellor, J.Chem.Soc. D, 507 (1970)

24 A.J.Kerr, Chem. Rev., 66, 465 (1966)

25 M.Julia, J.C.Chottard, J.J.Basselier, Bull.Chem.Soc.Fr., 3037 (1967); M.Julia, Pure Appl.Chem., 1 (1970)

26 H.C.McBay, O.Tucker, A.Milligan, J.Chem.Soc., 19. 1003 (1954)

212

27 M. Shibuya, K.Toyooka, S.Kubota, Tetr. Letters,25. 1171 (1984)
   Y.Koide et al., J.Antibiot., 33, 342 (1980), Y.Koide et al.,
   Chem.Pharm.Bull., 34, 4425 (1986)
28 Landort-Bornstein, Vol.13, Radical Reactions rates in Liquids,
   Part d, H.Fischer Ed., Springer Verlag, Berlin (1984)
29 Ref. 1b amd 1d.
30 A.Citterio, F.Minisci, M.Serravalle, J.Chem.Res., (S), 198,
   (M) 2174 (1981)
   A.Citterio, Org.Synth., 62, 67 (1984); Synthesis, 308 (1986)
31 A.Citterio, A.Cominelli, Gazz.Chim.It., in press
32 H.Cohen, D.Meyerstein, Inorg.Chem., 13, 2434 (1974), V.Gold,
   S.M.Pemberton, D.L.Wood, J.C.S.Perkin II, 1230 (1981)
33 A.Citterio, A. Arnoldi, A. Griffini, Tetrahedron, 38, 393
   (1982) and ref. 30.
34 It.Pat. 24421/A (1978), Belg.Pat. 877290 (1979)
35 E.J.Corey, S.G.Pyne, Tetr. Letters, 24 2821 (1983); G.Sterk,
   M.Kahn, J.Am.Chem.Soc., 107, 500 (1985), ref. 1a.
36 .Danieli, .Palmisano, unpublished results.

ONE-ELECTRON REDOX REACTIONS BETWEEN RADICALS AND MOLECULES. DOMINANCE OF INNER-SPHERE MECHANISMS

S. STEENKEN
Max-Planck-Institut für Strahlenchemie
D-4330 Mülheim, Federal Republic of Germany

ABSTRACT. In aqueous solution the electron transfer between (reducing) carbon-centered radicals or (oxidizing) heteroatom-centered inorganic radicals and organic molecules often proceeds by covalent bond formation between the radical and the molecule followed by heterolysis of the so-formed bond between the carbon and the heteroatom. It is the heterolysis step in which the actual electron transfer between the radical and the molecule takes place. This makes electron transfer a part of the area of (heterolytic) solvolysis reactions. Structure-activity relations for heterolysis of the radical-molecule adducts and thus the electron transfer between the adduct components can be rationalized in terms of the classical solvolysis concepts.

1. Introduction

Radicals $R^{\bullet}$, whose oxidation state is by definition in-between the two (stable) even-numbered systems R^+ and R^-, are prone to engage in one-electron transfer $(ET)^1$ reactions because this type of behavior constitutes the simplest way for the molecules to lose their radical nature, cf. eq 1:

$$(A^{-\bullet}+)\ R^+ \xrightleftharpoons[(+A)]{-e}\ R^{\bullet}\ \xrightarrow[(+D)]{+e}\ R^-\ (+D^{+\bullet}) \tag{1}$$

In order for this transfer to occur, there has to be an electron donor (D) or acceptor (A). If $R^{\bullet}$ and D or A are neutral species, electron transfer leads to the ionic products R^+, R^-, $A^{-\bullet}$, $D^{+\bullet}$, and the transistion state must therefore have some ionic character. The alternative to the outer-sphere ET mechanism of scheme (1) is a reaction in which the electron transfer occurs via an addition/elimination sequence:

$$R^{\bullet} + A \xrightarrow{k_{ad}} R\text{-}A^{\bullet} \xrightarrow{k_{hs}} R^+ + A^{-\bullet} \tag{a}$$

$$\tag{2}$$

$$R^{\bullet} + D \xrightarrow{k_{ad}} R\text{-}D^{\bullet} \xrightarrow{k_{hs}} R^- + D^{+\bullet} \tag{b}$$

213

F. Minisci (ed.), Free Radicals in Synthesis and Biology, 213–231.
© *1989 by Kluwer Academic Publishers.*

This scheme represents an *inner*-sphere path of overall electron transfer. The transition state for the addition step (rate constant k_{ad}) can be but does not necessarily have to be polar; however, the heterolysis (elimination) step (rate constant k_{hs}) must have a polar transition state. Therefore, in a situation in which the heterolysis of the adduct $RA^\bullet$ or $RD^\bullet$ is rate determining, the dependence on structural or environmental conditions of the rates of redox product formation will be similar for the outer-sphere (eq 1) and the inner-sphere (eq 2) case. The two types of mechanism can therefore not easily be distinguished by studying, e.g., substituent effects on R, A, or D or solvent effects.

In the following, examples will be presented for reactions that proceed according to scheme 2, with the radicals serving as electron-donors (eq 2a) or electron-acceptors (eq 2b). In the former case the radicals involved are *carbon*-centered and substituted at C_α by heteroatoms which provide - via their lone pairs - the necessary electron density ("nucleophilic" radicals[2]); in the latter case the radicals are heteroatom-centered, their oxidizing power being due to the higher electron affinity of the heteroatom as compared to carbon.

2. Carbon-centered radicals as electron donors (eq 2a).

2.1 NITROBENZENES AS OXIDANTS

Interest in the mechanism of reduction of nitro compounds is in part due to the (potential) importance of this class of chemicals as sensitizers in the radiotherapy of cancer.[3] It has long been recognized that many organic radicals produced by ionizing radiation, photolysis, or by transition metal ion catalyzed decomposition of peroxides eventually lead to the one-electron reduction of the nitro compounds. The rates and yields of these reactions increase with increasing reducing power of the radicals and with increasing oxidizing power (reduction potential) of the nitro compound.[3-6] This behavior is suggestive of an electron transfer mechanism, and the radiobiological effect of sensitizers has in fact been interpreted on this basis.[3,5] On the other hand, oxygen, still the best radiosensitzer, typically does *not* react by electron transfer but by addition,[7,8] and this in spite of the fact that electron transfer is often highly exergonic ($-\Delta G > 1$ eV).[9] The question thus arises whether the radiosensitizing action of nitro compounds is related to their ability to scavenge radicals by *addition*, a reaction type that is well documented for them,[4,10-12] and whether addition and the equally well documented electron transfer are just different aspects of a more general reaction mechanism. As pointed out below, the latter is in fact the case. The unified mechanism, which allows for addition as well as electron transfer, involves an ion pair type transition state (scheme 4).[6] Competition between ion combination (which leads to addition) and ion separation by solvation (which results in electron transfer) determines the product distribution, as will be seen from the following examples.

2.1.1. α-Hydroxyalkyl Radicals.

The simplest of this type of radical, $\overset{\bullet}{C}H_2OH$, reacts with nitrobenzenes in aqueous solution *exclusively* by addition, eq 3, to give the hemiacetal-type alkoxynitroxyl radical:[4,6,12]

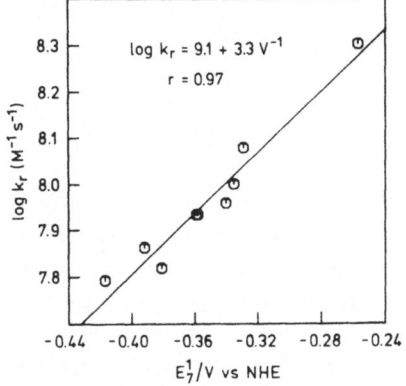

$$HOC\overset{\bullet}{H_2} \quad + \qquad \xrightarrow{\quad k_{ad}\quad} \qquad \qquad \text{(3)}$$

The rate constants k_{ad} and activation parameters for this reaction have been determined for systems with different substituents R.[6] The rate constants ($\simeq 10^8$ M^{-1} s^{-1}) increase with increasing reduction potential E_7^1 of the nitrobenzene according to a Marcus relation (Figure 1), which demonstrates that the addition reaction has electron transfer character.[6]

Figure 1. Dependence of log $k_r (\equiv k_{ad})$ for reaction in aqueous solution of $\overset{\bullet}{C}H_2OH$ with nitrobenzenes on their reduction potential E_7^1 (Marcus plot). From reference 6.

However, the slope of the Marcus line is only 3.3 V^{-1}, to be compared with 7–11 V^{-1} for a process that involves outer-sphere electron transfer.[3b,13] From this it may be concluded that in the transition state of the addition reaction electron transfer between the radical and the nitrobenzene is only partial.

Concerning the activation parameters for reaction (3), it is interesting that the activation *enthalpies* are as low as $\simeq 7$ kJ/mol, i.e. lower than that for diffusion in water ($\simeq 16$ kJ/mol). The less than diffusion controlled *rate constants* are therefore due to negative activation *entropies* ($\simeq -70$ J/molK). These have been explained as resulting from immobilization of water molecules, caused by the ionic transition state.[6]

If the nucleophilicity of the radical is increased by introduction of a methyl group at C_α, i.e. if one goes from $\overset{\bullet}{C}H_2OH$ to $CH_3\overset{\bullet}{C}HOH$, the slope of the Marcus line from the log k_{ad} vs E_7^1 plot increases to 4.5 V^{-1}, which indicates that there is more electron transfer character in the transition state than in the case of $\overset{\bullet}{C}H_2OH$.[6] In apparent contrast to this explanation is the experimental observation[6] that the activation entropies for the reaction of $CH_3\overset{\bullet}{C}HOH$ (which yields radical anion in addition to nitroxyl)[12] are less negative than in the case of $\overset{\bullet}{C}H_2OH$. This would be suggestive of *less* charge in the transition state. However, the two pieces of information can be reconciled by considering that in spite of a larger degree of electron transfer in the case of $CH_3\overset{\bullet}{C}HOH$ compared to $\overset{\bullet}{C}H_2OH$ the charge *density* in the former system is smaller than in the latter, due to the hyperconjugative effect of the methyl group. This argument is also useful in explaining why the reaction of $CH_3\dot{C}HOH$ with nitrobenzenes leads not only to nitroxyl radicals (analogous to eq 3) but also to nitrobenzene radical anions (cf. scheme 4): The cation $RC^+H(OH)$ produced by electron loss from $R\dot{C}H(OH)$ is less electrophilic if $R = CH_3$ than if $R = H$. The tendency for ion pair collapse (which leads to addition) will therefore be less pronounced for $R = CH_3$, giving the competing separation by solvent of the ions a greater chance, which leads to the electron transfer products.

(4)

It is interesting that the activation parameters for reaction of the α-*alkoxy*alkyl radicals $CH_3O\dot{C}H_2$ and $CH_2CH_2\dot{C}HO\dot{C}H_2$ are very similar to those of the related α-*hydroxy*alkyl radicals $HO\dot{C}H_2$ and $CH_3\dot{C}HOH$.[6] This indicates that in the transition state of the addition reaction deprotonation from the α-OH group is *not* taking place, in spite of the high charge density at C_α. Proton transfer to water molecules in the solvent shell would lead to an additional loss of entropy, due to the resultant immobilization of the water molecules. This conclusion is supported by the observation that k_{ad} for $\dot{C}H_2OH$ is the same in D_2O as in H_2O, i.e. the solvent kinetic isotope effect is 1, indicating that deprotonation in the transition state is not occurring.[6] If, however, deprotonation *was* occurring with concerted (full) hydration of the proton, the electron would be trapped on the nitrobenzene, since hydration of the proton is irreversible due to the huge exothermicity of this reaction (1135 kJ/mol). In other words, proton transfer would freeze electron transfer.

In the case of $CH_3\dot{C}HOH$, the reaction with the nitrobenzene does not stop at the nitroxyl stage. A spontaneous heterolysis reaction takes place in aqueous solution,[12] which results in the ultimate transfer of the (former) electron from the α-hydroxyalkyl radical to the nitrobenzene to give acetaldehyde, nitrobenzene radical anion, and H^+, eq 5.

TS

(5)

In contrast to the *formation* reaction of the nitroxyl, in the heterolytic *decomposition* of the nitroxyl deprotonation of the hemiacetal proton *does* take place, as judged by the kinetic isotope effect of $\simeq 2.2$ and by the very negative activation entropies. Electron-*donating* substituents R_D at the electron-donor part of the nitroxyl *increase* the rate of heterolysis, as do electron-*withdrawing* substituents at the electron-acceptor end of the molecule, in agreement with a push-pull mechanism of the S_N1 type.[12] Further support for S_N1 are a) the effect of introducing a second alkyl group at C_α which leads to an increase of heterolysis rates by the factor $\geq 10^3$, and b) the very drastic decrease of the heterolysis rates on decreasing the solvent polarity. The S_N1 mechanism has also been documented for the heterolysis of acetal type nitroxyl radicals produced by addition of α-*alkoxy*alkyl radicals to tetranitromethan[14] and to nitrobenzenes.[15]

As mentioned above, in the case of the nitroxyls from $\dot{C}H_2OH$ the spontaneous heterolysis in aqueous solution is slow ($\simeq 1s^{-1}$ at pH 4). However, if the hemiacetal proton is removed by reaction with base, the resulting radical anion heterolyzes with a rate constant $\geq 10^5 \ s^{-1}$, eq 6:[12]

$$+ \ CH_2O$$

The increase in the rate constant is obviously due to the enhancement of electron density at the acetalic carbon, or, expressed differently, by the base induced conversion of the bad electrofuge CH_2OH^+ into the good one CH_2O. From the point of view of electrofugacity, it is obvious that CH_3CHO and $(CH_3)_2CO$ are considerably better leaving groups than CH_2O.

2.1.2   5,6-Dihydropyrimidine-6-yl Radicals

These radicals can easily be produced by addition of the OH radical to (naturally occurring) pyrimidines or by H-abstraction by $\dot{O}H$ from 5,6-dihydropyrimidines, cf. eq's 7 and 8:

In both of these reactions $\dot{O}H$ is quite selective: $\geq 90\%$ of the reducing 6-yl radicals are formed.[16-18]

As shown by ESR and pulse radiolysis with optical and conductance detection, the pyrimidine-6-yl radicals react with nitrobenzenes exclusively by addition to the nitro group to give nitroxyl type radicals.[19a] The transition state for this addition reaction is highly polar, as judged by the strongly negative activation entropies (≈ -70 kJ/mol).[6] If the pH is < 6, the lifetime of the nitroxyls with respect to heterolysis is > ms. However, in basic solution the nitroxyls are rapidly converted into nitrobenzene radical anion and one-electron oxidized (hydroxy)pyrimidine. The reaction involves again heterolysis of the carbon-oxygen bond joining the pyrimidine with the nitrobenzene, made possible by the increased electron density at N-1 due to deprotonation at this site, cf. eq 9:[19a]

(9)

The electrofugal leaving group properties of the oxidized pyrimidine system are considerably improved by introducing a methyl group at C-6.[19b] In this case a *spontaneous* heterolysis takes place with rate constants k_{hs} between 10^3 and 10^5 s^{-1}, depending on the pyrimidine and on the substituent at C-5. The rate constants increase with increasing electron-deficiency of the nitrobenzene, as given by the substituent at the para-position, and the Hammett, Brönsted, and even Marcus relations are fulfilled for this heterolysis reaction.[19b]

As compared to the non-methylated pyrimidines, the 6-methyl-6-yl radicals react with nitrobenzenes not only by addition but also by formation of radical anion.[19b] Also in this respect they behave in a similar way as $CH_3\dot{C}HOH$ (see section 2.1.1). The explanation is again similar: The transition state is assumed to be ion-pair like. Collapse leads to addition, separation by solvent to electron transfer. A methyl group at C_α decreases the electrophilicity of the carbocation giving water molecules a chance to compete with ion combination. Hydration of the ions finalizes electron transfer.

2.2. O_2 AS AN OXIDANT

On the basis of its one-electron reduction potential, oxygen ($E_7^1 = -0.155$ V/NHE), is a much more powerful oxidant than nitrobenzene ($E_7^1 = -0.486$ V/NHE). In spite of this, O_2 does usually *not* react with radicals by electron transfer but by *addition*. This is true even in the case of the strong one-electron reductant $(CH_3)_2\dot{C}OH$ ($E_7^2 = -2.2$ V/NHE), where reaction with O_2 leads to the hydroxyperoxyl radical, cf. eq 10a:[20]

$$(CH_3)_2\overset{\centerdot}{C}OH + O_2 \xrightarrow[a]{k_{ad}} (CH_3)_2C(OH)\text{-}O_2^{\centerdot} \xrightarrow[b]{k_{hs}} (CH_3)_2CO + O_2^{-\centerdot} + H^+ \qquad (10)$$

The addition route eq 10a is favored over the hypothetical electron transfer although the driving force for the electron transfer is 2 eV, as obtained from the difference in the reduction potentials of O_2 and $(CH_3)_2\overset{\centerdot}{C}OH$. Obviously the activation barrier for electron transfer is higher than that for addition, and this reflects high bond and solvent reorganization energies for the ET process. Considerable solvent reorganization energies are expected due to the necessity to hydrate the proton produced on electron transfer. On the basis of an electron transfer/addition mechanism analogous to that suggested for nitrobenzenes (eq 4), the more strongly pronounced tendency of O_2 to add is understandable on the basis of the greater nucleophilicity of $O_2^{-\centerdot}$ compared to nitrobenzene radical anions, as a result of which ion pair collapse (of $>^+C(OH) ...O_2^{-\centerdot}$) is able to dominate over ion separation by solvation.

The adduct (the hydroxyperoxyl radical) undergoes spontaneous heterolysis of the carbon-peroxyl bond to yield acetone, H^+ and $O_2^{-\centerdot}$ (eq 10b), with the rate constant $k_{hs} = 670\ s^{-1}$.[20] If the hemiacetal proton is removed by reaction with base, the resulting α-oxyanion radical has sufficient electron density to "push out" the electron pair joining C_α and $O_2^{\centerdot}$, and this increases the heterolysis rate constant to $> 10^5\ s^{-1}$.[20] Introduction at C_α of a second OH or Oalkyl group has a similar effect, e.g.[21]

$$HC(OH)_2\text{-}O_2^{\centerdot} \longrightarrow HC(O)OH + O_2^{-\centerdot} + H^+; \quad k_{hs} \geq 10^6\ s^{-1} \qquad (11)$$

A dimethylamino group at C_α is at least equally efficient in stabilizing in the transition state the carbocation produced on heterolysis, at least as judged by the high rate constant for heterolysis of the corresponding peroxyl radical, cf. eq 12:[22]

$$(CH_3)_2NCH_2\text{-}O_2^{\centerdot} \longrightarrow (CH_3)_2\overset{+}{N}=CH_2 + O_2^{-\centerdot}; \quad k_{hs} \geq 10^6\ s^{-1} \qquad (12)$$

If the electron-donating power due to the lone pair on nitrogen is reduced by attachment of a carbonyl group to the nitrogen, $C\text{-}O_2^{\centerdot}$ heterolysis is considerably slowed down, as in the case of, e.g., the peroxyl radical produced by O_2 addition to 5,6-dihydrouracil-6-yl (eq 13; $k_{hs} < 10^3\ s^{-1}$). However, deprotonation of N(1)-H provides an electron push that makes $k_{hs} = 8.3 \times 10^4\ s^{-1}$.[16-18,23]

$$(13)$$

The isopyrimidine produced[23] is the one-electron oxidation product of the 6-yl radical. As pointed out above, it is formed by transfer of an electron to oxygen, not by an outer-sphere path but via addition/elimination.

These processes are analogous to those occurring on radical oxidation by nitro compounds and they have the same dependences on structure and solvent. Nitrobenzene radical anions appear to be better leaving groups than is $O_2^{-\bullet}$, and this is reflected by the pK_a values of the conjugate acids: $pK_a \leq 3.2$ for the nitrobenzene systems[24] and 4.8 for $HO_2^{\bullet}$.

An addition/elimination sequence with an ion-pair type transition state for the addition part has also been observed in the reaction of α-heteroatom substituted alkyl radicals with a quinone, anthraquinone-2,6-disulfonate.[9]

3. Heteroatom-centered radicals as electron acceptors (eq 2b)

3.1 $SO_4^{-\bullet}$

On the basis of the Hammett $\rho = -2.4$ for reaction with substituted benzenes, the $SO_4^{-\bullet}$ radical has been suggested to react by outer-sphere electron transfer.[25] However, although $SO_4^{-\bullet}$ is in fact a very strong oxidant $(E \simeq 2.5\text{-}3.1 \text{ V/NHE})$[1a,26], a Marcus type treatment of its reactivity[1a] indicates that inner-sphere processes are important in those cases where the substituent is less electron-donating than CH_3O. A search for SO_4^- adducts to benzene derivatives has, however, been unsuccessful so far. The potential SO_4^- adduct to benzene has a lifetime ≤ 100 ns,[27,28] and even in the case of benzonitrile (whose radical cation should be a much weaker electrofugal leaving group than that of benzene) the rate constant for heterolysis of the SO_4^- adduct is $\geq 5 \times 10^6 \text{ s}^{-1}$.[9] The products of the reaction of $SO_4^{-\bullet}$ with many benzenes are hydroxycyclohexadienyl radicals, whose formation can be explained by both an inner-sphere (addition/elimination, eq 14a) or by an outer-sphere path (eq 14b).

A clear-cut distinction between the inner- and outer-sphere paths can, however, be made in the case of reaction of $SO_4^{-\bullet}$ with simple alkenes. For instance, $SO_4^{-\bullet}$ reacts in aqueous solution with allyl alcohol exclusively by addition[28,29] via an ion pair type transition state.[30,31] If the SO_4^- adduct undergoes heterolysis at all, the rate constant for this reaction is $\leq 10^2 \text{ s}^{-1}$.[30] Howewer, if the electron density of the system is increased by introducing two geminal methyl groups at the C-C double bond, the SO_4^- adduct heterolyzes spontaneously with the rate constant $4 \times 10^4 \text{ s}^{-1}$,[30] as determined by conductance, cf. eq 15:

$$\text{(15)}$$

A second example is cyclohexene, where the SO_4^- adduct undergoes hydrolysis with $k_{hs} = 3.0 \times 10^4 \text{ s}^{-1}$ at $20°$,[9,31] cf. eq 16 (R=H).

$$\text{(16)}$$

The heterolysis rate is drastically increased by introducing a methyl group: For $R=CH_3$, $k_{hs} \geq 5 \times 10^7 \text{ s}^{-1}$.[28,31] As in the case of the allyl alcohols, the methyl effect indicates that the hydrolysis is of the S_N1 type. For $R=H$, the activation parameters for eq 16 have been determined to be $\Delta H^{\ddagger} = 17 \text{ kJmol}^{-1}$ and $\Delta S^{\ddagger} = -103 \text{ J(molK)}^{-1}$.[9] The strongly negative $\Delta S^{\ddagger}$ value for a decomposition reaction (with an intrinsic entropy gain) indicates that water molecules are immobilized in the ionic transition state of the heterolysis.

A further example for the addition/elimination sequence in one-electron oxidation by SO_4^- is its reaction with thymine derivatives. The position of highest electron density of the pyrimidine is C-5,[32] and it is therefore to be expected that the electrophilic[25] SO_4^- would, if possible, attack at this site.[33] However, interaction of SO_4^- with N-1 substituted thymines such as thymidine gives rise to only the (oxidizing) 6-hydroxy-5,6-dihydrothymine-5-yl radical,[33-35] which indicates that attack has occurred at C-6 rather than at C-5. It is reasonable to assume that the approach of SO_4^- to C-5 is sterically hindered by the methyl group at C-5 and that SO_4^- attaches to the next-best position, i.e. C-6. The adduct thus formed hydrolyzes with the spontaneous rate $k_{hs} = 3.2 \times 10^5 \text{ s}^{-1}$,[35] cf. eq 17 and Figure 2.

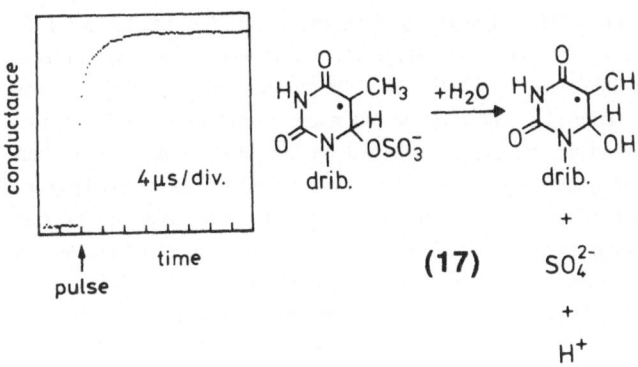

$$\text{(17)}$$

Figure 2. Conductance change on reaction of SO_4^- with 0.5 mM thymidine at pH 4.4. $[K_2S_2O_8] = 2 \text{ mM}$.

3.2 Cl_2^- $^\bullet$

This radical (E_7^2 = 2.1 V/NHE) is a weaker oxidant than SO_4^- $^\bullet$ (2.5-3.1 V/NHE) or $\dot{O}H$ (2.7 V/NHE) at pH 0)[36]. In aqueous solution Cl_2^- $^\bullet$ can be produced by reaction of SO_4^- $^\bullet$ (k = 2.1 x 10^8 M^{-1} s^{-1})[37] or $\dot{O}H$ with Cl^- to give $Cl^\bullet$ which is then scavenged by additional Cl^- to yield Cl_2^- $^\bullet$. The oxidation of Cl^- by $\dot{O}H$ requires the presence of H^+. It proceeds by an addition/elimination sequence with the elimination step consisting of an H^+ induced dehydration[38] (see section 3.3.1).

In spite of its lower oxidation potential as compared to SO_4^- $^\bullet$ and $\dot{O}H$, Cl_2^- $^\bullet$ tends to react more by what appears to be electron transfer and less by addition.[29,39] However, it was possible to find an example for one-electron oxidation of an organic substrate by Cl_2^- $^\bullet$ via addition/elimination. The radical $(CH_3)_2\dot{C}CH_2Cl$, whose room temperature rate of heteroysis has been determined to be 3.5 x 10^4 s^{-1} (after formation by H-abstraction from iso-butylchloride),[40] was produced by pulse irradiating an aqueous solution containing 10 mM $S_2O_8^{2-}$, 20 mM Cl^- and saturated with isobutene and monitoring the conductance.[41] A first order production of H^+ was seen with the rate constant 3.1 x 10^4 s^{-1} at 20°, which can be considered equal within experimental error with that determined by the H-abstraction approach (eq 18c). The addition/elimination mechanism for oxidation of an alkene by Cl_2^- $^\bullet$ is thereby established (eq 18a,b):

$$k_{hs}=3 \times 10^4 s^{-1} \quad (18)$$

If the heterolysis rate constant for $(CH_3)_2\dot{C}CH_2Cl$ is compared with that ($\leq 10^3$ s^{-1}) for $(CH_3)_2\dot{C}CH_2OSO_3^-$ (formed by addition of SO_4^- $^\bullet$ to isobutene) it is obvious that - in terms of rapidity of formation of the product radical $(CH_3)_2\dot{C}CH_2OH$ - the stronger oxidant SO_4^- $^\bullet$ is less efficient than the weaker Cl_2^- $^\bullet$. This type of result alone would make an outer-sphere mechanism highly unlikely.[42] The result can, however, be easily explained on the basis of the inner-sphere mechanism eq 18a,b. Since step b is a heterolysis, the classical mechanistic solvolysis concepts can be applied in order to understand structure-activity relations, solvent effects, and dependences on leaving groups. The adducts of alkenes to $Cl^\bullet$ and SO_4^- $^\bullet$ differ with respect to the leaving group: Cl^- is a considerably better leaving group ($pK_a(HCl)$ = -7) than SO_4^{2-} ($pK_a(HSO_4^-)$ = 1.9). If the Brönsted catalysis law is applicable in this case and if the Brönsted α is assumed to be 0.5, the difference in the pK_a values of the conjugate acids translates into a heterolysis rate ratio of $10^{4.5}$ in favor of the chloride elimination. This example shows that in the inner-sphere electron transfer reaction 18a,b the *activity of the oxidant is not determined by its oxidizing power but by the leaving group ability of its conjugate redox partner.* This situation is analogous to the oxidation of *radicals* by the *molecular* oxidants O_2 and nitrobenzene (sections 2.1/2.2), where the weaker oxidant nitrobenzene is the more efficient one in leading to one-electron oxidized product because its radical anion is the better leaving group.

3.3 ȮH

The ȮH radical is probably one of the most reactive radicals. This refers to reactions with both organic and inorganic molecules. In aqueous systems it is produced by ionizing radiation, by photolysis or transition metal ion catalyzed decomposition of hydroperoxides and possibly even in enzyme catalyzed reactions. ȮH is an electrophile (Hammett ρ = -0.5 from reaction with substituted benzenes), and it is a very powerful oxidizing agent (E = 2.7 V/NHE at pH 0 and still 2.3 V at pH 7). In fact, in aqueous solution ȮH is the strongest of all thermodynamically stable one-electron oxidants, since more powerful oxidants are converted to ȮH by reaction with water, eq 19:

$$H_2O \xrightarrow{\ -e^-\ } H_2O^{+\bullet} \xrightarrow[k \simeq 10^{12}\ s^{-1}]{} \dot{O}H + H^+ \tag{19}$$

The ȮH radical is therefore the "sink" for oxidizing equivalents stronger than water.

However, of the many ȮH reactions with inorganic and organic substances so far studied,[43] there is little, if any, solid evidence for outer-sphere electron transfer mechanisms. Oxidative ȮH reactions can, however, be well understood in terms of addition/elimination processes. Since some representative cases have already been reviewed,[9,44] in the following the emphasis will be placed on examples involving "biomolecules" rather than attempting a complete coverage of the area. A few inorganic systems will be touched upon.

3.3.1   Halides and pseudohalides

Except fluoride, the halides can be converted to the corresponding halogen atoms by reaction with ȮH in aqueous solution.[38] This reaction, which proceeds by addition and not by electron transfer, is probably one of the earliest examples for an inner-sphere mechanism in oxidation reactions of ȮH. The fate of the OH adducts $HOX^{-\bullet}$ depends on the oxidizability of X^- (i.e. the reduction potential of $X^\bullet$), or, using a related parameter, on the nucleofugal leaving group properties of X^-. In the case of Br^-, which is an excellent leaving group ($pK_a(HBr) = -8$) and a fair reductant ($E(Br^\bullet) = 2$ V/NHE), the OH adduct $HOBr^{-\bullet}$ undergoes a *spontaneous* heterolysis, eq 20b:[45]

$$
\text{HO}^{\bullet} + \text{X}^- \rightleftharpoons \text{HO} - \text{X}^{\bar{\bullet}}
\begin{array}{l}
\xrightarrow{\quad b \quad} \text{HO}^- + \text{X}^{\bullet} \\[2ex]
\underset{-H^+}{\overset{+H^+}{\rightleftarrows}} \quad \text{H}_2\overset{+}{\text{O}} - \text{X}^{\bar{\bullet}} \longrightarrow \text{H}_2\text{O} + \text{X}^{\bullet}
\end{array}
\tag{20}
$$

In the case of Cl$^-$, however, spontaneous heterolysis is *not* observed, and only after protonation of the nucleofugal leaving group OH$^-$ to give H$_2$O is the heterolysis of the adduct (step 20c) fast enough to compete with its *homolysis* (step reverse 20a) or with the bimolecular decay of the radical(s).[46] The function of the proton is thus seen to consist in converting the bad leaving group OH$^-$ (pK$_a$(H$_2$O) = 15.7) into the excellent one H$_2$O (pK$_a$(H$_3$O$^+$) = –1.7).

The mechanism of oxidation of the pseudohalide SCN$^-$ by $\overset{\bullet}{\text{O}}$H is similar to that of Br$^-$.[47]

3.3.2 Metal ions

The one-electron oxidation of metal ions by $\overset{\bullet}{\text{O}}$H also proceeds by an addition/elimination sequence.[48] An example is given in eq 21. Electron transfer from the metal to $\overset{\bullet}{\text{O}}$H is finalized by proton assisted dehydration. *Re*hydration of the oxidized metal is rapid as a result of the larger electrophilicity of the metal in the higher oxidation state.

$$
\text{Ag}^+ + \,^{\bullet}\text{OH} \longrightarrow \text{AgOH}^{+\bar{\bullet}} \underset{+H_2O,\, -H^+}{\overset{+H^+,\, -H_2O}{\rightleftarrows}} \text{Ag}^{2+}
\tag{21}
$$

3.3.3 Organic systems

With systems containing double bonds, $\overset{\bullet}{\text{O}}$H typically reacts by addition. In spite of high rate constants, $\overset{\bullet}{\text{O}}$H can be quite selective in these reactions.[49] In the case of pyrimidine bases, attachment occurs at C–5 with $\geq$ 80% probability to give the reducing 6-yl radicals,[16,17] cf. eq 22:

One-electron oxidation of uracil by ȮH (22)

Oxidation of the pyrimidine system takes place only after base induced conversion of one or both keto groups into the corresponding enolate groups.[16] This results in an increase of the electron density sufficient for the C(5)–OH bond to heterolyze. Mechanistically, the essential feature is the enhancement of the leaving group properties of the electrofuge by base induced deprotonation.

The reciprocal to the base induced improvement of the leaving group qualities of the *electro*fuge by its *de*protonation is the strengthening of the *nucleo*fugal leaving group by its *protonation*. Equations (20) and (21) are examples for this principle from *in*organic systems. Whereas the deprotonation results in the heterolyzing electron pair being *pushed* out, by protonation the electron pair is being *pulled* out.

Examples for H$^+$ assisted one-electron oxidation by ȮH are 2-phenylalkanols, where sidechain fragmentation is observed that originates from intermediate radical cations.[50-52] In Figure 3 is shown the time-dependent conversion of the OH-adduct to 1-ethyl-2-phenylethanol to yield the benzyl radical, eq 23.

226

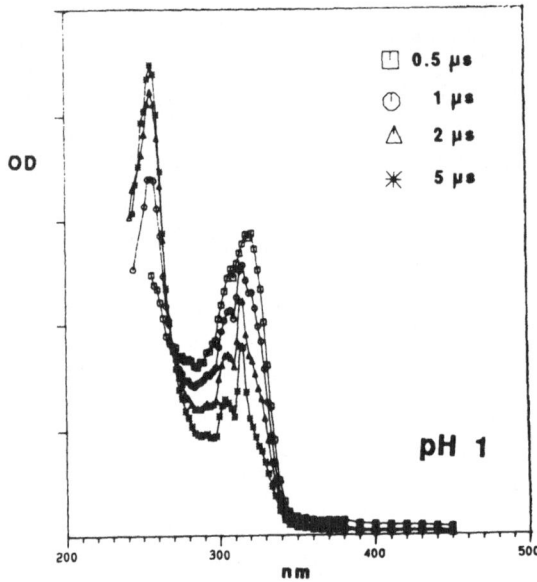

Figure 3. H^+ induced conversion of the OH adduct(s) of 1-ethyl-2-phenylethanol (1 mM) at pH 1.

The times are after the 20 ns 248 nm laser pulse. $[H_2O_2] = 0.1$ M. The initial spectrum is that of the OH adduct(s), the final one that of $Ph\overset{\cdot}{C}H_2$. From ref 62.

The ease of dehydration of an OH adduct *increases* with *increasing* electron density of the system.[53] Conversely, the activity of a radical cation with respect to electrophilic reaction with water to give a hydroxycyclohexadienyl type radical (an OH adduct) *increases* with *decreasing* electron density. In addition to changing the electron density by deprotonation (cf. eq 22), the electron density can of course be modified by substituents. For the benzene series the dependence on substituent of the propensity to hydrate/dehydrate of the radical cations/OH adducts is shown in scheme 24: whereas in aqueous solution the radical cation of benzene undergoes quantitative and irreversible hydration in ≤ 20 ns (k(hydration) $\geq 5 \times 10^7$ s^{-1}), the OH adduct of anisole is quantitatively converted to the radical cation by reaction with H^+ (k(hydration) $\leq 10^2$ s^{-1}).[44] In the case of phenol the OH adduct undergoes a *spontaneous* dehydration reaction ($k_{hs} \simeq 10^3$ s^{-1}).[54] A further increase in the electron density such as that occurring on ionization of the phenolic OH makes dehydration (which now consists in elimination of OH$^-$) become very rapid: $k_{hs} \geq 10^7$ s^{-1}.[44,54] The situation can be summarized by stating that increases in electron density increase the electro*fugacity* with respect to OH$^-$, decreases in electron density increase the electro*philicity* of the radical cation.

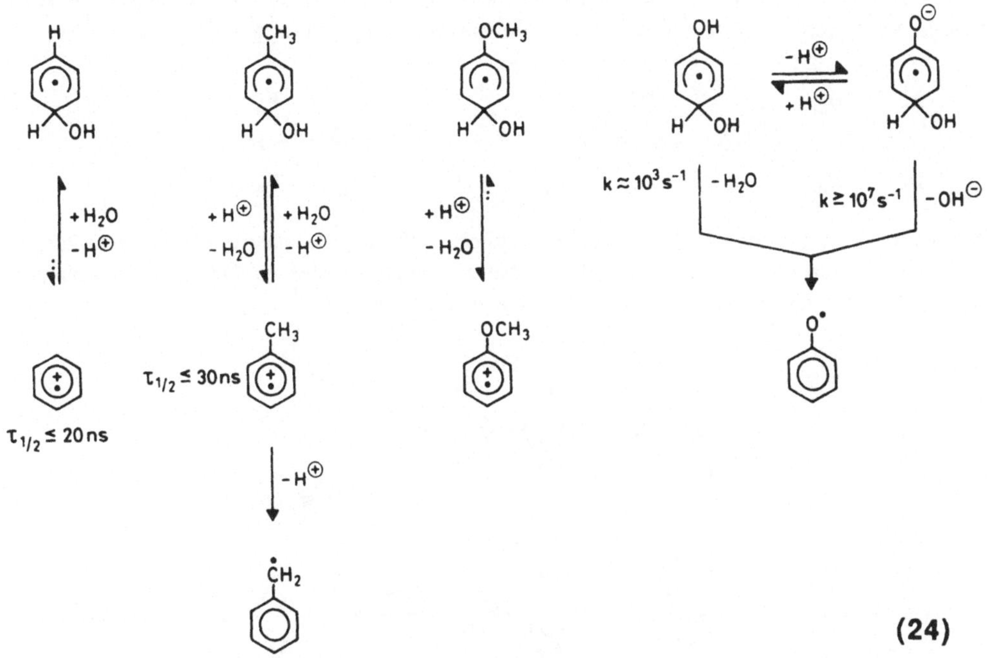

(24)

On this basis one expects dehydrations to become noticeable even at short times (e.g. $\leq$ ms) only if the electrofuge is "sufficiently" electron-rich. In addition to the examples already quoted (see scheme 24) and OH adducts of aniline[55] and N,N-dimethylaniline,[56] the C-4 OH adducts of *purines* undergo spontaneous dehydration to give one-electron oxidized purines,[57,58] or their deprotonated analogues cf., e.g., eq 25:

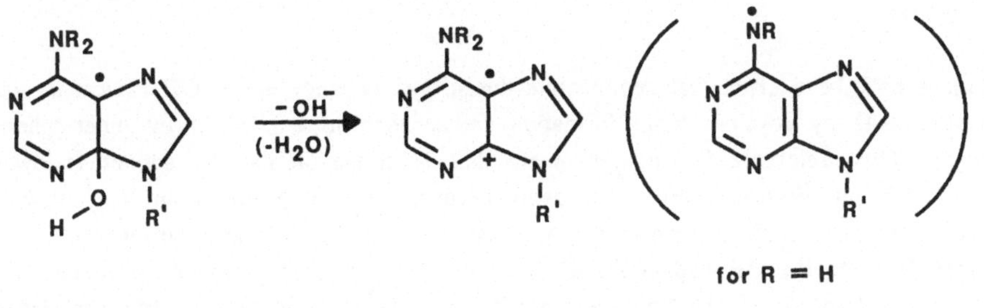

(25)

for R = H

The rate constants for the dehydration reactions increase with increasing electron-donating power of the substituents R and R' (for the effect of substituents at C-6. the Hammett $\rho^+ = -3.0$). The substituents also have an effect on the rate constants for $\dot{O}H$ *addition* ($\rho^+ = -0.9$), from which it may be concluded that the transition state for addition of $\dot{O}H$ is similar to that of its elimination (as OH^-), i.e. both have ion pair character.[58]

The OH adducts to alkenes are aliphatic counterparts of aromatic OH adducts. The same radicals are accessible by H-*abstraction* from the corresponding saturated systems, cf. eq 26:

$$(26)$$

Depending on the electron density of the system and on the nature of the substituents, one-electron oxidation by $\dot{O}H$ is possible via the acid-induced route eq 27a, the base catalyzed path eq 27b, or even by spontaneous elimination of water, eq 27c:[59]

$$(27)$$

These dehydration reactions, like their aromatic counterparts, lead to an inversion of the redox character of the radicals. The OH adducts typically are reducing, whereas after dehydration the radicals are oxidizing.[44] The reactions shown in eq 27 are of relevance for the radical chemistry of sugars.[8]

Concluding remarks

Several examples have been given that demonstrate the strong tendency of the $\dot{O}H$ radical to add to electron donors D by covalent bond formation rather than to oxidize D by outer-sphere electron transfer. This preference for addition means that the transition state for additon is lower in energy than that for electron transfer, probably because addition profits from bond *making* whereas electron transfer requires entropically expensive bond and solvent reorganization.

$\dot{O}H$ addition leads to the OH adduct HO-D$^{\bullet}$. If this reaction is to ultimately yield electron transfer products, heterolysis of the bond joining HO and D$^{\bullet}$ has to occur. However, since OH$^-$ is a bad leaving group (as reflected by the high pK$_a$ (15.7) of its conjugate acid H$_2$O), the rate of the spontaneous heterolysis is typically very low (e.g., $< 10^2$ s^{-1}). One-electron oxidation of D by $\dot{O}H$ can, however, be accomplished by operating on the leaving group abilities of the adduct components, HO- and D$^{\bullet}$. Protonation of HO- yields H$_2$O$^+$- (pK$_a$ (H$_3$O$^+$ = – 1.7) whose nucleofugal leaving group qualities are orders of magnitude better than that of HO-. Conversely, *de*protonation from $D^{\bullet}$ improves its electrofugacity to a similar extent.

The summarize, the reactivity of the powerful oxidant $\overset{\bullet}{O}H$ is characterizable by the bad leaving group properties of its reduction product, OH^-. If it is assumed that the addition reaction has electron transfer character (as evidenced, e.g., by the Hammett ρ of -0.5 for substituted benzenes), in other words, if addition proceeds via an ion pair type transition state, completion of electron transfer is prevented by the high nucleophilicity of OH^- as a result of which ion combination (which yields addition) beats ion solvation (which gives electron transfer).

4. Addition/elimination (inner-sphere) versus outer-sphere electron transfer

It has been demonstrated that a common mechanism for one-electron redox reactions between radicals and molecules is the inner-sphere type addition/elimination process. The preference for addition compared to electron transfer is the result of transition state stabilization by bond *making*. In the addition step the transition state often acquires electron transfer character and it is for this reason that the nucleophilicity of the reduced oxidant and the electrophilicity of the oxidized reductant are important in determining to which extent solvation of the (quasi) ion pair can compete with ion combination. Nucleo- and electrophilicity of the adduct components are also of great importance in determining the further fate of the adduct. It is these properties that can be influenced by, e.g., substitution or (de)protonation.

The relationship between leaving group properties and reduction potential is shown is eq's 28 and 29:

$$A - D \; \rightleftharpoons \; A^{\bar{\cdot}} + D^+ \tag{28}$$

$$A^{\bullet} + e^{-\bullet} \; \rightleftharpoons \; A^{\bar{\cdot}} \tag{29}$$

Eq 28 defines the (heterolytic) leaving group abilities, eq 29 the (homolytic) reduction potential. The essential difference is that in eq 28 $A^{\bar{\cdot}}$ donates an electron *pair*, whereas in eq 29 $A^{\bar{\cdot}}$ donates a *single* electron.[60] On this basis it is understandable that the reduction potentials are not necessarily related in a simple way to the leaving group abilities. In any inner-sphere or "bonded"[61] electron transfer mechanism the leaving group properties of the conjugate redox partners are of great importance in determining the behavior of a redox system.

References and Notes

1. For a general discussion of electron transfer processes see a) Eberson, L. *Electron Transfer Reactions in Organic Chemistry*, Springer, Berlin 1987. b) Cannon, R.D. *Electron Transfer Reactions, Butterworths, London* 1980.

2. Minisci, F.; Citterio, A. *Adv. Free Radical Chem.* 1980, **6**, 65.

3. a) Adams, G.E.; Breccia, A.; Rimondi, C. (Eds.) *Advanced Topics in Hypoxic Cell Radiosensitization*, Plenum, New York 1982. b) Wardman, P.; Clarke, E.D. In *New*

230

Chemo- and Radiosensitizing Drugs; Breccia, A.; Fowler, J.F., Eds., *Edizione Scientifiche* : Lo Scarabeo, Italy, 1985, p. 21.

4. McMillan, M.; Norman, R.O.C. *J. Chem. Soc.* B 1968, 590.
5. Wardman, P. in *Radiation Chemistry: Principles and Applications*, Verlag Chemie, Weinheim 1987, p. 565, and references in this article.
6. Jagannadham, V.; Steenken, S. *J. Am. Chem. Soc.* 1988, **110,** 2188.
7. Willson, R.L. *Int. J. Radiat. Biol.* 1970, **17,** 349.
8. von Sonntag, C. *The Chemical Basis of Radiation Biology*, Taylor and Francis, London 1987.
9. Steenken, S. in: *Free Radicals: Chemistry, Pathology and Medicine, Vol. 3*, Rice-Evans, C.; Dormondy, T. (eds.), *Richelieu Press, London* 1988, p. 51.
10. Janzen, E.G.; Gerlock, J.L. *J. Am. Chem. Soc.* 1969, **91,** 3108.
11. Sleight, R.B.; Sutcliffe, L.H. *Trans. Faraday Soc.* 1971, **67,** 2195.
12. Jagannadham, V.; Steenken, S. *J. Am. Chem. Soc.* 1984, **106,** 6542.
13. Marcus, R.A. *Annu. Rev. Phys. Chem.* 1964, **15,** 155.
14. Eibenberger, J.; Steenken, S.; Schulte-Frohlinde, D. *J. Phys. Chem.* 1980, **84,** 704.
15. Steenken, S. manuscript in preparation.
16. Fujita, S.; Steenken, S. *J. Am. Chem. Soc.* 1981, **103,** 2540.
17. Hazra, D.K.; Steenken, S. *J. Am. Chem. Soc.* **105,** 4380.
18. Schuchmann, M.N.; Steenken, S.; Wroblewski, J.; von Sonntag, C. *Int. J. Radiat. Biol.* 1984, **46,** 225.
19. a) Steenken, S.; Jagannadham, V. *J. Am. Chem. Soc.* 1985, **107,** 6818. b) Jagannadham, V.; Steenken, S. *J. Phys. Chem.* 1988, **92,** 111.
20. Bothe, E.; Behrens, G.; Schulte-Frohlinde, D. *Z. Naturforsch.* 1977, **32b,** 886.
21. Bothe, E.; Schuchmann, M.N.; Schulte-Frohlinde, D.; von Sonntag, C. *Photochem. Photobiol.* 1978, **28,** 639; Bothe, E.; Schulte-Frohlinde, D. *Z. Naturforsch.* 1980, **35b,** 1035.
22. Das, S.; Schuchmann, M.N.; Schuchmann, H.-P.; von Sonntag, C. *Chem. Ber.* 1987, **120,** 319.
23. Al-Sheikley, M.I.; Hissung, A.; Schuchmann, H.P.; Schuchmann, M.N.; von Sonntag, C; Garner, A.; Scholes, G. *J. Chem. Soc. Perkin Trans.* 2 1984, 601.
24. Grünbein, W.; Fojtik, A.; Henglein, A. *Z. Naturforsch.* 1969, **24,** 1336. Grünbein, W.; Henglein, A. *Ber. Bunsenges. Phys. Chem.* 1969, **73,** 376.
25. Neta, P.; Madhavan, V.; Zemel, H.; Fessenden, R.W. *J. Am. Chem. Soc.* 1977, **99,** 163.
26. Eberson, L. *Adv. Phys. Org. Chem.* 1982, **18,** 79.
27. Neta, P.; Madhavan, V.; Zemel, H.; Fessenden, R.W. *J. Am. Chem. Soc.* 1977, **99,** 163.
28. Steenken, S. unpublished results.
29. Davies, M.J.; Gilbert, B.C. *J. Chem. Soc. Perkin Trans.* 2 1984, 1809.
30. Steenken, S. in *Radiation Research, Proceedings 8th Internat. Congr. Radiation Research, Edinburgh* 1987; Vol. 2, Fielden, E.M.; Fowler, J.F.; Hendry, J.H.; Scott, D. (eds), Taylor and Francis, London 1987, p. 84.
31. Koltzenburg, G.; Bastian, E.; Steenken, S. *Angew. Chem.* 1988, in press.
32. Pullman, B.; Pullman, A. *Quantum Biochemistry*. Interscience, New York 1963.
33. a) Behrens, G.; Hildenbrand, K.; Schulte-Frohlinde, D.; Herak, J.N. *J. Chem. Soc. Perkin Trans.* 2 1988, 305. b) For further discussion of $SO_4^{-\bullet}$ reactions with uracils see the contribution of D. Schulte-Frohlinde and K. Hildenbrand in this issue.

34. O'Neill, P., Davies, S.E. *Int. J. Radiat. Biol.* 1987, **52**, 577.

35. Deeble, D.J.; von Sonntag, C.; Steenken, S. unpublished results.

36. Schwarz, H.A.; Dodson, R.W. *J. Phys. Chem.* 1984, **88**, 3643; Kläning, U.K.; Sehested, K.; Holcman, J. *J. Phys. Chem.* 1985, **89**, 760.

37. Chawla, O.P.; Fessenden, R.W. *J. Phys. Chem.* 1975, **79**, 2693.

38. For a review see Fornier de Violet, P. *Rev. Chem. Intermediates* 1981, **4**, 121.

39. Hasegawa, K.; Neta, P. *J. Phys. Chem.* 1978, **82**, 854.

40. Koltzenburg, G.; Behrens, G.; Schulte-Frohlinde, D. *J. Am. Chem. Soc.* 1982, **104**, 7311; ibid. 1983, **105**, 5168.

41. Steenken, S.; Koltzenburg, G. unpublished results.

42. Unless the electron transfer was taking place in the Marcus "inverted region".

43. For a review see Buxton, G.V.; Greenstock, C.L.; Helman, W.P.; Ross, A.B. Critical Review of Rate Constants for Reactions of Hydrated Electrons, Hydrogen Atoms, and Hydroxyl Radicals ($\dot{O}H/O^{-\bullet}$) in Aqueous Solution, NSRDS–NBS, in press.

44. Steenken, S. *J. Chem. Soc. Faraday Trans. I* 1987, **83**, 113.

45. Matheson, M.S.; Mulac, W.A.; Weeks, J.L.; Rabani, J. *J. Phys. Chem.* 1966, **70**, 2092.

46. Jayson, G.G.; Parsons, B.J.; Swallow, A.J. *J. Chem. Soc. Faraday Trans. I* 1973, **69**, 1597; Pucheault, J.; Ferradini, C.; Julien, R.; Deysine, A.; Gilles, L.; Moreau, M. *J. Phys. Chem.* 1979, **83**, 330.

47. Ellison, D.H.; Salmon, G.A., Wilkinson, F. *Proc. Roy. Soc. A* 1972, **328**, 23.

48. O'Neill, P.; Schulte-Frohlinde, D. *Chem. Commun.* 1975, 387; Asmus, K.-D.; Bonifacic, M.; Toffel, P.; O'Neill, P.; Schulte-Frohlinde, D.; Steenken, S. *J. Chem. Soc. Faraday Trans. I* 1978, **74**, 1820.

49. see, e.g., Raghavan, N.V.; Steenken, S. *J. Am. Chem. Soc.* 1980, **102**, 3495.

50. Snook, M.E.; Hamilton, G.A. *J. Am. Chem. Soc.* 1974, **96**, 860.

51. Walling, C.; Zhao, C.; El-Taliawi, G.M. *J. Org. Chem.* 1983, **48**, 4910; Walling, C.; El-Taliawi, G.M.; Zhao, C. ibid. 1983, **48**, 4914; Walling, C.; El-Taliawi, G.M.; Amarnath, K. *J. Am. Chem. Soc.* 1984, **106**, 7573.

52. Gilbert, B.C.; Scarratt, C.J.; Thomas, C.B.; Young, J. *J. Chem. Soc. Perkin Trans. 2* 1987, 371.

53. Holcman, J.; Sehested, K. *Nukleonika* 1979, **24**, 887.

54. Land, E.J.; Ebert, M. *Trans. Faraday Soc.* 1967, **63**, 1181.

55. Christensen, H. *Int. J. Radiat. Phys. Chem.* 1972, **4**, 311; Ling Quin; Tripathi, N.R.; Schuler, R.H. *Z. Naturforsch. A* 1985, **40**, 1026.

56. Holcman, J.; Sehested, K. *J. Phys. Chem.* 1977, **81**, 1963.

57. O'Neill, P.; Davies, S.E. *Int. J. Radiat. Biol.* 1986, **49**, 937 and references therein; Vieira, A.J.S.C.; Steenken, S. *J. Am. Chem. Soc.* 1987, **109**, 7441.

58. Vieira, A.J.S.C.; Steenken, S. *J. Phys. Chem.* 1987, **91**, 4138.

59. Steenken, S.; Davies, M.J.; Gilbert, B.C. *J. Chem. Soc. Perkin Trans. 2* 1986, 1003.

60. It has been pointed out that single electron "shifts" may be important in overall heterolytic reactions (see Pross, A. *Acc. Chem. Res.* 1985, **18**, 212 and Shaik, S.S. *Progr. Phys. Org. Chem.* 1985, **15**, 197).

61. Littler, J.S. in: *Essays in Free Radical Chemistry. J. Chem. Soc.* Special Publication No. 24, London 1970, p. 383.

62. Ramaraj, R.; Steenken, S. unpublished material.

ELECTRON-TRANSFER PHOTOSENSITIZATION IN THE OXIDATION OF ALKYL AROMATICS AND HETEROAROMATICS

A. GALADI, M. JULLIARD and M. CHANON
Laboratoire de Chimie Inorganique Moléculaire - U.A. 126
Faculté des Sciences Saint Jérôme - 13397 - Marseille Cédex 13 - France.

I. INTRODUCTION

Photochemical excitation may induce intermolecular electron-transfer. Thus an electron donor-acceptor couple can be activated by light which then transforms an inert substrate by enhancing its redox properties. This is particularly significant if a very efficient process or a chain reaction follows the initial activation induced by electron-transfer.

A possible classification of the overall series of the catalytic like set of steps following a photoinduced electron-transfer follows (1-2):

Type 1 of photoinduced electron-transfer catalysis
This type of reaction is met in organic, organometallic and inorganic chemistry. Its most familiar form in organic chemistry is pictured by $S_{RN}1$ substitution. It covers the cases where the source of catalyst is one of the partners of the starting donor-acceptor couple. The transformation of the procatalyst usually requires a photochemical activation.

Type 2 of photoinduced electron-transfer catalysis
In this type the role of light is to transform an inert procatalytic substance introduced in a donor-acceptor couple or in a pure compound into a catalyst which, as in type 1, then does its work without further need of photons (in the theoretical situation were no termination reactions occur).

The difference from type 1 is that the procatalyst is not one of the reagents but an external agent added in small amounts.

Type 3 of photoinduced electron-transfer catalysis or redox photosensitization
This class is met when a photosensitizer in an excited state gives an electron or a hole to the substrate. This electron-transfer creates a reactive intermediate able to evolve to the products via a succession of steps. In one of the steps, the oxidized or reduced photosensitizer regains or releases one electron.Then it is ready to begin an another cycle. For this type 3, the reaction is usually not catalytic in terms of photons because at least one photon will be consumed for every molecule of product formed. An exception would be the electron-transfer photosensitization which initiates a chain reaction.

Photoinduced electron-transfer catalysis : connection with electrochemical processes
Type 2 reactions may alternatively be induced by an electrode. These electrode-triggered chains parallel type 2 photoreactions because the electrode functions as a third reagent added to the donor-acceptor couple.

A great number of electrochemical reactions of this type is now reported (3-4) and the unified approach of ETC catalysis (3) let expect many more. Some of the electrochemically induced chemical transformations now have their photochemical counterparts. Unfortunately, there are only a few studies whose goal is a quantitative comparison of results obtained photochemically and electrochemically under similar experimental conditions of substrate, temperature, concentration and solvent (5-6)

The goal of the following section is such a comparison for two different reactions : one is a typical $S_{RN}1$ reaction (type 1 or 2 depending upon the mode of initiation), the other is an electron-transfer photosensitization reaction applied to the activation of C-H bonds towards O_2.

233

F. Minisci (ed.), Free Radicals in Synthesis and Biology, 233–251.

II. CORRESPONDENCE BETWEEN ELECTROCHEMICALLY INDUCED AND PHOTOCHEMICALLY INDUCED ELECTRON-TRANSFER : PHOTOCHEMICAL SUBSTITUTION OF HALOARYL KETONES BY PHENYL THIOLATE ANION.

II. 1 General aspects

4-bromobenzophenone, 4-bromoacetophenone and its fluoro and chloro analogs, react in acetonitrile, dimethylsulfoxide or dimethylformamide (all poor hydrogen donors) with the phenylthiolate anion under thermal ($60°C$) or photochemical ($\lambda > 330$ nm) activation. Besides the substitution products, benzophenone or acetophenone are formed together with diphenyldisulfide. The yield and the rate of haloderivative consumption depend upon the irradiation time, the light intensity and the nucleophile concentration. Results are summarized in Table I.

$$X\text{-}Ph\text{-}CO\text{-}R \quad + \quad PhS^- \quad \longrightarrow \quad Ph\text{-}S\text{-}Ph\text{-}CO\text{-}R \quad + \quad Ph\text{-}S\text{-}S\text{-}Ph$$

Table I - Photosubstitution of 4-bromobenzophenone and 4-haloacetophenones $4\text{-}XC_6H_4COR$ by Phenylthiolate $PhS^- M^+$. PhSSPh was not accounted for mass balance.

R	X	Solvent	Nu	Nu/ArX	ArX unreacted %	ArNu %	ArH %	Reaction time (h) a)
C_6H_5	Br	DMF	$PhS^- Et_4N^+$	2	13.5	83	3	1.5
-	-	-	-	2	19	95.5	4.5	1
-	-	DMSO	-	5	1	97	3	3
-	-	CH_3CN	$PhS^- K^+$	5	10	86.5	3.5	1
-	-	-	$PhS^- Bu_4N^+$	5	58	39	3	0.5
-	-	-	$PhS^- Na^+$	5	0	96.5	3.5	7
-	-	DMSO	$PhS^- Bu_4N^+$	5	23	74	3	5
-	-	CH_3CH	$PhS^- K^+$	2	25	72	3	4
-	Cl	DMF	$PhS^- Na^+$	5	13	84	3	20
-	-	DMSO	$PhS^- Bu_4N^+$	5	23	74.5	2.5	12
-	F	DMF	$PhS^- Na^+$	5	5	92	3	30
-	-	DMSO	$PhS^- Bu_4N^+$	5	22	76	2	18

a) Medium pressure mercury lamp Hanau Q 81 - Pyrex vessel.

For the photochemical substitution, the incident light was filtered with a freshly prepared 2.5 x 10^{-2} M aqueous solution of phenylthiolate so that the nucleophile was not excited and the phenylthioarylketone was not photolyzed. In consequence, the diphenyldisulfide formed did not undergo photofragmentation and thus accumulated in the medium. So the photochemical activation (λ > 330-340 nm) excited only haloarylketones (PhS $^-$ absorbs at 300nm) and induced an electron transfer between the phenylthiolate Nu $^-$ anion (donor) and the ketone ArX (acceptor). We also determined that irradiation under otherwise identical reaction conditions in less polar solvents or without the presence of nucleophile induced a far slower homolysis of the C-X bond of 4-bromobenzophenone (10 % conversion after 50 h). In acetonitrile, the following byproducts are formed : 1-cyano-2-phenylethane (< 0.5 %) and tributylamine (< 0.5 %) if the tetrabutylammonium cation is the counter ion of the phenylthiolate anion.

In contrast to the electrochemical substitution (7-13) we did not observe any enhancement of the photochemical reactivity when using tetrabutylammonium phenylthiolate instead of the sodium or potassium salt.On the other hand, diphenyldisulfide was formed (10-20 % yield) in every photochemical experiment, in contrast to the reaction triggered with a cathode tuned to -1.8 V (7-9). Any diphenyldisulfide which forms in the electrochemical experiment cannot accumulate in the medium ; its reduction potential is -1.8 V vs SCE (14) and it is reduced at the cathode, thereby regenerating the original nucleophile :

$$\text{Ph-S-S-Ph} \xrightarrow{\quad e \quad} \text{Ph-S-S-Ph}^{\overline{\cdot}} \xrightarrow{\hspace{3cm}} \text{PhS}^- + \text{PhS}^{\cdot}$$

In addition, the intermediate PhS$^{\cdot}$ may also possibly be reduced to PhS $^-$ at the electrode.

II.2 Quantum yield

The observed quantum yield values are greater than 1, as reported in Table II, and agree with a $S_{RN}1$ mechanism. The reaction occurs only in polar solvents such as DMF, DMSO or acetonitrile.

Table II - Quantum yield Φdis. for the disappearance of the substrate of the reaction (360nm). Substrate concentration : 10^{-2} M ; nucleophile : PhS $^-$ Bu$_4$N$^+$; solvent : DMSO.

Substrate ArX	Nu / ArX	Irradiation time (min.)	Φ dis.
4-bromobenzophenone	2.5	3	3.5
-	2.5	5	3.36
-	10	1	6.2
4-bromoacetophenone	3	3	2.9
-	6	3	3.9

For the electrochemical substitution, the number of Faradays passed through the electrolytic cell per mole of starting material (Faradaic yield) was 0.2 (7-9). It should correspond to an overall yield of 0.2 photons per mole (Φ = 5) for the photocatalyzed process; the primary quantum yield is not known and may include some part of back electron-transfer. This overall quantum yield depends on the nucleophile concentration (section II.5). For the same ratio Nu/ArX = 10, we measured Φ as 6.2 for the initial rate of the reaction. Nevertheless, on a preparative scale, our own photochemical results suggest that the overall quantum yield for a high conversion process is < 5.

II.3 Overall quantum yield variation with irradiation time

The overall quantum yield for the disappearance of 4-bromobenzophenone decreases with the irradiation time of the substrate (Table III). A negative consequence is that it is difficult to completely consume the substrate (e.g.in Table I, note the irradiation time necessary to obtain 80 % conversion of the original haloarylketones).

Table III - Rate of the substrate disappearance with irradiation time.

Substrate : 4-bromobenzophenone 10^{-2} M ; nucleophile : PhS⁻ $Bu_4 N^+$

[Nu] / [ArX] = 2.5 ; solvent : DMSO ; pyrex filter + 1 cm aqueous solution PhS⁻ K^+ ;

Hanau Q.81 Lamp ; Intensity = 1.75 x 10^{-4} E / l. mn.

Irradiation time (min.)	Disappearance of the substrate (%)	Φ dis. (overall)
3	23.4	3.5
5	36.8	3.36
7.5	43.7	2.66
10	48.4	2.21
15	50.9	1.55
20	55.2	1.26

This self-inhibition of the reaction is probably not due to energy transfer quenching by one of the reaction products (ArNu or ArH), since 4-bromobenzophenone does not photosensitize the decomposition of benzophenone or 4-phenylthiobenzophenone. Thus 95 % yields of substitution product were obtained for an irradiation time of 1 h 30 mn.

Another explanation may result from the lowering of the rates of second order chain carrying steps versus the (probably) first order termination steps. This agrees with the same kinetic behaviour that we observed for the thermal reaction and for the photochemical reaction.

To minimize quantitative problems connected with self -inhibition, quantum yield variations with reaction parameters and with the inhibitory effects of electron acceptors were performed at very low conversion percentages (5-15 %).

II.4 Quantum yield variation as a function of intensity of irradiation and concentration of the nucleophile

The $S_{RN}1$ mechanism often corresponds to a chain reaction initiated by a photoinduced electron transfer. The dependence of the overall quantum yield on reactant concentration and absorbed light intensity has been treated kinetically by several authors. In the absence of a charge transfer complex between the donor and the acceptor, as in our case, and if all the incident light is absorbed by the substrate, Stranks and Yandell (15-16) and Tolbert (17) propose that the overall quantum yield is proportional to the nucleophile concentration Nu⁻ and to $I^{-1/2}$. The variation of Φ with these parameters provides further information on the nature of termination steps (18-19).

- Variation with light intensity

The results show that for 4-bromobenzophenone, Φ varies linearly with $I^{-1/2}$. Such a variation implicates quadratic termination steps, as was noted by Stranks and Yandell (15-16).

Adamson and Sporer observed the same behaviour for the photoinduced exchange of $PtBr_6^{2-}$ with radiolabelled Br⁻ (20).

This linear variation suggests termination steps involving two radicals or radical anions :

$$Ar^{\bullet} + Ar^{\bullet} \longrightarrow Ar - Ar \qquad [A]$$

$$Ar^{\bullet} + ArX^{\overline{\bullet}} \longrightarrow Ar^{-} + ArX \qquad [B]$$

$$Ar^{\bullet} + ArNu^{\overline{\bullet}} \longrightarrow Ar^{-} + ArNu \qquad [C]$$

Possibly followed by (SH : solvent)

$$Ar^{-} + SH \longrightarrow ArH + S^{-} \qquad [D]$$

Ar probably does not result from electron transfer between Ar + Nu , i.e.,

$$Ar^{\bullet} + Nu \longrightarrow Ar^{-} + Nu^{\bullet} \qquad [E]$$

We determined that the ArH yield does not increase when the nucleophile concentration increases. The reduction products ArH (benzophenone or acetophenone) form in every experiment. Therefore for photochemical stimulation, the termination steps involved are the same as those reported by Saveant's group (7-11) in electrochemical studies : [B] or [C] followed by [D]. Thus at this level, nothing unique apparently differentiates termination steps involved in the electrochemical and photochemical experiments. The photosubstitution of iodobenzene by potassium diethylphosphite follows a $S_{RN}1$ mechanism (21). It showed a dependence of initial rate on light intensity which was somewhat less than first-power but closer to it than to the half-power which prevails for many photocatalyzed radical reactions (22).

- Variation with nucleophile concentration

For a constant concentration of 4-bromobenzophenone, the value of the overall quantum yield is approximately linear with nucleophile concentration until a limiting value of Nu/Arx = 3 is reached.Tolbert (17) has proposed a limiting expression of the quantum yield for a chain process involving a linear relationship with nucleophile concentration. Stranks and Yandell (15-16), for the photoinduced electron exchange chain reaction between Tl^{III} and Tl^{I} , showed that the observed overall quantum yield was directly proportional to the concentration of each of the reactants. They also observed that, for higher concentrations, the quantum yields reached a plateau, in agreement with our experiments.

The analysis of Hoz and Bunnett (21) of the photosubstitution of iodobenzene by potassium diethylphosphite was complicated by the existence of a charge transfer complex (CTC) between the reactants. Dividing the raw quantum yield by the fraction of light absorbed by the substrate, these authors obtained a corrected value of Φ corresponding to the situation where the photons absorbed by iodobenzene are also the photons which stimulate the reaction. This corrected quantum yield was found to be linearly related to the nucleophile concentration. This result agrees with our results for the linear variation of Φ with nucleophile concentration. Because of the CTC, these authors were unable to conclude whether the photons absorbed by the charge transfer complex or by the substrate were responsible for the stimulation of the reaction. Therefore, they proposed that the rate law obtained resulted from a combination of the two phenomena.

We noted the same results as in the Bunnett's propositions: if no CTC is involved, the reaction is first order in nucleophile and zeroth order in substrate concentration.

II.5 Conclusion

This study of an electron transfer induced chain reaction, through electrochemical (7-11) and photochemical methods, illustrates both similarities and differences. The similarities are the same approximative yield, the same primary termination step, the same initiation step, and a comparable chain length involved.

The main difference is the production of small amounts of diaryldisulfide in the photochemical experiment, together with 3% of the reduction product. In the electrochemical study, small amount of reduction product were also formed. Nevertheless, this selectivity has been questioned (23). From a synthetic point of view this gives an advantage to the electrode process for this reaction. More examples are necessary to decide if this conclusion is general.

238

III. PHOTOINDUCED ELECTRON-TRANSFER CATALYSIS : REDOX PHOTOSENSITIZATION APPLIED TO THE ACTIVATION TOWARD OXYGEN OF ALKYL SIDE CHAIN AROMATIC COMPOUNDS.

Redox photosensitization emerges as a powerful activation mode.
Some examples from the literature (24) illustrate the high yields and the good selectivities which may be obtained.

In the foregoing example we have seen a photostimulated reaction where a chain follows the photoinitiation step. Thus there was no need for regeneration of the photosensitizer if any. In the third class : redox photosensitization, the first electron-transfer is exergonic ($\Delta G < 0$) since the photosensitizer is in an excited state. The reverse electron-transfer between the two resulting paramagnetic species (e.g. radical anion-radical cation) is also exergonic. In many cases, it cancels the first one so that there is no overall reaction. One way to by pass this back electron-transfer is to trap one of the transients resulting from the photosensitizing step. The activation of aromatic compounds with light and electron acceptors involves such a principle.

III. 1 General aspects
The redox photosensitized oxidation of substituted aromatic hydrocarbons leads to aromatic aldehydes with aromatic acids as secondary products.

Electron acceptor
Photosensitizer

Electron donor
Substrate

radical anion   **radical cation**   **radical**

III.2 Photosensitizer

In this study, the substrate is the electron donor. The electron-acceptor photosensitizers : 1-4 dicyanobenzene : **DCNB** or tetrachloro 1, 2, 4, 5-benzoquinone : **CIA**.display the following properties

Excited state energies :

$E_{o-o} (^1DCNB \rightarrow {}^1DCNB^*) = 98.6$ kcal / M

$E_{o-o} (CIA \rightarrow {}^3CIA^*) \quad = 62.3$ kcal / M

Excited state lifetime :

$\tau_s (^1DCNB^*) = 9.7$ ns

Redox potentials :
E (DCNB / DCNB$\dot{}$) = - 1.6 V (vs. SCE)
E (CIA / CIA$\dot{}$) = 0.02 V (vs. SCE)
SCE : standard calomel electrode.

III.3 Lamp and irradiation time

The used lamp was a UV medium pressure Mazda MAF 4OO W. With quartz vessel (**DCNB** sensitizer) the reaction was stopped after 15 min. With pyrex filter (**CIA** sensitizer) we irradiated for 90 min. These times correspond to 60-80 % substrate consumption.

III.4 Solvent

Acetonitrile is the most generally used solvent in electron-transfer photosensitized reactions and was retained for the present study after checking that it gives better results than benzene, methanol and DMSO.

Table IV : Solvent effect on the rate of the reaction for 1,4-dicyanobenzene photosensitized oxidation of p-xylene.

Solvent	Dielectric constant ε	Viscosity η (cP)	Substrate consumption %
C_6H_6	2.3	0.6	23
CH_3OH	32.7	0.54	28
CH_3CN	37.5	0.34	64
DMSO	46.7	2	0

400 W medium pressure mercury lamp, quartz vessel, 15 mn irrradiation.

III.5 Results

Cyano and nitro alkylaromatics do not react with oxygen when the substrate is photoactivated with an electron acceptor as redox photosensitizer.

Other substituents were tested which did not hinder the activation of the benzylic CH bond. For 15 mn irradiation, the substrate conversion varies from 63 to 83 % (Table V). The aldehyde ratios in the medium are then 7.5% to 49%.

$$\text{R-PhCH}_3 \xrightarrow[\text{CH}_3\text{ CN / 15 mn}]{\text{DCNB / O}_2 \text{ / hv / quartz}} \text{R-PhCHO + R-PhCOOH}$$

$$\mathbf{1} \qquad\qquad\qquad\qquad\qquad \mathbf{2} \qquad\quad \mathbf{3}$$

Table V : Consumption of the substrate and ratios of reaction products after 15 mn irradiation (medium pressure 400 W mercury lamp - quartz vessel) - Photosensitizer DCNB.

R	θ%a)	1 %	2 % b)	3 % b)
H	63	37	48	15
m-CH3	64	36	46	18
p-CH3	64	36	49	15
3,5-dimethyl	84	16	30	4
Ar = 6-methyl,2-naphthyl	84	16	15	8
p-Cl	82	18	16	2.5
p-OMe	83	17	7.5	1.5

a) θ : conversion of the substrate; b) percentage of identified products.

For 4-methyl anisole, the yield of product per mole of consumed substrate may be enhanced by irradiation through pyrex vessel with chloranil as photosensitizer since 4-methoxy benzaldehyde then is not photolyzed with **DCNB** in quartz.

$$\theta = 88 \% \qquad\qquad\qquad Rd = 74 \% \qquad 1 \% \qquad 1 \%$$

4-methoxy benzylchloride is formed since chloranil is slightly photolyzed (only 80-90 % ClA is recovered after the reaction). In this equation and the following, the preparative yield is the product of θ x Rd.

For 4-bromotoluene, if pyrex filter is not used the main reaction is the photolysis of the C-Br bond leading to toluene formation. By using chloranil photosensitizer with pyrex filter 4-bromo-toluene is oxidizided and a 57 % yield of 4-bromobenzaldehyde is obtained per mole of 4-bromotoluene; 42 % 4-chlorotoluene is formed as a secondary product.

$$\theta = 59\ \% \qquad\qquad Rd = 57\ \% \qquad 42\ \% \qquad 1\ \% \qquad 1\ \%$$

N,N-dimethyl-p-toluidine is oxidized with a significant yield only if pyrex filter is used. In this latter case the dicyano 1,4-benzene DCNB does not absorb light. The substrate is electronically excited and DCNB is no longer a photosensitizer but a recyclable electron-acceptor.

$$\theta\ = 99\% \qquad\qquad\qquad\qquad\qquad\qquad 34\%$$

III.6 Hints for a possible mechanism

The relative redox properties of the aromatic substrates studied and those of chloranil and dicyanobenzene in their fundamental and excited state are gathered in Fig 1.

The corresponding values of ΔG in the Rehm-Weller equation (29) are given in Table VI. These data show that the oxidation of the studied substrates by the photosensitizers in their excited state is thermodynamically feasible. This thermodynamic feasibility is not by itself sufficiently compelling to propose the formation of a radical cation ; indeed the lifetimes of the excited sensitizers could be too short or even if they are long enough one must keep in mind that some highly thermodynamically favoured electron-transfer have been reported to be overpassed by another type of reaction.(30-31).

$$E \left({}^{1}DCNB^{*} / DCNB^{\cdot} \right) = E \left(DCNB / DCNB^{\cdot} \right) + E_{DCNB \longrightarrow {}^{1}DCNB^{*}}$$

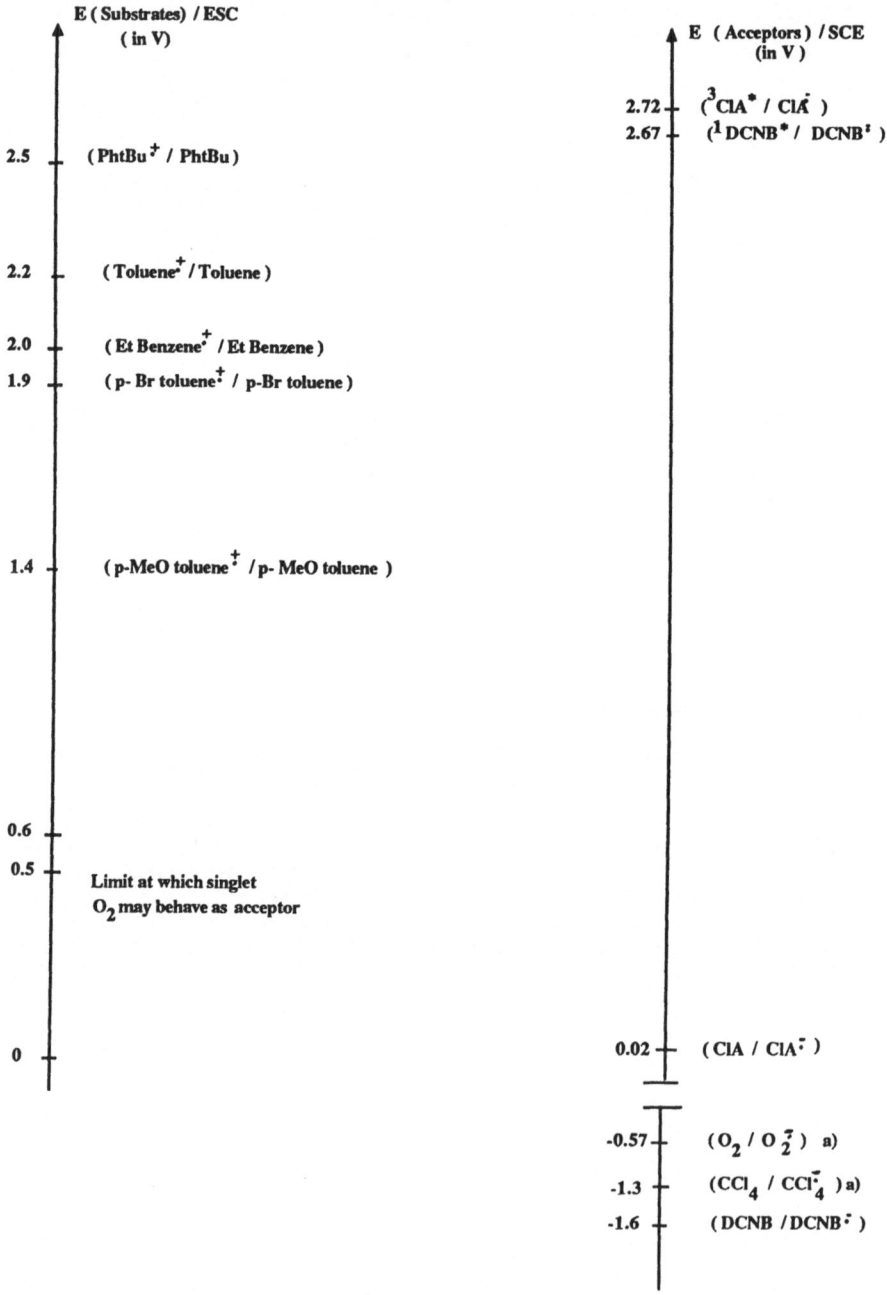

E (Substrates) / ESC (in V)

2.5 — (PhtBu^{+} / PhtBu)

2.2 — (Toluene^{+} / Toluene)

2.0 — (Et Benzene^{+} / Et Benzene)
1.9 — (p- Br toluene^{+} / p-Br toluene)

1.4 — (p-MeO toluene^{+} / p- MeO toluene)

0.6 —
0.5 — Limit at which singlet
O$_2$ may behave as acceptor

0 —

E (Acceptors) / SCE (in V)

2.72 — (^{3}ClA* / ClA$^{\cdot}$)
2.67 — (^{1}DCNB* / DCNB$^{\cdot}$)

0.02 — (ClA / ClA$^{\cdot}$)

-0.57 — (O$_2$ / O$_2^{\cdot}$) a)
-1.3 — (CCl$_4$ / CCl$_4^{\cdot}$) a)
-1.6 — (DCNB / DCNB$^{\cdot}$)

Fig. 1 Relative redox properties of the used sensitizers in their fundamental and excited states a) see ref. for more details .

L. Eberson   Electron Transfer   Verlag 1987

Table VI : ΔG values from Weller equation (29) calculated for photoinduced electron-transfers between excited photosensitizers and substrates.

Aromatics	Acceptor	Δ G / Kcal / mole
Toluene	(DCNB)*	-11
3-Me Toluene	(DCNB)*	-17.9
4-Me Toluene	(DCNB)*	-20.2
2-Cl Toluene	(DCNB)*	-13.3
4-Cl Toluene	(DCNB)*	-13.3
4-Br Toluene	(DCNB)*	-18
3-5-DiMe Toluene	(DCNB)*	-17.9
4-MeO Toluene	(DCNB)*	-29.7
4-MeO Toluene	(Chloranile)*	-30.7
4-CN Toluene	(DCNB)*	
4-NO$_2$Toluene	(DCNB)*	
4-NO$_2$Toluene	(DCNB)*	

Other experimental data back, however, this radical cation formation from the substrate.

Firstly, all the aromatic substrates irradiated under the same conditions but in a solution without sensitizer are not oxidized by O_2.

Secondly, the same aromatic substrates irradiated under the same conditions but using typical $^1O_2$ sensitizers (rose bengal).

Third, aromatic substrates such as 4-cyano, 4-nitro or 2-nitrotoluene, irradiated in the presence of chloranil or dicyanobenzene while O_2 is bubbled through the solution do not react ; values reported in Table VI show that, for these substrates, an electron-transfer would be endothermic.

Fourthly, an electron donor inhibits the reaction : 1,4-dimethoxybenzene **DMB** (E1/2 DMB$\ddagger$/DMB = 1.4 V vs SCE) strongly decreases the rate of photosensitized oxydation of p-xylene Xy (E1/2 Xy$\ddagger$/Xy = 1.8 V vs SCE) when added stoichiometrically to the medium.

Reaction	Xy consumption %	DMB consumption %	Aldehyde %
no DMB	64	-	76
2 x 10^{-4} M DMB	10	62	1

Fifthly, under comparable conditions, the rate of the photosensitized oxidation is slightly increased by addition of a cosensitizer : biphenyl **BP** (E1/2 BP^{+}/BP = 1.85 V vs SCE). When added stoichiometrically, it increases the conversion of the substrate for the **CIA** photosensitized oxydation of 4-methoxytoluene (E 1/2 MeO^{+}/MeO = 1.39 V vs SCE) from 88 % to 94 %.

This pattern of reactivity has been repeatedly reported in electron transfer sensitized oxidations (32-34)

Sixthly, the solvent effects gathered in Table IV are characteristic of an electron-transfer sensitized reaction (35-36). These experimental data added to the fluorescence quenching experiments performed by Albini (37) and Saito (38), strongly back the hypothesis of an electron transfer sensitization. One must note however, that the most compelling evidence for this first step of the mechanism : time resolved spectroscopic observation of the radical cation of the substrates is still missing.

What is the possible fate of this radical cation ?

The most striking property of alkyl aromatic radical cations is their acidity : the pKa of toluene radical cation has been estimated to be about -10 (39). The rate of deprotonation of toluene radical cation in weakly acidic aqueous solution is greater than 10^7 M^{-1}s^{-1}(40–41), but drops to 10^4 M^{-1}s^{-1} when pentamethylbenzene radical cation is considered.

In CH$_3$CN, these rates would certainly be higher, depending of course upon the attacking bases. The potential bases possibly present in the oxidizing medium presently studied are O$_2^{-}$ (pKa of conjugated acid = 13 in DMF) (42) Photosensitizer $^{-}$ (pKa of conjugated acid reported in ref 43-44; ArCH$_2$O$_2^{-}$ (pKa of conjugated acid = 12) (45-46)). The conditions to form a radical are therefore largely fullfilled :

$$Ar\text{-}CH_2\text{-}R \overset{+}{\cdot} \quad + \quad B \quad \longrightarrow \quad Ar\text{-}\overset{\bullet}{C}H\text{-}R \quad + \quad BH^{+}$$

and the benzyl radical may be trapped by quinones when O$_2$ is absent (37).

Such a radical, when formed, is expected to react at a diffusion limited rate (47) with O$_2$.

$$Ar\text{-}\overset{\bullet}{C}H\text{-}R \quad + \quad O_2 \quad \longrightarrow \quad \underset{\underset{O\text{-}O^{\bullet}}{|}}{Ar\text{-}CH\text{-}R}$$

Among the other patterns of reactivity expected for these radical cations one must recall the dimerization (48-49) which has trained several mechanistic controversies (50) in electrochemical works and the attack by nucleophiles. The analysis of products formed under our oxidation conditions does not show dimerization except with p-hydroxytoluene. The rate of hydratation of toluene radical cation is 5 x 10^8 M^{-1}s^{-1}. It shows that the aromatic nucleus, when ionized, is highly susceptible to nucleophilic attack.

Such a kind of nucleophilic reactivity could be thought of for the reaction of O$_2^{-}$with methyl viologen reported by Sawyer (42). Applied to aromatic radical cation it would give ring opening :

For some of the studied substrates the loss of substrate (Table IV) is not accompanied by the production of tars. For these substrates the attack of O$_2^{-}$ on the ring could possibly explain the material loss, however the DCNB photosensitized reaction of p-xylene with KO$_2$ crown solubilised in CH$_3$CN in the absence of O$_2$ does not yield p-methylbenzaldehyde (38). This point is presently under study.

When one reaches the stage of Ar-C(H)(R)OO˙ radicals the mechanistic discussion becomes more fuzzy.The experimental data which should help in making a mechanistic proposition are :
- Formation of benzaldehydes (primary alkyl arenes) or arylketone (secondary alkyl arenes).
- Greater oxidability under our conditions of benzylic alcohol in comparison with toluene (51).
- No consumption of the photosensitizer in most of the studied reactions (with exceptions for ClA).
- Slow only photochemical reaction between the substrate and the sensitizer when O$_2$ is absent.
- A radical trap, the galvinoxyl, stops the reaction. Less than stoichiometric (50 %) addition of galvinoxyl decreases the 4-methoxytoluene conversion from 84 % to 5 %.
- Few amount of CCl$_4$ (80 μl / 20 ml) accelerates the photosensitized oxidation of p-xylene : its conversion changes from 64 % to 79 %.
- Bibenzyl type of by products are not observed.
- Usually, at low conversion, the selectivity in aldehyde versus acid production under our conditions is higher than the one reported in autoxidation of aromatics.

One may recall that the splitting of hydroperoxides may be effected at 100° in sulfolane in 10 minutes by acid catalysis with amounts of pure acid varying from 0.1 % to 10 % by weight based on the hydroperoxide employed. The formed product, under these conditions are phenols and aldehydes (respectively 39 and 31 % for the case of benzyl peroxide) (52,53). As we observed no phenols among the products, we feel safe in discarding such a kind of reaction under our reaction conditions. Excited state of benzaldehyde could play a role in the production of acid because the use of a filter absorbing most of the light exciting benzaldehyde leads to a decrease in the production of acid.

At this point it is difficult to choose between the various overall schemes of transformation (or variations about them) shown in following schemes.

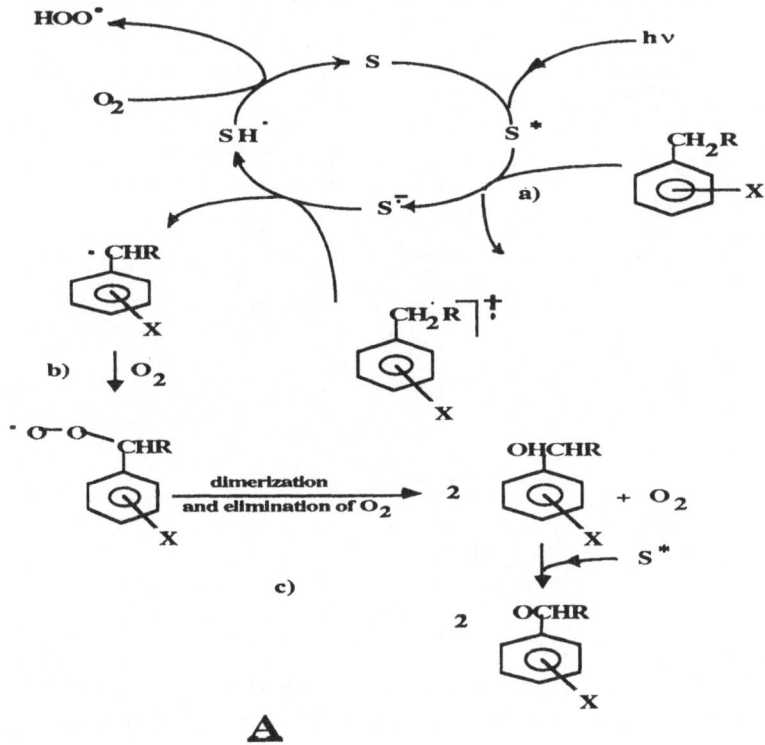

A

See caption under **B**.

246

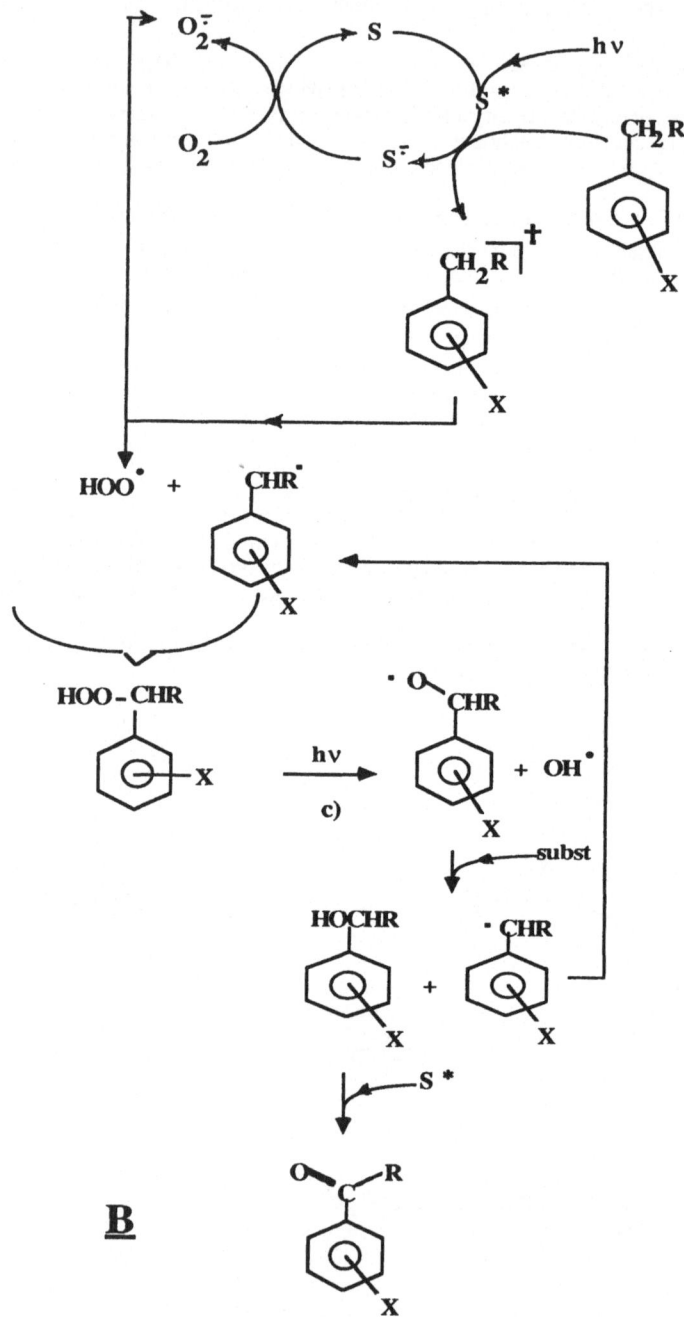

B

a) possibly via an exciplex (37).

b) alternative pathway, reaction with $O_2H\cdot$ formed in the regeneration of the sensitizer

c) other possible fate : reaction with $SH\cdot$ to produce $Ph\,CHR\,OOH$ and regenerate S.

d) O_2^- not detected by large excess of phenyl caprate (38).

Experiments are presently in progress to select which scheme is the most sensible keeping in mind that other non $^1O_2$ mechanisms are reported in the litterature as displayed in scheme :

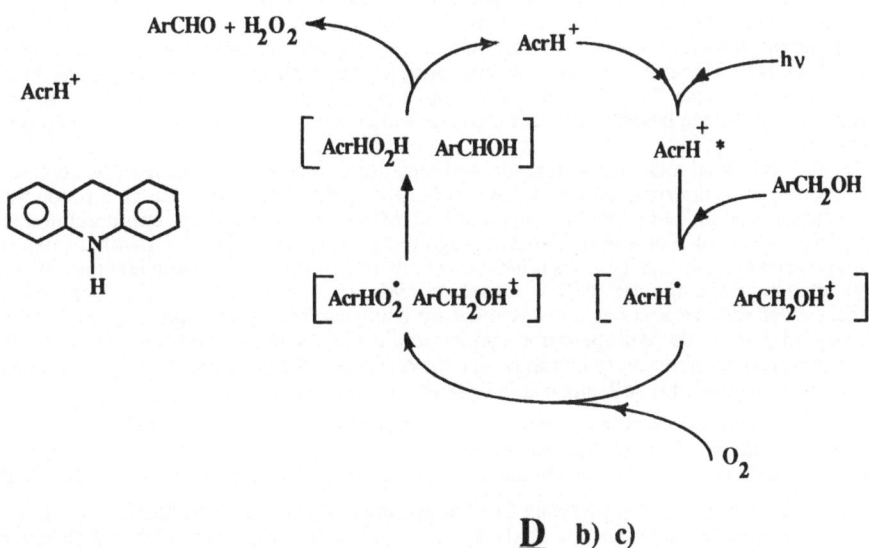

$\underline{C}$ a)

Other photocatalysed (54) mechanisms of oxidation are reported in the literature .

$\underline{D}$ b) c)

a) from ref (55, 56).
b) from ref (27).
c) in the original ref most of the steps are represented as reversible.

Finally one should stress that the photosensitized side chain oxidation of alkyl aromatics may take a completely different pathway. For example t-Bu benzene does not photooxidize in the presence of ClA and O_2 but does react if ClA is replaced by CCl_4 :

$$t\text{-Bu-}C_6H_5 \quad \xrightarrow[\text{O_2 \quad CH_3 CN}]{\text{hV \quad CCl_4}} \quad C_6H_5\text{-CO-CH}_3$$

Various explanations have been proposed (37) for the role of CCl_4 in such a reaction but the mechanism is here clearly different from the e.t. sensitization mechanism. In the case of 4-amino toluene the substituent becomes the center of reactivity :

The triggering step is probably one of deprotonation of the radical cation followed by dimerization

Although the products are reminiscent of those observed in the non photochemical oxidation of aromatic amines by $O_2^{\cdot}$ (57, 58) the mechanism involved here is different. Indeed the superoxide ion does not H abstract H-N for p-methylaniline as it does for p-naphthylamine.

III.7 Comparison of the synthetic value of electron-transfer photosensitized oxidation of alkyrarenes with other methods of preparation of arylaldehydes.

Reactions data banks count now more than 50.000 reactions (59), it should therefore become a duty for authors proposing a new synthetic method to compare it to those previously available for the selected target. This is not an easy task one must, on one hand, consider the industrial methods usually described in the Kirk Othmer encyclopedia of technology (60) , on the other hand one should also consider the laboratory scaled methods which bear some connection with the proposed method.

On the industrial side, benzaldehyde and substituted derivatives are mainly obtained as by-products during the oxidation of toluen to benzoic acid (60). Some are also produced by hydrolysis of benzal chloride but this method often yields products contaminated with chloroderivatives (60). Toluene oxidation at high temperature (170°C-220°C) with a mixture of gaseous oxygen and an inert gaseous diluent and with a conversion percentage less than 10 % has been claimed to yield a mixture of 30 % of benzaldehyde, 16 % of by-products (52). The method of preparing aldehydes and ketones later evolued by using preferably the splitting of primary or secondary alkylaromatic hydroperoxides under acidic (53) or basic conditions (61). We have selected some recent references which show that the oxidation of toluene under catalytic stimulation using O_2 or air as reagent is still aimed at in different countries (62, 65).

Another stream of research is based on the chemical generation of the radical cation of the substrate, this radical cation is then transformed by proton abstraction and the resulting radical reacted either with O_2 or with H_2O. On the side of O_2 as reagent one may cite Baciocchi's method (Table VII) which uses the strong oxidant Ce^{IV} ammonium nitrate and photostimulation to generate at room temperature the radical cation of alkylaromatics. The use of H_2O or another nucleophile as reagent apparently leads to a better selectivity in adehyde as shown by the results obtained by Minisci, Walling, Kreh (66, 67, 75, 76) or Shulpin (68) and Fujihara (photochemical stimulation added (69)). In this case a transition metal salt (usually Cu^{II}) is added to the mixture to generate the carbocation of the radical generated by the deprotonation of the radical cation.

Electro-organic reactions of industrial interest have been well developped during the last decade (70-74) and approximative correspondance principles let expect processes of oxidation for alkylaromatics based on electrochemistry. This is indeed the case and the best results have been obtained with indirect electroorganic syntheses (77). G. Kreysa and H. Medin report the indirect electrochemical oxidation of p-methoxy toluene to p-methoxybenzaldehyde with Ce^{4+}/Ce^{3+} as redox mediator system. Selectivity of p-methoxybenzaldehyde was optimized up to 98 % by factorial design of experiments (78). Kreh and al. have patented a process based on the same principle (75, 76). Another indirect method uses Mn^{3+} and OH· as electrochemically generated reagents able to provide high selectivity in benzaldehyde starting from toluene (79). Older direct (80-82) or indirect (83, 84) electrolytic methods have been patented.

These results, those gathered in Table VII and application of correspondance principles suggest that better yields and selectivity in aldehyde production should be obtained by electron transfer sensitization. We are presently exploring the possibilities offered by added transitior metal complexes, phase transfer catalysis, supported sensitizers and switch from O_2 to H_2O as a reagent.

Table VII : Comparison with Etard's reaction (CrO_2Cl_2 in CCl_4)

Substituent on Toluene	Etard's reaction a)	Anodic oxidation b)	Ce IV oxidation d)	e.t. photosens. oxidation g)
H	90		90 ; 24 e)	47
3-methyl	60		60	45
4-methyl	70-80%	86%	53 e)	48
2-chloro	52%			7
4-chloro	75%	78%	86	16
4-bromo	78%			33
3,5-dimethyl	6%			25
4-MeO	0%	78% , 98% c)	71% ; 81% f)	65
4-NO$_2$	70%			0

a) 85 ; b) 87; c) 78; d) 75, 76; e) 55, 56; f) metal transition catalysis. 66,86; g) This work

III.8 Conclusion

If we rest on the comparison of this redox photosensitized method of oxidation with other ones, one could conclude that it is not competitive. Nevertheless some substrates which are not oxidized with the Etard reagent lead to aldehydes with good yields in the photochemical method (e.g. 4-methylanisole).

We also underline that our system is to be optimized. Some improvements are now investigated :
- Use of light-filters to decrease the further oxidation of aldehyde
- Adsorption of the photosensitizer on SiO_2 or exchange resins to improve its stability and its ease of final separation.
- Search for new photosensitizers able to operate in near UV or visible region.

On a synthetic point of view, the method, earlier proposed by Saito, demonstrates the possibility of CH activation, at 25°C, with a cheap reagent : the air.

REFERENCES

(1) M. Julliard and M. Chanon, Chem. Rev. **83** (1983) 425.
(2) M. Julliard and M. Chanon, Chem. Scr. **24** (1984) 11.
(3) M. Chanon, Bull. Soc. Chim. Fr., **2** (1985) 209.
(4) M. Chanon, Acc. Chem. Res., (1987) 214.
(5) P. Nelleborg, H. Lund and J. Eriksen, Tetrahedron Lett.**26** (1985) 1773.
(6) M. Kojima, H. Sakuragi and K. Tokumaru, Bull. Chem. Soc. Jpn. **58** (1981) 521.
(7) C. Amatore and J.M. Saveant, J. Electroanal. Chem. Interfacial Electrochem. **126** (1981) 1.
(8) J.M. Saveant, Acc. Chem. Res., **13** (1980) 323.
(9) C. Amatore, J. Pinson, J.M. Saveant and A. Thiebault, J. Am. Chem. Soc. **103** (1981) 6930.
(10) J. Pinson and J.M. Saveant, J. Chem. Soc. Chem. Commun. (1974) 933.
(11) J. Pinson and J. M. Saveant, J. Am. Chem. Soc., **100** (1978) 1506.
(12) W. J. M. van Tilborg and C.J. Smit, Tetrahedron Lett., (1977) 3651.
(13) W.J.M. van Tilborg, C. J. Smit and J. J. Scheele, Tetrahedron Lett., (1978) 776.
(14) H. Tagaya, T. Aruga, O. Ito and M. Matsuda, J. Am. Chem. Soc. **103**, (1981) 5484.
(15) D.R. Stranks and J.K. Yandell, J. Phys. Chem., **73** (1969) 840.
(16) D. R. Stranks and J. K. Yandell, Exch. React. Proc. Symp. (1965) 83.
(17) L. M. Tolbert J. Am. Chem. Soc., **102** (1980) 6808.
(18) E.S. Huyser, "Free Radical Chain Reactions", Wiley Intersciences New York, N. Y. (1970) p 44.
(19) J.F. Bunnett "Investigation of Rates and Mechanisms of Reactions" 3rd ed., Part 1, E.S. Lewis Ed., Wiley, New York, N. Y. (1974) p 381.
(20) A. W. Adamson and A. H. Sporer, J. Am. Chem. Soc., 80 (1958) 3865.
(21) S. Hoz and J.F. Bunnett, J. Am. Chem. Soc. **99** (1977) 4690.
(22) E.S. Huyser "Free-Radical Chain Reactions" , Wiley Intersciences New York, N. Y. (1970) p 51.
(23) B. Helgee and V.D. Parker, Acta Chemica Scandinavica, **B 34**, (1980) 129.
(24) M. Chanon and L. Eberson, "Photochemistry of Homogeneous and Heterogeneous Gears Involving Electron Transfer Catalysis" in Photoinduced Electron Transfer, M.A. Fox and M. Chanon Eds., Elsevier, Amsterdam (1988), Vol. A, p 409.
(25) A. Albini and D.R. Arnold, Can. J. Chem., **56** (1978) 2985.
(26) M. Yasuda, T. Yamashita, T. Matsumoto, K.Shima and C. Pac, J. Org. Chem., **19** (1985) 3667.
(27) S. Fukuzumi, S. Kuroda and T. Tanaka, J. Chem. Soc. Chem. Commun., (1987) 120.
(28) K. Gollnick and A. Schnatterer, Tetrahedron Lett., **26** (1985) 173.
(29) D. Rehm and A. Weller, Isr. J. Chem. **8** (1970) 259.
(30) S. Steenken in "Free Radicals Chemistry, Pathology and Medecine" Vol 3, Rice Evans and Dormonty T. Eds., Richilieu Press London (1987).
(31) E. Bothe, G. Behrens and D. Schulte-Frolhinde, Z. Naturforsch. **32b** (1977) 886.
(32) A.P. Schaap, S. Siddiqui, D. Gagnon and L. Lopez, J. Am. Chem. Soc., **105** (1983) 5149.
(33) A.P. Schaap, L. Lopez, S.D. Anderson and S.D. Gagnon, Tetrahedron Lett., **23** (1982) 5493.
(34) A.P. Schaap, L. Lopez and S.D. Gagnon, J. Am. Chem. Soc., **105** (1983) 663.
(35) J. Santamaria,"Photoinduced Electron-Transfer", M.A. Fox and M.Chanon Eds., Elsevier, Amsterdam (1988), Vol. B, p 483.
(36) M. Julliard,"Photoinduced Electron-Transfer", M.A. Fox and M. Chanon Eds., Elsevier, Amsterdam (1988), Vol. B, p 216.
(37) A.Albini and S. Spreti, Z. Naturforsch., **41B** (1986) 1286.
(38) I. Saito, K. Tamoto and T. Matsuura, Tetrahedron Lett., **31** (1979) 2889.
(39) A.M. De Nicholas and D.A. Arnold, Can. J. Chem., **60** (1982) 2165.
(40) K. Schested, J. Holemen and E.J. Hart, J. Phys. Chem., **81** (1977) 1363.
(41) K. Schested and J. Holeman, J. Phys. Chem., **82** (1978) 651.
(42) D.T. Sawyer, E.J. Nanni and J.L. Roberts, "Electrochemical and Spectrochemical Studies of Biological Redox Compounds" K.M. Kadish Ed., Adv. Chem. Ser. n° 121, ACS 1982 p 585.
(43) S.L. Mattes and S. Farid, J. Am. Chem. Soc. **108** (1986) 7356.

(44) F.D. Lewis and J.R. Petisce, Tetrahedron, **42** (1986) 6207.
(45) I.M. Kolthoff and A.I. Medalia, J. Am. Chem. Soc. **71** (1949) 3789.
(46) A.J. Everett and G.J. Minkoff, Trans. Faraday Soc. **49** (1953) 410.
(47) B. Maillard, K.U. Ingold and J.C. Scaiano, J. Am. Chem. Soc., **105** (1983) 5095.
(48) R.O.C. Norman, C.B. Thomas and P.J. Ward, J. Chem. Soc. Perkin Trans I, **23** (1973) 2914.
(49) S. Uemara, T. Ikeda, S. Tanaka and M. Okano, J. Chem. Soc. Perkin Trans II, **10** (1979) 2574.
(50) O. Hammerich and V.D. Parker, Adv. Phys. Org. Chem., **20** (1984) 55.
(51) C. Walling, D.M. Camaioni and S.S. Kim, J. Am.Chem. Soc., **100** (1978) 4814.
(52) J. Bonnard and G.Poilane, Rhone Poulenc, US Patent 3,387,036 (June 4, 1968).
(53) Jouffret, Rhone Poulenc Textile, US Patent 3,948,995 (April 6, 1976).
(54) F. Chanon and M. Chanon in "Photocatalysis" E. Pelizetti and N. Serpone Eds., Wiley, New York, N. Y., (1988).
(55) E. Baciocchi, T. Del Giacco, C. Rol and G.V. Sebastiani, Tetrahedron Lett., **26** (1985) 3353.
(56) G.V. Sebastiani, Tetrahedron Lett., **26** (1985) 3353.
(57) I. Saito, T. Matsuura and K. Inoue, J. Am. Chem. Soc., **105** (1983) 3200.
(58) G. Crank and M.I.H. Makin, Aust. J. Chem., **37** (1984) 845.
(59) R. Barone and M. Chanon, "Computer Aids to Chemistry", G.Vernin and M. Chanon Ed., Horwood-Wiley, Chichester, New-York, 1986, p19.
(60) A.E. Williams, Encyclopedia of Chemical Technology, Kirk-Othmer, Wiley, New-York, 1978, p736.
(61) J.C. Brunic, M. Costantini, N. Crenne, M.Jouffret, (Rhône-Poulenc)U.S.Pat.3, 658, 875, Apr 25, 1972.
(62) M. Nakada, C. Miura, K. Ogawa, R. Hayashida, S. Fukushi, M. Hirota, T. Ishi, Kenkyu Hokoku-Chiba Kogyo Daigaku, Riko-hen, **32**, 27(1987).
(63) B. Jonson, R. Larsson and B. Rebenstorf, J. Catal., **102**, 29 (1986).
(64) A. Monaci, Gazz. Chim. Ital., **116**, 339 (1986).
(65) P.J. Van der Berg, K. Van der Wiele, J.J.J. Den Ridder, Int. Congr. Cat. 5Proced.) 8th, 1984, 5, V393, Verlag Chemie Weinheim.
(66) P. Maggioni, F. Minisci, Chimica e l'Industria, **61**, 101(1979).
(67) C. Walling, C. Zhao, G.M. El Taliani, J. Org. Chem., **48**, 4910 (1983).
(68) G.B. Shulpin, P. Lederer and E. Macova, Izv. Akad. Nauk. SSR, Ser. Khim., 2638 (1986).
(69) M. Fujihara, Y. Satoh, T. Osa, Bull. Chem. Soc. Jpn, **55**, 666 (1982).
(70) D. Pletcher, Chem. and Ind., 358 (1982).
(71) D. Degner, "Techniques of Chemistry", Ed. N.L. Weinber and B.V. Tilak, J. Wiley, New-York, 1982, p 251 .
(72) R. Clarke, A.T. Kunh and E. Okoh, Chem. in Britain, **11**, 2, 59 (1975).
(73) J. Chaussard, l'Actualité Chimique, 29 (Nove. 1982).
(74) J.H. Wagenknecht, Angew. Chem. Int. Ed., **25**, 683 (1986).
(75) R.P. Kreh, R.M. Spotnitz and J.T. Lundquist, Tet. Letters, **28**, 1067 (1987).
(76) R.P. Kreh, R.M. Spotnitz (W.R. Grace and Co) U.S. US 4, 647, 349 March 3 1987.
(77) E. Steckhan, Angew. Chem. Inter. Ed., **25**, 683 (1986).
(78) G. Kreysa, H. Medin, J. Appl. Electrochemistry, **16**, 757 (1986) and references therein.
(79) J.J. Jow, A.C. Lee and T.C. Chou, J. Appl. Electrochem., **17**, 753 (1987).
(80) Ger. Offen 2 848 397 (1980).
(81) Ger. Offen 2 851 732 (1980).
(82) Ger. Offen 2 912 058 (1980).
(83) US Patent 4 212 711 (1980).
(84) Ger. Offen 2 435 985 (1975).
(85) K.B. Wiberg, "Oxidation in Organic Chemistry", K.B. Wiberg Ed., Vol 5A, p 94.
(86) F. Minisci, A. Citterio,,C. Giordano, Acc. Chem. Res., **16**, 27 (1983).
(87) I. Nishigushi, T. Hirashima , J.Org.Chem. , **50**, 539 (1985)

SELENIUM RADICAL IONS IN ORGANIC SYNTHESIS

M. Tiecco, L. Testaferri, and M. Tingoli
Istituto di Chimica Organica
Universita' di Perugia
06100 - Perugia
Italy

ABSTRACT. Selenium radical anions are suggested to be the reactive intermediates in the reductive cleavage of aryl alkyl, vinyl alkyl or vinyl acyl selenides; the aryl and vinyl selenolate anions thus obtained have been used to synthetize several types of selenium containing compounds. Radical cations are suggested to intervene in the oxydeselenenylation of aryl alkyl selenides promoted by oxidizing agents in hydroxylated solvents. This process has been used to effect the conversions of alkenes and vinyl bromides into dialkoxyalkanes and α-alkoxy acetals, and of alkynes and methyl ketones into α-keto ketals and α-keto acetals which occur by treatment with persulphate ions and catalytic amounts of diphenyl diselenide.

1. INTRODUCTION

Important developments in organic synthesis have been achieved in recent years by the use of organoselenium compounds which permit efficient manipulation of functional groups with great selectivity and under mild conditions.[1] A number of free radical selenium reactions have been proved to be of general synthetic utility particularly when employed in conjunction with other non radical selenium based functional group transformations.[2] Selenium radical ions have been scarcely investigated and their potential importance in organic synthesis is substantially unexplored. We report here several examples of useful conversions which can be suggested to proceed through selenium radical anion or radical cation intermediates.

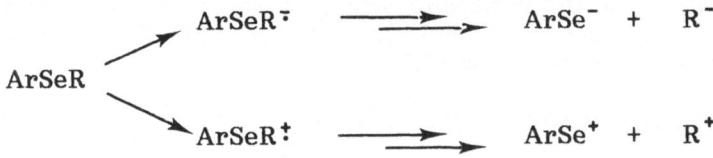

F. Minisci (ed.), Free Radicals in Synthesis and Biology, 253–262.

Reduction or oxidation of aryl alkyl selenides very likely produce radical anions or radical cations which suffer fragmentation at the alkyl-selenium bond to eventually afford aryl selenium anions or cations. In the first case this process has been used to effect selective syntheses of various types of selenium-containing compounds. In the second case, on the contrary, the formation of carbocations has been used to synthetize selenium-free compounds; in most cases these conversions require only catalytic amounts of the phenylselenium moiety precursor.

2. RADICAL ANIONS

The reductions of aryl alkyl selenides have been effected with sodium in HMPA or by adding the solution of sodium in HMPA to the solution of the selenide in DMF; in the latter case reaction conditions are much milder. We have employed the two methods to effect the cleavage of bis(alkylselenenyl)arenes and of vinyl alkyl and vinyl acyl selenides.

2.1. Aryl Alkyl Selenides

Under drastic conditions all the SeR groups are fragmented, whereas under mild conditions the reaction stops after the fragmentation of the first SeR group. The resulting arylselenolate solutions can be directly employed for further uses. By adding an alkylating agent we have prepared alkylselenenyl arenes which could not be obtained by direct nucleophilic aromatic substitution.[3]

$$Ar(SeR)_2 \longrightarrow Ar(SeR)Se^- \longrightarrow Ar(Se^-)_2$$

From intramolecular competitive experiments we have observed that the dealkylation of the SeR function occurs much more easily than those of the SR or OR functions. This property has been used to effect the selective dealkylation of the alkylselenenyl group in alkoxy and thioalkoxy substituted aryl alkyl selenides.[3]

$$Ar(SR)SeR_1 \longrightarrow Ar(SR)Se^- \qquad Ar(OR)SeR_1 \longrightarrow Ar(OR)Se^-$$

Several examples of the synthetic utility of these reactions to effect the selective dealkylation of aryl alkyl ethers, thioethers and selenoethers have been reviewed.[4]

2.2. Vinyl Alkyl and Vinyl Acyl Selenides

The reduction of vinyl alkyl as well as vinyl acyl selenides also gives the corresponding radical anions which suffer fragmentation to afford vinylselenolate anions. Interestingly, these anions are formed with complete retention of configuration and do not interconvert under the conditions employed. This is a peculiar property of the selenium compounds; the dealkylation of the corresponding sulphides, for instance, is not stereospecific:[5]

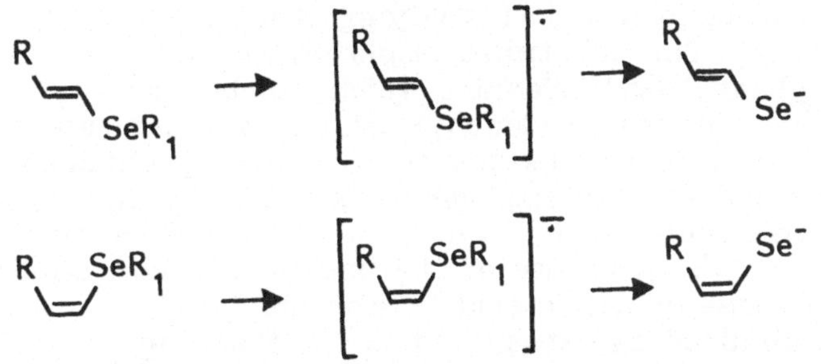

$R=Ph, PhS, PhSe$

$R_1=Alkyl, Acyl$

This configurational stability of the vinylselenolate anions has been fruitfully employed to effect a series of stereospecific syntheses of other vinylic selenium-containing compounds. Thus the addition of an alkyl, acyl or vinyl halide to the solution of the vinylselenolate anions produced vinyl alkyl,[6,7] vinyl acyl,[8] or divinyl selenides[9] in good yields; moreover, oxidation with iodine afforded divinyl diselenides.[8,9] These stereospecific conversions occurred equally well with both the Z- and the E-isomers.

An efficient, stereospecific synthesis of divinyl diselenides can be realized when vinyl acyl selenides are treated with catalytic amounts of an electron donor (Na or MeSNa). A chain mechanism involving vinyl acyl selenide radical anions and vinylselenenyl radicals has been proposed for this reaction (Scheme 1). The observed retention of configuration implies that the vinylselenenyl radicals also are configurationally stable.[9]

SCHEME 1

	VinSeCOR	$\xrightarrow{\text{e}}$	VinSeCOR$^{\bar{\cdot}}$	
	VinSeCOR$^{\bar{\cdot}}$	$\longrightarrow$	VinSe$^-$ + RCO$\cdot$	
VinSe$^-$ +	VinSeCOR	$\longrightarrow$	VinSeCOR$^{\bar{\cdot}}$ +	VinSe$\cdot$
	2 VinSe$\cdot$	$\longrightarrow$	VinSe-SeVin	

Vin = E - or Z-PhCH=CH, PhSCH=CH-, PhSeCH=CH-

3. RADICAL CATIONS

Recently we have undertaken an investigation with the aim of producing selenium radical cations and of studying their potential utility in organic synthesis. The substrates employed for this investigation were the phenyl alkyl selenides which can be easily produced from the addition of phenylselenenyl cations to unsaturated compounds. We found that these cations can be conveniently obtained from diphenyl diselenide and an oxidizing agent like the persulphate or nitrate anions, cerium ammonium nitrate, Cu^{++} and Pb^{+4} . When the reaction was carried out in the presence of the unsaturated compounds in alcohols, acetonitrile-water or acetic acid, the products of oxyselenenylation were obtained as exemplified in the following reactions:[10]

$$PhSeSePh + S_2O_8^{--} \longrightarrow 2PhSe^+ + 2SO_4^{--}$$

Alkenes, vinyl bromides, alkynes and methyl ketones easily gave the corresponding addition products which, in the presence of an excess of the oxidant afforded the products of oxydeselenenylation. In this way the entire process consisting in the production of the phenylselenenyl cations, in the oxyselenenylation and in the oxydeselenenylation steps can be effected in one-pot. The replacement of the phenylselenenyl group by the solvent very likely proceeds through the initial formation of radical cations.

3.1. Conversion of Alkenes into Dimethoxyalkanes

The results obtained with some alkenes in methanol are summarized in Scheme 2. The first step is stereospecific and gives the product of anti addition of the PhSe and MeO groups. If an excess of diphenyl diselenide is employed the reaction can be stopped at this stage. On the contrary, with catalytic amounts of PhSeSePh and an excess of persulphate ions, the addition compounds are rapidly consumed to afford a mixture of 1,2-and 1,1-dimethoxyalkanes. The 1,2-dimethoxy derivatives are formed with retention of configuration and the 1,1-dimethoxy are the result of a 1,2-shift of the phenyl group.[10]

SCHEME 2

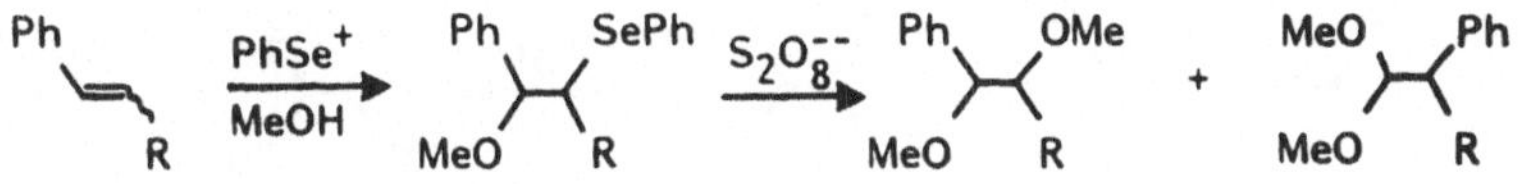

R=E-Ph	79[a]	34[c]		34
R=Z-Ph	72[b]	36[d]		46
R=E-Me	70[a]	36[a]		48

a: erythro; b: threo; c: meso; d: dl

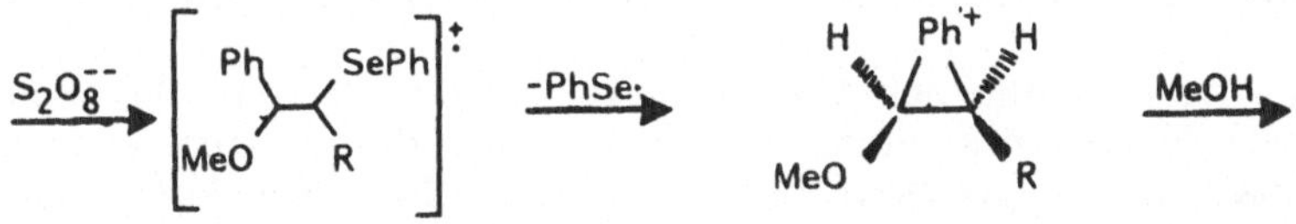

It can be suggested that the addition products are first transformed into the radical cations and that the fragmentation of the carbon-selenium bond is assisted by the participation of the phenyl group. An intermediate phenonium ion explains both the formation of the two types of the observed reaction products and the stereochemistry of the 1,2-dimethoxyalkanes. Phenylselenenyl cations are regenerated by oxidation of the radicals produced in the fragmentation step.

A similar reaction takes place with cholesterol also. In this case two diastereoisomeric dimethoxy derivatives were obtained. The oxidation of the addition product was carried out in CD_3OD and it was thus possible to demonstrate that the methoxy group partially migrates from the 5 to the 6 position. It might therefore be assumed that in this case the fragmentation of the radical cation is assisted by the methoxy group.[11]

3.2. Conversion of Vinyl Bromides into α-Alkoxy Acetals

Under similar conditions vinyl bromides (β-bromostyrene, bromostilbene, 1-bromo,1-octene) gave α-alkoxy acetals (Scheme 3). In these cases the addition products are the α-bromoselenides which are rapidly solvolized, through a selenium-stabilized carbocation, to give the α-alkoxyselenides. The radical cations of these compounds can easily fragment since an alkoxy-stabilized carbocation is formed.[12]

SCHEME 3

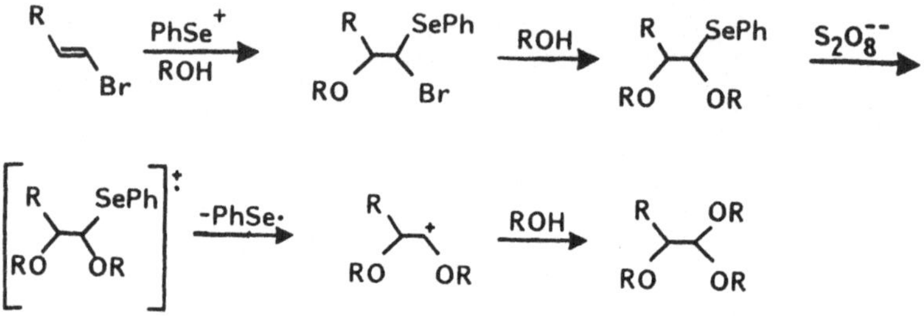

3.3. Conversion of Alkynes into α-Keto Acetals or α-Keto Ketals

In the case of alkynes the diphenyl diselenide must be used in stoichiometric amounts since the addition products decompose more slowly than in the previous examples and other reactions takes place in the absence of phenylselenenyl cations. The PhSeSePh, however, can be almost completely recovered from the final reaction mixture. Terminal alkynes give α-keto acetals and non terminal alkynes give α-keto ketals (Scheme 4). In the case of unsymmetrically substituted alkynes regioselectivity problems intervene and a mixture of the two possible α-keto ketals was obtained; the reaction mixtures were therefore treated with PTSA and the α-diketones were isolated. The products

SCHEME 4

$$R-\!\!\equiv\!\!-R_1 \longrightarrow R\text{-}COC(OMe)_2\text{-}R_1$$

PhCOCH(OMe)$_2$	87%	PhCOC(OMe)$_2$Ph	84%
PhCH$_2$COCH(OMe)$_2$	56%	PhCOCOC$_6$H$_4$CMe$_3$-m	79%
n-C$_4$H$_9$COCH(OMe)$_2$	51%	PhCOCOC$_6$H$_4$Br-o	70%
n-C$_6$H$_{13}$COCH(OMe)$_2$	67%	PhCOCOC$_6$H$_4$F-p	71%
		C$_3$H$_7$COC(OMe)$_2$C$_3$H$_7$	51%

obtained from these experiments are collected in Scheme 4, where reaction yields are also indicated.[10] Whenever a methylene group was present in the α-position of the starting alkyne (3-phenyl-1-propyne, 1-hexyne, 1-octyne, 4-octyne) small amounts (5-10%) of β-phenyl-selenenyl-α-keto acetals were also formed. Similar results were obtained when the reaction was carried out in ethanol. Using acetonitrile-water as solvent the unprotected α-dicarbonyl compounds were obtained. On the contrary, in ethylene glycol both the carbonyl functions were obtained in the protected form.

A possible interpretation of the reaction course in the case of alkynes is reported in Scheme 5. Two consecutive oxyselenenylation reactions are suggested to take place to afford product of the type $RC(OR)_2 C(SePh)_2 R_1$, which can also be isolated in some cases. The oxidation to radical cation is followed by fragmentation to give a selenium-stabilized carbocation which react with the solvent to afford an α,β,β'-trialkoxyalkyl phenyl selenide. Further oxidation to radical cation and fragmentation gives an oxygen-stabilized carbocation; a tetraalkoxy alkane is thus formed. In the acidic reaction medium this crowded compound gives the more stable α-keto acetal or α-keto ketal.

SCHEME 5

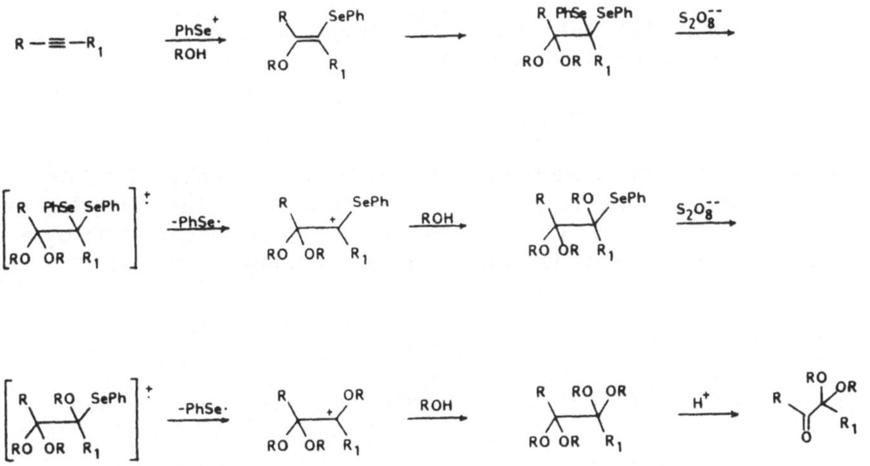

3.4. Conversion of Methyl Ketones into α-Keto Acetals

Finally the oxidation of diphenyl diselenide in methanol was carried out in the presence of enolizable ketones. The substrates investigated so far were the aryl, vinyl and alkyl methyl ketones. In the latter case the reaction took different courses as a function of the structure of the alkyl group.

3.4.1. Aryl Methyl Ketones

Using an excess of persulphate anions and catalytic amounts of PhSeSePh several aryl methyl ketones were easily converted into the corresponding α-keto acetals:[10]

$$\text{ArCOMe} \xrightarrow[\text{MeOH}]{(\text{PhSe})_2 , S_2O_8^{--}} \text{ArCOCH(OMe)}_2$$

This reaction seems to be of general application, good yields being obtained with phenyl (76%), 4-tolyl (90%), 4-nitrophenyl (62%), 2-hydroxyphenyl (61%), 4-biphenyl (80%), 2-naphthyl (92%), 2-thienyl (88%) and 3-thienyl (65%) methyl ketones. The addition of phenylselenenyl cations occurs on the enolic form and it is suggested to proceed towards the formation of the dimethoxy, diphenylselenenyl compounds identical to those obtained from terminal acetylenes:

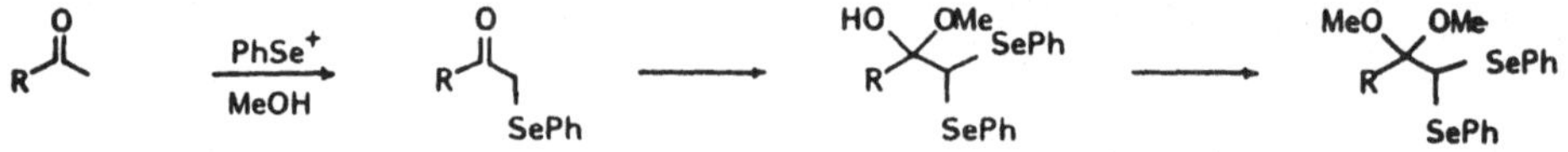

The reaction can then take the same course suggested in the case of the alkynes and indicated in Scheme 5.

3.4.2. Alkyl and Vinyl Methyl Ketones

Similar good results were obtained with alkyl methyl ketones whenever the α-carbon was not a methylene. Thus the following products were obtained, with the yield indicated in parentheses, from cyclohexyl and i-propyl methyl ketone and from pregnenolone acetate:

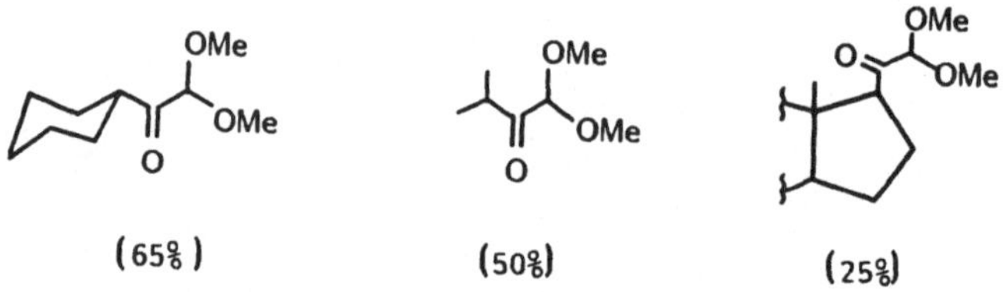

α,β-Unsaturated methyl ketones can be made to react at the methyl group without touching the carbon-carbon double bond. Some examples of the obtained products are reported below:

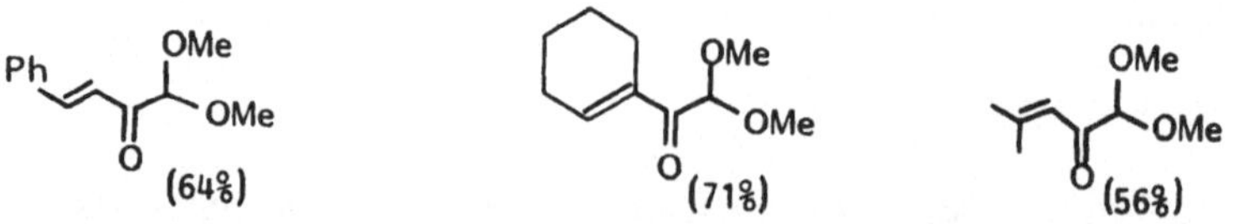

On the contrary when the α-carbon was a methylene the reaction required stoichiometric amounts of diphenyl diselenide since the phenylselenenyl group was incorporated to give the β-phenylselenenyl-α-keto acetals (Scheme 6).

SCHEME 6

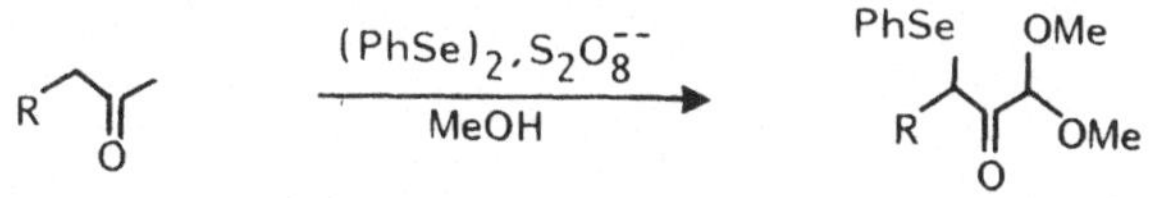

$$R = \quad Me \quad Et \quad i\text{-}Pr \quad n\text{-}Pr \quad n\text{-}Bu$$
$$Yield = \quad 52 \quad 59 \quad 77 \quad 62 \quad 62$$

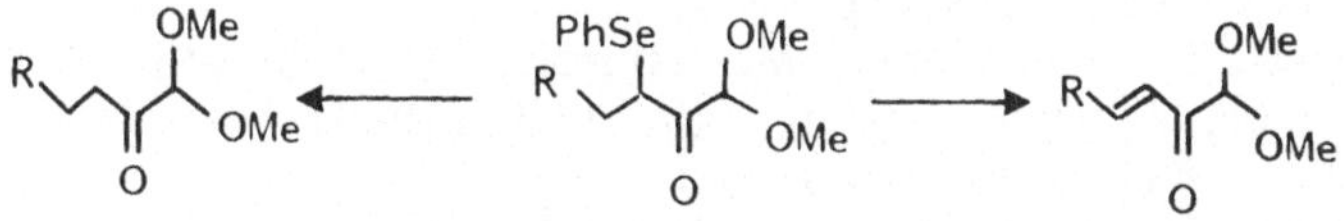

These compounds can be easily deselenenylated by treatment with acid in methanol in the presence of styrene which traps the phenylselenenyl cations; alternatively, oxidation with hydrogen peroxide in methanol affords the β,γ-unsaturated-α-keto acetals.

Finally some experiments were carried out in order to trap the intermediate carbocations intramolecularly and to obtain cyclic products. For this purpose the reactions with catalytic amounts of diphenyl diselenide and excess persulphate anions were carried out with β-hydroxyalkyl methyl ketones. Although in low yields the expected cyclization products shown below were indeed formed.

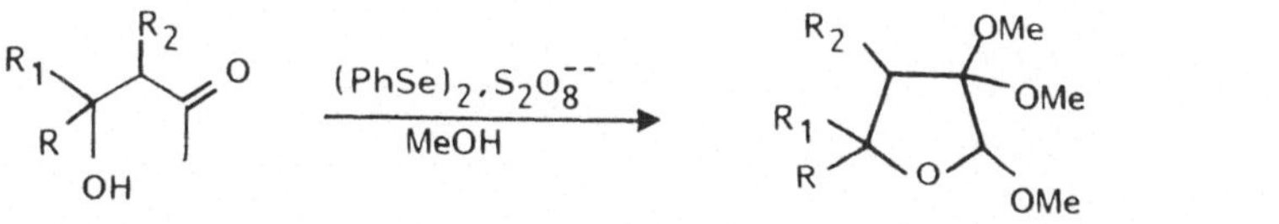

$$R=R_1=R_2=H \quad 25\% \qquad R=R_1=H, R_2=Me \quad 25\% \qquad R=R_1=Me, R_2=H \quad 45\%$$

4. CONCLUSIONS

Although the formation of aryl alkyl selenide radical ions was not directly proved, the reaction conditions employed and the structure of the reaction products obtained suggest that they can be the reactive intermediates in several processes. Both the radical anions and the radical cations evolve by fragmentation at the alkyl-selenium bond. The nature of the alkyl group does not have substantial influence in the case of radical anions. With radical cations, on the contrary, the experimental data so far available seem to indicate that the fragmentation process occurs only in the case in which a stabilized carbocation can be formed.

262

Acknowledgments. Financial support from the CNR, Rome, Progetto Strategico "Processi di Trasferimento Monoelettronico", and Ministero della Pubblica Istruzione, Italy, is gratefully acknowledged.

References

1) C. Paulmier, "Selenium Reagents and Intermediates in Organic Synthesis", Pergamon Press, Oxford, 1986.
2) T. G. Back in D. Liotta, "Organoselenium Chemistry", John Wiley & Sons, New York, 1987, p. 325.
3) M. Tiecco, L. Testaferri, M. Tingoli, D. Chianelli, and M. Montanucci, J. Org. Chem., 48, 4289 (1983).
4) M. Tiecco, Synthesis, in press.
5) M. Tiecco, L. Testaferri, M. Tingoli, D. Chianelli, and M. Montanucci, J. Org. Chem., 48, 4795 (1983).
6) M. Tiecco, L. Testaferri, M. Tingoli, D. Chianelli, and M. Montanucci, Tetrahedron Lett., 26, 2225 (1985).
7) L. Testaferri, M. Tiecco, M. Tingoli, and D. Chianelli, Tetrahedron, 41, 1401 (1985).
8) L. Testaferri, M. Tiecco, M. Tingoli, and D. Chianelli, Tetrahedron, 42, 4577 (1986).
9) L. Testaferri, M. Tiecco, M. Tingoli, and D. Chianelli, Tetrahedron, 42, 63 (1986).
10) M. Tiecco and coworkers, unpublished results.
11) P. Ceccherelli and coworkers, unpublished results.
12) M. Tiecco, L. Testaferri, M. Tingoli, D. Chianelli, and D. Bartoli, Tetrahedron, in press.

A SYNTHETICALLY USEFUL SOURCE OF ALKYL AND ACYL RADICALS

Catherine Tailhan and Samir Z. Zard*
Laboratoire de Synthese Organique
Ecole Polytechnique
91128 Palaiseau
France

ABSTRACT.   Alkyl and acyl xanthates are useful precursors of alkyl and acyl radicals. With acyl xanthates, the radical chain process can be initiated by visible light, and capture of the intermediate acyl radical leads to various carbonyl derivatives. In some cases, the acyl radical extrudes carbon monoxide to give an alkyl radical which in turn can be trapped by an electrophilic olefin. Alkyl xanthates in contrast require U.V. lighting but the reaction sequence can sometimes be initiated with visible light if small amounts of S-benzoyl-O-ethyl xanthate are incorporated into the medium.

Kinetic studies on various thiocarbonyl derivatives have shown that the addition of carbon and tin centered radicals to the carbon-sulphur double bond is a very fast reaction [1]. Nevertheless, at least with some substrates, this radical addition step can be reversible. We recently found this to be the case in the radical decarboxylation of thiohydroxamate esters of carboxylic acids [2] as well as in the Barton deoxygenation of secondary alcohols via their xanthate derivatives [3].

$$R^\bullet \; + \; S{=}C\Big\langle \quad \rightleftharpoons \quad R{-}S{-}\overset{\bullet}{C}\Big\langle$$

The surprising ease with which the fragmentation of the alkylthiyl-alkyl adduct radicals takes place suggested to us a simple chain process for generating alkyl and acyl radicals under mild conditions [4]. Our conception , applied to a xanthate, is outlined in scheme 1.

According to this scheme, a carbon radical R˙ produced from a xanthate 1 following chemical or photochemical initiation [5] will react very rapidly with its precursor to give a symmetrical radical 2. If R' is chosen such that rupture of the C-O bond is difficult [6] (R'= primary alkyl or aryl), then 2 can only fragment to give back R and the starting xanthate. This sequence of reactions is therefore degenerate and synthetically sterile. However, by adding a radical trap to the system such as an olefin, for example, it should be possible to trigger a succession of radical additions and fragmentations leading ultimately to compounds of type 3. The fact that a new carbon-carbon bond is created in the process has considerable synthetic implications [7]. Of course, the olefinic trap can be part of the starting xanthate resulting in the formation of cyclic products. Furthermore, the degenerate character of the reaction of the carbon radical with its progenitor effectively removes what otherwise would have been a serious competing pathway. Premature quenching of intermediate carbon radicals with tributylstannane for example, is a constant nuisance in cyclisations using tin based systems [8].

F. Minisci (ed.), Free Radicals in Synthesis and Biology, 263–268.
© 1989 by Kluwer Academic Publishers.

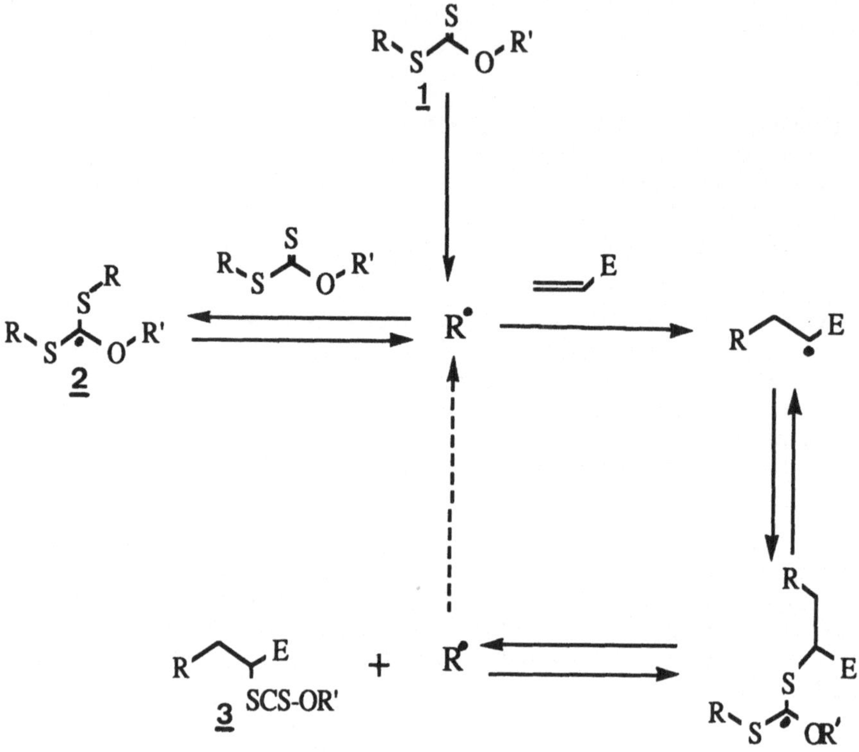

Scheme 1

This proposition was put to test by irradiating with U.V. light a solution of benzyl xanthate **4** and N-methyl- maleimide **5a**. The major product, isolated in about 40% yield was indeed the expected adduct **6**. The modest yield appeared to be due to the fact that **6** is also a xanthate absorbing in more or less the same region of the spectrum. This lack of discrimination causes the formation of side-products and diminishes the efficiency of the process.

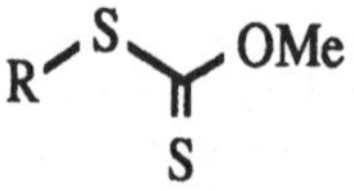

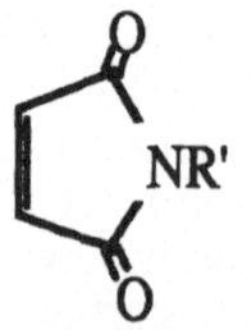

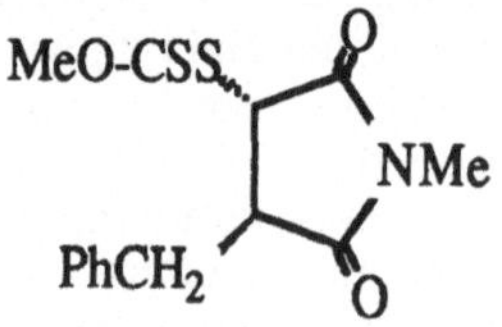

4 , R= PhCH$_2$-   **5a** , R'= Me   **6**
23, R= CH$_2$=CH-CH$_2$-   **5b** , R'= PhCH$_2$-
25, R= 4-(BnO-)-3-(MeO-)C$_6$H$_3$CH$_2$-
27, R= PhCOCH$_2$-

In order to circumvent this complication, at least in the initial phase of this study, we turned to the little known acyl xanthates RCO-SCSOR' 7 . These are yellow substances and therefore absorb at a different wavelength than simple alkyl xanthates. Indeed many years ago, a pioneering work by Barton and co-workers [9] revealed that on irradiation with U.V. or even visible light, these compounds underwent a decarbonylative rearrangement to give an ordinary xanthate as shown in the equation below:

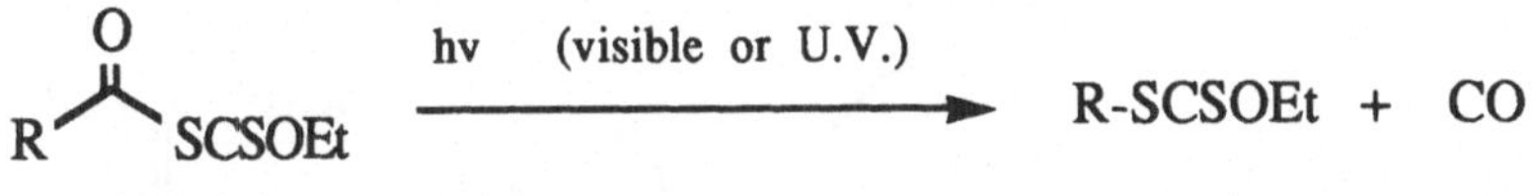

This transformation was found to occur smoothly when R is a tertiary or benzylic group. In contrast a slow or no reaction took place in the case primary alkyl or aryl derivatives.

In the light of the above discussion, a simple rationalisation of these observations may be formulated as shown in scheme 2 .Thus, photochemical initiation yields small amounts of acyl radicals which may loose carbon monoxide irreversibly to give an alkyl radical. The latter then reacts with the starting acyl xanthate to give an alkyl xanthate and an acyl radical thereby propagating the chain. In contrast, the reaction of the acyl radical with 7 is degenerate and accounts for the apparent inertness of aryl derivatives since aroyl radicals (ArCO) do not extrude carbon monoxide easily. By analogy with the mechanistic manifold displayed in scheme 1, incorporation of an external trap should allow the capture of the intermediate acyl radical or even the alkyl radical in cases where the decarbonylation step is too rapid. These expectations were readily borne out in practice.

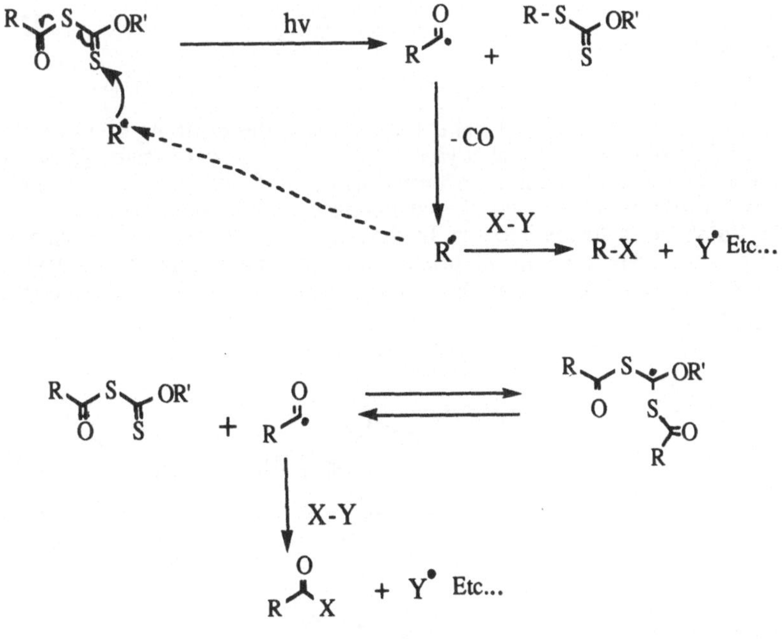

Scheme 2

The benzoyl derivative 8 , easily prepared from benzoyl chloride and potassium ethyl xanthate, is , as previously reported, impervious to irradiation with visible light (tungsten lamp) in boiling toluene. However in the presence of allyl acetate, which is only a modest radical trap, a smooth reaction occurs leading to adduct 9 in about 60% yield. The xanthate group can be eliminated either by treatment with a combination of triethylamine and methyl iodide or more conveniently by heating with copper powder and distilling the product as formed under reduced pressure. The somewhat labile 10 may thus be obtained in 87% crude (50% purified) yield.

As expected, the intramolecular variant was even more efficient. The salicylic derivative 11 and its sulphur analogue 12 were rapidly converted into the corresponding cyclic products 13 and 14 in 70% and 93% yield respectively. As in the example above, compound 13 gave the known [10], reactive exocyclic enone 15 in 40% yield on distillation from copper powder.

11 , X= O
12 , X= S

13 , X= O
14 , X= S

15

When decarbonylation of the acyl radical is fast, it is the resulting alkyl radical that is trapped. Thus, irradiation of the phenylacetyl and pivaloyl derivatives 16 and 17 in the presence of N-benzyl maleimide 5b afforded adducts 18 and 19 in 70% and 63% yield respectively. Similarly, compound 20 was obtained in 45% yield using the less reactive phenyl vinylsulphone as the olefinic trap. In these examples, a true competition exists since, unlike most of the other steps in the process, the decarbonylation of the acyl radical is irreversible. Consequently the olefin has to be quite reactive in order to divert efficiently the normal sequence of events.

16 , R= PhCH$_2$-
17 , R= t-Bu-

18 , R= PhCH$_2$-
19 , R= t-Bu-

20

The sensitivity of the acyl xanthates to visible light suggested to us the possibility of improving our first experiment using benzyl xanthate **4** by catalysing it with a small amount of an acyl xanthate such as **8**. As shown in scheme 3, this contrivance would allow the overall radical addition to be accomplished using visible light instead of the less discriminating U.V. irradiation. Thus the benzoyl radicals produced by photolysis of **8** will add reversibly to xanthate **4** to give an intermediate radical **21** that can then fragment, again reversibly, into a benzyl radical and acyl xanthate **22**. As the latter is a congener of **8**, the catalytic activity is therefore restored. The benzyl radical so created can now enter the synthetically useful cycle by interacting with the olefinic trap.

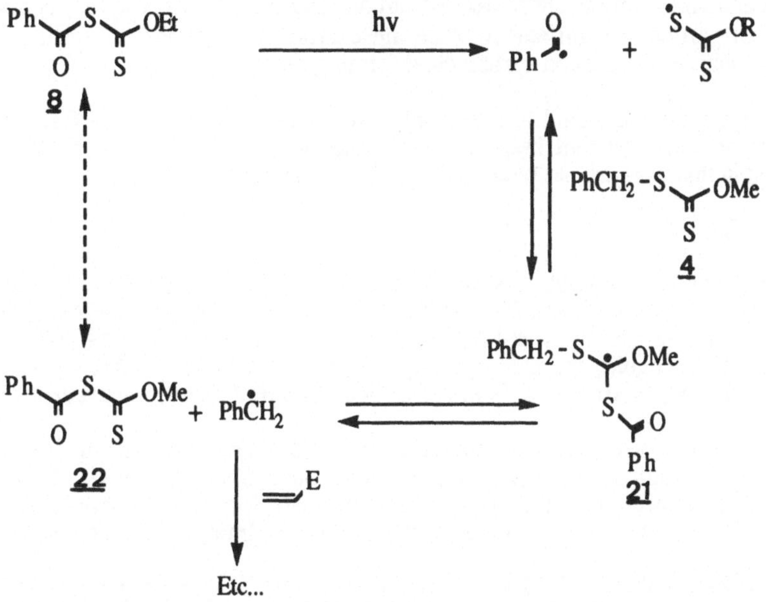

Scheme 3

In the event, shining visible light from a tungsten lamp onto a refluxing solution of xanthate **4**, N-methyl maleimide **5a**, and a small amount of **8** (ca 10%) in toluene afforded the same adduct **6** in a delightful yield of 77% instead of the 40% initially obtained. No reaction took place in the absence of the acyl xanthate. By a similar modification, compounds **24** and **26** were prepared from the corresponding xanthates **23** and **25** in 52% and 50% yield respectively.

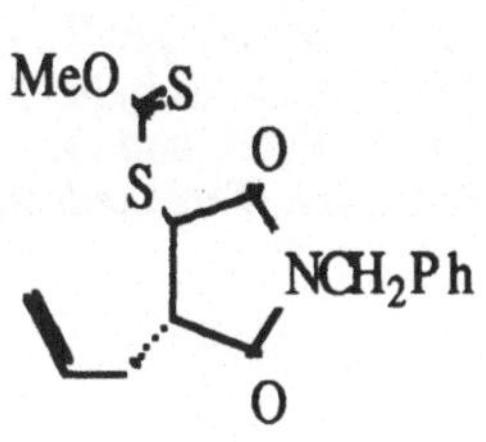

24 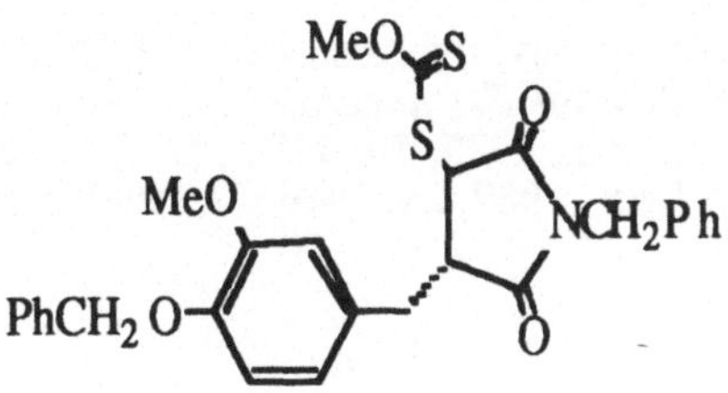   **26** 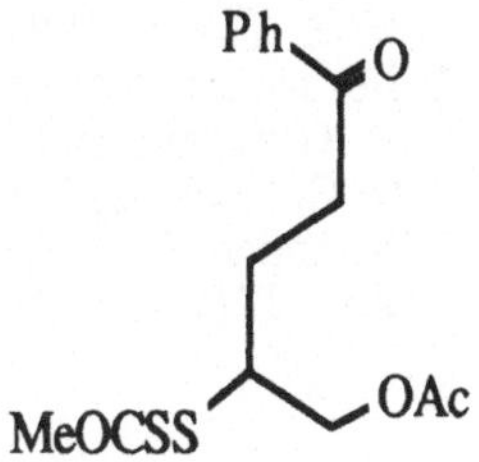   **28**

268

This mechanistically interesting approach appears however to be limited to the production of stabilised alkyl radicals. Thus the fragmentation of intermediate 22 (or its analogues in the other examples) in the desired sense was biased by choosing benzylic or allylic radicals as the "leaving group". One other class of radicals that may be generated by this procedure are those stabilised by carbonyl groups. For example, phenacyl radicals were easily produced from xanthate 27 and captured by allyl acetate to give derivative 28 in 71% yield. The electrophilic nature of the phenacyl radical implies the use of an olefin with opposite polar characteristics. It is interesting to note that in this case the use of U.V. light as initiator, and in the absence of acyl xanthate 8 gave the same adduct 28 but in only 40% yield.

It is clear from these preliminary results that alkyl and acyl xanthates are synthetically convenient and highly promising sources of carbon centered radicals. The easy access to acyl radicals is of particular significance since these little used species [11] furnish directly carbonyl derivatives in a carbon-carbon bond forming process.

Acknowledgments: We would like to thank Professors D. H. R. Barton, F.R.S. and J-Y. Lallemand for their constant help and encouragement. P. Delduc and F. Mestre have collaborated in this work as part of an undergraduate research project.

REFERENCES

1. D. Forrest and K.U. Ingold, *J. Amer. Chem. Soc.*, 1978, *100*, 3868 ; J.C . Scaiano and K.U. Ingold, *J. Amer. Chem. Soc.*, 1976, 98, 4727 and references therein ; J.C . Scaiano, J. P. -A Tremblay and K.U. Ingold, *Can. J. Chem.*, 1976, *54*, 3407 ; D. Forrest, K.U. Ingold, and D.H.R. Barton, *J. Phys. Chem.*, 1977, *81*, 915.
2. D. H. R. Barton, D. Bridon, I. Fernandez-Picot, and S. Z. Zard, *Terahedron*, 1987,*43*, 2733.
3. D. H. R. Barton, D. Crich, A. Lobberding, and S. Z. Zard, *Tetrahedron*, 1986,*42*, 2329
4. P. Delduc, C. Tailhan, and S. Z. Zard, *J. Chem. Soc., Chem. Commun.*, 1988, 308.
5. Xanthates have been reported to be photosensitisers for the polymerisation of vinyl monomers : M. Okawara, T. Nakai, Y. Otsuji, and E. Imoto, *J. Org. Chem.*, 1965, *30*, 2025.
6. D. H. R. Barton and S. W. McCombie, *J. Chem. Soc., Perkin Trans. 1* , 1975, 1574 ; D. H. R. Barton, W. B. Motherwell, and A. S. Stange, *Synthesis*, 1981, 743 ; W. Hartwig, *Tetrahedron*, 1983, *39*, 2609.
7. B. Giese, "Radicals in Organic Synthesis: Carbon- Carbon Bond Formation", Pergamon Press, Oxford, 1986.
8. L. J. Johnstone, J. Lusztyk, D. D. M. Wayner, A. N. Abeywickreyma, A. L. J. Beckwith, J. C. Scaiano, and K. U. Ingold, *J. Amer. Chem. Soc.*, 1985, *107*, 4594 and references there cited ; A. L. J. Beckwith and C. H. Schiesser, *Tetrahedron*, 1985, *41*, 3925.
9. D. H. R. Barton, M. V. George, and M. Tomoeda, *J. Chem., Soc.*, 1962, 1967.
10. F. E. Ward, D. L. Garling, R.T. Buckler, D. M. Lawler, and D. P. Cummings, *J. Med. Chem.*, 1981, *24*, 1073.
11. For recent synthetic work on acyl radicals, see : E. J. Walsh Jr., J. M. Messinger II , D. A. Grudoski, and C. A. Allchin, *Tetrahedron Lett.*, 1980, *21*, 4409 ; P. Gottschalk and D. C. Neckers, *J. Org; Chem.*, 1985, *50*, 3498 ; D. J. Coveney, V. F. Patel, and G. Pattenden, *Tetrahedron Lett.* , 1987, *28*, 5949 ; V. Patel and G. Pattenden, *Tetrahedron Lett.* , 1988, *29*, 707.

CHLORINATION BY HYPOCHLOROUS ACID. FREE-RADICAL VERSUS ELECTROPHILIC REACTIONS

F. Fontana, F. Minisci, E. Vismara
Dipartimento di Chimica del Politecnico
Piazza Leonardo da Vinci 32, 20133 Milano
Italy

G. Faraci, E. Platone
ENIRICERCHE, Via Maritano 26
S. Donato Milanese (Milano)
Italy

ABSTRACT - The chlorination at room temperature of alkanes, aromatics, alkylaromatics, olefins with hypochlorous acid shows two competitive processes in the absence of free-radical initiators: a free-radical chain chlorination and an electrophilic process, whenever this last is structurally possible. The factors affecting this competition, the mechanism of the free-radical chain involved, the selectivity and the synthetic potentiality of this new procedure of free-radical chlorination are discussed.

The use of hypochlorous acid for electrophilic chlorinations is well known[1] from long time. Its utilization for free radical chlorinations is less known, whereas numerous other free-radical chlorinating agents have been used[2]. Particularly t-butyl hypochlorite, structurally similar to hypochlorous acid, has been largely utilized[3] for free-radical chain chlorinations (Scheme 1).

$$t\text{-BuO}\cdot + R\text{-H} \longrightarrow t\text{-BuOH} + R\cdot$$

$$R\cdot + ClOBu\text{-}t \longrightarrow R\text{-Cl} + t\text{-BuO}\cdot$$

Scheme 1

A similar chain process was in principle possible with

269

F. Minisci (ed.), Free Radicals in Synthesis and Biology, 269–282.

hypochlorous acid (Scheme 2)

$$HO. + R-H \longrightarrow H_2O + R.$$
$$R. + ClOH \longrightarrow RCl + .OH$$

Scheme 2

Four factors, however, could negatively affect the process of
the Scheme 2: i) the low solubility of HOCl in organic media;
ii) the equilibria of HOCl in aqueous solution; iii) the low se
lectivity in hydrogen abstraction by the OH radical; iv) the
competition of the electrophilic chlorination when it was
structurally possible.

It was, in any case, interesting to investigate the poten
tiality of the use of the hypochlorous acid as agent of free-
radical chlorination, also in view of possible practical appli
cations due to the large and cheap commercial availability of
the reagent.

We have found that the free-radical chlorination of hydro
carbons can be smoothly carried out with hypochlorous acid at
0°-50°C in a two-phase system: one phase formed by commercial
aqueous solution of sodium or calcium hypochlorite (5-6%) with
an equivalent amount of an acid stronger than HOCl, (HCl, H_2SO_4
CH_3COOH etc.) and the other phase by the organic substrate as
such or in suitable solvent. The free-radical chlorination occurs in the absence of initiators and in the dark. To verify
if fortuitous impurities could act as initiators of the free-
radical chain chlorination, sodium, potassium and calcium hypochlorites were also prepared from pure reagents, but the results were identical in all cases.

In order to understand the synthetic and mechanistic aspects of this new procedure of free-radical chlorination, we
have investigated the behaviour of a variety of alkylbenzenes
for what concerns the selectivity of the side-chain chlorination and the competition with the nuclear chlorination.

Benzene and naphtalene were also investigated. Cyclohexane and 2,3-dimethylbutane were particularly studied among the
alkanes in the attempt to individuate the free-radical species
involved from the inter- and intramolecular selectivity. Cyclohexene and 1-cyanocyclohexene were investigated among the olefins.

The results of Table I show the range of the synthetic
applications of this procedure. An excess of reagent has
been utilized in order to convert generally more than 80% of

the alkylbenzenes in the absence of solvents. No reaction occurs with strongly deactivated alkylbenzenes, such as nitrotoluenes, and only nuclear chlorination takes place with strongly activated alkylaromatics, such as methoxytoluenes.

TABLE I - Chlorination of alkylaromatics by HOCl (Substrate: HOCl 1:3)

Substrate	T°C	Conversion %	Reaction product (%) based on converted substrate
Toluene	20-30	98.0	$PhCH_2Cl$ (65.7); $PhCHCl_2$ (16); o+pCl-C_6H_4-Me (2)
o-Cl-Toluene	40	94.3	o-Cl-C_6H_4-CH_2Cl(85.1); o-Cl-C_6H_4-$CHCl_2$ (11.9)
p-Cl-Toluene	40	88.0	p-Cl-C_6H_4-CH_2Cl (85.4); p-Cl-C_6H_4CHCl$_2$ (11.8)
mCl-Toluene	40	96.0	m-Cl-C_6H_4CH$_2$Cl (83.2); m-Cl-C_6H_4CHCl$_2$ (11.8)
p-Br-Toluene	40	95.4	p-Br-C_6H_4CH$_2$Cl (88.3); m-Cl-C_6H_4CHCl$_2$ (10.1)
p-F-Toluene	40	87.4	p-F-C_6H_4CH$_2$Cl (91.2); p-F-C_6H_4CHCl$_2$ (6.5)
o-CN-Toluene	50	70.2	o-CN-C_6H_4CH$_2$Cl (91)
p-CN-Toluene	50	72	p-CN-C_6H_4CH$_2$Cl (90)
m-CN-Toluene	50	78	m-CN-C_6H_4CH$_2$Cl (93)
m-Bu-benzene	10-30	100	α-chloro (58); β + γ + δ (35)
Diphenyl-methane	10	89	α-chloro (80); benzophenone(20)
p-Ac-C_6H_4-CH_3	50	85	p-Ac-C_6H_4CH$_2$Cl (82)
p-Toluic acid	50	77	p-ClCH$_2$-C_6H_4COOH
m-Xylene	0	75	Benzyl chlorides (70); Nuclear chloroderivatives (30)
m-Xylene	20-45	80	Benzyl chlorides (25); Nuclear chloroderivatives (75)
o-Xylene	0	87	Benzyl chlorides (78); Nuclear chloroderivatives (21)
p-Xylene	2-7	89	Benzyl chlorides (73); Nuclear chloroderivatives (24)
o-Nitrotoluene	60	0	No reaction
p-Nitrotoluene	60	0	No reaction
p-Methoxy-toluene	20	100	3-Cl-4-Meo-toluene (98)

With toluene only traces of nuclear chlorination occur at room temperature; the reaction is moderately exothermic (the temperature increases from 20 to 40°) and benzal chloride is a significant by-product of the reaction with 97% conversion of toluene.

With xylenes under the same conditions the nuclear chlorination prevails on the benzylic chlorination, but the temperature has a marked effect on the competition: at 0°C the selectivity is reversed and the benzylic chlorination is prevailing.

If electronwithdrawing substituents, such as halogens, CN, COOR, COR, deactivate the aromatic ring a somewhat higher temperature (40-50°C) is necessary and the best synthetic results are obtained because no nuclear chlorination occurs and a high selectivity of benzylic monochlorination is obtained.

Diphenylmethane leads to 1-chlorodiphenylmethane (80%) and benzophenone (20%); this last clearly arises from hydrolysis of the 1,1-dichlorodiphenylmethane initially formed. n-Butylbenzene gives 58% of benzylic chlorination and 35% chlorination of the β, γ and δ positions of the butyl group.

The chlorination of α-chloro-m-xylene leading to α,α'-m-dichloroxylene is of considerable interest, because of the industrial importance of this intermediate[4]. Actually, the chlorination is more selective than with m-xylene for what concerns the benzylic/nuclear competition and that agrees with the electron availability of the aromatic ring, but the selectivity of the benzylic chlorination is not high (eq.1)

$$(1)$$

The effect of the solvents on the nuclear/benzylic chlorination for toluene and m-xylene are reported in Table II. With toluene the highest selectivity in benzylic chlorination is obtained in the absence of solvents; the nuclear chlorination increases with the polarity of the solvent and becomes dominant with polar solvents, such as acetonitrile.

With m-xylene the behaviour is less linear; polar solvents, such as acetonitrile, THF, ethyl acetate give high prevalence of nuclear chlorination. Other solvents, however, such as CH_2Cl_2, CCl_4, 1,2-DCE, 1,3-DCB, nitrobenzene, increase considerably the benzylic chlorination compared with that of m-xylene in the absence of solvents. The behaviour in these cases is opposite to that observed with toluene; the highest selectivity of benzylic chlorination is observed in CS_2.

It is noteworthy the fact that no benzylic chlorination is observed by using Cl_2 under the same conditions utilized in the Tables I and II.

TABLE II - Solvent effect in the chlorination of toluene and m-xylene by HOCl

Substrate	Solvent	T°C	Benzyl : Nuclear chlorination
Toluene	–	20–40	40 : 1
"	CH_2Cl_2	20–40	10 : 1
"	1,2-DCE	20–36	15 : 1
"	1,3-DCB	20–35	18 : 1
"	Acetonitrile	20–40	Only nuclear chlorination
"	Anisole	20–40	Only o- and p-chloroanisoles
m-Xylene	–	0–2	2.3 : 1
"	–	20–45	1 : 3
"	$CHCl_2$	20–40	2 : 1
"	CCl_4	20–35	5 : 1
"	1,2-DCE	20–42	1 : 1.5
"	1,3-DCB	20–30	2.2 : 1
"	CS_2	20–32	7 : 1
"	Acetonitrile	20–40	1 : 60
"	Ph-NO_2	20–32	3 : 1
"	THF	20–48	1 : 25
"	EtAc	20–35	1 : 10

The reaction easily takes place also with benzene leading to hexachlorocyclohexane and no significant amount of chlorobenzene is formed. On the contrary naphtalene leads to α-chloronaphtalene in 65% yield and 35% of a complex mixture of polychlorinated products (probably products of addition-chlorination as with benzene).

Cyclohexane gives cyclohexyl chloride as the only reaction product at low conversion; it has been utilized in competition with toluene (Table III) in order to determine the relative rates and to individuate the free-radical species involved in the reaction.

TABLE III - Relative rates in the chlorination of
toluene and cyclohexane

Chlorinating agent	T°C	Reactivity C_6H_{12} : $Ph-CH_3$	Reactivity C_6H_{12} : $Ph-CH_3$ per hydrogen
HOCl	10	12.16	3.04
"	80	8.12	2.03
Cl_2	10	8.43	2.17
"	80	15.00	3.75

For the same reason the tertiary/primary C-H selectivity was investigated in the chlorination of 2,3-dimethyl-butane under various conditions and in comparison with photo-chlorination by Cl_2 (Table IV)

A comprehensive view of these results clearly suggests that a free-radical chain chlorination (alkane and benzylic chlorination, addition to benzene) is working in competition with electrophilic chlorinations, whenever these last are structurally possible.

The first problem, which arises from these results, concerns the mechanism of the initiation of the radical chain. The fact that hypochlorite from different origin, also from pure reagents, gives always the same results indicates that the initiation is not due to fortuitous impurities, but it is intrinsic of the reacting system. The reaction takes place in the dark with the same rate, indicating that the initiation step is thermal and not photochemical.

Now the possible species present in the reacting system are those shown by the equilibria of eqs.2, 3

$$2HOCl \rightleftharpoons Cl_2O + H_2O \qquad (2)$$

$$HOCl + HCl \rightleftharpoons Cl_2 + H_2O \qquad (3)$$

TABLE IV - Chlorination of 2,3-dimethyl butane
with Cl_2 (photochlorination) and HOCl
at 30-40°C

Chlorinating agent	Solvent (ml)	Tertiary/primary selectivity
Cl_2	-	3.56
HOCl	-	5.04
Cl_2	CCl_4 (8.77)	3.78
"	CCl_4 (3.5)	3.75
HOCl	"	4.19
Cl_2	CCl_4 (3.5) Benzene (0.9)	13.22
HOCl	"	15.35
Cl_2	Benzene (6.06)	31.35
HOCl	"	34.54
Cl_2	CS_2 (6.8)	89.05
HOCl	"	88.03

a) 10 mmol of 2,3-dimethylbutane were used in
   all cases.

Molecular chlorine is certainly not involved in the
initiation step because the experiments with this reagent
in the dark under the same conditions only lead to the
electrophilic chlorination of benzene and toluene. Thus
the species responsible of the initiation must be HOCl or
Cl_2O.
We think that two mechanisms of initiation of the free-
radical chains can be envisaged:
i) A thermal homolysis of a Cl-O bond; in this case only
   species with a bond energy less than 25 Kcal/mol are suitable at the moderate temperatures used in the reaction.
   The most probable species is Cl_2O, in which the high electronegativity of the three atoms weakens the strength of
   the Cl-O bond (eq.4)

$$Cl_2O \longrightarrow Cl. + .OCl \tag{4}$$

ii) An electron-transfer process between the substrate and
    the initiating species; also for this interaction the

most probable species is Cl_2O (eq.5), due to its high electron affinity.

$$R-H + Cl_2O \longrightarrow R\cdot + ClO^- + Cl\cdot + H^+ \qquad (5)$$

A similar process very likely occurs in the spontaneous free-radical fluorination of hydrocarbons by molecular fluorine (eq.6), due to the high redox potential of this last

$$R-H + F_2 \longrightarrow R\cdot + F^- + F\cdot + H^+ \qquad (6)$$

On the other hand, it is known[5] that Cl_2O can chlorinate hydrocarbons by a free-radical chain process and spontaneous initiation, even if the mechanism of this last has not been yet elucidated.

Thus we think that the free-radical initiation is due to Cl_2O formed according to the equilibrium of eq.2 and that probably both eqs.4 and 5 contribute to the initiation step. With substrates of high electron availability, such as m-xylene, at low temperature (0°C) the electron-transfer process is probably the important initiation step (eq.7)

$$(7)$$

Deactivated substrates, such as cyanotoluenes, require higher temperatures ($\dot{\div} 50°C$) for the benzylic chlorination because the corresponding aromatic radical cations are much less stable compared with that of m-xylene and the homolysis of the O-Cl bond according to eq.4 can become important as initiation step.

On the ground of the species present in the reacting systemn (HOCl, Cl_2O, Cl_2) several propagation steps can be considered (eqs.8-10)

$$R\cdot + ClOH \longrightarrow RCl + \cdot OH \qquad (8)$$

$$R\cdot + Cl_2O \longrightarrow RCl + \cdot OCl \qquad (9)$$

$$R\cdot + Cl_2 \longrightarrow RCl + Cl\cdot \qquad (10)$$

Now dichlorine monoxide is present in small amount in the reacting system because of the unfavourable equilibrium of eq.2; it can act as initiator of the free-radical chains, but its participation in the propagation step (eq.9) appears to be less important. That is supported by the fact that Cl_2O smoothly chlorinates deactivated substrates, such as nitrotoluenes[5], which do not react with HOCl. This last, on the other hand, is present in the aqueous phase and its contribution to the propagation chain (eq.8) occurring in the organic phase is not favoured. Moreover, the formation of hexachlorocyclohexane from benzene suggests that molecular chlorine and chlorine atom are involved in the propagation steps (eqs.11-13)

$$(11)$$

$$(12)$$

$$\text{etc.} \qquad (13)$$

The tertiary/primary selectivity in the chlorination of 2,3-dimethylbutane (Table IV) confirms the participation of chlorine atom as important hydrogen abstracting species. The selectivity is somewhat higher with HOCl compared to that of photochlorination by Cl_2 in the absence of solvents or in CCl_4. However, solvents such as benzene or CS_2 well-known[6] for the marked effect on the selectivity by chlorine atoms, give similar results in the photochlorination by Cl_2 and chlorination by HOCl.

The relative rates of hydrogen abstraction from toluene and cyclohexane (Table III), determined by the competitive method, are more intriguing above all for what concerns the

temperature effect: cyclohexane is more reactive than toluene
with both Cl_2[7] and HOCl; however, an increase of temperature
increases the selectivity in the photochlorination with Cl_2
and decreases the selectivity with HOCl. These results would
suggest that the chlorine atom is not the only chain-carrying
species and that a superimposition of different chains better
explains the behaviour. In any case an important contribution of the mechanism of the propagation is similar to the
Goldfinger mechanism in the bromination by N-bromosuccinimide
(Scheme 3) and it involves a chlorine atom chain-carrying
species (Scheme 4)

$$R-H + Br. \longrightarrow R. + HBr$$

$$R. + Br_2 \longrightarrow R-Br + Br.$$

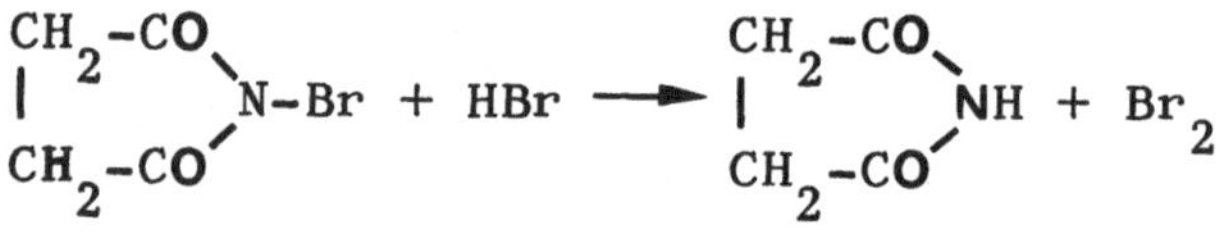

$$\begin{matrix} CH_2-CO \\ | \\ CH_2-CO \end{matrix} N-Br + HBr \longrightarrow \begin{matrix} CH_2-CO \\ | \\ CH_2-CO \end{matrix} NH + Br_2$$

Scheme 3

$$R-H + Cl. \longrightarrow R. + HCl$$

$$R. + Cl_2 \longrightarrow RCl + Cl.$$

$$HOCl + HCl \longrightarrow Cl_2 + H_2O$$

Scheme 4

The overall selectivity is not much different from that
of photochlorination by Cl_2; the moderate differences can be
ascribed to superimpositions of other propagating chains
(eqs. 8,9).

An advantage of this new procedure is purely practical:
hypochlorite is more practical to handle than molecular chlorine. Another practical advantage is the fact that no initiator is necessary to carry out the chlorination; the reaction is spontaneous at moderate temperature (0°-50°C). A further advantage is the same observed in the free-radical bromination by N-bromosuccinimide compared with Br_2, that is the
steady-state concentration of Br_2 is very low, the N-bromosuccinimide acting as reservoir of Br_2. Similarly, HOCl in
the aqueous phase acts as reservoir of Cl_2, whose stationary concentration in the organic phase is very low

during the reaction. That is favourably reflected when an electrophilic process is a competitive process because the reactions of alkyl radicals with Cl_2 (eq.10) are very fast, close to the diffusion controlled limit, whereas the rates of the electrophilic chlorinations by Cl_2 can range within very wide limits. Thus the low stationary concentration of Cl_2 generally favours the free-radical process. A consequence with moderately activated alkylbenzenes, such as toluene and xylenes, is the temperature effect on the free-radical/electrophilic competition: with HOCl, in which the stationary concentration of Cl_2 is low, the free-radical process is depressed by an increase of temperature. In the photochlorination by Cl_2 (relatively high concentration generally) the behaviour is opposite.

A further consequence is the fact that the chlorine atom can abstract benzylic atoms from alkylbenzenes, but it can also add to the aromatic ring (eq.14)

(14)

Path (a) is irreversible, whereas path (b) is reversible and the reversibility is strongly affected by the temperature and the concentration of Cl_2. An increase of temperature and a decrease of Cl_2 concentration favours the reversibility and therefore the benzylic chlorination. Thus the low stationary concentration of Cl_2 with HOCl is more suitable for the benzylic chlorination at low temperatures.

All the other conditions being equal the ratio benzylic/nuclear chlorination of alkylaromatics decreases by the electron availability of the aromatic substrate. A high electron availability favours both the initiation step of the free-radical chain, if this occurs according to an electron-

transfer process (eq.7), and the benzylic hydrogen abstraction (all the possible abstracting species are electrophilic).
However, the influence of the electron availability of the alkyl aromatics is much higher in favouring the electrophilic chlorination of the aromatic ring than the steps of the free-radical chain. The consequence is that electronrich alkyl-aromatics, such as methoxytoluenes, only give electrophilic chlorination; electron-poor alkylaromatics, such as halo-toluenes, cyanotoluenes, acyltoluenes and toluic acids only give benzylic chlorination and more deactivated substrates, such as nitrotoluenes, do not react.

The behaviour appears quite similar with olefins. Cyclohexene is more reactive than toluene towards the electrophilic chlorination, whereas the allylic C-H bond is nearly as reactive as the benzylic one; thus the electrophilic addition prevails (90%) on the free-radical substitution (10%).
With 1-cyanocyclohexene, in which the electrophilic process is depressed, mainly the free-radical substitution takes place; the process is not selective and a full analysis of the reaction products was not accomplished. However, similar results were obtained with HOCl and in the photochlorination by Cl_2 (the same peaks in similar ratios in g.l.c.), once again supporting the fact that the chlorine atom is an important chain-carrying species. A similar trend is observed in the competition between radical addition to the aromatic ring and electrophilic substitution. Thus benzene only gives free-radical addition and no significant amount of chloro-benzene, whereas the more reactive naphtalene (the partial rate factor for the electrophilic chlorination of the α-position in naphtalene is 9.1×10^4) mainly gives electrophilic substitution.

The effect of the solvents is explained by the fact that the electrophilic substitution is strongly favoured by a polar medium. However, the results with m-xylene (Table II) indicate that the situation is more complex and the overall rationalization is complicated by the fact that a hetero-geneous system is operating, in which initially the initiator and the chlorinating agent are in the aqueous phase, whereas chlorination takes place in the organic phase. A further factor arises if an electron-transfer initiation is working as suggested with m-xylene (eq.7); a cage collapse of the geminate radical pair can lead to nuclear chlorination (eq.15)

$$\left[\begin{array}{c} \text{CH}_3 \\ \text{CH}_3 \end{array} , \text{ Cl}^{\cdot} , \text{ ClO}^{-} \right] \rightarrow \begin{array}{c} \text{CH}_3 \\ \text{CH}_3 \\ \text{Cl H} \end{array} \cdot ^{-}\text{OCl} \longrightarrow \begin{array}{c} \text{CH}_3 \\ \text{CH}_3 \\ \text{Cl} \end{array} \cdot \text{ HOCl} \qquad (15)$$

The efficiency of the initiation of the radical chains is strictly connected with the diffusive separation of the radical pair of eq.7. Thus all the factors which influence this separation also contribute to determine the overall selectivity between nuclear and benzylic chlorination. This type of chlorination by HOCl has been developped[8] in the early 1983. Immediately afterwards it has been reported[9] that during the oxidation by inorganic hypochlorites in basic medium (pH 7.5-9) and in the presence of phase-transfer catalyst the chlorination of alkanes, alkenes and alkylbenzenes can occur at some extent. No substantial reaction takes place under these conditions in the absence of phase-transfer catalyst. However, the selectivities obtained in this case are quite different from those observed with our procedure: it is different the intermolecular selectivity between toluene and cyclohexane and the intramolecular selectivity in dimethylbutane; moreover, benzene does not substantially react (only 1% of chlorobenenzene is formed). It has been proposed that the epoxidations and chlorinations observed under phase-transfer catalysis proceed both by a free-radical mechanism involving the ClO. radical. However, the efficiency of chlorination by this system is poor: high conversions can be obtained only by a large excess of reagent.

Conclusion

Hypochlorous acid, well-known as agent of electrophilic chlorination, can be utilized under suitable, mild conditions also for free-radical chain chlorination in the absence of initiators and in the dark. It takes place in a two-phase system and it is a practical procedure for benzylic chlorination of moderately activated (toluene, xylenes) or deactivated (halo-, carboxy-, cyano-, acyltoluenes etc.) alkyl aromatics. A high activation (i.e. alkyl anisoles) or a high deactivation (i.e. nitrotoluenes) prevent from benzylic chlorination.

REFERENCES

1. "Aromatic Substitution. Nitration and Halogenation".
    P.B.D. De La Mare', J.H. Ridd, Butterwords, London 1959,
    pag.16.
    "Electrophilic Substitutions in Benzenoid Compounds",
    R.O.C. Norman, R. Taylor, Elsevier, London, 1965, pag.119.
2. M.L. Poutsma, Methods in Free-Radical Chemistry, vol.1,
    79, 1969.
3. C. Walling, B.B. Jacknow, J.Am.Chem.Soc. 1960, 82, 6108,
    6113.
4. F. Minisci, E. Platone, G. Serboli, U.S.P. 4.423.263
    (15/7/1980).
5. F.D. Marsh, W.B. Fornham, D.J. Sam, B.E. Smart, J.Am.Chem.
    Soc. 1982, 104, 4680.
6. G.A. Russell, J.Am.Chem.Soc., 1958, 80, 4987.
7. G.A. Russell and H.C. Brown, J.Am.Chem.Soc. 1955, 77,
    4578.
8. G. Faraci, E. Platone, F. Minisci, Ital.Pat. 1.161.227
    (1983).
9. H.E. Fanouni, S. Krishnan, D.G. Kuhn, G.A. Hamilton,
    J.Am.Chem.Soc. 1983, 105, 7672.

INDUCED DECOMPOSITION OF PEROXYCOMPOUNDS IN SYNTHESIS : FREE RADICAL FUNCTIONALIZATION OF CROWN ETHERS

B. MAILLARD[*](a), M.J. BOURGEOIS (b), R. LALANDE (b) , E. MONTAUDON (b)

(a) Laboratory of Organic and Organometallic Chemistry, CNRS UA 35
(b) Laboratory of Applied Chemistry
University of Bordeaux I, 351, cours de la Libération,
F-33405 TALENCE-Cedex (FRANCE).

ABSTRACT . Free radical additions of commercial 12-crown-4, 15-crown-5 and 18-crown-6 to various unsaturated peroxyesters and peroxides allowed their functionalization : acetonyl, 2,3-epoxypropyl, 2-oxacyclopentylmethyl, 3-oxo-2-oxacyclopentylmethyl and 3-oxo-2,4-dioxacyclopentylmethyl groups were grafted onto the heterocycle. Polyfunctionalization could be obtained by changing the ratio crown ether/peroxycompound.

1 . INTRODUCTION

The importance of crown ethers in various fields ([1]) is increasing every day with the discovery of new applications. In the last few years numerous papers ([2]) dealing with the preparation of armed crown ethers have been published.

Our studies on the synthetic potentialities of the induced decomposition of unsaturated peroxy-compounds prompted us to apply this methodology to access such heterocycles. Indeed, in our previous work ([3]), we succeeded in the functionalization of the dioxane that could be considered as the first member of the family of crown ethers deriving from ethylene glycol (Figure 1, n = 1).

Several of these compounds are similarly formed by the following mechanism ([3a], [4], [5]) :

   - Addition to the double bond :

$$Z^{\bullet} + CH_2=CH \text{ wwwwwwwOOt-Bu} \longrightarrow ZCH_2\overset{\bullet}{CH}\text{wwwwwwwOOt-Bu}$$

   - Intramolecular homolytic substitution :

$$ZCH_2\overset{\bullet}{CH}\text{wwwwwwwOOt-Bu} \longrightarrow ZCH_2\overset{\bullet}{CH}\text{wwwwwwO} + {}^{\bullet}Ot\text{-Bu}$$

283

F. Minisci (ed.), Free Radicals in Synthesis and Biology, 283–292.
© 1989 by Kluwer Academic Publishers.

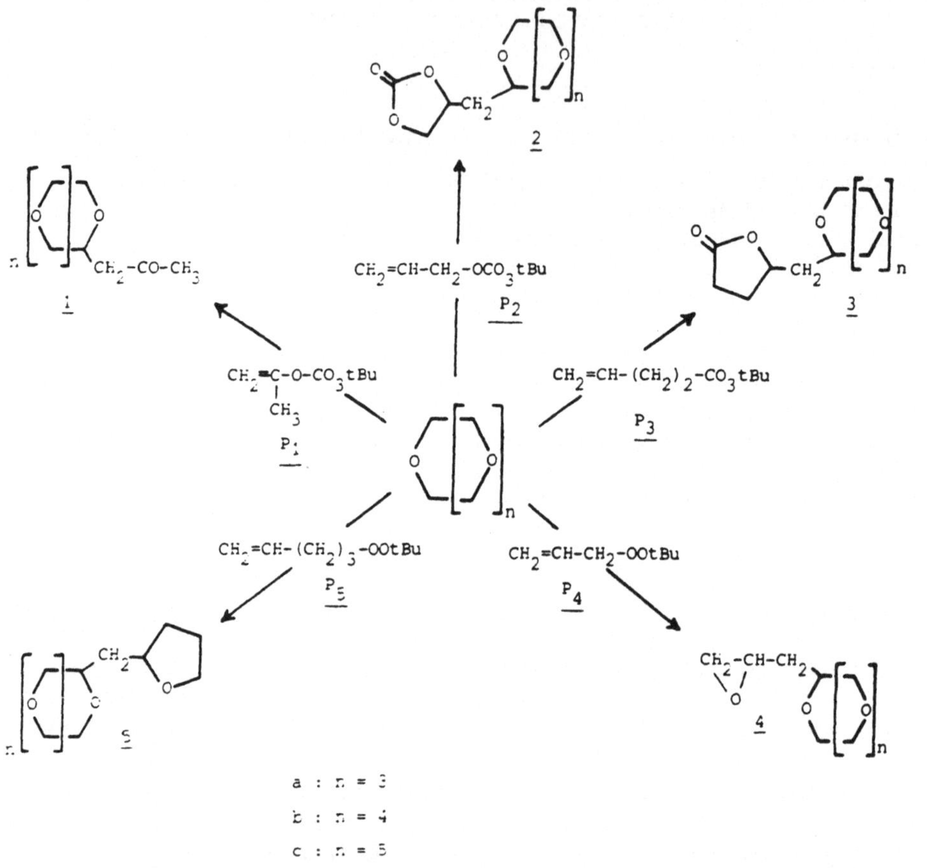

284

Figure 1 . Homolytic functionalization of ethylene glycol derivatives

- Regeneration of the initial radical :

$$t\text{-}BuO^{\cdot} + ZH \longrightarrow Z^{\cdot} + t\text{-}BuOH$$

The mechanism of formation of the ketone from O-isopropenyl O,O-t-butyl percarbonate has not yet been elucidated ; it could involve three similar steps (addition, β-elimination, regeneration of Z·) or two steps (the addition and the elimination being concerted). However from a synthetic point of view this is not important ; the total reaction occurs as a chain :

$$Z^{\cdot} + CH_2=C\!\!\!\!\begin{array}{c} R \\ \end{array}\!\!\!\!\sim\!\!\sim\!\!\sim A\text{-}OOt\text{-}Bu \xrightarrow{\qquad} \begin{cases} \xrightarrow{R=H} ZCH_2CH\sim\!\!\sim\!\!\sim A\text{-}O \\ \xrightarrow{R=CH_3} ZCH_2C(O)CH_3 + CO_2 \\ \sim\!\!\sim A=OC(O) \end{cases} + {}^{\cdot}Ot\text{-}Bu$$

with ZH regenerating Z·

Because previous work ([6]) showed that hydrogen abstraction from crown ethers by t-butoxy radical was efficient enough to initiate its homolytic addition to captodative olefins, we decided to start a study with these commercially available compounds.

2 . RESULTS

The free radical functionalization of 12-crown-4 by various peroxy-compounds (figure 1, n=3) could be performed with yields in the range 30-60 %, relative to the unsaturated compound, the crown ether being used in excess (ether/unsaturated peroxy-compound = 10). However, it must be noticed that this solvent was easily removed by distillation and could be reused without further purification. The conditions of the reaction (initiator and temperature) were chosen according to a low spontaneous decomposition of the peroxy-compound involved in the reaction and an easy access to the initiator : the ratios of these two last reactants have been taken as precedently determined as the optimal ones ([3]). All these variables are summarized in table I.

Taking into account the fairly good reactivity of epoxides and ketones towards various reactants, the compounds 4a and 1a could be valuable precursors of armed 12-crown-4 ethers. This remark prompted us to extend the induced decomposition of O-isopropenyl O,O-t-butyl percarbonate and allyl t-butyl peroxide by free radicals generated from 15-crown-5 and 18-crown-6 (table I). We obtained comparable yields for the same functionalization of the three ethers in all the examples ; only the ketone deriving from the 18-crown-6 was isolated with a lower yield than from its two homologs : no explanation could be advanced for the observed difficulty in the purification of this adduct.

TABLE I – Monofunctionalization of Crown Ethers

Unsaturated Peroxy Compound (a) P_x	Initiator	Molar Ratio P_x-Init.	T°C	t	Functionalized Crown ether(b) (yield(c))		
$CH_2=\underset{CH_3}{C}-OCO_3tBu$ $\underline{P_1}$	Diethyl perdicarbonate	1-0.2	60	12h	$\underline{1a}$ (45)	$\underline{1b}$ (47)	$\underline{1c}$ (30)
$CH_2=CH-CH_2-OCO_3tBu$ $\underline{P_2}$	Benzoyl peroxide	1-0.1	80	24h	$\underline{2a}$ (30)		
$CH_2=CH-(CH_2)_2-CO_3tBu$ $\underline{P_3}$	Benzoyl peroxide	1-0.1	80	24h	$\underline{3a}$ (38)		
$CH_2=CH-CH_2-OOtBu$ $\underline{P_4}$	t-Butyl peracetate	1-0.1	110	12h	$\underline{4a}$ (60)	$\underline{4b}$ (55)	$\underline{4c}$ (50)
$CH_2=CH-(CH_2)_3-OOtBu$ $\underline{P_5}$	t-Butyl peracetate	1-0.5	110	12h	$\underline{5a}$ (55)		

(a) Crown ether/unsaturated peroxy compounds = 10

(b) Compound n_a deriving from 12-crown-4
    Compound n_b deriving from 15-crown-5
    Compound n_c deriving from 18-crown-6

(c) Calculated relative to the unsaturated peroxy compound

TABLE II – Polyepoxypropanation of 12-crown-4

Molar Ratio[a]		Yield[b]			Total yield of
Crown Ether	P_4	Monoepoxyde 4a	Diepoxyde 42a	Triepoxyde 43a	functionalization[c]
2	1	50	10	–	58
2[d]	1	–	60	3	51
1	2	23	15	5	33
1	2[e]	30	23	10	47
1	3[e]	18	22	14	32

(a) t-butyl peracetate = 0.1

(b) calculated relative to the crown consommed

(c) calculated relative to the unsaturated peroxy compound

(d) Crown used is 4a

(e) P_4 and t-butyl peracetate are added slowly with an automatic syringe (0.66 ml/h)

As polyfunctionalized crown ethers would be valuable compounds in polymer chemistry we decided to extend the work in this way. Polyepoxypropanation of 12-crown-4 was obtained by changing the ratio of the reactants (table II). No trials have been done to determine the relative ratios of the various isomers of 42a and 43a that could be obtained in these reactions.

Acetonylation of 4a to give 14a could be achieved by its free radical addition to O,O-t-butyl O-isopropenyl percarbonate with a yield of 36% relative to the crown ether 4a consumed.

In conclusion, we wish to underline the easy functionalization of commercially available crown ethers through their homolytic addition to unsaturated peroxy-compounds ; in particular 2,3-epoxypropanation and acetonylation of 12-crown-4, 15-crown-4 and 18-crown-6 could be a good way to start the building of an arm for such heterocycles.

3 . MATERIALS

3.1 - Unsaturated Peroxycompounds P_x

These products were obtained in the same way as in our previous work ([3]).

3.2 - Initiators

t-butyl peracetate was prepared according to ([7]).

Diethylperdicarbonate was obtained by reaction of ethyl carbono-choridate with sodium peroxide ([8]).

Benzoyl peroxide was obtained from the commercial stabilized product by water elimination (extraction with chloroform, drying of the organic phase by sodium sulfate, elimination of the solvent under vacuum).

3.3 - Crown ethers

Starting materials are commercially available.

Monofunctionalized crown ethers are described in tables III and IV.

Polyfunctionalized crown ethers are identified on the basis of mass and NMR spectra.

In the polyepoxypropanation reaction mono-, di- and trifunctional compounds were identified by GC-MS under chemical ionization by ammonia (VG Micromass 16 F) : monoepoxide 4a (M+1 = 237, M+18 = 254), diepoxyde 42a (M+1 = 293, M+18 = 310), triepoxyde 43a (M+1 = 349, M+18 = 356).

Diepoxide 42a had been isolated by distillation (Kugelrohr 165°C/0.05 torr) meanwhile triepoxide 43a could never been obtained pure : it contained around 10 % of 42a.

Crown 42a NMR ^{1}H (BRUKER AC 200, CD$_3$COCD$_3$) : 1.3-1.8, m, 4H (CH-CH$_2$-CH) ; 2.4-2.5, m, 4H (CH-CH$_2$) ; 2.9-3.05, m, 2H(CH$_2$-CH) ; 3.35-4.0, m, 14H (H of crown).

Crown 14a NMR ^{1}H (BRUKER AC 250, CDCl$_3$) : 1.5-1.9, m, 2H (CH-CH$_2$-CH) ; 2.1, s, 3H(CO-CH$_3$) ; 2.3-2.8, m, 4H (CH-CH$_2$; CH$_2$-CO) ; 2.9-3.1, m, 1H (CH-CH$_2$) ; 3.3-4.1, m (H of crown).

TABLE III - Monofunctional 12-Crown-4

Compound	b.p. °C/Torr	n_D^{20}	^{1}H-NMR (CCl$_4$/TMS) δ(ppm) (b)	^{13}C-NMR (CDCl$_3$/TMS) δ(ppm) (c)
1a $12\text{-}O_4\text{-}\overset{3}{C}H_2\text{-}\overset{2}{C}O\text{-}\overset{1}{C}H_3$	115/0.1	1.4710	2.1, s, 3H(H$_1$) ; 2.4-2.6, m, 2H(H$_3$) ; 3.4-4.1, m, 15H(crown)	C$_1$: 31.0 ; C$_3$: 45.8 ; CH$_2$(crown): 70.0-70.4-70.7-70.9-73.1 ; CH(crown): 75.2 ; C$_2$: 207.1
2a $12\text{-}O_4\text{-}\overset{4}{C}H_2$	170-180/0.01 (a)	1.4850	3.3-5, m	C$_4$: 35.4-36.6 ; C$_2$ and CH$_2$(crown): 69.2-69.8-70.0-70.3-70.5-70.8-72.6-72.9 ; C$_3$ and CH(crown): 74.6-75.0-75.2-75.5 ; C$_1$: 155.2
3a $12\text{-}O_4\text{-}\overset{5}{C}H_2$	170-180/0.01 (a)	1.4860	1.5-2.7, m, 6H(H$_5$,H$_3$,H$_2$) ; 3.5-4.9, m,16H(crown,H$_4$)	C$_2$,C$_3$: 28.2-28.6 ; C$_5$: 37.4-38.5 ; CH$_2$(crown) : 69.4-70.3-70.6-71.4-73.0-73.5 ; C$_4$ and CH(crown): 75.9-76-77.8 ; C$_1$: 177.0
4a $12\text{-}O_4\text{-}\overset{3}{C}H_2\text{-}\overset{2'}{C}H\text{-}\overset{1}{C}H_2$	110/0.05	1.4730	1.2-1.75, m, 2H(H$_3$) ; 2.2-3, m, 3H(H$_1$, H$_2$) ; 3.3-4, 15H (crown)	C$_3$: 34.8-35.5 ; C$_1$: 46.8-47.5 ; C$_2$: 49.3-49.6 ; CH$_2$(crown) : 69.7-70.3-70.4-70.7-70.8-71.0-71.1-73.5-74 ; CH(crown) ; 77.4
5a $12\text{-}O_4\text{-}\overset{5}{C}H_2$	150/0.01 (a)	1.4775	1.15-2.2, m, 6H(H$_2$,H$_3$,H$_5$); 3.25-4.3, m, 18H(crown, H$_1$, H$_4$)	C$_2$,C$_3$: 25.6-25.7-31.74-31.85 ; C$_5$: 37.5-38.4 ; C$_1$: 67.5-67.6 ; CH$_2$(crown): 69.6-70.3-70.4-70.5-70.6-70.8-70.9-71-73.5-74.5 ; C$_1$ and CH (crown): 75.8-76.1-77.1-77.3

(a) Kugelrohr distillation
(b) Recorded on a Bruker WP 60 CW Spectrometer . H$_x$ protons are bound to the carbon numbered X.
(c) Recorded on a Bruker WP 90 Spectrometer (22,63 MHz) (Broad-band ^{1}H decoupled).

TABLE IV - 2.3-Epoxy propanation and Acetonylation of 15-Crown-5 and 18-Crown-6

Compound	b.p. °C/torr	n_D^{20}	^{1}H-NMR (CCl$_4$/TMS) (c) δ(ppm)	^{13}C-NMR (CDCl$_3$/TMS) (d) δ(ppm)
1b $15\text{-}O_5\text{-}\overset{3}{C}H_2\text{-}\overset{2}{C}O\text{-}\overset{1}{C}H_3$	120-125/0.01	1.4720	2.1, s, 3H(H$_1$) ; 2.5-2.8, m, 2H (H$_3$) ; 3.4-4.1, m, 19H(crown)	C$_1$: 30.6 ; C$_3$: 46.0 ; CH$_2$ (crown) : 70.0-70.1-70.3-70.6-70.8-72.7 ; CH(crown) : 75.2 ; C$_2$: 206.9
1c $18\text{-}O_6\text{-}\overset{3}{C}H_2\text{-}\overset{2}{C}O\text{-}\overset{1}{C}H_3$	155/0.001 (b)	1.4720	2.1, s, 3H(H$_1$) ; 2.4-2.7, m, 2H(H$_3$) ; 3.4-4.1, m, 23H(crown)	C$_1$: 30.9 ; C$_3$: 45.9 ; CH$_2$ (crown) : 69.6-70.5-70.8-73 ; CH(crown) : 75.1 ; C$_2$: 207.3
4b $15\text{-}O_5\text{-}\overset{3}{C}H_2\text{-}\overset{2}{C}H\text{-}\overset{1}{C}H_2\backslash O$	125-130/0.1	1.4750	1.3-1.8, m, 2H(H$_3$) ; 2.25-3.1, m, 3H(H$_1$,H$_2$) ; 3.25-4, m, 19H(crown)	C$_3$: 34.8-35.5 ; C$_1$: 46.5-47.2 ; C$_2$: 49.2-49.4 ; CH$_2$(crown) : 69.5-70.1-70.3-70.5-70.7-70.8-70.9-73.2-73.6 ; CH(crown):77.2
4c $18\text{-}O_6\text{-}\overset{3}{C}H_2\text{-}\overset{2}{C}H\text{-}\overset{1}{C}H_2\backslash O$	155/0.001 (b)	1.4752	1.3-1.9, m, 2H(H$_3$) ; 2.15-3.1, m, 3H(H$_1$,H$_2$) ; 3.3-4.1, m, 23H (crown)	C$_3$: 34.35-35.0 ; C$_1$: 46.3-46.9 ; C$_2$: 48.9-49.1 ; CH$_2$ (crown) : 68.9-69.4-70.2-70.4-73.2-73.6 ; CH(crown) : 76.5

(a) (b) (c) see table I

4 . METHODS

4.1 - General procedure of the functionalization under static conditions :

The mixture of the reactants was heated in a flask fitted with a condenser for t hours at a temperature of T°C (table I). The excess crown ether was then distilled off and the residue distilled at reduced pressure (distillation apparatus or Kugelrohr according to the boiling point of the substituted crown ether). If necessary, a complementary purification was performed by liquid-solid chromatography (SiO_2 ; solvent : diethyl ether-acetone).

4.2 - General procedure under dynamic conditions:

The mixture of allyl t-butyl peroxide and t-butyl peracetate was added with an automatic syringe (0.66 ml/h) to the crown ether heated (110°C) under agitation in a flask fitted with a condenser. The medium was maintained for 12 hours at 110°C after the end of the addition. The functionalized and starting crown ethers were then separated by distillation with a Kugelrohr apparatus.

The purity of the compounds was checked by gas chromatography (3% OV-1 on Chromosorb WHP 80/100 mesh - Intersmat ICG 112 F).

LITERATURE

(1)- Kopolow S., Hogen Esch T.E., Smid J., _Macromolecules_ 1973, 6, 133.
  - Ungaro R., El Haj B., Smid J., _J. Am. Chem. Soc._ 1976, 98, 5198.
  - Kimura K., Maeda T., Shono T., _Anal. Lett._ 1978, A11, 821.
  - Ikeda I., Katayama T., Okahara M., Shono T., _Tetrahedron Lett._ 1981, 22, 3615.
  - Maeda T., Ouchi M., Kimura K., Shono T., _Chem. Lett._ 1981, 1573.
  - Ikeda I., Emura H., Yamamura S., Okahara M., _J. Org. Chem._ 1982, 47, 5150.

(2) See e.g. :

  - Nakatsuji Y., Nakamura T., Okahara M., Dishong D.M., Gokel G.W., _J. Org. Chem._ 1983, 48, 1237.
  - Jarvis B.B., Vrudhula V.M., Dishong D.M., Gokel G.W., _J. Org. Chem._ 1984, 49, 2423.
  - Kaifer A., Echegoyen L., Gokel G.W., _J. Org. Chem._ 1984, 49, 3030.
  - Elben U., _Liebigs Ann. Chem._, 1985, 210.
  - Belohradsky M., Stibor I., Holy P., Zavada J., _Coll. Czech. chem. Commun._, 1987, 52, 2500.
  - Nakatsuji Y., Nakamura T., Yonetami M., Yuya H., Okahara M., _J. Amer. Chem. Soc._ 1988, 110, 531.

292

(3)a – Jahouari, R., Maillard B. , Filliatre C., Villenave J.J.,
      Synthesis 1982, 760.

   b – Montaudon E., Rakotomanana F., Maillard B., Bull. Soc. Chim. Fr.
       1985, 198.

   c – Maillard B., Kharrat A., Rakotomanana F., Montaudon E.,
       Gardrat C., Tetrahedron, 1985, 41, 4045.

   d – Kharrat A., Gardrat C., Maillard B., Bull. Soc. Chim. Belg.
       1986, 95, 535.

   e – Agorrody M., Montaudon E., Maillard B., to be published.

(4) Maillard B., Kharrat A., Gardrat C., Nouv. J. Chim. 1987, 11, 7.

(5) Bourgeois M.J., Maillard B., Montaudon E., Tetrahedron 1986, 42,
     5309.

(6) Beaujean M., Mignani S., Merenyi R., Janousek Z., Viehe H.G.,
    Kirch M., Lehn J.M., Tetrahedron 1984, 40, 4395.

(7) Bartlett P.D., Hiatt R.R., J. Am. Chem. Soc. 1958, 80, 1398.

(8) Strain F., Bissinger W.E., Dial W.R., Rudof H., Dewitt B.J.,
    Stevens H.C., Langston J.H., J. Am. Chem. Soc. 1950, 72, 1254.

Robert Louw
Center for Chemistry and The Environment
Gorlaeus Laboratories, Leiden University
P.O. Box 9502, 2300 RA Leiden,
The Netherlands

ABSTRACT. Gas-phase thermal conversion ("pyrolysis"), important in the chemical industry, can also be very useful in the laboratory, for example for the synthesis of compounds with small rings, difficult to prepare otherwise. High temperature pyrolysis of hydrocarbons results in carbon and hydrogen, but milder reaction gives redistribution (cf cracking of mineral oils); free radicals play a major part in such reactions. Examplified by CH_4 ($\longrightarrow C_2H_4$?) and benzene ($\rightarrow$ biphenyl), hydrocarbon pyrolysis is briefly discussed on a thermochemical basis. A comparison is made with hydrogenolysis, and with combustion. The backgrounds of soot formation are summarized, and the fate of O, Cl, etc - when present in the starting material is outlined. Combustion of Cl-containing materials entails generation (and emission) of toxic Cl_x-DBF's and -"dioxins", and chlorophenols are likely precursors. Model studies have shown that DBF's and dioxins arise via combination of two (chloro)phenoxy radicals.

1. INTRODUCTION

Most research chemists are accustomed to study reactions in solution, which are hoped for to take place at 350 ± 50 K. However, in the (petro)chemical industry numerous reactions and plants run at elevated temperatures and in the vapour phase - without solvents, eventually with the aid of (heterogeneous) catalysts. Despite the increased importance of entropy terms - making synthesis by combination, build up, less likely at higher temperatures, many large-scale operations involving group transformation are, at least in part, thermolytic reactions. Examples include refining mineral oils, vinyl chloride production from ethylene, and various processes for making useful derivatives of propene: allyl chloride/epichlorohydrin, perchlorination/chlorinolysis to tetrachloroethylene, or ammonoxidation to acrylonitrile. Beyond doubt, in gas-phase chlorinations, as in the two general types of high-temperature reactions practised on a gigantic scale: cracking and combustion, free radicals are crucial intermediates.
Remarkably, gas phase thermal reactions are not very popular as research subjects in university laboratories; as with free

F. Minisci (ed.), Free Radicals in Synthesis and Biology, 293–302.
© *1989 by Kluwer Academic Publishers.*

radical reactions in general, there is a prejudice amongst chemists, and students, that these types of chemistry are messy and of little value. Inspection of R.F.C. Brown's monograph[1] learns, however, that a pyrolysis set-up, consisting of standard laboratory devices, and whose operation requires no special skills, can be of great help in synthesizing a variety of organic compounds. Especially useful is FVT, flash-vacuum-thermolysis: nothing else, in principle than distillation in vacuo through a hot tube rather than a condenser. By rearrangement and/or fragmentation one can obtain, for example, exotic small ring, strained, compounds with good yields in single-step operations, where "traditional" solution chemistry would require a multistep approach. A better known category comprises olefin-forming (molecular) 1,2-elimination, from esters, bromides, sulphoxides etc. These reactions are endothermal, typically, by 10-20 kcal/mol, but entropy gain make them go to "completion" at, say, 400° C. (At ambient temperature this situation is reversed, and, at the right conditions, one can add HX to an alkene to obtain RX).

2. HYDROCARBON PYROLYSIS

Writing eq. (1) correctly is easy, but even a first year undergraduate will suspect, I guess, that thermolysis of natural gas will fail to give useful yields of ethylene. Indeed, below 1000 K thermodynamics is against him and you, and only above, say, 1500 K, entropy is important enough to counterbalance the huge endothermicity of over 50 kcal/mol. Alas, reaction (2) then is more favourable, substantiating also that at high temperatures, hydrocarbons are thermodynamically unstable relative to the elements, carbon and hydrogen.

One may add air, and - incomplete - combustion will generate heat needed for conducting pyrolysis, as well as the additional products, carbon oxides and water. This operation is, in principle, also beneficial for making C_2H_4 from CH_4. Reaction (3) is thermochemically pleasing, but the only snags are to find the right experimental conditions (catalyst), and to prevent C_2H_4 from further oxidation. Those researchers who will succeed in passing the "20/80" line (20% CH_4 conversion, 80% selectivity to C_2H_4) in an economically attractive fashion, will catch attention anyhow. Meanwhile, pyrolysis (cracking) of mixtures of - mostly paraffinic - hydrocarbons is practised to make C_2H_4, and the entropy-driven β-splitting of alkyl radicals which leads to ethylene is the thermodynamic counterpart of its polymerisation - whether free-radical, or otherwise - at lower temperatures (and, often higher pressures). Another feature is "redistribution" to get thermodynamically more stable molecules (more branched hydrocarbons, aromatisation, etc.), entailing better petrol/gasoline properties.

Turning to benzene as starting material, its pyrolysis is a useful method of making biphenyl. The thermal reaction sets in

Equation	$\Delta H°$ 1000K (kcal/mol)	logKp (unit atm, 1000 K)
(1) $2\ CH_4 \rightleftharpoons C_2H_4 + 2\ H_2$	52	− 4.2
(2) $CH_4 \longrightarrow C(s) + 2\ H_2$	21	+ 1.0
(3) $2\ CH_4 + O_2 \longrightarrow C_2H_4 + 2\ H_2O$	−66	+16.0
(4) $2\ C_6H_6 \rightleftharpoons (C_6H_5)_2 + H_2$	7	− 1.0
(5) $C(s) + H_2O \rightleftharpoons CO + H_2$	32	+ 0.4
(6) $2\ HCl + \frac{1}{2}\ O_2 \rightleftharpoons H_2O + Cl_2$	−14	− 0.4

around 500°C and the mechanism involves phenylation of benzene rather than combination of phenyl radicals[2]. In one atmosphere of hydrogen, equilibrium (4) can be attained around 700°C (1000 K), comprising ca. 1% of biphenyl. Cracking reactions - to give CH_4 and C_2's - and formation of soot, then, are insignificant[3], but become increasingly more important at higher temperatures[4]. (Hydro)pyrolysis of benzene also leads to small proportions of "PAH's" (polycyclic aromatic hydrocarbons), including terphenyls, naphthalene, phenanthrene and flouranthene. Formation of naphthalene can be rationalised in various ways, and which is most prominent will depend on the composition of the hydrocarbon mixture. In the presence of sufficient C_2H_4/C_2H_2, the route from benzene/phenylacetylene (and/or radical derivatives thereof), another combination with C_2, and ring closure, seems logical[5]. In some cases, benzyne, C_6H_4, may be intermediate; reaction with benzene should be followed by loss of C_2H_2[6]. The entry "2 x C_5" seems to be important in pyrolysis of cyclopentadiene or phenol[7]. As mentioned before aromatic hydrocarbons (also) arise from pyrolysis of nonaromatic hydrocarbons. One entry to benzene nuclei is Diels-Alder type condensation involving alkenes/alkynes and dienes/enynes, at least in part via biradical intermediates[8]. This pathway may merge with the so-called acetylene growth reaction, which has $C_4H_3\cdot + C_2H_2 \longrightarrow$ linear $C_6H_5\cdot \longrightarrow$ phenyl as crucial steps[5,9].
Aromatisation can occur at temperatures well below 500°C, and in those cases combination of allylic radicals to 1,5-hexadienes, H abstraction thereof, and ring closure seems to be important.[7,10]

Anyway, continued condensation/ring closure - eventually assisted by particulate matter, and/or the reactor wall - can give way to larger molecules of low volatility, and an outbreak of sooting. In turn, if time and temperature permit, soot is subject to further stripping, eventually to the ultimate stage: carbon.

3. FUNCTIONALIZED MOLECULES

When thermolysing compounds with elements other than C and H, their fate is of great theoretical and practical interest. To begin with oxygen containing molecules, H_2O usually is a prominent product, whereas - depending on structure - CO and/or CO_2 can also be formed, whether directly, or at high temperatures, indirectly from H_2O and carbonaceous material:

(5) $H_2O + C(s) \longrightarrow CO + H_2; CO_2 + C(s) \longrightarrow 2\ CO$

Chlorinated derivatives will yield HCl - in line with thermochemistry, Cl_2 formation is negligible if the H/Cl ratio in the feed is $\geq$ 1. Hence, methylene chloride "cleanly" gives 2 HCl + C(s), but CCl_4 per se obviously cannot give but Cl_2 and C_2Cl_6/C_2Cl_4, and at high temperatures, C(s) + 2 Cl_2.
Pyrolysis under conditions which keep organic structures grossly intact will be governed by kinetic principles. The degree of "mineralisation" of heteroatoms will then depend on the type of bonding in the feed. So, alkyl chlorides possessing β-hydrogen easily loose HCl below 600°C[11], but (poly)chlorobenzenes and PCB's are stable then, and will give off little HCl even at 700°C.[12] In phenol pyrolysis, CO and H_2O are the major oxygen containing products, and dibenzofuran (DBF) is part of the complex reaction mixture.[13] By analogy, chlorinated phenols will give rise to the notorious, - toxic - (poly)chlorinated DBF's. Note that thermal hydrogenolysis (pyrolysis in an atmosphere of H_2) results in a relatively rapid, and clean displacement of Cl[14a], OH[14b], etc., bound to benzene. This chemistry involves H atoms - present in about equilibrium concentration with H_2 at elevated temperatures - which displace and/or abstract such substituents.
Biphenyls, DBF's and (chlorinated) dibenzo-p-dioxins - are also "de-substituted" by H. Thermal hydrogenolysis, therefore, holds promise as a method for treatment of (hazardous) organic wastes, including those with complex chemical composition, and high contents of chlorine, etc.[14c]

4. ABOUT COMBUSTION

If combustion of hydrocarbons is conducted at ideal conditions
(excess of oxygen, T > 1200°C, t > 1 second, adequate mixing) it
should run to completion, and one will obtain H_2O and CO_2 without
even traces of CO and of other PIC's (products of incomplete
combustion). In principle, hazardous wastes should give the same
results; chlorinated compounds will not only yield HCl, but also
some Cl_2 -depending on conditions- due to the (equilibrium)
reaction (6).

However, even well-operating waste incineration facilities
emit traces of chlorinated PIC's. These result mostly from de
novo synthesis, which is also subject to fly ash catalysis.[15] In
other words, in efficient combustors it is the element
composition of the feed rather than its molecular buildup that
counts; so "clean" fuels such as polyethylene, with added HCl,
lead to (poly)chlorinated benzenes, DBF's and dioxins just as
more harmful starting materials do[16].

4.1. On the formation of DBF's and "dioxins".

Incomplete combustion of (chlorinated) biphenyl can lead to
substantial yields of (chlorinated) DBF's (that is why a PCB-fire
can be catastrophic). Also, the statement, that
pyrolysis/combustion of phenols also yields DBF's -and probably
dioxins as well- will hardly be contradicted. Recalling that
aromatisation is an important thermal reaction, and, that in
oxidising environments, direct or subsequent production of
phenols will be likely, their importance can not be
overestimated. There is however, very little mechanistic
knowledge behind the balanced equations of type (7).

(7)   2 C_6H_5OH DBF   (+ H_2O + H_2)

"dioxin"   (+ 2 H_2)

We are presently studying the mechanisms of vapour-phase
pyrolysis and slow combustion of phenols - including rates,
products and structural effects. Controlled slow combustion of

phenol at ca. 600°C can give over 70% yields of DBF, but dioxin
is not observed. When starting with benzene, biphenyl, phenol and
DBF are important products - in addition to CO/CO_2 - their ratios
depending on conditions: feed/oxygen ratios,
conversion/temperature.

Interestingly, the pyrolytic conversion of phenol to DBF can
also be effected by adding suitable free-radical generating
species. Whereas phenol _per se_ gives negligible conversion (in a
nitrogen atmosphere) at 450-500°C, 10-100 s, addition of a few
percent of nitromethane (which dissociates smoothly into $\cdot CH_3$ and
NO_2) leads to DBF, typically, in 10-20% molar yield on $MeNO_2$.
This observation strongly suggests that phenoxy radicals are
intermediates. Formation of $C_6H_5O\cdot$ is certain, as anisole as well
as o- and p-cresol (combination products of $C_6H_5O\cdot$ and $\cdot CH_3$) are
also formed, together with o- + p-nitrophenol (from $C_6H_5O\cdot$ and
NO_2). Evaluation of product compositions aids in quantitative
interpretation of DBF formation (_vide infra_).

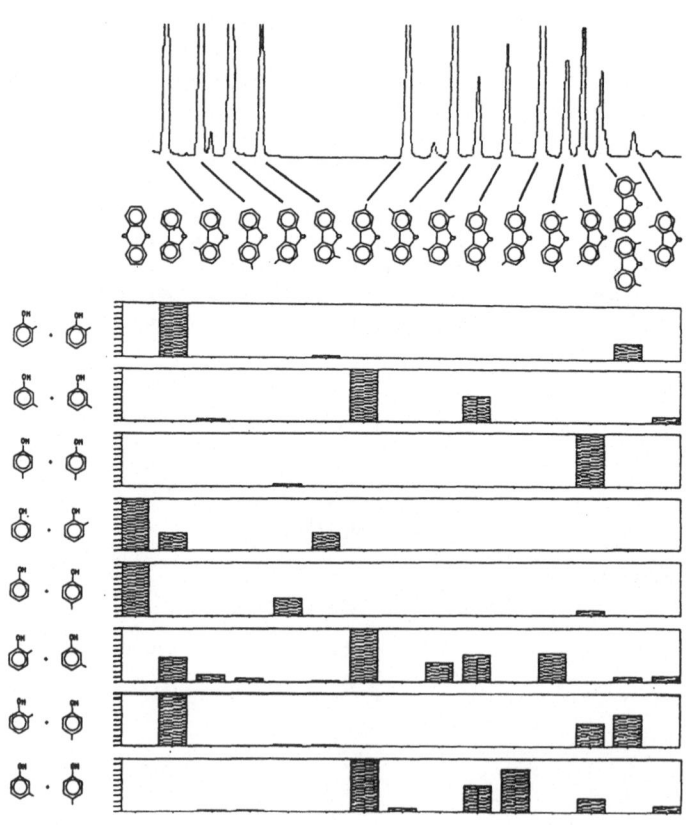

There is more to learn when using substituted phenols, such as o-, m- and p-chlorophenol. Whether subject to slow combustion or to free-radical assisted thermolysis, each individual isomer gives characteristic results. For example, slow combustion of o-chlorophenol (455°C, 100 s) led to
4,6-Cl$_2$-DBF, 4-Cl$_1$-DFB, DBF, dioxin, and Cl$_1$-dioxin, in the ratio 100:3.3:0.5:2.2:1.4. The other chlorophenols gave different isomers of
Cl$_x$-DBF, and no dioxins. Slow combustion of chlorobenzene resulted in a "combined" product pattern, not unlike that observed when starting with mixtures of phenol + o-/m-/p-chlorophenol. A typical GC spectrum is given next, together with an overview of products obtained from selected co-pyrolyses of phenols. (Cl substituents are indicated by dashes).

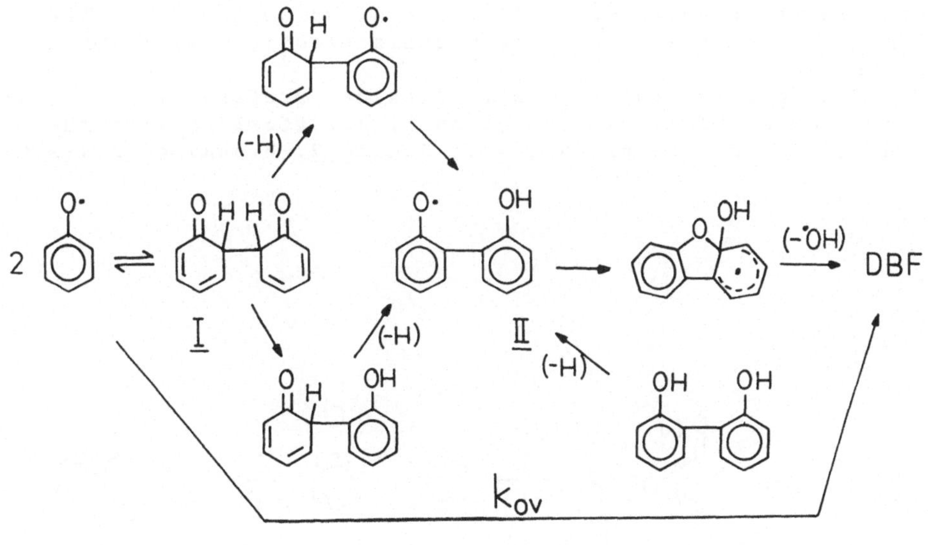

Scheme 1

The results on DBF's are in full accord with a mechanism involving combination of two (chloro)phenoxy radicals; dimerisation, via O-O, O-C, or C-C coupling wil be reversible, but when starting from dimer I, conversion into II can be envisaged, which forms DBF by expelling ·OH (Scheme 1). In line with this interpretations, o-, o'- dihydroxybiphenyl gives > 95% conversion and an > 80% yield of DBF when pyrolysed with 8% $MeNO_2$ at 490°C in nitrogen.

Product formation from phenol/$MeNO_2$ pyrolysis allows us to derive [$C_6H_5O·$] and k_{ov}, the overall rate constant for DBF formation. In a typical example [495°C, 95 s., inflow molar ratios phenol (1), $MeNO_2$ (0.09), N_2 (1.6)] the yield of DBF was 18%m on $MeNO_2$, and the DBF: $C_6H_5OCH_3$: C_2H_6 product ratio 1:0.061:0.046. The rate of formation of C_2H_6 translates into [$CH_3·$] = 4.8 x 10^{-9}M. Accepting k = 1.2 x 10^8 $M^{-1}s^{-1}$ for combination of ·CH_3 with $C_6H_5O·$ to give anisole[17], [$C_6H_5O·$] ~ 220 [·CH_3], or ca. 10^{-6} M. As the rate of formation of DBF is close to 10^{-6} M.s^{-1}, k_{ov} [$C_6H_5O·$]^2 = 10^{-6}, or k_{ov} ~ 10^6 $M^{-1}s^{-1}$. This is a factor 10^2-10^3 lower than that for the primary step giving dimer I. Proper thermokinetic analysis has revealed, that, in this example, [I]$_{eq.}$ ~ $10^{-5.5}$ M. Assuming conversion of I into II is rate determining in DBF formation, the requisite removal of H could be performed by 10^{-6} M of $C_6H_5O·$ with a rate constant of ca. 10^6 $M^{-1}s^{-1}$ - for example, with the parameters, log k = 9 - 10.5/2.3 RT. NO_2 may also play a part, and in combustion, ·OH and/or O_2 are likely agents as well. Note that, in the special case of o, o'- dihydroxybiphenyl, the mechanism can be envisaged as a chain reaction, with ·OH as a carrier.

Only o-chlorophenol led to dioxins, preferably the chlorine-free congener. This can be rationalised, starting with the O-ortho C-combination product A of Scheme 2. Breaking of its C-Cl

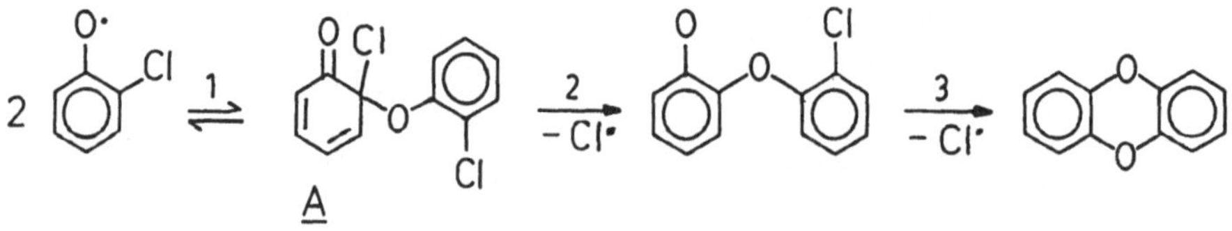

Scheme 2

bond needs only ca. 60 kcal/mol. If $[\underline{o}\text{-}ClC_6H_4O\cdot] = 10^{-6}$ M, $[A] = 10^{-7}$ M. Accepting $\log k_2/s^{-1} = 16 - 60/2.3$ RT, $\log v_2 = -8.0$ at 495°C. This is consonant with production of dioxin at the per cent level of DBF's.

Studies on other phenol derivatives will further help in identifying important structural parameters and detailing mechanism. Also, translations to more "practical" systems have to be made in order to determine the significance of mere gas-phase reactions in the formation, and survival, of DBF's and dioxins upon combustion.

ACKNOWLEDGEMENTS

The author is indebted to dr. Peter Mulder, drs. Jaap Joosting Bunk, drs. Jan G.P. Born, Jeff A. Manion, B.Sc., and students Isabel Arends, René Ophorst, Ronald IJsselstein, Roland Spronk and Willem Klinkenberg, whose efforts and results form the basis for this contribution. Financial support by the Dutch Ministry of Housing, Physical Planning, and Environmental Affairs (VROM) is also gratefully acknowledged.

REFERENCES

1. Brown, R.F.C., 'Pyrolytic Methods in Organic Chemistry', Acad. Press, **1980**.

2 a. Louw, R., and Lucas, H.J., *Recl.Trav.Chim.Pays-Bas*. **1973**, *92*, 55.
   b. Poutsma, M.L., 'A Review of the Thermolysis Studies of Model Compounds Relevant to Processing of Coal', **1987**, ORNL/TM-10637.

3. Louw, R., Dijks J.H.M., and Mulder, P., *Recl.Trav.Chim.Pays-Bas*, **1984**, *103*, 271.

4. Kiefer, J.H., Mizerka, L.J., Patel, M.R. and Wei, H.-C., *J.Phys.Chem.*, **1985**, *89*, 2013.

5. Harris, S.J., Weiner, A.M. and Blint, R.J., *Comb. and Flame*, **1988**, *72*, 91.

6. Fields, E.K. and Meyerson, S., *Adv.Phys.Chem.*, **1968**, *6*, 1.

7. Klinkenberg, W. and Louw, R., *CCE Special Report Series 87-01*, Leiden University **1987**.

8. Harper, C. and Heicklen, J., *Int.J.Chem.Kinet*, **1988**, *20*, 9.

9. Frenklach, M., Clary, D.W., Yuan, T., Gardiner, W., and Stein, S.E., *Comb.Sci.Technol.*, **1986**, *50*, 79.

10. Arends, I., Struijk, J., and Louw, R., unpublished observations.

11. Maccoll, A., *Adv.Phys.Chem.*, **1965**, *3*, 91.

12. Manion, J.A., Mulder, P., and Louw, R., *Environ.Sci.Techn.*, **1985**, *19*, 280.

13. Cypres, R., and Bettens, B., *Tetrahedron*, **1974**, *30*, 1253.

14 a.Manion, J.A., Dijks, J.H.M., Mulder, P., and Louw, R., *Recl.Trav.Chim. Pays-Bas*, **1988**, *107*, 434; b. Manion, J.A., and Louw, R., *J.Phys.Chem.*, submitted; c. Eur.Pat. 85.20136; U.S. 770.392.

15. Hagenmaier, H.P., Kraft, M., Brunner, M., and Haag, R., *Environ.Sci.Technol.*, **1987**, *21*, 1080.

16. Lahaniatis, E.S., Road, R., Bieniek, D., Klein, W., and Korte, F., *Chemosphere*, **1981**, *10*, 1321.

17. Lin, C-Y., and Lin, M.C., *Austral.J.Chem.*, **1986**, *39*, 723.

POLAR EFFECTS ON RADICAL ADDITION REACTIONS: AN AMBIPHILIC RADICAL

I. Beranek and H. Fischer
Physikalisch-Chemisches Institut der Universität
Winterthurerstrasse 190
CH-8057 Zürich
Switzerland

ABSTRACT. Absolute rate constants are presented for the radical $\cdot CH_2COOC(CH_3)_3$ to 19 mono- and 1,1-disubstituted ethenes in acetonitrile. They vary with the substituents in the range $4 \cdot 10^4 \le k \le 3 \cdot 10^6 M^{-1}s^{-1}$. Plots of $\log k$ vs ionization energies and electron affinities of the olefins yield U-shaped curves. This shows that both SOMO-LUMO and SOMO-HOMO interactions in the transition state lower the activation energy as expected from simple FMO-concepts.

1. INTRODUCTION

Polar effects of radical and substrate substituents on the rates of radical reactions have found considerable attention in recent years (1). They can lead to variations within many orders of magnitude, and their consideration is essential in the planning of successful syntheses, therefore (2). In much of our work (3) we have concentrated on the determination of absolute rate constants for prototype radical reactions by time resolved electron spin resonance, optical spectroscopy and muon spin rotation and have presented large series of data which allow rationalizations in terms of rather simple concepts. In the particular field of radical additions to olefins (4) we presented rate constants and their Arrhenius parameters for the reaction of the tert.-butyl radical $((CH_3)_3C\cdot)$ with 29 substituted ethenes which covered the range $6 \cdot 10^1 \le k_{300} \le 10^6 M^{-1}s^{-1}$. For mono- and 1,1- disubstituted ethenes addition occurs only at the unsubstituted carbon, and the substituent effects were in accord with a simple FMO model: tert.-Butyl has a very low ionization potential (6.92 ev (5)), i.e. a high lying SOMO which interacts in the transition state with the olefin's LUMO. For

F. Minisci (ed.), Free Radicals in Synthesis and Biology, 303–316.
© 1989 by Kluwer Academic Publishers.

substituents lowering the LUMO energy the addition is fast, whereas ethenes with high LUMO energies react reluctantly. A linear increase of logk and a decrease of the activation energy with increasing olefin electron affinities was found. This behaviour contrasts that of radicals with low lying SOMO'S as trifluoromethyl (6) and dicyanomethyl (7) where the rate constants increase with decreasing ionization potential of the olefin, i.e. the SOMO-HOMO interaction dominates.

Tert.-butyl and dicyanomethyl (IP > 10.87 e.V, that of $\cdot CH_2CN$ (8)) represent extreme cases of strongly nucleophilic and electrophilic radicals where one type of interaction dominates the other. If both are substantial then the FMO concept predicts for olefins with equal HOMO-LUMO gaps that plots of rate constants vs. ionization potentials and electron affinities show a minimum, i.e. the radical should react fast with both electron deficient and electron rich ethenes and slow with unsubstituted ethene (8). Such a behaviour is known for certain cycloaddition reactions (8) but has not yet been demonstrated for radicals.

To explore its existence we have chosen the radical $\cdot CH_2COOC(CH_3)_3$ whose ionization potential is expected to fall between that of $\cdot CH_3$ (9.84 eV(5)), a weakly nucleophilic species (9), and $\cdot CH_2CN$ (10.87 eV (8)) which is electrophilic. The radical was generated photochemically and its rate of addition to ethenes was measured near room temperature in acetonitrile solution. The temperature dependence could not yet be measured because the rates of addition are rather high and reach for several cases the limits of our experimental capabilities.

2. PROCEDURES AND MECHANISTIC ASPECTS

Our experimental arrangements for steady state and time resolved ESR-spectroscopy and the methods for data analysis have been published in detail previously ((3) and references therein). For steady state spectroscopy the tert-butoxicarbonylmethyl radical $CH_2COOC(CH_3)_3$ was produced photochemically by (a) Norrish Type I cleavage of the symmetrically substituted ketone 0.1 M in acetonitrile,

$$(CH_3)_3COOCCH_2COCH_2COOC(CH_3)_3 \rightarrow (CH_3)_3COOCCH_2\dot{C}O$$
$$+ (CH_3)_3COOC\dot{C}H_2$$

$$(CH_3)COOCCH_2\dot{C}O \rightarrow CO + (CH_3)_3COOC\dot{C}H_2$$

(1)

(b) Br-abstraction by $(Et)_3Si\cdot$, generated from tert.- butoxiradicals and $(ET)_3SiH$ in acetonitrile containing 0.4 M tert.-butyl-α-bromoacetate, 10% by volume silane and 20% di-tert-butylperoxide, and (c) photolysis of tert.-butyl-α-bromoacetate 0.4 M in acetonitrile or 3-methyl-3-pentanol. All methods yielded clean spectra of the radical $(2H_\alpha:21.30$ Gauss, g = 2.0034) in accord with previous findings (10). To investigate its mode of addition to olefins 22 different mono-or 1,1-disubstituted ethenes were added in 0.02 to 0.4 M concentrations. Clean spectra of the adduct radicals $(CH_3)_3COOCCH_2CH_2CH_2\dot{C}XY$ were observed for most cases which shows that the radical adds highly predominantly at the unsubstituted end of the double bond as expected (9,11). One example is given in Fig.1, and coupling constants of adducts are collected in Table I. For several styrenes the spectra were not completely analyzed but the large number of sharp ESR-lines ensured the benzylic character of the adduct radicals. Allylchloride and allylamine gave no observable adducts, and allylalcohol lead to the formation of $CH_2 = CH\dot{C}HOH$. Consequently, these three compounds were not used in kinetic runs. For several other olefins with allylic hydrogens photolysis of the ketone lead to the appearance of weak lines of $((CH_3)_3COOCCH_2)_2\dot{C}OH$ presumably by a photoreduction reaction, but the corresponding allyl type radicals could not be detected.

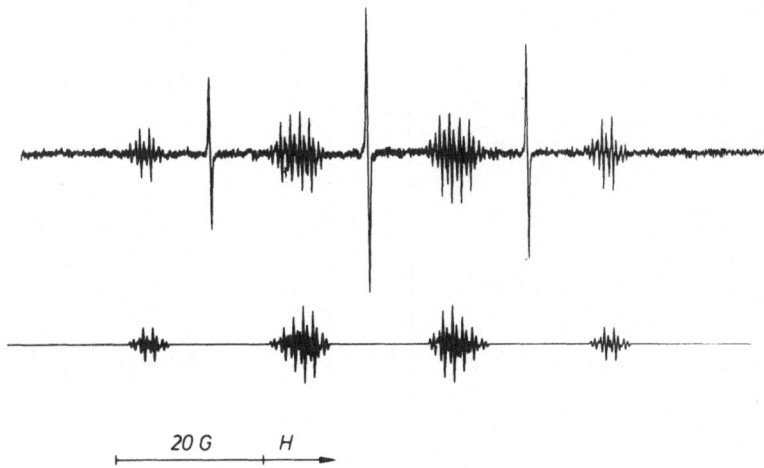

20 G   H

Figure 1. ESR-spectrum (top) of $(CH_3)_3COOC\dot{C}H_2$ (triplet) and $(CH_3)_3COOCCH_2CH_2\dot{C}HOCOCH_3$ observed during photolysis of the ketone in acetonitrile containing vinylacetate. Lower part: Simultation of the spectrum of the adduct radical.

Table I. Coupling Constants of Radicals $(CH_3)_3COOCCH_2CH_2\dot{C}XY$ (in ACN near room temperature, in Gauss)

X	Y	H_α	H_β	$3H_\beta$	$2H_\gamma$	other
H	CN	20.3	23.3			N: 3.4
H	$SiCl_3$	20.7	25.1			
H	$SiMl_3$	20.1	24.6			$9H_\gamma$: 0.4
H	CH_2CN	22.1	25.3,23.7			
H	CH_2CH_3	22.1	$25.3^{a)}$		$0.24^{b)}$	
H	$C(CH_3)_3$	21.3	27.2			$9H_\gamma$: 0.6
H	CH_2SiMe_3	20.7	24.4,18.4			
H	$COOCH_3$	20.5	25.1			$3H_\delta(CH_3)$:1.3
H	OEt	14.0	18.7		0.5	$2H_\gamma(OEt)$:1.5
H	OCOMe	18.9	21.5		0.6	$3H_\delta(CH_3)$:1.3
CH_3	$C(CH_3)_3$		16.8	22.7		
CH_3	$COOCH_3$		15.8	22.0		$3H_\delta(CH_3)$: 1.3
CH_3	Cl		12.7	22.7		^{35}Cl:2.0
CH_3	OCH_3		15.6	20.3	0.7	$3H_\gamma(OCH_3)$:1.5
CH_3	$OCOCH_3$		18.9	22.3		
Cl	Cl		11.7			$2^{35}Cl$:3.3

a) 2 CH_2 - groups

b) 5 γ - protons

The acyl radical of reaction (1) could not be detected directly nor could its adduct to the olefins. Obviously its decarbonylation is quite fast at room temperature ($k_D > 10^5 \text{s}^{-1}$) though a theoretical estimation using the methods of ref. (12) gave $k_D = 8.10^2 \text{s}^{-1}$. This needs further studies, though the fast decarbonylation is supported by the fast that ESR spectra taken in the μs-range with a direct detection technique (13) showed a pure multiplet type electron spin polarization of $(CH_3)_3COOC\dot{C}H_2$ and no indication of the presence of the acyl species.

GLC-analysis of the products formed by photolysis of the ketone in acetonitrile revealed the symmetric dimer $((CH_3)_3COOCCH_2)_2$ as the main product. Further products were $(CH_3)_3COOCCH_3$ and $(CH_3)_3COOCCH_2CH_2CN$ in yields of 5% and 1-3% relative to that of main product. This shows that the radicals formed in reaction (1) self-terminate by coupling and that side reactions involve hydrogen abstraction from the solvent and presumably from the ketone. They are of minor importance, however. Reaction products of the acyl radical could not be found.

In kinetic runs the radicals were generated by ketone photolysis in acetonitrile containing the olefins in $5 \cdot 10^{-5}$ to $4 \cdot 10^{-3}$ M concentrations. For fast reacting olefins special precautions as high flow rates and reduction of the photolysis zone ensured that substrate depletion did not exceed about 15%. Typically 100'000 to 200'000 concentration vs time profiles were accumulated for the center line of $(CH_3)_3COOC\dot{C}H_2$ and averaged. Two methods for data analysis were applied. The first assumes an only weakly perturbed second order decay and is valid if the pseudo-first-order reaction contributes to the overall radical decay to less then about 15% (3). For fast reacting olefins this condition could not be met. In these cases the more complex procedure of fitting numerically integrated forms of the appropriate rate laws to the time profiles (3) was used. For the addition to $CH_2 = C(CH_3)OCH_3$ both methods of analysis were applied and yielded results which were identical within the error limits.

Fig.2 shows a kinetic trace for solutions containing trimethylallylsilane as olefin and the resulting fit to numerically integrated rate laws. The individual addition rate constants were obtained from measurements employing different olefin concentrations. Plots of the inverse pseudo-first order lifetimes $\tau_1^{-1} = k \cdot [S]$ vs $[S]$ were fairly linear (Fig.3). The first order contribution for $[S] = 0$ is due to recations of the radical with the solvent acetonitrile and with the parent ketone. In separate experiments the rate constants for these hydrogen transfer reactions were determined as $k(CH_3CN) = (2.0 \pm 0.1) M^{-1} s^{-1}$ and

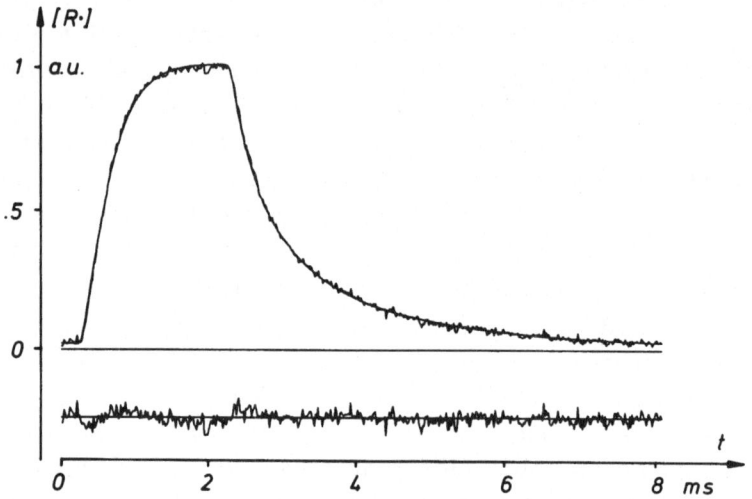

Figure 2. Concentration (arb.units) vs time for $\cdot CH_2COOC(CH_3)_3$ in solution containing trimethylallylsilane. Lower trace: Residual of the fit.

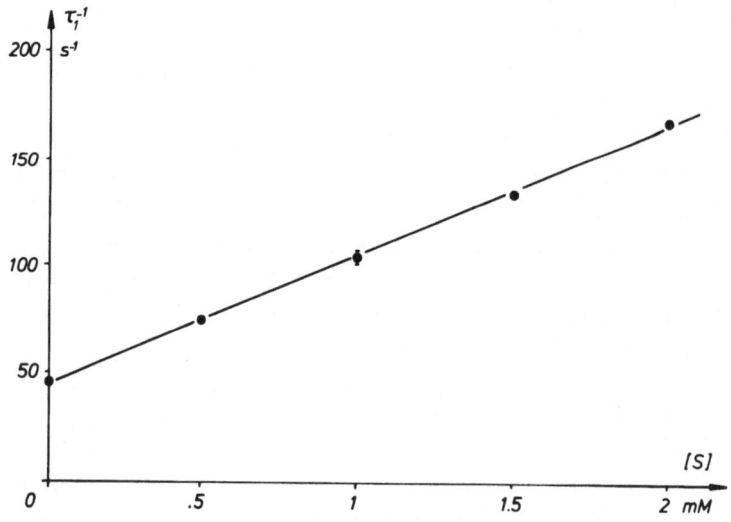

Figure 3. Pseudo-first order plot for the reaction of $\cdot CH_2COOC(CH_3)_3$ with vinyl acetate.

k(ketone) = $(100 \pm 11)M^{-1}s^{-1}$. Further, the self-termination rate constant of $\cdot CH_2COOC(CH_3)_3$ in acetonitrile at 295K was found to be $2k_t = (7.9 \pm .7) \cdot 10^9 M^{-1}s^{-1}$, i.e. probably diffusion controlled.

3. RATE CONSTANTS FOR THE ADDITION REACTION

Table II gives the rate constants for the addition of $\cdot CH_2COOC(CH_3)_3$ to 19 olefins together with the concentration range employed, the number n of individual experiments, the ionization potentials and electron affinities of the olefins.

Table II: Rate constants for the addition of $\cdot CH_2COOC(CH_3)_3$ to olefins $CH_2 = CXY$ at $(296\pm1)K$ in acetonitrile.

X	Y	c[mM]	n	$k \cdot 10^{-3}[M^{-1}s^{-1}]$	IP[eV[a)]	EA[eV][b)
H	CN	.1-.2	6	512 ± 47	10.95	-.21
H	$SiCl_3$	.05-.2	6	890 ± 90[c)]	10.79	-.52
H	CH_2CN	.2-3.2	12	38 ± 2	10.56	-1.31
H	$COOCH_3$	.1-.2	6	450 ± 33	10.52	-.49
CH_3	$COOCH_3$	.05-.2	6	1250 ± 95[c)]	10.28	-.38
Cl	Cl	.2-.4	5	243 ± 11	10.00	-.76
H	$Si(CH_3)_3$	.6-2	12	84 ± 3	9.80	-1.14
H	$OCOCH_3$	.5-2	12	60 ± 2	9.77	-1.19
CH_3	Cl	.5-1.5	9	132 ± 3	9.76	-1.44
CH_3	$OCOCH_3$	.5-1.5	7	82 ± 3	9.74	-1.51
H	CH_2CH_3	.4-1.5	8	50 ± 2	9.59	-1.90
H	$C()CH_3)_3$	1.5-3.0	6	38 ± 1	9.45	-1.73
H	OCH_2CH_3	.4-.8	5	134 ± 5	9.15	-2.24
CH_3	$C)CH_3)_3$	.4-.8	6	156 ± 9	9.02	-1.77
H	$CH_2Si(CH_3)_3$	.5-1.5	9	82 ± 2	9.00	-1.72

310

CH$_3$	OCH$_3$	.4-.8	11	121 ± 8	8.75	-2.48
H	Ph	.05-2	8	1950 ± 120[c]	8.50	-.25
CH$_3$	Ph	.05-.2	11	3500 ± 100[c]	8.35	-.23
Ph	Ph	.1	1	≥3000[c]	8.00	+.36

a) Ref (14)
b) Ref (15)
c) From numerical fits, other data from fits of analytical functions.

The addition rate constants cover the range 38'000 ≤ k ≤ 3'500'000 M^{-1}s^{-1}, i.e. vary by about a factor 10^2 with substitution. For tert.-butyl they vary from 60 M^{-1}s^{-1} to 2'400'000 M^{-1}s^{-1}, i.e. cover a much larger range (4a). To most olefins $\cdot$CH$_2$COOC(CH$_3$)$_3$ adds faster than tert.- butyl as may be expected for a primary radical. Also, the apparent lower selectivity correlates reasonably with the higher reactivity. However, a more detailed comparison of the rate constants of the two radicals reveals further interesting differences: To olefins like acetonitrile, 1,1-dichloroethene and acrylic esters which react with tert.-butyl with very high rate constants $\cdot$CH$_2$COOC(CH$_3$)$_3$ adds equally fast or slightly slower by factors of 2 to 5. Olefins reacting with tert.butyl with intermediate rate constants ($5\cdot10^3$ to $5\cdot10^5$M^{-1}s^{-1}) react with $\cdot$CH$_2$COOC(CH$_3$)$_3$ about 10 to 50 times faster. This factor increases to several hundreds for electron rich olefins like vinylethers which react slow ($<5\cdot10^3$M^{-1}s^{-1}) with tert. butyl. That is, $\cdot$CH$_2$COOC(CH$_3$)$_3$ adds fast both to electron deficient and electron rich olefins but slower to intermediate cases.

In discussions of rate constants of addition or abstraction reactions of intermediates with substrates in the frame of FMO concepts it has become customary to use measurable molecular quantities as ionization potentials and electron affinities as a basis for correlations (4,7,8,16). For olefins the ionization potential serves as a measure of the HOMO energy and the electron affinity measures the LUMO energy. For additions of radicals whose SOMO energies fall within the HOMO-LUMO gap both polar SOMO-HOMO and SOMO-LUMO interactions lower the transition state energy and increase the rate constants. In nucleophilic reactions SOMO-LUMO interactions dominate whereas electrophilic reactions are governed by SOMO-HOMO interactions. Substitution of the olefins changes both HOMO and LUMO energies. For a strongly nucleophilic radical with high SOMO energy, i.e. low ionization potential the rate constant then increases with increasing electron affinity (decreasing LUMO energy) and shows no correlation with the ionization potential. For a strongly electrophilic radical with low SOMO energy the rate constant increases with decreasing ionization potential (increasing HOMO energy). For the intermediate case high rate constants are expected both for olefins with high electron affinities and olefins with low ionization potentials. If the HOMO-LUMO-gap IP-EA is constant for a series of olefins and if the substituents act solely via polar effects on the activation energy then plots of logk vs either IP and EA should yield curves exhibiting a minimum. Figs. 4 and 5 show these plots together with reasonable curves. Though the general behaviour is as expected for our radical which has an intermediate ionization potential of about 10.5 eV several remarks are appropriate: (1) The scatter of the data is considerable. Part of it may be due to experimental errors and uncertainties in the electron affinities and ionization potentials. Further, there may be influences of the substituents on the A-factors which were not measured. (2) For most of the olefins IP-EA is in the narrow range of 10.6 to 11.5 eV, i.e. approximately constant. A linear correlation for these gave

$$IP = (10.90 \pm 0.14) + (0.87 \pm 0.09) \cdot EA \quad \text{in eV, } r = 0.9303$$

which confirms the constant energy gap. (3) The phenylsubstituted olefins have a much smaller gap and should not be used in a correlation with the other olefins in a strict sense. They are expected to react fast with all radicals and it will be difficult to separate SOMO-HOMO from SOMO-LUMO contributions. (4) Allylcyanide has a particularly large HOMO-LUMO gap of 11.87 eV and its rate constant is low whereas methylmethacrylate has a relatively low gap (10.66 eV) and a high rate constant. This indicates that both SOMO-HOMO and SOMO-LUMO interactions are important, and this probably holds for all olefins. Only for the extreme cases of high electron affinities-high ionization potentials and low electron affinities-low ionization potentials SOMO-LUMO and SOMO-HOMO interactions dominate the other. To be specific for our case: We believe that the ambiphilic $\cdot CH_2COOC(CH_3)_3$ radical reacts fast with acrylonitrile, trichloro vinylsinale and the acrylates because their electron affinities are high whereas it reacts fast with vinylethers and dialkylsubstituted olefins

312

because of their low ionization potentials.

Parallel to this study Giese et al (17) have measured relative rate constants for the addition of cyclohexyl, dimethyl malonyl, cyanomethyl and ethoxicarbonylmethyl radicals to seven α-substituted styrenes mainly at 110°. The effects of substitution show that cyclohexyl reacts in a SOMO-LUMO and the malonyl radical reacts in a SOMO-HOMO dominated way. Cyanomethyl and $\cdot CH_2COOCH_2CH_3$ behave ambiphilic as the radical studied here. However, the polar effects were less pronounced and lead to variations of relative rates by only a factor 5. In a comparison of the relative reactivities of cyclohexyl and cyanomethyl towards diphenylethene and other styrenes Giese found a pronounced influence of radical stabilizing effects for cyanomethyl addition, and concluded that in such borderline cases where polar effects are weak radical stability effects may become of major importance. It would be interesting to know the variation of IP and EA in Giese's styrene series to quantify the polar effects more clearly and possibly achieve a separation from radical stability factors.

ACKNOWLEDGEMENT

We thank Dr. M. Allan, University of Fribourg for providing electron affinities needed in this work, Prof. B. Giese, Technische Hochschule Darmstadt, for the communication of results prior to publication and discussions, D. Rügge for help with the analysis procedures and acknowledge financial support by the Swiss National Foundation for Scientific Research.

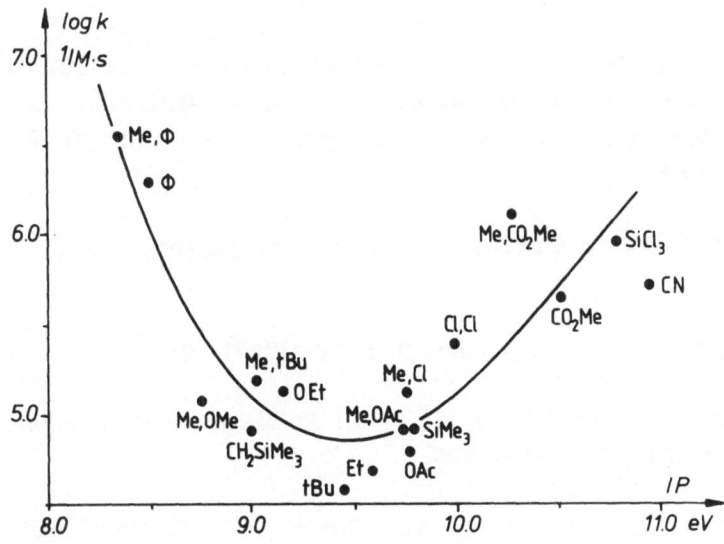

Figure 4. Log k₃₀₀ versus Ionization Potentials.

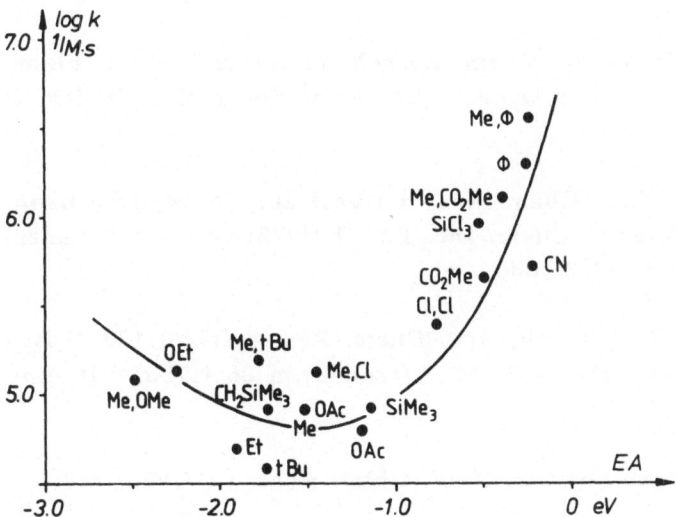

Figure 5. Log k₃₀₀ versus Electron Affinities.

314

REFERENCES

(1) (a) F.Minisci, A. Citterio, Adv. Free Radical Chem. 6 (1980) 65,
     (b) "Substituent Effects in Radical Chemistry", H.G.Viehe, Z. Ja-
     nousek, R. Merènyi, eds. NATO ASI Series, C, Vol. 189, D. Reidel,
     Dortrecht, 1986.

(2) B. Giese "Radicals in Organic Synthesis", Pergamon Press;
     Oxford, 1986.

(3) H. Fischer, H. Paul, Acc. Chem. Res. 20 (1987) 200.

(4) (a) H. Fischer, in ref. (1b), p.123ff. (b) K. Münger, H. Fischer,
     Int. J. Chem. Kin. 17 (1985) 809.

(5) F.A. Houle, J.L. Beauchamp, J. Am. Chem. Soc. 101 (1979) 4067.

(6) A.P. Stefani, Fluor. Chem. Rev. 5 (1971) 115.

(7) K. Riemenschneider, H.H. Bartels, R. Dornow, E. Drechsel-Grau,
     W. Eichel, H. Luthe, Y.M. Matter, W. Michaelis, P. Boldt, J. Org.
     Chem. 52 (1987) 205.

(7) (a) K. Fukui, Fortschr. Chem. Forsch. 15 (1970) 1. (b) I. Fleming,
     "Frontier Orbitals and Organic Chemical Reactions", Wiley, New
     York, 1978.

(8) (a) K.N. Houk, Acc. Chem. Res. 8 (1975) 361, (b) R. Sustmann,
     R. Schubert, Angew. Chem. Int. Ed. 11 (1972) 840, (c) R. Sustmann
     H. Trill, ibid. 11 (1972) 838.

(9) J.M. Tedder, J.C. Walton, Acc. Chem. Res. 9 (1976) 183, Adv.
     Phys. Org. Chem. 18 (1978) 61, Tetrahedron 36 (1980) 701, Angew.
     Chem. 94 (1982) 433.

(10) Wu Lung-min, H. Fischer, Helv. Chim. Acta, 66 (1983) 138.

(11) B. Giese, Angew. Chem. 95 (1983) 771.

(12) J.K. Vollenweider, H. Paul, H. Fischer, Int. J. Chem. Kin. 18, (1986) 791.

(13) F. Jent, H. Paul, H. Fischer, Chem. Phys. Lett., in press.

(14) (a) K.N. Houk, L.L. Munchausen, J. Am. Chem. Soc. 98 (1976) 937. (b) B. Albrecht, Zürich, priv. comm. (c) M.A.M. Meester, H. van Dam, D.J. Stufkens, A. Oskam, Inorg. Chim. Acta 20 (1976) 155, (d) K. Watanabe, T. Nakayama, J. Mottl, J. Quant. Spectr. Rad. Transfer, 2 (1962) 369, (e) R.F. Lake, H. Thompson, Proc. Ray. Soc. London (A) 317 (1970) 187, (f) R. Sustmann, H. Trill, Tretrah. Lett. (1972) 4271, (g) K. Wittel, H. Bock, Chem. Ber. 107 (1974) 317, (h) U. Weidner, A. Schweig, J. Organomet. Chem. 39 (1972) 261, (i) D. Nelson, M.J.S. Dewar. H.M. Buscheck. E. McCarthy, J. Org. Chem. 44 (1979) 4109 4109, (k) M.J.S. Dewar, S.D. Worley, J. Chem. Phys. 50 (1969) 654, (l) M. Bachiri, G. Mouvier, P. Carlier, J.E. Dubois, J. Chim. Phys. 77 (1980) 899, (m) H. Friege, M. Klessinger, J. Chem. Res. S (1977) 208, (n) W.C. Herndon, J. Am. Chem. Soc. 98 (1976) 887 and references cited therein.

(15) (a) M. Allan, Fribourg, priv. comm. (b) Ref. (4a) and references therein, (c) L. Wojnarovits, G. Földiak, J. Chromatogr. 206 (1981) 511.

(16) (a) R. Paltenghi, E.A. Ogryzlo, K.D. Bayes. J. Phys. Chem. 88 (1984) 2595, (b) Y.M.A. Naguib, C. Steel, S.G. Cohen, M.A. Young, J. Phys. Chem. 91 (1987) 3033.

(17) B. Giese, private communication.

NOTE ADDED IN PROOF

At the NATO workshop R. Merényi, kindly pointed out earlier evidence for an ambiphilic behaviour of the isobutyronitril radical $(CH_3)_2\dot{C}CN$ in its addition to p,p'- substituted diphenylenes and α, p-substituted styrenes (F. Lahousse, R. Merényi, J.R. Desmurs, H. Allaime, A. Borghese and H.G. Viehe, Tetrah. Letters 25 (1984) 3823). The rather small variation of relative rate constants with substitution by a factor of about seven is in agreement with Giese's recent work. For 5 diphenylenes (6a,b,g,i,k) the original data support the dominance of SOMO-HOMO-interactions, whereas the influence of SOMO-LUMO-interactions for compounds (6a,b,c,d,e,f) was much less pronounced than in our case.

THE EFFECT OF TEMPERATURE IN THE REACTIONS OF ARENESULPHONYL RADICALS

CARLOS M. M. DA SILVA CORRÊA,
MARIA DANIELA C. M. FLEMING,
and MARIA AUGUSTA B. C. S. OLIVEIRA
Centro de Investigação em Química (INIC)
Faculdade de Ciências
4000 PORTO – Portugal

ABSTRACT The effect of substituents and temperature on the reactivity of
a) arylacetylenes towards arenesulphonyl radicals and
b) arenesulphonyl chlorides towards substituted phenyl radicals,
was studied. Arenesulphonyl radicals were prepared from photolysis of arenesulphonyl iodides and substituted phenyl radicals from arenediazonium tetrafluoroborates. Relative reactivities correlate with Hammett and Brown substituent constants. Relative Arrhenius parameters were determined and its effects on the relative rates analised.

1. Introduction

The addition of sulphonyl iodides to unsaturated compounds is a very convenient process to get iodosulphones in high yield.[1] Iodosulphones can be easily dehydrohalogenated to give unsaturated sulphones or functionalized *via* nucleophilic substitution.[1b,2] The addition proceeds quickly under visible light and is consistent with the free radical chain mechanism (1-3).

$$ArSO_2I + h\nu \longrightarrow ArSO_2^{\bullet} + I^{\bullet} \qquad (1)$$
$$ArSO_2^{\bullet} + M \longrightarrow ArSO_2\text{-}M^{\bullet} \qquad (2)$$
$$ArSO_2\text{-}M^{\bullet} + ArSO_2I \longrightarrow ArSO_2\text{-}M\text{-}I + ArSO_2^{\bullet} \qquad (3)$$

where M is an olefinic or acetylenic compound.

We have been engaged in the study of structure-reactivities relationships in reactions (2) and (3). Althoug the addition (2) is a fast process, the chain propagation (3) is much more fast;[3] being so, we have been able to measure relative reactivities of addition to unsaturated compounds (substituted

F. Minisci (ed.), Free Radicals in Synthesis and Biology, 317–324.
© *1989 by Kluwer Academic Publishers.*

styrenes [4] and phenylacetylenes [5]) by competition experiments (4 – 5) by using the equation (6),

$$\begin{cases} ArSO_2^{\bullet} + XM \longrightarrow \text{Radical adduct} & \{k_X\} & (4) \\ ArSO_2^{\bullet} + HM \longrightarrow \text{Radical adduct} & \{k_H\} & (5) \end{cases}$$

$$k_X/k_H = (\log [XM] - \log[XM]_0) / (\log[HM] - \log[HM]_0) \qquad (6)$$

where X is the substituent in the unsaturated compound.

Relative rates of halogen abstraction have been measured [3b,6] based on competition rections (7 – 8) by using equations (9) and (13).

$$\begin{cases} Y\text{-}C_6H_4SO_2Hal + R^{\bullet} \longrightarrow R\text{-}Hal + Y\text{-}C_6H_4SO_2^{\bullet} & \{k_{Hal}\} & (7) \\ A\text{-}Z + R^{\bullet} \longrightarrow R\text{-}Z + A^{\bullet} & \{k_Z\} & (8) \end{cases}$$

$$k_{Hal} / k_Z = \log\{1 - ([R\text{-}Hal]/[Y\text{-}C_6H_4SO_2Hal]_0)\} / \log\{1 - ([R\text{-}Z]/[A\text{-}Z]_0)\} \quad (9)$$

Hal = Cl, Br, and I ; R = Ph , PhCH$_2$, Cl$_3$C , p-MeOC$_6$H$_4$, and p-NO$_2$C$_6$H$_4$;

A-Z = compound with suitable Z atoms to be abstracted (Sulphonyl halide, Ph$_3$C-Cl, and CH$_3$CN).

2. Addition of Tosyl Radicals to Arylacetylenes

2.1. SELECTIVITIES

The relative rates of addition of tosyl radicals to substituted phenyl-acetylenes, X-C$_6$H$_4$C≡CH (X = p-MeO, p-Me, H, p-Cl, m-NO$_2$), have been determined at several temperatures (T/°C = -25, 0, 25, 50, 70). Plots of log(k_X/k_H) versus · σ^+ gave good correlations showing the importance of polar effects. A typical plot is given in **Fig.1**. At all the temperatures examined, phenylacetylenes with electron-donating substituents react faster. Results of competition experiments at several temperatures are summarized in **Table 1**. A plot of the ρ^+ values against 1/T gave a concave upward curve with the minimum near 25 °C.

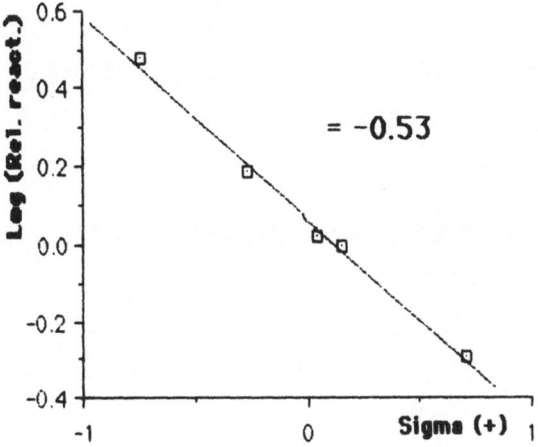

Fig.1. Plot of $\log(k_X/k_H)$ versus σ^+ for the addition of p-Me-$C_6H_4SO_2I$ to X-$C_6H_4C\equiv CH$ at 25 °C.

Table 1 – Relative reactivities, ρ^+ values, and goodness of fit for the addition of p-Me-$C_6H_4SO_2^\bullet$ to X-$C_6H_4C\equiv CH$ in CCl_4, at several temperatures.

X	Temperature / °C				
	−25	0	25	50	70
p-MeO	3.15	3.00	2.83	2.46	2.10
p-Me	1.64	1.55	1.44	1.42	1.32
H	(1)	(1)	(1)	(1)	(1)
p-Cl	1.25	1.01	0.94	0.97	0.92
m-NO$_2$	0.67	0.55	0.48	0.48	0.49
ρ^+	−0.46	−0.51	−0.53	−0.48	−0.43
r	0.97	0.99	1.00	1.00	1.00
+t·s (90%)	0.13	0.08	0.05	0.04	0.03
+t·s (95%)	0.17	0.10	0.06	0.05	0.04

2.2. ACTIVATION PARAMETERS

Relative Arrhenius parameters have been determined from plots of $\log(k_X/k_H)$ *versus* $1/T$. Results are given in **Table 2**.

Table 2-Relative Arrhenius parameters and goodness of fit for the addition of $p\text{-Me-}C_6H_4SO_2^{\bullet}$ to $X\text{-}C_6H_4C\equiv CH$

X:	$p\text{-MeO}$	$p\text{-Me}$	H	$p\text{-Cl}$	$m\text{-NO}_2$
$\ln(A_X/A_H)$:	-0.20	-0.27		-0.83	-1.63
$s_{\ln}$:	0.27	0.08		0.26	0.28
$\pm t \cdot s_{\ln}(90\%)$:	0.54	0.15		0.52	0.57
$\pm t \cdot s_{\ln}(95\%)$:	0.69	0.20		0.66	0.73
A_X/A_H:	**0.82**	**0.77**	1	**0.44**	**0.20**
$E_{aH}-E_{aX}$ / kcal:	**0.69**	**0.37**	0	**0.49**	**0.58**
s_E:	0.16	0.04		0.15	0.16
$\pm t \cdot s_E(90\%)$:	0.32	0.09		0.30	0.33
$\pm t \cdot s_E(95\%)$:	0.41	0.11		0.38	0.42
r:	0.93	0.98		0.88	0.90
k_H/k_X (if $\Delta E_a = 0$):	1.2	1.3		2.3	5.0
k_X/k_H (if $A_X/A_H = 1$):	3.2	1.9		2.3	2.7
Overall effect :	**E_a**	**E_a**		**E_a, A**	**A**

Results show that all the substituents decrease the pre-exponential factor; the retardation ability, how far as $\Delta S^{\ddagger}$ is concerned, is:

$$m\text{-NO}_2 > p\text{-Cl} > p\text{-Me} > p\text{-MeO} > H.$$

In the respect of $\Delta H^{\ddagger}$, all the substituents increase the reactivity:

$$p\text{-MeO} > m\text{-NO}_2 > p\text{-Cl} > p\text{-Me} > H.$$

No correlation between ΔE_a and σ^+ (or σ) could be obtained; $\log(A_X/A_H)$ correlate poorly with σ^+ ($r = 0.82$) and σ ($r = 0.93$). There is no correlation between ΔE_a and $\log(k_X/k_H)$

3. Reaction of Arenesulphonyl Chlorides With Aryl Radicals

3.1. SELECTIVITIES

Substituted phenyl radicals were produced by iododediazoniation of arenediazonium tetrafluoroborates with tetrabuthylammonium iodide, in acetonitrile, following Sing and Kumar [7] (equation 10).

$$XC_6H_4N_2^+ + I^- \longrightarrow XC_6H_4^{\bullet} + N_2 + I^{\bullet} \qquad (10)$$

The relative reactivities of chlorine abstraction by p-methoxyphenyl radicals were measured in competition experiments based on reactions (11) and (12) by using equation (13), following Tilset and Parker.[8]

$$MeOC_6H_4^{\bullet} + \begin{cases} YC_6H_4SO_2Cl \xrightarrow{\ k_Y\ } MeOC_6H_4Cl + YC_6H_4SO_2^{\bullet} \quad (11) \\ \qquad\qquad\qquad\qquad p\text{-Chloroanisole} \\ \\ CH_3CN \xrightarrow{\ k_0\ } MeOC_6H_5 + {\overset{\bullet}{C}}H_2CN \qquad (12) \\ \qquad\qquad\qquad\quad \text{Anisole} \end{cases}$$

$$k_Y/k_0 = (\,[\,CH_3CN\,]_0/[\,YC_6H_4SO_2Cl\,]_0\,).(\,[\,MeOC_6H_4Cl\,]/[\,MeOC_6H_5\,]\,) \quad (13)$$

Equation (13) was established by assuming a constant ratio of the concentrations of CH_3CN and $YC_6H_4SO_2Cl$ during all the reaction time [this was obtained by using a large excess of CH_3CN (the solvent) and $YC_6H_4SO_2Cl$ relatively to the total amount of aryl radicals generated]. Arenesulphonyl chlorides with electron-withdrawing substituents in the ring react faster, according with the nucleophilic character of the p-methoxyphenyl radical. At all the temperatures examined, plots of $\log(k_Y/k_0)$ against σ and σ^+ gave quite good correlations. The results are summarized in **Table 3**.

Table 3. Relative reactivities (k_Y/k_0), ρ values, and goodness of fit for chlorine abstraction (k_Y) from $YC_6H_4SO_2Cl$, and hydrogen abstraction (k_0) from CH_3CN, by $p\text{-MeOC}_6H_4^\bullet$ radicals.

Y	Temperature / °C				
	0	10	25	40	60
p-MeO	80.4	64.1	49.9	44.9	34.1
p-Me	87.0	73.1	64.0	50.3	32.1
H	87.1	71.5	61.7	50.4	39.6
p-Cl	103.0	93.0	72.9	54.4	49.5
m-NO$_2$	117.1	108.2	77.8	68.6	51.1
ρ	0.17	0.23	0.17	0.17	0.21
r	0.98	0.96	0.86	0.98	0.90
±t.s (95%)	0.05	0.10	0.14	0.06	0.15
±t.s (90%)	0.04	0.07	0.11	0.04	0.12
ρ^+	0.12	0.16	0.13	0.12	0.15
r	0.93	0.92	0.93	0.95	0.85
±t.s (95%)	0.07	0.10	0.08	0.06	0.13
±t.s (90%)	0.05	0.08	0.06	0.05	0.10

3.2. ACTIVATION PARAMETERS

Plots of $\ln(k_Y/k_0)$ against $1/T$ yielded the relative Arrhenius parameters E_0-E_Y and A_Y/A_0 from which the relative Arrhenius parameters for the chlorine abstraction from $YC_6H_4SO_2Cl$ and from $C_6H_5SO_2Cl$ (E_H-E_Y and A_Y/A_H) were obtained. Results are given in **Table 4**

At all the temperatures examined the order of decreasing reactivity is
$$m\text{-NO}_2 > p\text{-Cl} > H > p\text{-Me} > p\text{-MeO}.$$
As in the case of the addition reaction, the same general trends are present.

All the substituents decrease the A factor, decreasing the reaction rate; the retardation ability of the substituents is:

$$p\text{-Me} > p\text{-MeO} > m\text{-NO}_2 > p\text{-Cl} \approx H$$

How far as the activation energy is concerned, all the substituents increase the rate:

$$p\text{-Me} > m\text{-NO}_2 > p\text{-MeO} > p\text{-Cl} > H$$

Table 4. Relative Arrhenius parameters and goodness of fit for the reaction of $p\text{-MeOC}_6H_4^{\bullet}$ radicals with $YC_6H_4SO_2Cl$ and CH_3CN.

Y:	p-MeO	p-Me	H	p-Cl	m-NO$_2$
$\ln(A_Y/A_0)$:	-0.20	-0.76	0.21	0.24	0.11
$s_{\ln}$:	0.33	0.58	0.15	0.40	0.35
$\pm t \cdot s_{\ln}(90\%)$:	0.67	1.17	0.31	0.80	0.70
$\pm t \cdot s_{\ln}(95\%)$:	1.71	2.87	0.80	2.06	1.79
A_Y/A_0:	0.82	0.47	1.24	1.27	1.12
A_Y/A_H:	**0.66**	**0.38**	**(1)**	**1.02**	**0.91**
r:	0.99	0.98	1.00	0.99	0.99
$E_{a0}-E_{aY}$ / kcal:	+2.47	+2.86	2.30	+2.39	+2.54
s_E:	0.20	0.34	0.09	0.24	0.20
$\pm t \cdot s_E(90\%)$:	0.40	0.69	0.18	0.48	0.41
$\pm t \cdot s_E(95\%)$:	0.50	0.88	0.23	0.61	0.52
E_H-E_Y / kcal:	**+0.17**	**+0.56**	**(0)**	**+0.09**	**+0.24**
k_H/k_Y (if $\Delta E_a = 0$):	1.5	2.6		1.0	1.1
k_Y/k_H (if $A_Y/A_H = 1$):	1.3	2.6		1.2	1.5
Overall effect :	A	E$_a$. A		E$_a$	E$_a$

Results show that the interpretation of these relative reactivities must take entropies and enthalpies of activation into consideration. For instance, either m-nitrobenzenesulphonyl chloride or p-chlorobenzenesulphonyl chloride are more reactive than benzenesulphonyl chloride due to enthalpic reasons; however, the greater reactivity of benzenesulphonyl chloride as compared with p-methoxybenzenesulphonyl chloride is the result of pre-exponential factors. With toluene-p -sulphonyl chloride both effects seem compensate each other. Results also show that no linear relationship exists between ΔE_a and $\ln(k_Y/k_H)$, ΔE_a and σ^+ (or σ), $\ln(A_Y/A_H)$ and σ^+ (or σ), and ρ^+ (or ρ) and $1/T$.

4. Conclusions

The changes in the rate of addition of tosyl radicals to arylacetylenes (as well as the chlorine abstraction from arenesulphonyl chlorides by p-methoxyphenyl radicals) when the ring substituents are changed are quite small and correlate well with substituent polar constants . Relative rates can not be related with activation energies or pre-exponential factors separately. Pre-exponential factors play a very important part in the process and may be the rate determining factor.

5. References

1 a) P. Skell and J. H. McNamara, *J. Am. Chem. Soc.*, 1967, **79,** 85; P. Skell.R. C. Woodworth, and J. H. McNamara, ibid., p.1253; b) C. M. M. da Silva Corrêa and W. A. Waters, *J. Chem. Soc. (C),* 1968, 1874; c) W. E. Truce and G. C. Wolf, *J. Org. Chem.,* 1971, **36,** 1727.
2 J. Barluenga, J.M. Martinez-Gallo, C. Nájera, F. J. Fañanás, and M. Yus, *J. Chem. Soc., Perkin Trans. 1,* 1987, 2605.
3 a) C. Chatgilialoglu, *J. Org. Chem.,* 1986, **51,** 2871; b) C. M. M. da Silva Corrêa and M. A. B. C. S. Oliveira, *J. Chem. Soc., Perkin Trans. 2,* 1987, 811.
4 C. M. M. da Silva Corrêa and W. A. Waters, *J. Chem. Soc., Perkin Trans. 2,* 1972,1575; C. M. M. da Silva Corrêa, M. A. B. C. S. Oliveira, M. D. C. M. Fleming and M. P. F. Gonçalves, *Rev. Port. Quim.,* 1973, **15,** 100.
5 C. M. M. a Silva Corrêa e M. D. C. M. Fleming, *J. Chem. Soc., Perkin Trans.2,* 1987,103.
6 C. M. M. da Silva Corrêa and M. A. B. C. S. Oliveira, *J. Chem. Soc., Perkin Trans. 2,* 1983,711.
7 P. R. Sing and R.Kumar, *Aust. J. Chem.,* 1972, **25,** 981.
8 M. Tilset and V. D. Parker, *Acta Chem. Scand,* 1982, **B 36,** 281.

SYNTHESIS OF NEW 5-NITROIMIDAZOLES HIGHLY ACTIVE AGAINST ANAEROBES BY SUBSTITUTION REACTIONS WHICH PROCEED VIA RADICAL AND RADICAL-ANION INTERMEDIATES

M.P. Crozet, O. Jentzer and P. Vanelle
Aix-Marseille University ; CNRS UA 10⁹ - Box 562
13397 Marseille Cedex 13 - France

ABSTRACT. Electron-transfer substitution reactions which proceed via radical and radical-anion intermediates ($S_{RN}1$) were extended to heterocyclic electrophiles of the nitroimidazole series and to heterocyclic nucleophiles prepared from 3-nitrolactams. New 5-nitroimidazoles bearing a trisubstituted ethylenic double bond in position 2 were obtained in good yields. Biological assays have shown that several of these compounds are highly active against anaerobes in vivo. The best compound was found to be greater than 40-fold more active than metronidazole (Flagyl*).

INTRODUCTION.

In 1966 researchers at Rhône-Poulenc first published their results on the synthesis of metronidazole (1-hydroxyethyl-2-methyl-5-nitroimidazole) and its therapeutic trials for the treatment of trichomoniasis and amoebiasis (1).

A chance observation that metronidazole was effective in the treatment of Vincent's stomatitis led to a study of its activity against a variety of obligate anaerobes (2) Since these early investigations, the usefulness of metronidazole has been firmly established and metronidazole is the current drug of choice for the treatment of several protozoal diseases as well as for treating infections due to anaerobic bacteria (3,4,5). Since 1963 many thousands of nitroimidazoles have been synthesized by different industrial groups but these massive efforts to develop newer agents, which might offer advantages over the well-established metronidazole, have resulted in the introduction of a few new nitroimidazoles.

It was also in 1966 that N. KORNBLUM and G.A. RUSSELL published the mechanism of a new type of substitution process at a saturated carbon (6). These reactions which proceed via a chain sequence, in which radical anions and free radicals are intermediates, are called $S_{RN}1$ reactions, with the term, $S_{RN}1$, being coined by J.F. BUNNETT and co-workers in 1970 (7). The $S_{RN}1$ reactions are noteworthy for providing novel and powerful means of synthesis as they occur readily under mild conditions to give excellent yields of pure products.

Our initial studies, which were aimed at the synthesis

325

F. Minisci (ed.), Free Radicals in Synthesis and Biology, 325–333.
© *1989 by Kluwer Academic Publishers.*

of novel 5-nitroimidazoles using the $S_{RN}1$ reaction of simple nitronate anions with 1-methyl-2-chloromethyl-5-nitroimidazole (8) have been extended to 3-nitrolactam anions. This paper describes our studies of the application of the $S_{RN}1$ mechanism to the synthesis of new biologically active 5-nitroimidazoles bearing a trisubstituted ethylenic double bond in the 2-position which is conjugated with a lactam group.

RESULTS AND DISCUSSION.

Many nucleophiles participate in $S_{RN}1$ reactions but few examples of heterocyclic nitronate anions are known. During the last 30 years H. FEUER and co-workers have developed a systematic study of alkyl nitrate nitrations (9) and recently reported the sucessful introduction of a nitro group into the position 3 of N-alkyl lactams (10). By using lithium diisopropylamide, prepared _in situ_ from diisopropylamine and butyllithium, as the base, tetrahydrofuran as the solvent, and n-propyl nitrate as the nitrating agent the resulting lithium salt of 1-methyl-3-nitro-2-pyrrolidone 2 was readily converted to 2 in 53% yield by first diluting the reaction mixture with water and then acidifying with acetic acid (scheme 1).

Scheme 1. Preparation of 1-methyl-3-nitro-2-pyrrolidone 2.

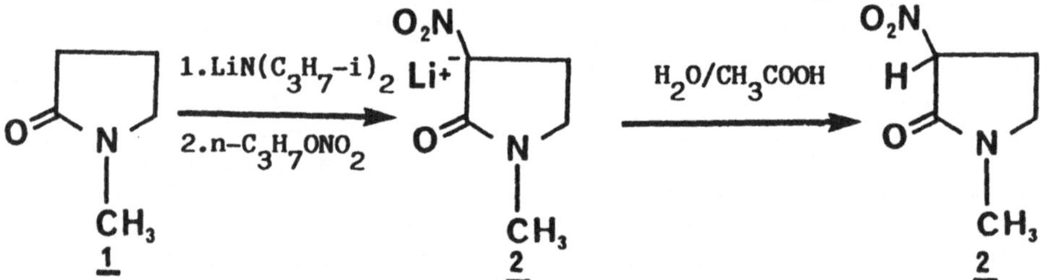

We have found that the lithium salt of 2 can be isolated as a yellow solid in 70% yield by evaporating water before acidification. This salt after titration was used for further reactions. By this procedure other N-alkyllactams were readily converted into the corresponding lithium salts of N-alkyl-3-nitrolactams (scheme 2)

Scheme 2. N-alkyl-3-nitrolactam lithium salts prepared.

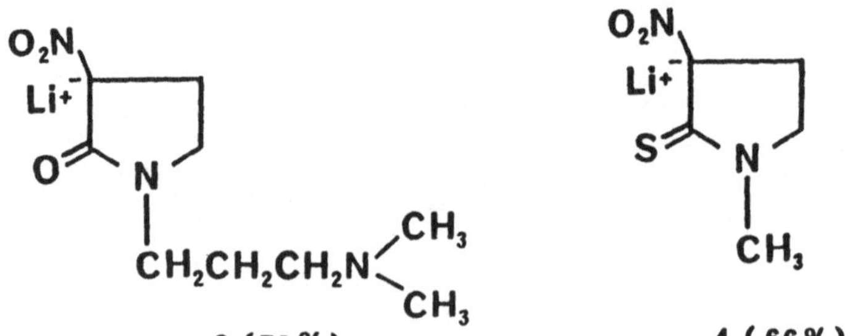

3 (70%)   4 (66%)

5 (57%)   6 (62%)   7 (66%)

To the best of our knowledge the reactivity of these salts with alkylating agents has not been studied. By employing **2** as a model compound different experimental conditions conductive to $S_{RN}1$ reactions (inert atmosphere, light catalysis) were used. The results are tabulated in Table 1.

Table 1. $S_{RN}1$ reactions between 1-methyl-3-nitro-2-pyrrolidone anion **2** and 1-methyl-2-chloromethyl-5-nitroimidazole[a] **8**.

Entry	Solvent	Time	Conditions	% Yield[b]			
				9	**10**	**11**	**12**
1	THF	16h		0	18	tars	
2	CH$_3$OH	2h		0	60	tars	
3	CH$_3$CH$_2$OH	1h		0	50	tars	
4	DMF	20h	1 equiv. **2**	10	0	tars	
5	DMSO	6h		15	11	tars	
6	H$_2$O/CH$_2$Cl$_2$	7h	0.1 equiv. Bu$_4$NBr	35	3	1	10
7	H$_2$O/C$_6$H$_5$CH$_3$	28h	0.1 equiv. Bu$_4$NBr	85	6	0	0
8	H$_2$O/C$_6$H$_5$CH$_3$	28h	0.1 equiv. Bu$_4$NCl	85	6	0	0

[a] All reactions were carried out under argon with fluorescent lamps (2 x 60 W) with **8** (1 equiv.), **2** (2 equiv.) unless otherwise stated. [b] % Yields of pure isolated materials are based on **8**.

The results of Table 1 show that experimental conditions have a dramatic influence on the yield of **9**. In THF, almost only untractable tarry matters were formed. In alcohols, besides tars, only the reduction product **10** was isolated. In dipolar solvents, with one or two equivalents of **2**, disappointing yields of **9** were obtained. At this stage of the study, it was found that **9** in the presence of the anion **2** gave tars, and hence it was necessary to find a way to avoid any contact between the final product **9** and the nitronate anion. Phase-transfer catalysis conditions using water and toluene, solvents in which **9** is partially soluble have allowed us to reach 85% yield of pure isolated **9**.

Proofs for the $S_{RN}1$ mechanism for these reactions were not studied but these results for the reaction between **8** and **2** are in close agreement with those obtained between **8** and simple nitronate anions for which the $S_{RN}1$ mechanism has been studied in full details (8). We suggest the following $S_{RN}1$ mechanism for the C-alkylation of anion **2** (eq. 1-4) with a base-promoted elimination of nitrous acid giving **9** (eq. 5) (scheme 3).

Scheme 3. The $S_{RN}1$ mechanism for the reaction between 1-methyl-2-chloromethyl-5-nitroimidazole **8** and 1-methyl-3-nitro-2-pyrrolidone lithium salt **2**.

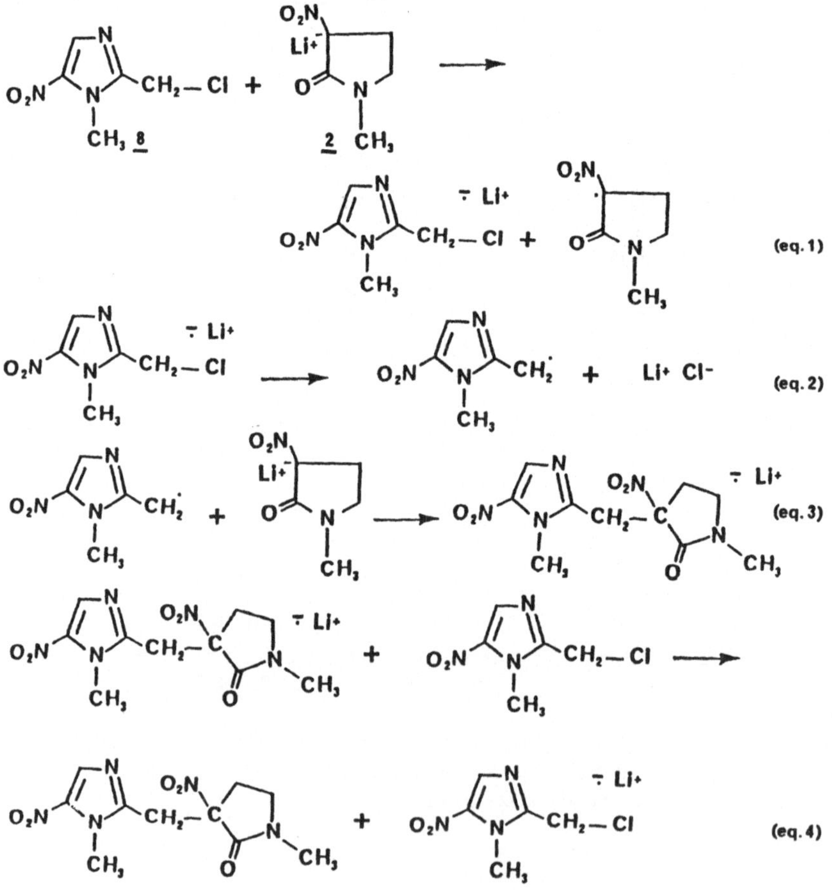

Scheme 3 (Continued)

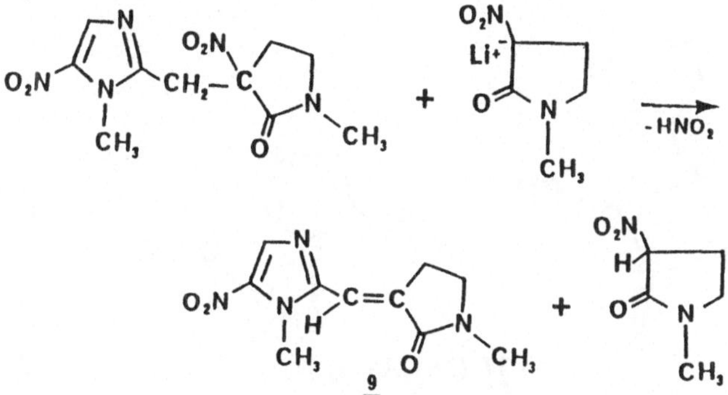

(eq.5)

The use of the best experimental conditions (Table 1 entry 7) for the reactions of the different other heterocyclic nitronate anions have afforded the new 5-nitroimidazoles reported in scheme 4.

Scheme 4. New 5-nitroimidazoles with lactam group.

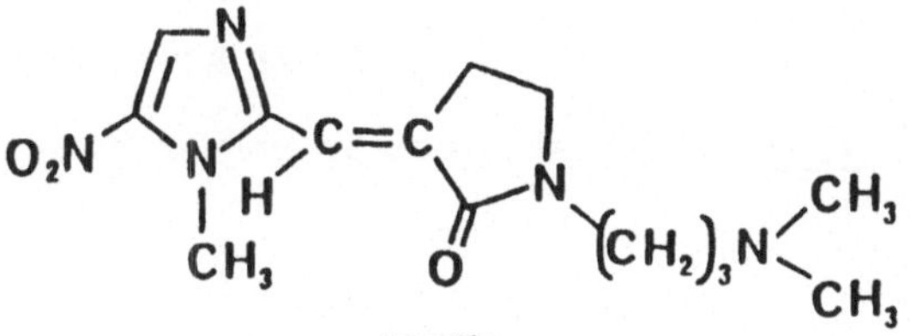

13 (36 %)

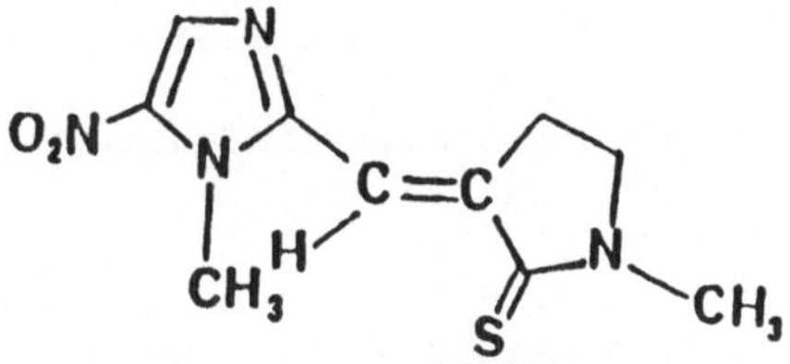

14 (15 %)

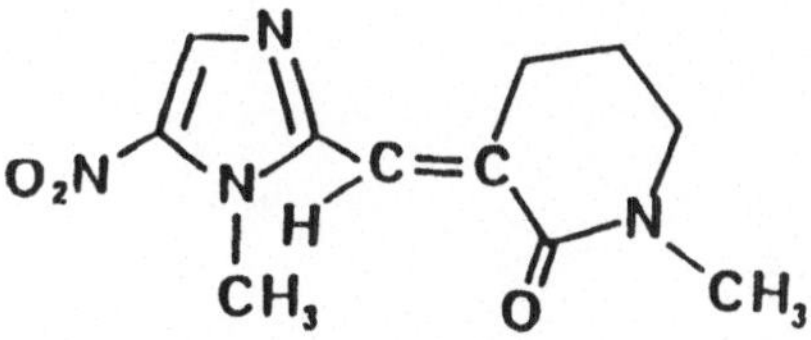

15 (44 %)

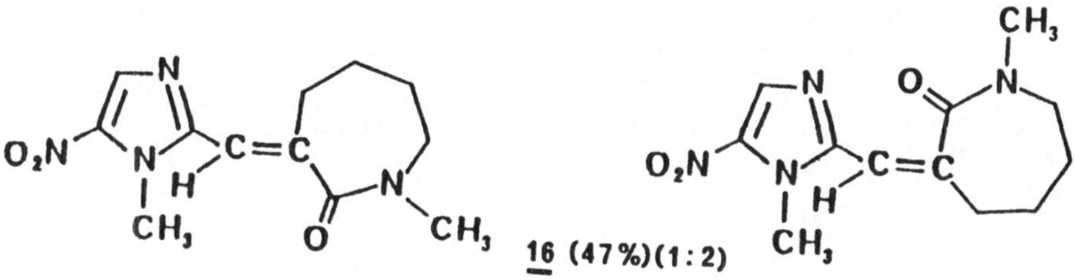

16 (47 %)(1 : 2)

Scheme 4 (Continued)

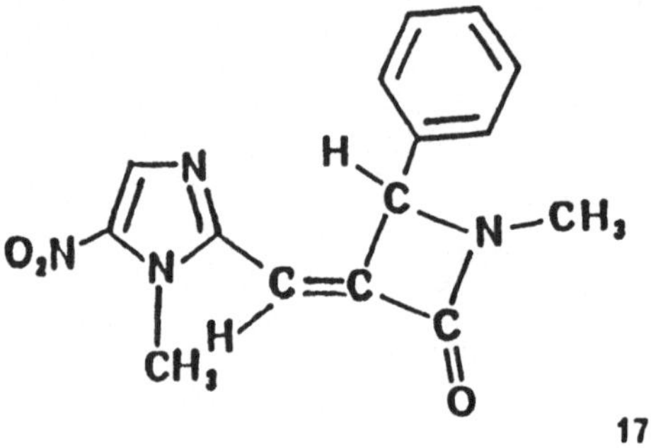

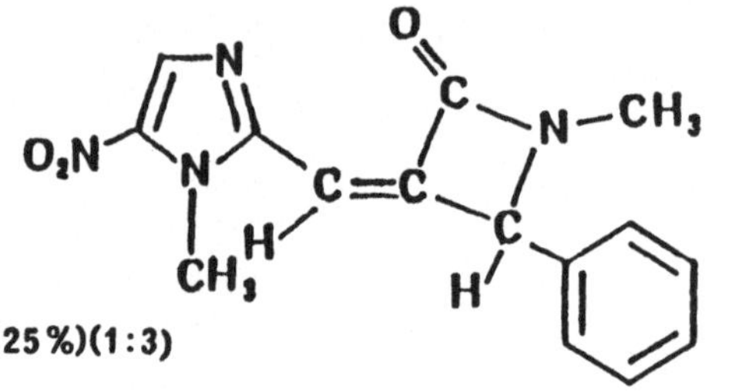

$\underline{17}$ (25%)(1:3)

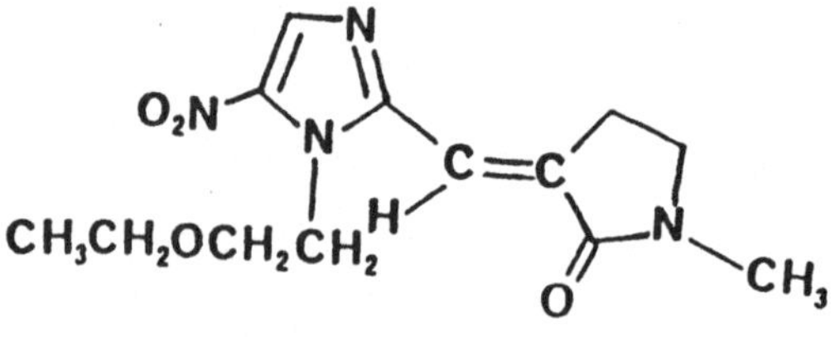

$\underline{18}$ (41%)

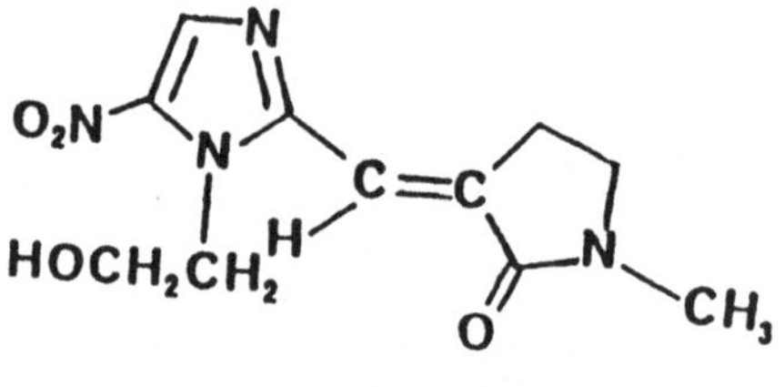

$\underline{19}$ (40%)

Table 2. Antimicrobial activity, _in vivo_ against Clostridium perfringens
(IP 2794)

Nitroimidazole	Number	Relative ED$_{50}$ (metronidazole = 1)[a]	
		orally	subcutaneously
Metronidazole	8823 RP	1	1
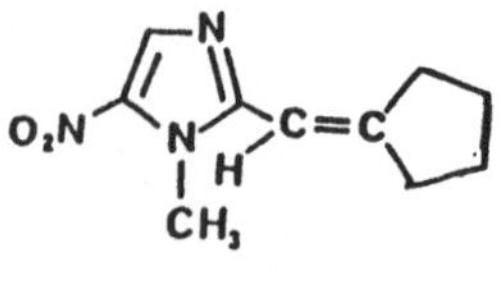	$\underline{20}$[b]	1	N.D.
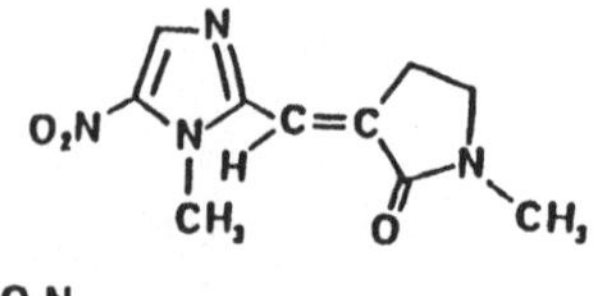	$\underline{9}$	27	43
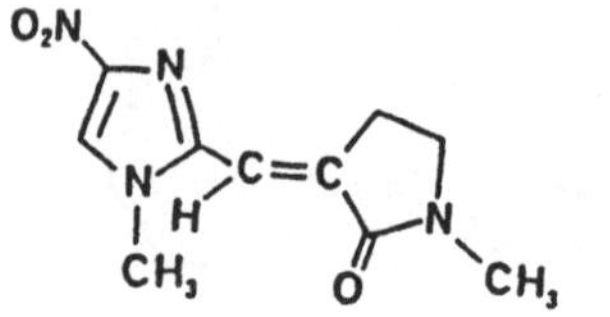	$\underline{21}$[c]	0.1	N.D.

Table 2 (Continued)

Compound		
13	15	8
14	17	23
15	17	23
18[d]	7	2.5
19[e]	1	10

[a] The figures express the ratio ED_{50} (compound)/ED_{50}(metronidazole) ; [b] See : reference 8a; [c] See reference 8c for the preparation of 4-nitro imidazoles with ethylenic group at position 2 ; [d] prepared from 1-methyl-2-chloromethyl-4-nitroimidazole in 75% yield. [e] Prepared from 1-ethoxyethyl-2-chloromethyl-5-nitroimidazole. Prepared from 1-acetoxyethyl-2-chloromethyl-5-nitroimidazole.

BIOLOGICAL RESULTS AND DISCUSSION.

The _in vitro_ antibacterial activity of the compounds was investigated with 33 anaerobic strains. The results have shown that the compounds are very active. Compound **9** was found to show the most potent activity being, for example, 125-fold more effective against Bacteroides fragilis VPI 9032 than metronidazole. The detailed pharmacology of these compounds will be published elsewhere.

For the determination of the _in vivo_ activity against Clostridium perfringens (IP 2794), the compounds were administered orally and subcutaneously to mice. The data reported in Table 2 represent the ratio between the ED_{50} value of the compounds and that of metronidazole. Actually, the Ed_{50} of metronidazole in these experimental conditions can vary from 0.5 to 30 mg/kg.

The data reported in Table 2 indicate that all the

332

compounds with a lactam group which have been tested <u>in vivo</u> are more active than metronidazole when the nitro group is at position 5 of the imidazole ring. **9** is the most active compound. The substitution of cyclopentane in position 2 by N-methylpyrrolidone increases the activity by a factor of 27 orally and of more than 43 subcutaneously. Concerning the effects of the substituent on the nitrogen atom of the pyrrolidone, the size of the lactam or the introduction of a sulfur atom, it can be seen that these modifications result in a marked lowering of activity. An examination of the influence of the substituent in position **1** leads to the conclusion that a methyl group increases the activity. <u>In vivo</u> a similar sequence has been observed (1) for 2-methyl-5-nitroimidazoles, with the choice of metronidazole resulting in a more favorable therapeutic index.

These compounds were assayed for their antiviral and antitumor activities. They have been found to be inactive.

CONCLUSION.

Our results show that heterocyclic nitronate anions prepared from 3-nitrolactams are able to act as nucleophiles in the $S_{RN}1$ reactions of 1-substituted-2-chloromethyl-5-nitroimidazoles. The resulting C-alkylation products are not isolated a base-promoted nitrous acid elimination giving the final products bearing a trisubstituted ethylenic double bond in position 2. The nature of the solvent affects the yields of these products which give tars with nitronate anions. Phase-transfer catalysis conditions in which the final products are partially soluble in water and also in the organic solvent gave good or excellent yields. This new class of 5-nitroimidazoles appears of interest because it possesses a fair degree of activity and a very low toxicity.

Despite the fact that these new compounds have to our knowledge, one of the strongest activities known , <u>in vivo</u> , against anaerobic bacteria, as with metronidazole and other 5-nitroimidazole drugs they give a positive Ames test. Even if the conclusions drawn from mutagenicity studies of nitroimidazoles using the Ames test are questionable (4), the presence of the nitro group in these compounds is a defect for the FDA registration.

The first challenge concerning the synthesis of new 5-nitroimidazoles which are much more active <u>in vivo</u> than metronidazole has been taken up but a bigger challenge lies ahead : that of finding new non-mutagenic 5-nitroimidazoles (11) or new active compounds without a nitro group.

ACKNOWLEDGEMENTS.

This work was supported by The Centre National de la Recherche Scientifique and Rhône-Poulenc Recherches. It is a pleasure to acknowledge the Departement Biologie of Rhône-Poulenc Santé at Vitry sur Seine for all biological studies and the Departement Analyses for mass spectra , ultra-violet spectra and NMR studies using NOE measurements. We thank J.-F. Sabuco for his collaboration in the preparation of 6- and 7-membered compounds. We gratefuly acknowledge helpful discussions with J.-C. Aloup, Dr. M. Barreau (RP Santé) and Professor J.-M. Surzur.

REFERENCES.

1 - Cosar, N. ; Crisan, C. ; Horclois, R. ; Jacob, R.N. ; Robert, J. ; Tchelitcheff, S. ; Vaupré, R. Arzneim-Forsch. 1966, **16**, 23.

2 - Shinn, D.L.S. Lancet, 1962, **1**, 1191.

3 - Nitroimidazoles : Chemistry, Pharmacology and Clinical Applications, Breccia, A. ; Cavalleri, B. ; Adams, G.E., Ed. Plenum Press, New York, 19832, Vol.**42**.

4 - Edwards, D.I. ; "Modes and Mechanisms of Microbial Growth Inhibitors" in Antibiotics, Hahn, F.E. ; Ed. ; Springer-Verlag, Berlin, 1983, Vol. **VI**, pp. 121-135.

5 - Nair, M.D. ; Nagarajan, K. "Nitroimidazoles as Chemiotherapeutic Agents" in Progress in Drug Research, Jucker, E. ; Ed. ; Birkhauser Verlag, Basel, 1983, Vol. **27**, pp. 163-252.

6 - (a) Kornblum, N. ; Michel, R.E. ; Kerber, R.C. J. Am. Chem. Soc. 1966, **88**, 5660 and and 5662. (b) Russell, G.A. ; Danen, W.C. J. Am. Chem. Soc. 1966, **88**, 5663.

7 - Bunnett, J.F. ; Kim, J.K. J. Am. Chem. Soc, 1970, **92**, 7463.

8 - (a) Crozet, M.P. ; Surzur, J.M. ; Vanelle, P. ; Ghiglione, C. ; Maldonado, J. Tetrahedron Lett. 1985, **26**, 1023. (b) Crozet, M.P. ; Archaimbault, G. ; Vanelle, P. ; Nouguier, R. Tetrahedron Lett. 1985, **26**, 5133. (c) Crozet, M.P. ; Vanelle, P. ; Jentzer, O. ; Maldonado, J. C.R. Acad. Sci., Ser. II, 1988, **306**, 967.

9 - Feuer, H. "Alkyl nitrate nitrations" in The Chemistry of amino, nitroso and nitro compounds and their derivatives, Supplement F, Part 2, Patai, S. ; Ed., Wiley, New York, 1982, Chap. 19, pp. 805-847.

10 - Feuer, H. ; Panda, C.S. ; Hou, L. ; Bevinakatti, H.S. Synthesis, 1983, 187.

11 - Walsh, J.S. ; Wang, R. ; Bagan, E. ; Wang, C.C. ; Wislacki, P. ; Miwa, G.T. J. Med. Chem, 1987, **30**, 150.

ELECTRON SPIN RESONANCE STUDIES OF THE REACTIONS OF $^\bullet$OH AND $SO_4^{\bullet-}$ RADICALS WITH DNA, POLYNUCLEOTIDES AND SINGLE BASE MODEL COMPOUNDS

D. Schulte-Frohlinde and K. Hildenbrand
Max-Planck-Institut für Strahlenchemie
Stiftstr. 34
D433 Mülheim/Ruhr
FRG

ABSTRACT. Results of studies by esr spectroscopy and pulse radiolysis on the reactions of $^\bullet$OH and $SO_4^{\bullet-}$ in aqueous solution with nucleobases, nucleosides, nucleotides and polynucleotides of the pyrimidine series are reported. The esr parameters of a large number of radicals are presented and several of the chemical pathways leading to their formation are discussed. Upon reaction of $SO_4^{\bullet-}$ with ribose nucleosides and nucleotides a fast transfer of the radical site from the base to the sugar is observed. Tentatively a base radical cation is assumed to be the highly reactive intermediate formed by reaction of $SO_4^{\bullet-}$ with the base moieties. A relatively long-lived radical cation is identified from the reaction of $SO_4^{\bullet-}$ with tetramethyluracil. No transfer reaction is observed with the 2'-deoxyribose-derivatives. Differences in sugar ring puckering may explain these results. Upon reaction of $^\bullet$OH radicals with polyuridylic acid [poly(U)] the C(5)-OH adduct radical of the uracil moiety was observed at neutral pH but a sugar radical at pH 3.5. This result is in agreement with pulse conductivity measurements showing a 100-fold increase in the rate of strand break formation of poly(U) upon lowering the pH. From esr spectroscopic parameters of the sugar radical of poly(U) it can be concluded that H abstraction occurs from position 2' of the sugar by base radicals. This is evidence for the C(2')-pathway leading to strand break formation in poly(U) as postulated earlier on the basis of pulse radiolysis work. Results obtained with purines are briefly discussed. With polyadenylic acid [poly(A)] and with DNA long-lived radicals ($t_{1/2}$ > 10 s) are observed which are inert to oxygen.

1.INTRODUCTION

Damage of deoxyribonucleic acid (DNA) in cells has been found to be the major cause for mutation, cancer and cell deactivation under the influence of chemicals, UV-light and high-energy irradiation. Among the various kinds of lesions observed, double-strand breaks are the ones most severe and most difficult to repair. The chemical details of the pathways to strand break (sb) formation, e.g. cleavage of the sugar-phosphate bonds, is therefore a matter of considerable concern. For reviews see ref. (1).

From the structure of DNA (Fig. 1) it is obvious that sugar damage is a prerequisite for sb formation. However, the results obtained from reaction of hydroxyl radicals ($^\bullet$OH) with polyuridylic acid [poly(U)], a single-stranded ribose-homopolynucleotide, were rather surprising. It

335

F. Minisci (ed.), Free Radicals in Synthesis and Biology, 335–359.
© *1989 by Kluwer Academic Publishers.*

336

was shown that the $^\bullet$OH radicals add to the nucleobase in the primary step and that the OH-adduct radicals of the base act as precursors of sb formation.[2]

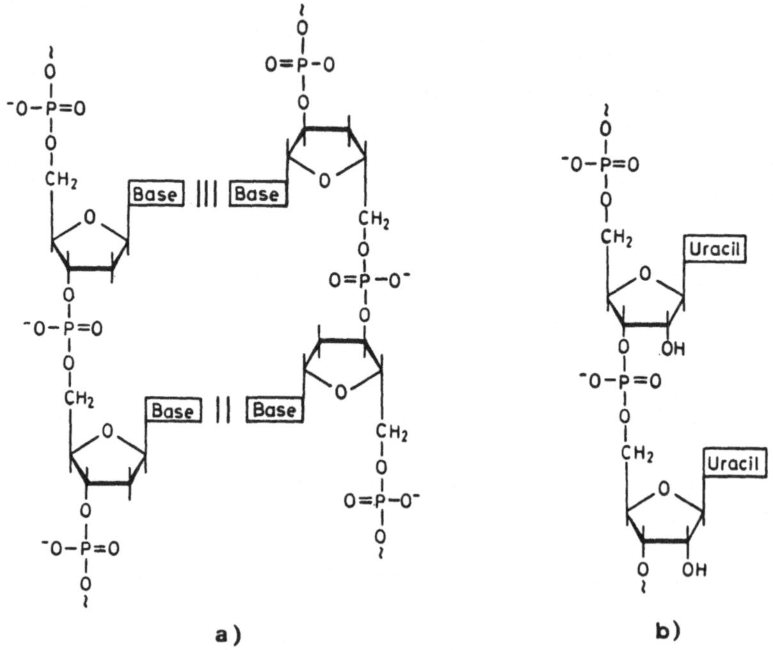

Figure 1. Schematic views of a segment of DNA (a) and of poly(U) (b)

This led to the conclusion that there exists a pathway of transfer of the radical site from the base to the sugar moieties within the macromolecule. This discovery prompted various studies with the goal to learn more about this kind of intramolecular reactions. Some of the questions are: At which position does the addition of $^\bullet$OH radicals to the nucleobases appear? Does the transfer of radical site occur only in polynucleotides or also in mononucleotides? Is the transfer of the site of the free spin also found in deoxyribopolynucleotides? Are radicals of nucleobases other than uracil also able to interact with the sugar residues? Is the rate and site of transfer of the free spin dependent on conformational properties of the polynucleotides?

The situation is complicated by the fact that by reaction with H atoms or with $^\bullet$OH different radicals may be formed from the same nucleobase and H atoms in five different positions are available for abstraction from the sugar residues.

In order to gain information on these problems studies have been carried out not only on polynucleotides but also on model compounds such as nucleobases, mononucleosides and -nucleotides.

The radical reactions were studied by pulse radiolysis with optical detection, with pulse conductivity and by *in situ* esr spectroscopy in aqueous solution. All results reported in this paper have been obtained under anoxic conditions.

It is known that the damaging effect of ionizing radiation on nucleic acids can be described by two mechanisms, known as the 'direct' and the 'indirect' effect. The main damaging species in the 'indirect effect' is the $^\bullet$OH radical formed by radiolysis of H_2O whereas by the 'direct effect' solvated electrons, radical cations and radical anions, especially of the nucleobases, are formed in the primary step. In order to try to mimic, to some extent, the 'direct effect' by a chemical method we reacted sulfate radical anions ($SO_4^{\cdot-}$) with the substrates. It was shown that this species gave rise to esr spectra of radical cations of aromatic compounds.[3,4] Although in most cases there is no rigorous proof for formation of radical cations by reaction of nucleobases with $SO_4^{\cdot-}$ this method has been found to be rather rewarding, especially in esr studies in aqueous solution where a variety of unexpected radical reactions could be detected.

In the pulse radiolysis experiments the aqueous solutions were saturated with N_2O. Under these conditions $^{\cdot}OH$ radicals are generated according to reactions (1) and (2).

$$H_2O \longrightarrow {}^{\cdot}OH, e_{aq}^-, H^{\cdot}, H^+, H_2O_2, H_2 \qquad (1)$$

$$e_{aq}^- + N_2O + H_2O \longrightarrow {}^{\cdot}OH + N_2 + OH^- \qquad (2)$$

When solutions of peroxodisulfate are exposed to the electron beam, $SO_4^{\cdot-}$ is produced in reaction (3).

$$e_{aq}^- + S_2O_8^{2-} \longrightarrow SO_4^{\cdot-} + SO_4^{2-} . \qquad (3)$$

The esr experiments were carried out by *in situ* photolysis of solutions containing the substrates and either H_2O_2 or $S_2O_8^{2-}$. The radical reactions were induced by photoreactions (4) and (5)

$$H_2O_2 + h\nu \longrightarrow 2\,{}^{\cdot}OH \qquad (4)$$

$$S_2O_8^{2-} + h\nu \longrightarrow 2\,SO_4^{\cdot-} \qquad (5)$$

The first part of the paper is dedicated to the radical reactions of pyrimidine derivatives starting with the pyrimidine bases uracil, thymine and cytosine

uracil   *cytosine*   *thymine*

and their methylated homologues. Then we present results on ribose- and 2'-deoxyribose-nucleosides (uridine, cytidine, dU, dC and thymidine) and on the 3'-nucleotides [uridine-3'-phosphate, (3'-UMP); 2'-deoxyuridine-3'-phosphate, (3'-dUMP)]. In the second part the corresponding radical reactions with the derivatives of the purine bases guanine and adenine are briefly reviewed. Finally, the implications of these model studies for strand breakage of poly(U) and poly(dU), of the adenine polynucleotides poly(A) and poly(dA) and of DNA are discussed.

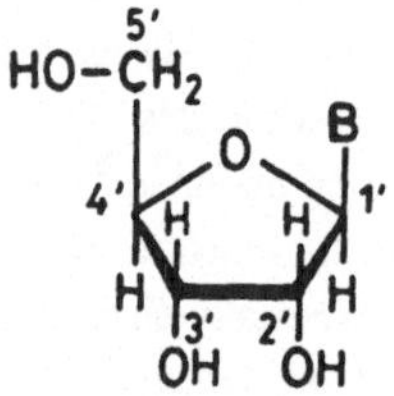

B = uracil: uridine
B =cytosine: cytidine

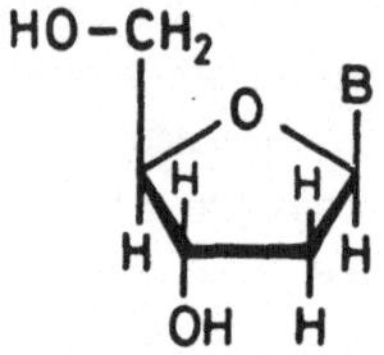

B = uracil: 2'-deoxyuridine (dU)
B = cytosine: 2'-deoxycytidine(dC)
B = thymine: thymidine

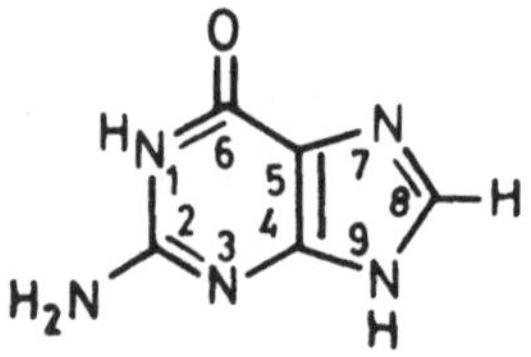

guanine

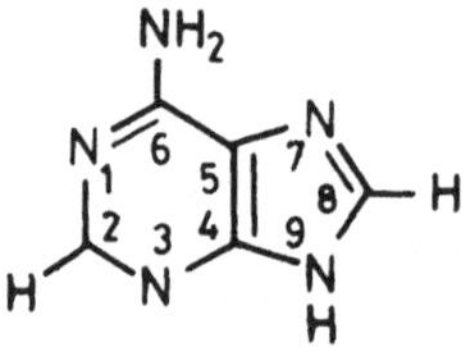

adenine

2. REACTION OF $^{\bullet}$OH AND SO$_4^{\bullet -}$ WITH PYRIMIDINE DERIVATIVES; SITES OF REACTIONS

2.1. REACTIONS OF $^{\bullet}$OH AND SO$_4^{\bullet -}$ WITH URACIL

It is generally accepted that $^{\bullet}$OH radicals add to the C(5)-C(6) double-bond of uracil [reaction (6)]. The isomer distribution of the OH-adduct radical was determined by pulse radiolysis with

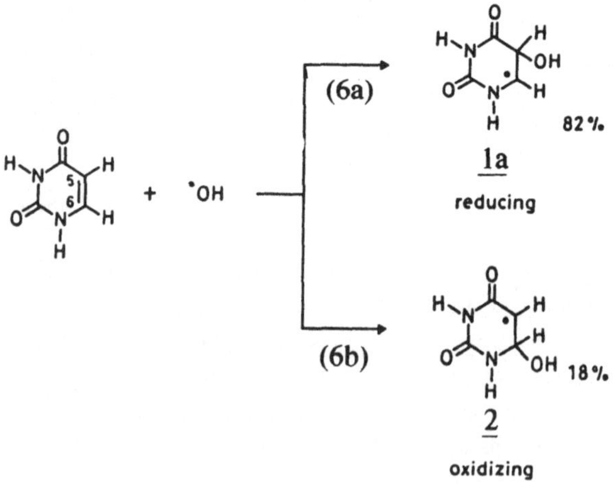

(6a)

82%

1a

reducing

(6b)

18%

2

oxidizing

optical detection utilizing the different redox properties of the radicals.[5] The C(5)-OH adduct 1a reduces tetranitromethane (TNM) and the C(6)-OH radical 2 oxidizes N,N,N',N'-tetramethyl-p-phenylenediamine (TMPD). A ratio of ∼ 8:2 was reported for addition of ˙OH to C(5) vs. addition to C(6) (see Table 1).

In agreement with these results esr spectra generated by reaction of ˙OH with uracil in acidic solutions (pH 3-7) indicate the predominance of the C(5)-OH radical 1a.[6,7]

In alkaline solution a totally different esr spectrum was generated by reaction of ˙OH with uracil. (For esr parameters see Tables 2 and 3). The original interpretation was that a radical formed by ˙OH addition to C(6) undergoes base-catalyzed ring opening.[8] However, this has been criticized because with SO4˙⁻ an identical esr spectrum was found by Bansal and Fessenden[9] which was assigned to radical 3 with high spin-density at N(1) and C(5) formed in reaction (7). It was concluded that a base catalyzed dehydration of the C(5)-OH adduct in alkaline solutions takes place [reactions (8) and (9)] rather than ring opening.

Table 1. Sites of reaction of ˙OH with pyrimidine bases[5,10], methylated pyrimidine bases[11] and poly(U)[12]. Yields in per cent.

Substrate	C(5)	C(6)	methyl group or sugar
uracil	82	18	-
thymine	60	30	10
cytosine	87	10	-
1,3-dimethyl-uracil	74	19	-
1,3-dimethyl-thymine	59	-	-
1-methyl-cytosine	87	8	-
poly(U)	70	23	7

Formation of radical $\underline{3}$ presumably involves addition of $SO_4^{\cdot-}$ to C(5) [reaction (10)] followed by release of a proton and SO_4^{2-}. It is not known whether the sulfate dianion and the proton are eliminated simultaneously [reaction (11)] or in two separate steps [reactions (12) and (13)] via the radical cation $\underline{5}$. In view of recent findings this addition-elimination mechanism seems to be more favourable than one-electron oxidation as suggested originally[9] for reaction of $SO_4^{\cdot-}$ with uracil.

The OH-adducts $\underline{1a}$ and $\underline{2}$ are not produced. Obviously, deprotonation of the sulfate adduct $\underline{4}$ [reaction (11)] or of the radical cation [reaction (13)] are too fast to allow the competition of hydrolysis reactions (14) and (15).

Reaction of $\cdot$OH with thymine and cytosine has been studied by pulse radiolysis. In thymine, besides addition to the double-bond there is also H abstraction from the methyl group. At high pH where $O^{\cdot-}$ is the reacting species, the latter gains in importance[9] and at pH 13.5 the esr spectra are dominated by the signals of the allyl radical $\underline{6}$.

$$6$$

As shown by Novais and Steenken[13] reaction of $SO_4^{\bullet-}$ with thymine results in the esr spectrum of radical 7 (Table 3), a homologue of the N(1)-uracilyl radical 3. Its formation may be described by the reaction scheme discussed for uracil [reactions (10) - (13)].

2.2. REACTION OF $^{\bullet}OH$ AND $SO_4^{\bullet-}$ WITH N(1)-METHYLATED PYRIMIDINES

Methyl substitution of N(1) of uracil changes the situation because deprotonation is not possible. Therefore, with both species, $^{\bullet}OH$ and $SO_4^{\bullet-}$, the OH-adduct radicals are formed [reactions (16) and (17)]. The isomer distribution of the OH adducts obtained by reaction of $^{\bullet}OH$ with 1,3-dimethyluracil is similar to that of uracil, the C(5)-OH adduct being formed in considerable excess over the C(6)-OH adduct[11] (see Table 2). Pulse radiolysis experiments showed that the C(5)-OH adduct was also the main product of the reaction of $SO_4^{\bullet-}$ with 1,3-dimethyluracil[14] in aqueous solution [reaction (17)]. It was found[14] to be present already 1 μsec after the radiolysis pulse with a yield of more than 90%. In esr measurements using in situ photolysis of peroxodisulfate to generate $SO_4^{\bullet-}$ the C(5)-OH adducts of N(1)-methylated uracils were rather difficult to detect.[15] On the ms time-scale of the esr experiment, with peroxodisulfate concentrations of > 3 mM the C(5)-OH adducts of 1-methyluracil and 1,3-dimethyluracil are oxidized, the isomeric C(6)-OH adducts formed in side-reactions are enriched in the course of a chain reaction[14] and dominate the esr spectra. Only with low peroxodisulfate concentrations of < 3 mM was it possible to detect the C(5)-OH adduct radical 8 of 1-methyluracil.

R₁ = CH₃; R₂ = H: 1-methyluracil
R₁ = R₂ = CH₃ : 1,3-dimethyluracil

After formation of the sulfate adduct $\underline{9}$ the following two pathways for $SO_4^{\cdot-}$-induced formation of OH adduct radicals of N-methylated uracils have to be considered:

i) The sulfate adduct $\underline{9}$ hydrolyzes to form the OH-adduct radicals $\underline{8}$ and $\underline{11}$ via the S_N2 reaction (18) or ii) The sulfate dianion is eliminated [reaction (19)] and subsequent hydrolysis of the radical cation $\underline{10}$ in the S_N1 reaction (20) leads to $\underline{8}$ and $\underline{11}$.

From N(1)-methylthymine no esr signals were obtained by reaction with $\cdot$OH. However, with $SO_4^{\cdot-}$ an esr spectrum was generated which could be unambiguously assigned to the C(6)-OH radical $\underline{12}$. Characteristic parameters are the large quartet splitting of 2.26 mT and the OH splitting of 0.037 mT. Radical $\underline{12}$ is formed by reactions similar to those discussed for N(1)-methyluracil. By the presence of the C(5)-methyl group reaction of H_2O with C(5) is hindered and therefore the C(5)-OH isomer is missing.

In general, the sulfate adducts and the radical cations of the N(1)-methylated uracils and thymines are too short-lived to be detected by esr and also by pulse radiolysis. Only in the case of tetramethyluracil was it possible to detect a short-lived transient[16] (λ_{max} = 410 nm, see Fig. 2) which was assigned to the radical cation $\underline{13}$. The radical cation disappeared by hydrolysis [reaction (22)] under formation of the OH-adduct(s) $\underline{14}$ (λ_{max} = 525 nm). Increase in proton concentration due to reaction (22) was monitored by conductivity measurements (inset in Fig. 2). The rates of increase in conductivity, of increase in concentration of OH-adduct(s) and of decrease in concentration of the radical cation were identical ($t_{1/2} \sim 10$ μs)

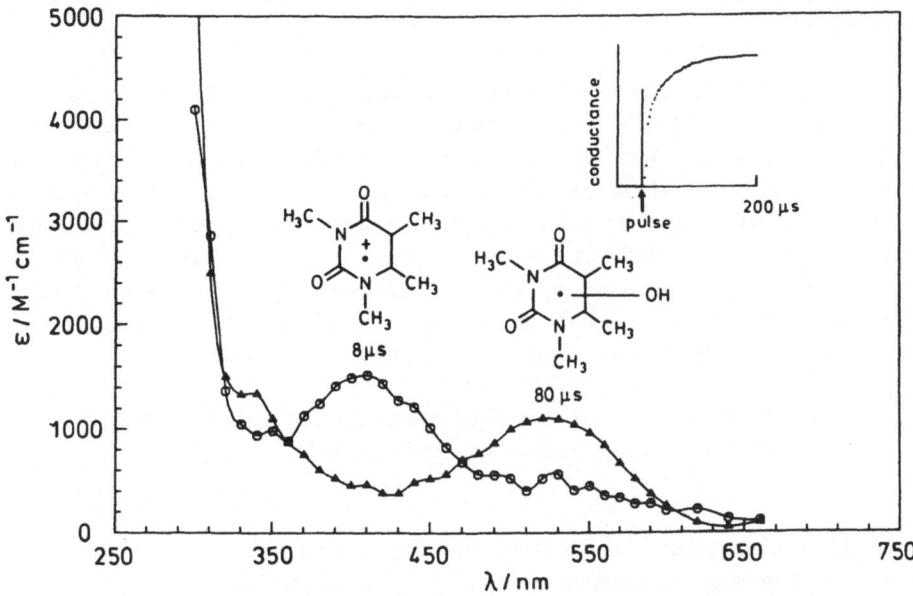

Figure 2. Transient spectra obtained upon irradiating a solution of tetramethyluracil (0.2 mM), $K_2S_2O_8$ (10 mM) and tert.-butanol (60 mM) with a laser pulse (248 nm), pH 4.5 [according to ref. (16)]. The first transient (o) was observed 8 μs and the second one (▲) 80 μs after the laser pulse. In the inset the increase of conductivity after the laser pulse is shown.

No esr spectra were detected upon reaction of $^\bullet$OH with N(1)-methylcytosine. However, by *in situ* photolysis of a solution containing the substrate, peroxodisulfate and 1% acetone 5 equidistant groups of signals with a g factor of 2.0035 were observed.[17] The spectrum was consistent with two similar triplet splittings of 1.175 mT and 1.16 mT, each of them due either to a nitrogen or 2 equivalent protons. The pattern of the individual groups of lines indicated two further small triplets of 0.18 and 0.11 mT and a proton splitting of 0.09 mT. The C(4)-aminyl structure 17 expected[18] upon deprotonation of the cytosine radical cation 16 should give rise to a large doublet splitting[19] ($a_{NH} > 1$ mT) which was not found. Possibly, the spectrum has to be assigned either to a tautomeric form of the aminyl radical 17 or to a radical produced in secondary reactions from 17.

2.3. REACTIONS OF $^\bullet$OH AND $SO_4^{\bullet-}$ WITH PYRIMIDINE NUCLEOSIDES AND NUCLEOTIDES

2.3.1. *Site of reaction.* From γ-radiolysis and pulse radiolysis Deeble et al.[12] were able to demonstrate that in poly(U) 93% of the $^\bullet$OH radicals add to the base moiety and only 7% abstract H atoms from the sugar residues. It may be assumed that this result is representative for reaction of $^\bullet$OH with single base model compounds. As far as the site of reaction of $SO_4^{\bullet-}$ with nucleosides and nucleotides is concerned we have to rely on kinetic data. Rates of addition of $SO_4^{\bullet-}$ to the nucleobases[14] are of the order of $\sim 5 \times 10^9$ $dm^3mol^{-1}s^{-1}$ whereas H abstraction from alcohols and ethers by $SO_4^{\bullet-}$ is one to two orders of magnitude slower.[20] Therefore, we expect that the majority of the $SO_4^{\bullet-}$ radicals adds to the bases whereas H abstraction by $SO_4^{\bullet-}$ from the sugar moieties is negligible.

2.3.2. *Esr spectra.* By reaction of $^\bullet$OH with uracil nucleosides and nucleotides the spectra of the C(5)-OH-6-yl adducts were detected (radicals 1a - 1e in Table 2).

Table 2. Hyperfine couplings of the radicals obtained upon reaction of $^\bullet$OH with uracil and some of its derivatives at pH 7, 4°C. g factors are 2.0028

1a - 1e

Substrate	Radical	Hyperfine Splittings (mT)		
		a(H-α)	a(H-β)	a(other)
uracil[7]	1a	2.14	1.83	0.9(1N) 0.9(1N)
uridine[17]	1b	2.15	1.85	0.3(H1')
3'-UMP[21]	1c	2.18	1.88	0.3(H$_1$')
dU[17]	1d	2.20	1.88	0.28(H1')
poly(U)[21]	1e	2.15	1.85	0.33(H$_1$')

Table 3. Hyperfine splittings and g factors of the radicals obtained upon reaction of SO$_4^{\cdot-}$ with pyrimidine bases, nucleosides and nucleotides

Substrate	Radical	Hyperfine splittings (mT)			g
		a(H-α)	a(H-β)	a(other)	
uracil[9]	3	-	-	0.528(1N) 0.073(1N) 1.576(1H) 0.137(1H) 0.033(1H)	2.00433
thymine[13]	7	-	-	0.470(1N) 0.075(1N) 1.962(3H) 0.058(1H) 0.031(1H)	2.00408
1-methyl-uracil[15]	8	2.04(1H)	1.83(1H)	0.09(1N) 0.048(1N) 0.09(CH$_3$-1)	2.0028
1-methyl thymine[15]	12	2.26(3H) (α-CH$_3$)	1.51(1H)	0.058(1N) 0.011(1N) 0.150(H-3) 0.037(OH-6)	2.0031
1-methyl cytosine[17] dC IPC	18	-	-	1.175[a] 1.16[a] 0.18[a] 0.10[a] 0.09(1H)	2.0035

a) triplets due to one nitrogen or two equivalent protons

Table 3, continued

Substrate	Radical[a]		Hyperfine splittings (mT)			g
			a(H-α)	a(H-β)	a(other)	
dU[17]	1d	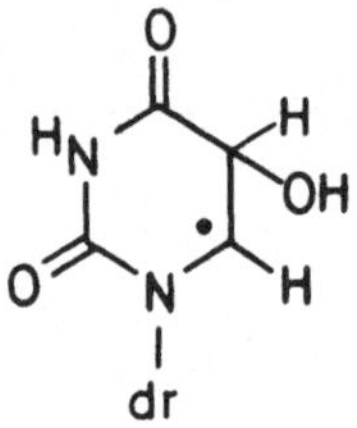	2.20(1H)	1.88(1H)	0.28(H-1')	2.0028
3'-dUMP[21]	19		2.18(1H)	1.88(1H)	0.30(H-1')	2.0028
thymidine[17]	20	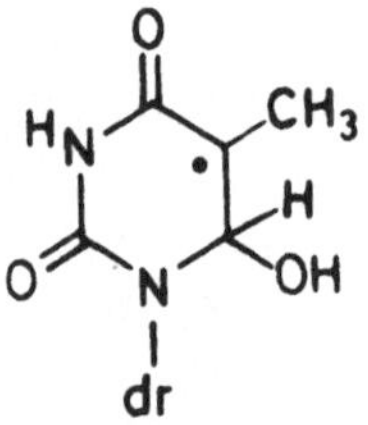	2.23(3H) (α-CH₃)	1.125(1H)		2.0032
uridine[17] cytidine	21a	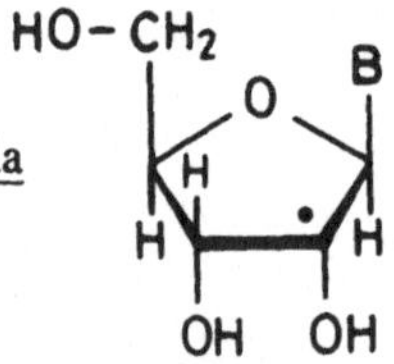	not detected			
uridine[17] cytidine	22	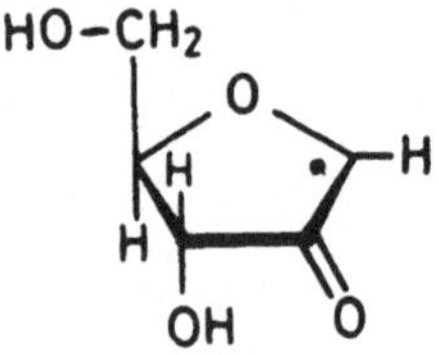	1.36(1H)		0.54(1H) 0.25(1H,	2.0049
3'-UMP[21]	23	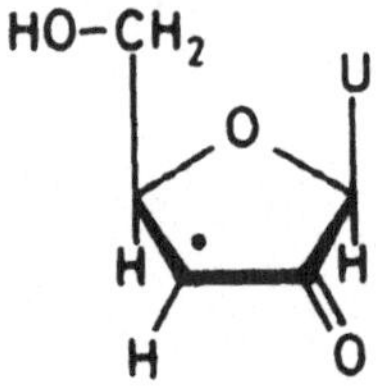	1.86(1H)	3.65(1H)	0.31(1H) 0.08(2H)	2.0042

a) dr = 2'-deoxyribose

Table 3, continued

Substrate	Radical[a]		a(H-α)	a(H-β)	a(other)	g
				Hyperfine splittings (mT)		
uridine[17]	24 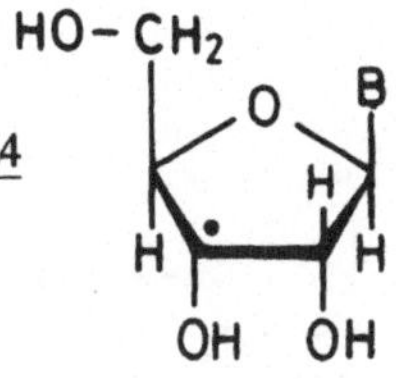			2.65(1H) 2.00(1H)	0.175(1H) 0.07(1H)	2.0033
uridine[17] cytidine	25a 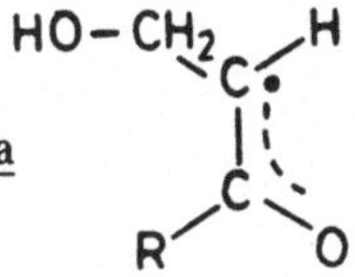		1.87(1H)	2.62(2H)	0.13(1H)	2.0043
uridine[17] cytidine	25b 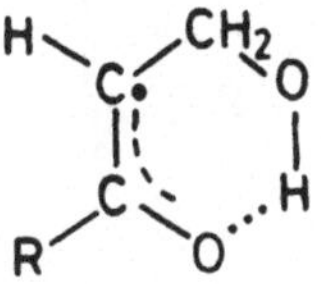		1.81(1H)	2.83(2H)	0.14(1H)	2.0045
poly(U)[21]	34 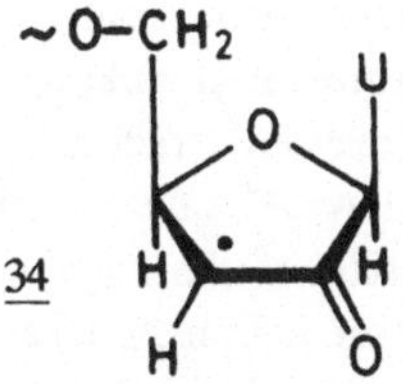		1.875(1H)	3.35(1H)	0.25(1H) 0.05(1H)	2.0044

a) R = -CH(OH)-CH(OH)-B

In contrast, the sulfate adducts are too short-lived to be detected by esr. Depending on the sugar residues they undergo two different types of reactions. The products generated from deoxyribose-derivatives are reminescent of those of N(1)-methylated pyrimidine bases. The spectra obtained from dU, 3'-dUMP and thymidine (assigned to radicals 1d, 19 and 20) are similar to those of the base radicals 8 and 12 of 1-methyluracil and 1-methylthymine.

The spectrum of the radical derived from dC was identical with that generated from 1-methylcytosine. We assume that the base radicals of the deoxyribose-nucleosides are formed by the reactions already discussed for the N(1)-methylated pyrimidines [reactions (18) - (20) for uracil derivatives and (23) - (25) for cytosine derivatives].

For the ribose-derivatives the situation is more complex. The spectra originate from heterolytic decay of the C(2')- (not detected by esr) and the C(3')-hydroxyalkyl radicals.

a) sugar radicals formed by heterolytic decay of the C(2')-hydroxyalkyl radical 21: Reaction of $SO_4^{\cdot-}$ with uridine and cytidine in the pH range 2-9 yielded an esr spectrum with g =2.0049 and three doublet splittings of 1.36, 0.54 and 0.25 mT which is characteristic for radical 22 with the free spin adjacent to both, carbonyl and oxygen functions.

(27)

HO−CH₂ ... 21a OH OH (26a) ... HO−CH₂ ... OH O⁻ (26b) ... HO−CH₂ ... OH O −H + B⁻ 22

The C(2') radical 21b derived from 3'-UMP decays by elimination of the phosphate group thus giving rise to the 2'-oxo-3'-yl radical 23.

(29)

HO−CH₂ ... 21b O OH (28a) PO_3^{2-} ... O O⁻ (28b) PO_3^{2-} ... HO−CH₂ ... H O 23 + $HO-PO_3^{2-}$

Elimination of the base moiety from radical 21a and of the phosphate group from 21b was independent of pH values in the region 2-9. This is in agreement with results by Fitchett et al.[22] who report rapid fragmentation at pH 2-8 of the C(2)-hydroxyalkyl radicals derived from ribose-5-phosphate and from inosine and of the C(2')-hydroxyalkyl radical derived from adenosine. Reasons for the rapidity of the elimination even at low pH values might arise from the strain of the ring or overlap of the β-C-O bond and the orbital of the unpaired electron or both. Presumably, in neutral and alkaline solutions fragmentation of radical 21 takes place at the stage of the deprotonated form [reactions (26) and (28)] whereas in acidic solutions the reaction may occur at the stage of the hydroxyalkyl radical as such [reactions (27) and (29)].

HO−CH₂ ... 24 OH OH (30) −H⁺ ... HO−CH₂ ... O⁻ OH (31) +H⁺ ... HO−CH₂ OH ... ·CH CH−B ... C−CH ... O OH 25

b) sugar radicals formed by heterolytic decay of the C(3')-hydroxyalkyl radical 24; sugar ring opening: In acidic solutions (pH 1-4) of uridine besides the signals of 22 a spectrum with doublets of 2.00, 2.65, 0.175 and 0.07 mT and g = 2.0033 was identified. These parameters are similar to those of the α-hydroxyalkyl radicals formed by H abstraction from six-membered sugar rings.[23,24] Therefore, the spectrum was assigned to the C(3')-hydroxyalkyl radical 24. In neutral and alkaline solutions of uridine and cytidine two new, strongly overlapping spectra with much wider spread appeared. They are consistent with radicals possessing one α-proton, 2 equivalent β-protons and one γ- proton. The spectra are assigned to the Z and E isomers of the open chain carbonyl conjugated radical 25 formed by heterolytic cleavage of the C(4')-oxygen bond of the sugars [reactions (30) and (31)].

2.3.3. *Influence of the electronic properties of the leaving groups on the rate of heterolytic decay of the α-hydroxyalkyl radicals.* It is well-known that rates of heterolytic decay of α-hydroxyalkyl radicals are strongly influenced by the electronic properties of the leaving groups L.[25] A measure for the 'nucleofugacity' is the pK$_a$ (H-L) of the conjugate acid. The pK$_a$ values of 7.19 for the phosphate monoanion, 9.5 for N(1)-H of the nucleobases and 15.7 for H$_2$O explain why elimination of the nucleobase in radical 21a is faster than elimination of the C(3')-OH group and why elimination of the phosphate group in radical 21b is faster than elimination of the base. For the same reasons due to the electron withdrawing effect of the nucleobase B elimination of the $^-$O-CH-(B)-CH(OH) -moiety, i.e. ring opening, is faster than elimination of the C(2')-OH group in radical 24.

2.3.4. *Mechanism of intramolecular H atom transfer.* If we accept that the primary attack of SO$_4^{\cdot-}$ is directed towards the base moieties we have to explain how the site of the free spin is transferred from the base radical to the sugar moieties and why this reaction takes place in the ribose- and not in the deoxyribose-derivatives. The most obvious reason for formation of sugar radicals from ribose-nucleosides seems to originate from activation of H(2'). This explains, for example, why SO$_4^{\cdot-}$ abstracts H(2') from D-ribose but not from 2-deoxy-D-ribose.[23,24] Accordingly, reactive base radicals formed with SO$_4^{\cdot-}$ might be able to react with H(2') in ribose-nucleosides and -nucleotides under formation of radical 21. In agreement with the experimental results this pathway is not feasible in the deoxyribose-derivatives.

The reasons for abstraction of H(3') in the ribose- but not in the deoxyribose- derivatives are less evident. The electronic environment of H(3') is similar in the two classes of compounds. It should be mentioned, however, that there are conformational differences which may be relevant for the H-transfer. It is known that the furanose rings exist in a variety of non-planar conformations which interconvert by pseudorotation.

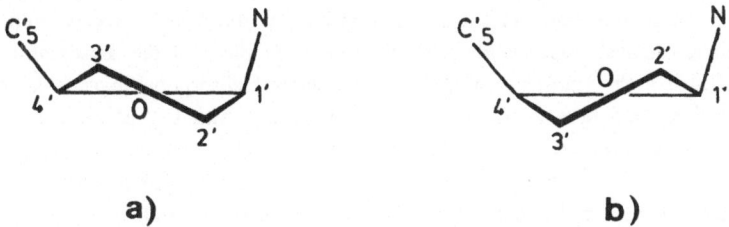

a)   b)

Figure 3. Schematic drawings of average sugar ring conformations favoured in ribose-nucleosides (a) and in deoxyribose-nucleosides (b)

Nmr data[26] show that in ribose-nucleosides and -nucleotides the furanose rings prefer conformations with C(3') in endo position relative to C(5') ('N-type' conformations, see Fig. 3a) and, thus, closer to the base moiety. In the deoxyribose-nucleosides and in 2',3'-O-isopropylidene-uridine (IPU) and -cytidine (IPC) the conformational equilibrium is shifted towards structures with C(3') in exo position ('S-type' structures, see Fig. 3 b). From this situation it is conceivable that fast intramolecular H abstraction from C(3') is more favourable in the ribose- than in the deoxyribose-nucleosides and in the 2',3'-O-isopropylidine-derivatives (see also Table 4).

B = uracil : IPU
B = cytosine: IPC

One might argue that for the same reasons H abstraction from C(2') should be hindered in the ribose-nucleosides. This was not observed. Possibly the effect of sugar ring puckering is not strong enough to overcompensate the favourable spatial conditions allowing rapid intramolecular H transfer from C(2') to the base moiety.

In the case of uridine reaction of $SO_4^{\bullet-}$ with the uracil moiety may result in 4 different base radicals, namely the sulfate adduct, the radical cation and the two OH-adducts. [The structures of these species are analoguous to those of radicals 9, 10, 8 and 11 derived from N(1)-methyluracil]. Upon reaction of $^{\bullet}$OH with uridine only the C(5)-OH-6-yl radical 1b was detected by esr whereas sugar signals did not contribute to the spectrum. Obviously the C(5)-OH adduct of the base is not able to react with the sugar residues on the ms time scale of the esr experiment. The same is probably true for the isomeric C(6)-OH-5-yl radical which may be formed in minor amounts. Both of the remaining species, the sulfate adduct and the base radical cation are too short-lived to be identified by esr or even by pulse radiolysis. We would expect that H transfer to the sulfate adduct is too slow to account for the experimental results which points to the radical cation as the reactive intermediate.

2.3.5. *Pulse radiolysis studies of the reaction of* $SO_4^{\bullet-}$ *with pyrimidine nucleosides.* The differences in reaction products of $SO_4^{\bullet-}$ with ribose- and deoxyribose-nucleosides are corroborated by pulse radiolysis. The absorption spectra of the transients generated by reaction of $SO_4^{\bullet-}$ with ribose-nucleosides were quite different from those of the deoxyribose-nucleosides. As shown in Table 4 the transients of dU and dC showed absorption maxima at λ =380 and 405 nm, respectively and decayed by second order processes with $t_{1/2}$ of more than 100 μs. A transient with similar properties was obtained from IPU. The transients of dU and IPU are assigned to the C(5)-OH-6-yl adduct radicals, that of dC might be due to radical 17 or 18. The transient spectra of the ribose-derivatives showed two maxima and decayed by first order reactions with $t_{1/2}$ of 3.5 μs (uridine) and 11μs (cytidine). Essentially the same transients have been detected independently by Fujita and co-workers.[27] The data listed in Table 4 suggest a relation between the furanose ring conformations, the type of transients detected by pulse radiolysis and the type of radicals detected by esr. However, at the present moment, the structures of the transients of the ribose-compounds are not known and it still has to be clarified whether they play a role in the formation of the sugar radicals.

Table 4. Reaction of SO$_4^{\cdot -}$ with pyrimidine nucleosides. Comparison of pulse radiolysis data[a], esr results and furanose conformations.

Substrate	absorption maxima [nm]	esr results	preferred furanose con- formation
dU	380(1000)[b]		
dC	405(1000)	base radicals	'N-type'
IPU	380(1000)		
IPC	-		
uridine	325(4500) 500(1000)	sugar radicals	'S-type'
cytidine	345(4000) 530(1800)		

a) the transients of dU, dC and IPU decay by second order reactions with $t_{1/2}$ of > 100 µs; those of uridine and cytidine decay by first order rate laws with $t_{1/2}$ of 3.5 µs and 11 µs; b) values in brackets are extinction coefficients [M^{-1} cm^{-1}];

3. REACTION OF $^{\cdot}$OH and SO$_4^{\cdot -}$ WITH PURINE DERIVATIVES

In contrast to the radiolysis of pyrimidines which, except for the case of cytosine, is fairly understood, the knowledge on the radical reactions of purine derivatives is less profound. For reviews see ref. (1d) and ref. (28).

3.1. REACTION OF $^{\cdot}$OH WITH GUANINE DERIVATIVES.

Reaction of $^{\cdot}$OH with 2'-deoxyguanosine (dG) and dG-5-monophosphate has been studied by O'Neill and co-workers.[18,29-31] In pulse radiolysis experiments intermediates with oxidizing and reducing properties in a ratio of ~ 1:1 were detected. The following reaction scheme was proposed.

Assignment of the redox properties is discussed by O'Neill[29] and by Steenken.[28]

3.2. REACTION OF $^{\bullet}$OH WITH ADENINE DERIVATIVES

The addition of $^{\bullet}$OH to substituted adenines, to adenosine and to N^6,N^6-dimethyladenosine has been investigated by pulse radiolysis by Vieira and Steenken[32,33] [reaction (33)].

Both, A4OH$^{\bullet}$ and A8OH$^{\bullet}$ undergo unimolecular transformation, namely dehydration [reaction (34)] and ring opening [reaction (35)].

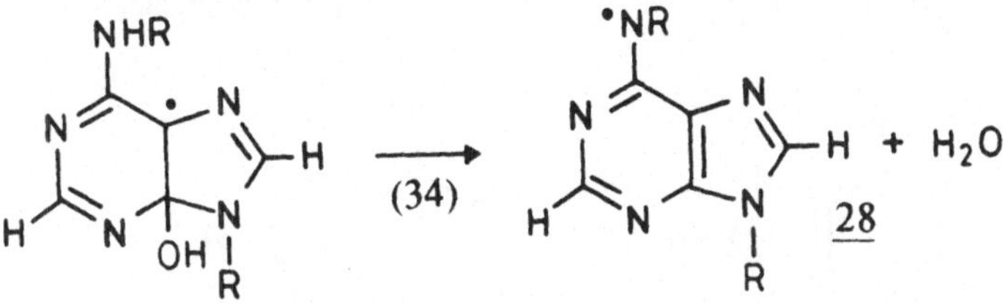

(34)

28 + H_2O

(35)

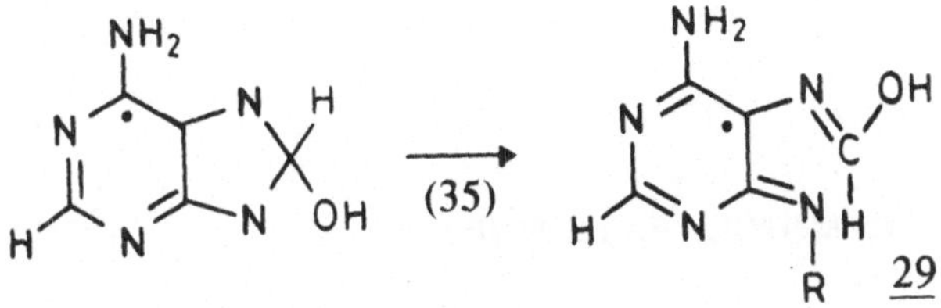

29

Relative rates of these reactions and activation parameters have been studied in detail by Vieira and Steenken.[32,33]

Gilbert and co-workers[22,34] reported on esr spectra generated by reaction of $^\bullet OH$ with adenosine and its 5'-monophosphate (5'-AMP). The $^\bullet OH$ radicals were produced by reaction of Ti^{3+} with H_2O_2

$$Ti^{3+} + H_2O_2 \longrightarrow Ti^{4+} + {}^\bullet OH + OH^- \tag{36}$$

A drastic difference was observed for reaction of $^\bullet OH$ with adenosine and 5'-AMP as compared to uridine. Whereas with uridine the C(5)-OH-adduct 1b was produced, from the adenine derivatives sugar radicals were generated by H abstraction from C(2') and C(4') (radicals 30 and 31). It was shown that the C(2') radical 31 decays by heterolytic elimination of the adenine moiety under formation of radical 22. The reactions are similar to those already discussed for uridine [reactions (26) and (27)].

30

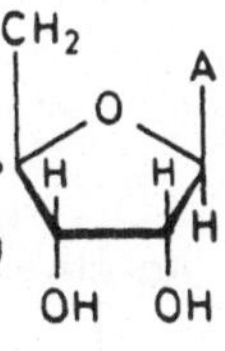

31

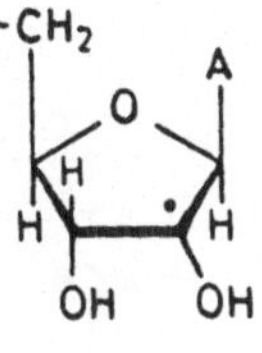

3.3. REACTION OF $SO_4^{\bullet-}$ WITH PURINES

According to O'Neill and Davies[18] the transient detected in pulse radiolysis upon interaction of dG with $SO_4^{\bullet-}$ is either the deprotonated radical cation 32 or the C(4)-OH adduct 33 produced by hydrolysis of the radical cation. On interaction of 2'-deoxyadenosine (dA) with $SO_4^{\bullet-}$ the resulting intermediate is assigned to a one-electron oxidized and deprotonated species (according to structure 28) as shown by redox-titration[29-31] and conductivity measurements.[32]

32   R   33   OH   R

4. STRAND BREAKAGE OF HOMOPOLYNUCLEOTIDES

Radical reactions with single base compounds were carried out with the goal to gain information on the mechanisms of radiation induced DNA damage in aqueous solution. Homopolynucleotides like e.g. poly(U), poly(dU), poly(A) and poly(dA) provide useful models to study sb reactions.

4.1. STRAND BREAKAGE OF POLY(U) AND POLY(dU)

The yield of $^{\bullet}$OH induced sb of poly(U) is much higher[2,35] than of DNA.[36] It was shown that the $^{\bullet}$OH radicals add to the uracil residues and that in a sequence of reactions the radical sites are transferred from the base to the ribose residues within the macromolecule. Heterolytic decay of the sugar radicals results in sb formation. In principle, heterolytic cleavage of the sugar-phosphate bond can originate either from C(2') or from C(4') radicals [reactions (37) and (38), respectively].

Esr studies on single base compounds showed that the reaction of $^{\bullet}$OH with uridine and 3'-UMP yielded the spectrum of the C(5)-OH adduct radicals 1b and 1c of the uracil moieties

(Table 2) whereas with $SO_4^{\bullet-}$ sugar radicals were produced. Similar results were obtained for poly(U), namely generation of the C(5)-OH adduct $\underline{1e}$ with $^{\bullet}$OH at neutral pH and sugar radical $\underline{34}$ with $SO_4^{\bullet-}$ in the pH range 2-9. This means that the C(5)-OH adducts of the single base compounds as well as of poly(U) are stable on the esr time-scale whereas for the base radical cation formed by $SO_4^{\bullet-}$ there exists a pathway for rapid transfer of the radical site to the sugar moiety. The structure of sugar radical $\underline{34}$ proves that the C(2')-mechanism contributes to the $SO_4^{\bullet-}$-induced breakage of the poly(U) chain.

In this context it should be mentioned that in our experiments the freely diffusing radicals of the single base compounds show lifetimes of several milliseconds whereas the lifetimes of the radicals of homopolynucleotides are in the range of seconds. This means that transfer of radical sites to the sugars in mononucleosides and -nucleotides induced by $SO_4^{\bullet-}$ occurs in the submillisecond range, otherwise the precursors of the sugar radicals would be detected. This fast reaction points to a radical cation as intermediate. The C(5)-OH adducts of the single base compounds are not able to abstract H atoms from the sugar residues in the millisecond time range and the C(5)-OH adduct of poly(U) shows no detectable reaction with the sugar residues on the time scale of seconds at neutral pH.

However, in acidic solutions at pH < 4, upon interaction of $^{\bullet}$OH with poly(U) the spectrum of the sugar radical $\underline{34}$ was detected but no longer the C(5)-OH adduct of the nucleobase. Thus, under these conditions the rate of transfer of the radical site to the sugar moieties is significantly larger than 1 s^{-1}. This result is in agreement with pulse conductivity studies by Bothe and Schulte-Frohlinde.[35] It was found that the rate constant of $^{\bullet}$OH-induced sb of poly(U) increased from < 1 s^{-1} at pH 6.5 to 100 s^{-1} at pH 3.4. In view of the results with $SO_4^{\bullet-}$, especially taking into account the relatively long-lived radical cation $\underline{13}$ of tetramethyluracil formed with $SO_4^{\bullet-}$, we postulate that the increase in the rate constant for the $^{\bullet}$OH-induced strand break formation of poly(U) on lowering the pH is due to the formation of the uracil radical cation. It is generated from the C(5)-OH-6-yl radical $\underline{1e}$ by protonation and release of a water molecule. The radical cation should be more efficient in abstracting H atoms from the sugar residues than the parent C(5)-OH-6-yl radical. Therefore, it is suggested that the rate-determining step of the $^{\bullet}$OH-induced strand breakage of poly(U) at neutral pH is transfer of the radical site to the sugar whereas at lower pH it is the generation of the radical cation.

As shown by Adinarayana et al.[37] the yield of sb of poly(dU) was considerably lower than of poly(U) [G_{sb} =0.042 μmolJ^{-1} for poly(dU) and 0.24 μmolJ^{-1} for poly(U)]. This was rather unexpected because the primary structures of the macromolecules are similar and neither of the polynucleotides shows base stacking at 25^0C.[38] These differences can be explained from the model studies presented above showing that transfer of radical sites from base radicals to sugar residues is favoured in ribose- as compared to dexyribose-nucleotides of the pyrimdine series. Possibly differences in sugar ring puckering in the ribose- and deoxyribose- derivatives are responsible for this effect. As far as the behaviour of the deoxyribose-nucleotides is concerned, we agree with studies by Simic and co-workers[39] showing that the base-OH-adduct radical of thymidine does not react with its own sugar residue.

4.2. STRAND BREAKAGE OF POLY(A), POLY(dA) AND DNA.

Compared to the data on single base compounds of uracil the information on adenine derivatives is relatively poor. Esr studies by Gilbert and co-workers[22,34] showed that by reaction of $^{\bullet}$OH with adenosine and 5'-AMP C(2')- and C(4') sugar radiclas were produced. Therefore, both mechanisms discussed for poly(U), i.e. the C(2')- and the C(4')-mechanism, may be also of relevance for poly(A). The high yield of sb [G_{sb} =0.23 μmolJ^{-1} for poly(dA)] shows that base radicals are involved in the sb formation of poly(dA). The rate constant[40] for the addition of $^{\bullet}$OH radicals to adenine is 5 x 10^9 dm^{3}mol^{-1}s^{-1} whereas the rate constants for H abstraction from sugars are $\sim$ 1 x 10^9. Therefore, a base to sugar transfer of the radical site contributes to sb formation also in the case of poly(dA). The yield of sb of poly(A) was G_{sb} =0.044 μmolJ^{-1}. This means that the influence of the structures of the sugar residues on the sb is opposite for the adenine than for the uracil polynucleotides. The explanation for this effect may arise from the fact that, contrary to poly(U) and poly(dU), both adenine polynucleotides show base stacking, poly(dA) possessing a lower amount of secondary structure than poly(A).[41,42] Therefore, the segmental mobility of the poly(dA) macromolecular chain is expected to be higher than of poly(A) and the rate of encounter of base radicals and sugar residues is larger. An interesting result has been obtained by comparison of the rate constants of sb of poly(dA) and single-stranded DNA as a function of pH.[37] The pH dependence and the absolute rate constants are virtually identical. This may indicate that in DNA sb formation induced by $^{\bullet}$OH radicals is occurring via adenine radicals.

5. TIME-RESOLVED ESR SPECTROSCOPY OF PURINE RADICALS IN POLY(A) AND DNA.

As discussed above reactions of SO$_4^{\bullet -}$ (and of $^{\bullet}$OH at pH < 4) with poly(U) yielded the well-resolved esr spectrum of sugar radical $\underline{34}$.

The results of the reaction of SO$_4^{\bullet -}$ with poly(A) were quite different (Fig. 4) showing one broad signal ($\Delta \nu_{1/2}$ = 0.7 mT).

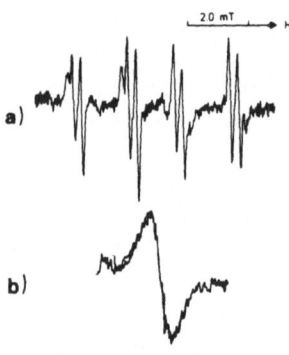

Figure 4. Esr spectra obtained by in situ photolysis of solutions containing $K_2S_2O_8$ (6 mM) and a) poly(U) (1 mM) or b) poly(A) (1 mM)

Whereas in presence of oxygen the signals of the sugar radical of poly(U) disappeared at the expense of a peroxyl signal (g =2.015) the spectrum of poly(A) did not change upon addition of oxygen. Obviously it is due to an oxidizing radical of the adenine moiety. The g factor of 2.004 shows that it is not a peroxyl radical. Since upon interaction of SO$_4^{\bullet -}$ with N^6-methyladenosine the nitrogen centered radical $\underline{28}$ was formed[32] with a yield close to 100% we assign the spectrum of poly(A) tentatively to this structure. A similar signal was obtained by reaction of SO$_4^{\bullet -}$ and also of $^{\bullet}$OH with DNA from calf-thymus. The decay kinetics of these radicals were characterized by time-resolved esr spectroscopy using a chopper-device.[43] The data could be fitted with sums of exponential functions whereas second order rate laws give unsatisfactory results. The decay parameters are given in Table 5. The decay times range from several seconds for the fast components to some hundred seconds for the slow components.

Table 5. Decay characteristics of the purine radicals of poly(A) and calf-thymus DNA.

Substrate	primary radical	decay parameters	
poly(A)	$SO_4^{\bullet -}$	2 ± 1 s	$(60\%)^{a)}$
		50 ± 10 s	(40%)
DNA	$SO_4^{\bullet -}$	2 ± 1 s	(30%)
		30 ± 10s	(30%)
		470 ± 50s	(40%)
DNA	$^{\bullet}OH$	18 ± 5s	(50%)
		170 ± 20s	(50%)

a)Values in brackets give the percentage to which the decay component contributes to the total intensity.

We expect that the decay characteristics are a function of the segmental mobility of the macromolecules which depend on the kind of the neighbouring bases. Taking this into account the differences for poly(A) and DNA are not very large. Upon saturating the solutions with oxygen the decays did not change significantly in agreement with the assumption that the radicals are oxidizing. It is obvious from the lifetimes that the radicals are rather unreactive. Whether or not they attack the sugar residues is not known.

6. CONCLUSIONS.

Two unexpected conclusions have been drawn from this work. Firstly: the formation of a highly reactive intermediate in the reactions of $SO_4^{\bullet -}$ with uridine, 3'-UMP and poly(U) which is not found with $^{\bullet}OH$ radicals at neutral pH. This intermediate is most probably the base radical cation. Secondly: the strong influence of the conformation of the sugar moieties (sugar ring puckering) on the yield of the transfer of the radical site from the base to the sugar residue.

358

7. REFERENCES

1 a) G. Scholes in *Photochemistry and Photobiology of Nucleic Acids*, Vol 1, S.Y. Wang, Ed., Academic, New York, (1976), 521
   b) G. Scholes in *Effects of Ionizing Radiation on DNA*, J. Hüttermann, W. Köhnlein, R. Teoule and A.J. Bertinchamps, Eds., Springer Verlag, Berlin, (1978), 153
   c) C. von Sonntag, *The Chemical Basis of Radiation Biology*, Taylor and Francis, London, (1988)
   d) J. Cadet and M. Berger, Int. J. Rad. Biol., 47, (1985), 127
   e) C. von Sonntag and H.-P. Schuchmann, Int. J. Rad. Biol., 49, (1986), 1
2 D.G.E. Lemaire, E. Bothe and D. Schulte-Frohlinde, Int. J. Rad. Biol., 45, (1984), 351
3 S. Steenken, P. O'Neill and D. Schulte-Frohlinde, J. Phys. Chem., 81, (1977), 26
4 P. O'Neill, S. Steenken and D. Schulte-Frohlinde, J. Phys. Chem., 79, (1975), 2773
5 S. Fujita and S. Steenken, J. Am. Chem. Soc., 103, (1981), 2540
6 C. Nicolau, M. McMillan and R.O.C. Norman, Biochim. Biophys. Acta 174, (1969), 413
7 J.K. Dohrmann and R. Livingston, J. Am. Chem. Soc., 93, (1971), 5363
8 P. Neta, Rad. Res., 49, (1972), 1;
9 K.M. Bansal and R.W. Fessenden, Rad. Res. 75, (1978), 497
10 D.K. Hazra and S. Steenken, J. Am. Chem. Soc., 105, (1983), 4380
11 M. Al-Sheikhly and C. von Sonntag, Z. Naturf. 38b, (1983), 1622
12 D.J. Deeble, D. Schulz and C. von Sonntag, Int. J. Rad. Biol., 49, (1986), 915
13 H.M. Novais and S. Steenken, J. Phys. Chem., 91, (1987), 426
14 H.-P. Schuchmann, D.J. Deeble, G. Olbrich and C. von Sonntag, Int. J. Rad. Biol., 51, (1987), 441
15 G. Behrens, K. Hildenbrand, D. Schulte-Frohlinde and J.N. Herak, J. Chem. Soc. Perkin Trans. II, (1988), 305
16 F. Fockenberg, S. Steenken and D. Schulte-Frohlinde, manuscript in preparation
17 K. Hildenbrand, G. Behrens, J.N. Herak and D. Schulte-Frohlinde, J. Chem. Soc. Perkin Trans. II, submitted
18 P. O' Neill and S.E. Davies, Int. J. Rad. Biol., 52,(1987), 577
19 a) P. Neta and R.W. Fessenden, J. Phys. Chem., (1974), 78, 523
   b) K.V.S. Rao and M.C.R. Symons, J. Chem. Soc. A, (1971), 2163
20 H. Eibenberger, S. Steenken, P.O'Neill and D. Schulte-Frohlinde, J. Phys. Chem., 82, (1978), 749
21 K. Hildenbrand and D. Schulte-Frohlinde, manuscript in preparation
22 M. Fitchett, B.C. Gilbert and M. Jeff, Phil. Trans. R. Soc. London, B 311, (1985), 517
23 B.C. Gilbert, D.M. King and C.B. Thomas, J. Chem. Soc. Perkin Trans II, (1981), 1186
24 J.N. Herak and G. Behrens, Z. Naturf., 41c, (1986), 1062
25 G. Behrens and G. Koltzenburg, Z. Naturf., 40c, (1985), 785
26 D.B. Davies in *Progress in Nucl. Magn. Res. Spectrosc.*, Vol. 12, (1978), 135
27 S. Fujita, private communication
28 S. Steenken in *Pulse Radiolysis of Irradiated Systems*, Tabata,Y., Ed., CRC Press, New York, (1988)
29 P. O'Neill, Rad. Res., 96, (1983), 198

30 P. O'Neill and P.W. Chapman, Int. J. Rad. Biol., 47, (1985), 71

31 P. O'Neill in *Oxidative Damage and Related Enzymes*, Life Chem. Rep. Suppl. Ser. 2, G. Portilio and J.V. Bannister, Eds., Chur: Harwood, (1984), 337

32 A.C.J.S. Vieira and S. Steenken, J. Phys. Chem. 91, (1987), 4138

33 A.C.J.S. Vieira and S. Steenken, J. Am. Chem. Soc. 109, (1987), 7441

34 M. Fitchett and B.C. Gilbert, Life Chem. Rep. 3, (1985), 57

35 E. Bothe and D. Schulte-Frohlinde, Z. Naturf., 37c, (1982), 1191

36 A. Bopp and U. Hagen, Biochim. Biophys. Acta 209, (1970), 320

37 M. Adinarayana, E. Bothe and D. Schulte-Frohlinde, Int. J. Rad. Biol., submitted

38 e.g. W. Saenger, *Principles of Nucleic Acid Structure*, Springer Advanced Texts in Chemistry, C.R. Cantor, Ed., Springer-Verlag, (1984), p. 311

39 M.G. Simic and S. Jovanovic in *Mechanisms of DNA Damage and Repair*, M.G. Simic, L. Grossman and A.D. Upton, Eds., Plenum Press, New York, (1986), p. 39-50

40 Farhataziz and A.B. Ross, Nat. Stand. Ref. Data System, Nat. Bur. Stand. Vol. 59 (NSRDS-NBS-59)

41 F.E. Evans and R.H. Sarma, Nature (London) 263, (1976), 567

42 C.S.M. Olsthoorn, L.J. Bostelaar, J.H. van Boon and C. Altona, Eur. J. Biochem. 112, (1980), 95

43 D. Schulte-Frohlinde, G. Behrens and A. Önal, Int. J. Rad. Biol. 50, (1986), 103

DETECTION OF FREE RADICALS IN BIOCHEMISTRY BY ELECTRON SPIN RESONANCE SPECTROSCOPY

J. Z. Pedersen*, G. Musci° and G. Rotilio*
*Dept. of Biology, "Tor Vergata" University of Rome
Via O. Raimondo, 00173 Rome, Italy
°Dept. of Biochemical Sciences and CNR Center for
Molecular Biology, "La Sapienza" University of Rome
Piazzale A. Moro 5, 00185 Rome, Italy

ABSTRACT. Free radical studies in biochemistry are mainly concerned with the detection and characterization of the radical species, rather than with the determination of the molecular details of the radical reaction. Two examples of electron spin resonance spectroscopy studies are described: the production of radicals from the favism-causing pyrimidine glucoside vicine, and the detection of radicals from anthracycline antitumor drugs in human blood. The results demonstrate the importance of working under conditions as close as possible to the ones found in intact biological systems.

INTRODUCTION

Free radical mechanisms have become of increasing interest in biochemistry, in a development parallel to the one seen in pure organic chemistry. However, the type of free radical research differs considerably between the two fields. Whereas the chemist will pursue the molecular mechanism and the underlying principles governing the reaction, the main task in biochemistry will often be the detection and identification of the radical species. The biochemist is faced with the problem of measuring radicals occurring at low natural abundancies in systems with a very complex composition, furthermore these systems (e.g. cells and tissues) have a complex structural organization of a kind never encountered in pure chemistry. To circumvent these obstacles it is necessary to work with simplified model systems where the radical reactions take place under conditions suitable for experimental investigations. However, with this type of work it always remains a problem to establish to what extent the observed findings will be valid also for the intact, complex system. There is always the possibility

361

F. Minisci (ed.), Free Radicals in Synthesis and Biology, 361–372.
© 1989 by Kluwer Academic Publishers.

that the relative importance of individual reaction rates etc. could be
very different under natural conditions, or that the model system might
in fact directly produce artificial results.

In this paper we present results from two different lines of free
radical research, working with electron spin resonance determination
and characterization of naturally occurring unstable radicals. The
first example concerns the different radicals formed by enzymatic or
chemical degradation of vicine, one of the compounds causing favism.
The second example deals with the radicals of anthracyclines, a group
of drugs with potential antitumor activity, and in particular with
the first detection of this type of radicals in human blood. The work
will be described here with special emphasis on the kind of constraints
and problems occurring in biochemical studies of radical reactions.

ENZYMATICALLY AND CHEMICALLY GENERATED DIVICINE RADICALS

Vicine and the related compound convicine are pyrimidine glucosides
found in broad beans (Vicia faba). Upon ingestion these compounds are
enzymatically hydrolyzed, presumably by a ß-glucosidase, to form the
aglycones divicine and isouramil respectively (1,2). Under physiological conditions both of the aglycones are unstable and undergo rapid
oxidation by molecular oxygen in a series of reactions involving O_2^-
and H_2O_2 (3-5). It is assumed that the reaction products can be
re-reduced by some suitable reductant, e.g. red blood cell glutathione,
so that a redox cycle is formed, consuming oxygen and cellular re-
ductants and producing potentially harmful oxygen derivatives.

In some individuals affected by a deficiency of the enzyme
glucose-6-phosphate dehydrogenase, which is normally responsible for
a major part of the NADPH production in the cells, the action of the
redox cycle may lead to an acute hemolytic crisis known as favism.
This condition is characterized by depletion of the cell glutathione
pool and impairment of the functioning of several key enzymes in the
red blood cells (6-9).

The details of the redox cycle as well as the steps in the
subsequent metabolism of divicine and isouramil are not very well
known; the oxidized products decompose rapidly, and their charac-
terizations by standard analytical methods have so far not been
reported. One of the problems of research in this field is the high
concentration of ß-glucosidase necessary, the presence of the enzyme
prevents the spectrophotometric observation of the behaviour of the
radicals. It has therefore been customary to produce divicine
chemically by acid hydrolysis of vicine in boiling HCl. In this way
the enzyme step is avoided and the divicine reactions can be followed

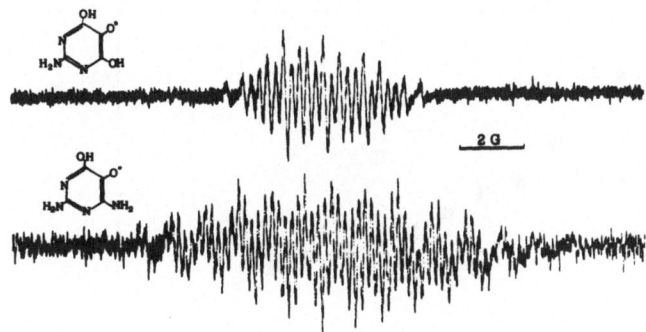

Figure 1. High resolution spectra of the radicals produced by chemical
(upper spectrum) or enzymatic (lower spectrum) hydrolysis of vicine.
The enzymatic preparation was made by incubation of 5 mg ß-glucosidase
with 10 mM vicine in 50 mM phosphate buffer at pH 5.0 for 30 min at room
temperature. Acid hydrolysis was done by heating 0.1 mmol vicine in
1.0 M HCl for 10 min at 100 °C. In both cases the reaction took place
in 1 ml nitrogen-flushed solution kept under vacuum in a Thunberg cell;
before the measurements the pH was adjusted to 7.0 by admixing a small
volume of NaOH stored in the sidearm of the cell. The cell was opened
under a stream of nitrogen, and the sample was transferred rapidly in
a reproducible way to a standard flat quartz ESR cell. Measurements
were done with a Bruker ESP-300 spectrometer using the following
instrument settings: 2mW power, 0.1-0.25 G modulation, 10 ms time
constant, 10 s scan time; 16 scans were accumulated for the upper
spectrum, 102 scans for the lower.

optically. Almost all published studies on the reaction mechanism have
applied this method, the enzymatic cleavage has only been used in a
few cases (8-10).

Although it had long been assumed that the autoxidation of the
aglycones involved the formation of radicals, early attempts to detect
such radicals by electron spin resonance (ESR) spectroscopy turned out
unsuccessful (3). Only recently did Albano and co-workers succeed in
obtaining an ESR spectrum of a radical formed after enzymatic hydrolysis
of vicine. However, their spectral resolution was too low to reveal
any information about the identity of the radical; besides the reported
stability seemed inconsistent with the reaction rates otherwise found
(3-5).

To resolve these problems we have made a series of experiments in
which the radicals formed by enzymatic as well as by chemical hydrolysis
have been characterized by their ESR spectra (12). As can be seen from

364

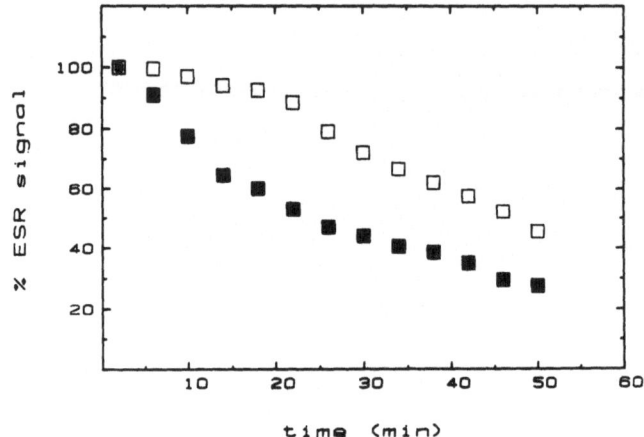

Figure 2. Decay curves at pH 7.0 of the divicine ( ) and deamino
divicine ( ) radicals. Disappearance of the ESR signal caused by
diffusion of air into the flat section of the aqueous sample cell was
measured using 10 G modulation and 50 mW microwave power. Under these
conditions the signals from both radicals were composed of a single
large line, ideal for kinetics experiments.

Figure 1, the two methods give quite different spectra. Obviously these
two types of vicine hydrolysis do not yield the same products, in
contrast to what has been assumed hitherto. By spectral simulation it
could be shown that the radical formed following acid hydrolysis is a
deamino derivative; evidently the harsh reaction conditions apart from
releasing the glucose moiety at position 5 also causes the loss of the
aminogroup at position 4. An exact assignment of the spectrum from the
enzymatically prepared radical has not been possible so far, but it is
certain that the radical retains all four nitrogens present in vicine.
Thus there can be little doubt that this is indeed the genuine divicine
radical.

The abovementioned conclusions were supported by studies of the
pH dependency of the radical spectra. For the divicine radical spectral
changes seem to reflect the existence of a pK value around 9, which was
no longer observed in spectra from the deamino divicine radical (12).
From the high resolution ESR spectra of the enzymatically formed radical
two other important observations have been made. Firstly, the spectra
are clearly isotropic and characteristic of a radical moving freely in
solution; no binding to the ß-glucosidase is seen. Secondly, the
spectral characteristics did not change with time, indicating that only
a single radical species is involved in the vicine oxidation
process (13).

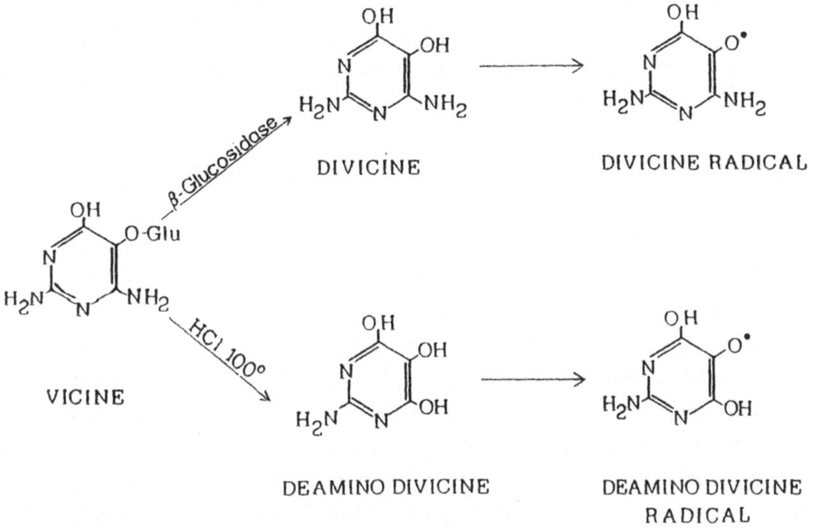

Figure 3. Reaction scheme outlining the formation of the two different radicals from vicine. The details of the scheme as well as the fate of the radicals are still unknown.

An immediate consequence of these findings was that the great majority of the studies on vicine metabolism has been done with an artificial intermediate that does not occur under physiological conditions. It now became important to establish to what extent the deamino derivative would behave as the real divicine. In Figure 2 is shown the time dependent disapperance of the ESR signals due to the reaction of the radicals with oxygen. The two radicals showed markedly different decay curves, in particular during the initial phase. The autoxidation of divicine and isouramil has been suggested to take place as outlined in the following reaction sequence (4):

$$RH_2 + O_2 \quad -- \quad RH^{\bullet} + O_2^{-} + H^{+} \qquad (1)$$

$$RH_2 + O_2^{-} + H^{+} \quad -- \quad RH^{\bullet} + H_2O_2 \qquad (2)$$

$$RH^{\bullet} + O_2 \quad -- \quad R + O_2^{-} + H^{+} \qquad (3)$$

in analogy with the mechanism previously described for adrenaline (14). It is possible that this set of equations is valid for both divicine and deamino divicine, but with the relative importance of the individual steps being different. In Figure 3 is shown a reaction scheme for the formation of the two radicals. The production of O_2^{-} as an intermediate and H_2O_2 as a product have been demonstrated for the oxidation of what is now known to be the deamino divicine (3-5). However, the details of the metabolism of divicine still remains to be clarified. The ability to observe directly the presence of the radical by its ESR spectrum will be a useful tool for the forthcoming research in this area.

ANTHRACYCLINE RADICAL GENERATION BY RED BLOOD CELLS

The anthracycline antibiotics represent another group of compounds, the most prominent members being adriamycin and daunomycin, where the formation of a radical is assumed to be an important aspect of their biological activity. These drugs are efficient antitumor agents; their use in cancer chemotherapy, however, is limited by their cardiac toxicity. In spite of many efforts no clear picture of the mode of action of the anthracyclines has emerged; from the bewildering number of observations made it appears that these compounds can interact in several ways with cellular systems. Two cases in particular have been much studied: The intercalation of the drugs into DNA with the consequent inhibition of the functioning of the nucleic acids, and the unusual redox properties of adriamycin and daunomycin, which are both dihydroxy-substituted anthraquinone derivatives (for an overview of earlier literature see ref. 15).

The involvement of a radical mechanism in anthracycline metabolism has been shown by the demonstration of the formation of reactive oxygen species using the spin trapping method (16-21). Since the anthracyclines are easily reduced to their semiquinone forms it has been possible to detect the semiquinone radicals directly in biological systems by ESR. In this way the radical could be observed in tumor cells (22-24), in microsomal cytochrome P-450 preparations (16,18,25-29) and in mitochondrial preparations (30-31). The spectrum is always observed as a single line without any fine splitting, this is probably a result of the tendency of the anthracyclines to form aggregates in aqueous solutions, even at very low concentrations (32,33).

The complexity of living cells is often a serious problem when trying to single out the various effects caused by a drug. It is therefore useful to study a simplified system or a model structure. One such simple system is the red blood cell, it has a comparatively simple and much-studied metabolic machinery, and is without nucleic acids; an important feature for anthracycline studies. Furthermore there is at the moment considerable interest in the application of red blood cells as drug carriers for clinical use; isolated cells can be loaded in various ways with anthracyclines or other drugs and re-introduced into the circulatory system, whereafter the drugs are released slowly in the organism (34).

In a recent report (35) it was shown that isolated red blood cell membranes could give a radical ESR signal upon incubation with the anthracycline carminomycin. This observation prompted us to examine further the interaction between anthracyclines and red blood cells, using traditional biochemical methods as well as ESR experiments; only the results from the latter shall be discussed here. During the experiments the daunomycin radical was seen for the first time in intact human blood (Figure 4a). As expected the radical was found to be extremely sensitive to oxygen (16,23,36). No ESR signal was ever seen in air saturated samples and not even in samples flushed with nitrogen. Only after substituting the O_2 bound to hemoglobin with CO could the free radical accumulate to reach measurable concentrations. Flushing the sample with carbon monoxide was done directly in the flat cell via the bottom capillary access, since any attempt to transfer the sample would inevitably lead to exposure to air. Following a lag time of 10-15 min the ESR signal would appear slowly, and the steady accumulation of the radical could be seen over a period of several hours. Introduction of air would cause the immediate disappearance of the signal.

The same radical signal could be obtained with samples of red blood cells and of isolated erythrocyte membranes (Figure 4b,c); no daunomycin radicals were ever seen in samples of blood plasma. On the contrary, often the characteristic two line signal of ascorbate was seen in plasma samples (Figure 4d). In whole blood samples the ascorbate signal was never seen together with the semiquinone radical. These findings point to the erythrocyte membrane as the site of the reductive generation of anthracycline semiquinones, in accordance with the conclusions of Bartosz and co-workers (35). Under similar conditions also adriamycin gave ESR visible radicals in whole blood samples, although the signal was somewhat weaker than that observed for daunomycin, a tendency which is known from other systems (37).

Based on the initial observations, experiments were made to determine the localization of the daunomycin radical. No ESR signal was found in samples of red blood cell homogenates or isolated erythrocyte cytosol (Figure 4e,f). This is not surprising, given the high amounts of reducing compounds (e.g. glutathione and ascorbate) present inside the intact cell. These samples were found even to be able to abolish the signal of radicals generated artificially by incubation of the daunomycin with the enzyme xanthine oxidase; this enzyme will give rise to a strong radical production as it reacts directly with the anthracycline. The susceptibility of anthracycline radicals to thiol reduction is well known (36,38), so it can be concluded that the ESR signal observed is caused by radicals located outside the red blood cells. However, this again implies that the diffusion of the radicals

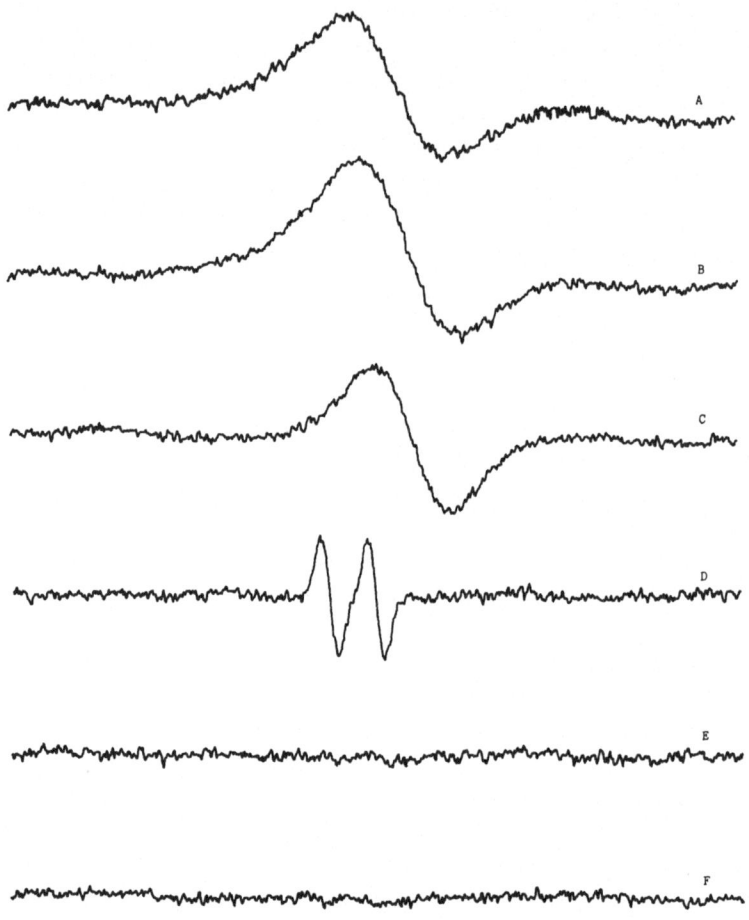

Figure 4. ESR spectra of daunomycin radicals formed by whole blood (a), intact red blood cells (b) and isolated erythrocyte membranes (c). In spectra of blood plasma the two-line ascorbate radical signal could sometimes be observed (d). No ESR signals were seen in samples of cell homogenates (e) or membrane-free cell homogenates (f). Red blood cells and isolated membranes were suspended in a standard PBS buffer at concentrations comparable to whole blood conditions. 5 mM glucose was added to blood and cell samples, whereas 5 mM NADPH served as electron source for the isolated membranes. All measurements were made at room temperature with a Bruker ESP-300 instrument using 40 G scan width, 1 G modulation, 25 mW microwave power, 82 ms time constant and a scan time of 84 s. Scans were accumulated for 1 h, apart from the experiment with isolated membranes where 2.5 h of scan accumulation was necessary.

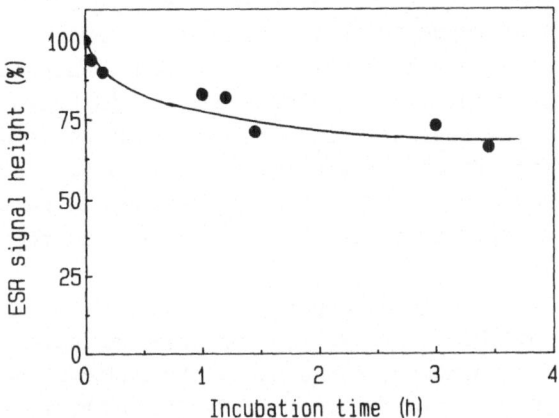

Figure 5. Demonstration of the generation of O_2^- during the reduction of daunomycin by red blood cells. A 50% hematocrit suspension of erythrocytes was incubated with 1 mM daunomycin and and 0.05 mM of the Cu^{2+}-enzyme superoxide dismutase. The size of the characteristic Cu^{2+} ESR signal from the enzyme decreases due to the reaction with O_2^-, and will under steady state conditions reach a level close to 50% of the signal from the oxidized enzyme (40).

across the cell membrane must be slow, slow enough to allow for the accumulation of radicals outside the cells for a period of several hours, and too slow to make it possible for the radical to be formed inside the cells and then diffuse out without being reduced to the non-paramagnetic hydroquinone species. In agreement with this conclusion is a recent report, stating that in anthracycline-loaded red blood cells the passive diffusion out of the cells to reach equilibrium is a process taking two days (34). Also results from other cell systems indicate that diffusion of anthracyclines through the cell membrane requires considerable time (39).

Although we cannot determine by the ESR technique if daunomycin is involved in a redox cycle inside the red blood cells, we can conclude that the radicals observed are generated on the outside of the erythrocyte membrane. Further support for this result comes from the experiment with superoxide dismutase reported in Figure 5. It demonstrates that the reduction of daunomycin by red blood cells generates a flux of O_2^-, which can drive externally added superoxide dismutase towards the steady state condition; considering the known efficient O_2^--scavenging activity inside the erythrocytes, it is unlikely that the flux of O_2^- observed has internal origins.

Our conclusion drawn from this work is that daunomycin is reduced by an electron donor at the outside of the red blood cell membrane; the

semiquinone radical formed reacts rapidly with oxygen to generate superoxide. From other lines of research there are several indications for
the presence of a reducing enzyme localized on the outer surface of the
erythrocyte cell membrane, but very little conclusive is known about
its nature (for an overview see ref. 41). It remains to be established to
what extent the cytotoxic effects of the anthracyclines can be explained
by reactions at the outer surface of the cell membrane. In fact, recent
experiments have shown that immobilized anthracyclines can mimic the
action of the free drugs without being able to enter the cells, thus
indicating the membrane as the main target (42,43). As demonstrated
here, ESR measurements of the anthracycline radicals can give useful
information about the drug-membrane connection. The radical spectra in
Figure 4 do in fact contain a small immobilized component; whether this
is caused by membrane-bound radicals or by aglycone formation (16) will
be the subject of forthcoming studies.

ACKNOWLEDGEMENTS

This work was partially supported by the CNR special project "Oncologia".
JZP was supported by a grant from the Danish Natural Sciences Research
Council.

REFERENCES

1. Mager J, Glaser G, Razin A, Izak G, Bien S and Noam J (1965)
    Biochem. Biophys. Res. Commun. 20, 235-240.
2. Mager J, Chevion M and Glaser G (1980) in Toxic Constituents of
    Plant Foodstuff (Liener LI, ed.) 2nd edn., pp 265-294, Academic
    Press, New York.
3. Chevion M, Navok T, Glaser G and Mager J (1982) Eur. J. Biochem.
    127, 405-409.
4. Winterbourn CC, Benatti U and De Flora A (1986) Biochem. Pharmacol.
    35, 2009-2015.
5. Musci G, Mavelli I and Rotilio G (1987) Biochim. Biophys. Acta
    926, 369-372.
6. De Flora A, Benatti U, Guida L, Forteleoni G and Meloni T (1984)
    Blood 64, 294-297.
7. Benatti U, Guida L, Grasso M, Tonetti M, De Flora A and
    Winterbourn CC (1985) Arch. Biochem. Biophys. 242, 549-556.
8. Mavelli I, Ciriolo MR, Rossi L, Meloni T, Forteleoni G, De Flora A,
    Benatti U, Morelli A and Rotilio G (1984) Eur. J. Biochem.
    139, 13-18.

9. Mavelli I, Ciriolo MR and Rotilio G (1985) Biochim. Biophys. Acta 847, 280-284.

10. De Flora A, Benatti U, Morelli A and Guida L (1983) Biochem. Int. 7, 281-290.

11. Albano E, Tomasi A, Manuzzu L and Arese P (1984) Biochem. Pharmacol. 33, 1701-1704.

12. Pedersen JZ, Musci G and Rotilio G (1988) submitted to Biochemistry.

13. Musci G, Pedersen J and Rotilio G (1988) in Proceedings on Medical, Biochemical and Chemical Aspects of Free Radicals (Yoshikawa T, ed.) Elsevier, Amsterdam, in press.

14. Misra HP and Fridovich I (1972) J. Biol. Chem. 217, 3170-3178.

15. Gianni L, Corden BJ and Myers CE (1983) in Reviews in Biochemical Toxicology (Hodgson E, Bend J and Philpot R, eds.) pp 1-82, Elsevier/North Holland, New York.

16. Kalyanaraman B, Perez-Reyes E and Mason RP (1980) Biochim. Biophys. Acta 630, 119-130.

17. Lown JW and Chen H-H, (1981) Can. J. Chem. 59, 390-395.

18. Gutierrez PL, Gee MV and Bachur NR (1983) Arch. Biochem. Biophys. 223, 68-75.

19. Bannister JV and Thornally PJ (1983) FEBS Lett. 157, 170-172.

20. Doroshow JH and Davies KJA (1986) J. Biol. Chem. 261, 3068-3074.

21. Sinha BK, Katki AG, Batist G, Cowan KH and Myers CE (1987) Biochemistry, 26, 3776-3781.

22. Sato S, Iwaizumi M, Handa K and Tamura Y (1977) Gann 68, 603-608.

23. Bozzi A, Mavelli I, Mondovi B, Strom R and Rotilio G (1981) Biochem. J. 194, 369-372.

24. Peskin AV, Konstatinov AA, Zbarsky IB, Ledeney AN and Ruuge EK (1987) Free Rad. Res. Commun. 3, 47-55.

25. Bachur NR, Gordon SL and Gee MV (1977) Mol. Pharmacol. 13, 901-910.

26. Bachur NR, Gordon SL and Gee MV (1978) Cancer Res. 38, 1745-1750.

27. Bachur NR, Gordon SL, Gee MV and Kon H (1979) Proc. Natl. Acad. Sci. USA 76, 954-957.

28. Berlin V and Haseltine WA (1981) J. Biol. Chem. 256, 4747-4756.

29. Nohl H and Jordan W (1983) Biochem. Biophys. Res. Commun. 114, 197-205.

30. Davies KJA, Doroshow JH and Hochstein P (1983) FEBS Lett. 153, 227-230.

31. Davies KJA and Doroshow JH (1986) J. Biol. Chem. 261, 3060-3067.

32. Chaires JB, Dattagupta N and Crothers DM (1982) Biochemistry 21, 3927-3932.

33. De Flora A, Benatti U, Guida L and Zocchi (1986) Proc. Natl. Acad. Sci. USA 83, 7029-7033.

34. Schreiber J, Mottley C, Sinha BK, Kalyanaraman B and Mason RP (1987) J. Am. Chem. Soc. 109, 348-351.
35. Peskin AV and Bartosz G (1987) FEBS Lett. 219, 212-214.
36. Zweier JL (1985) Biochim. Biophys. Acta 839, 209-213.
37. Zweier JL, Gianni L, Muindi J and Myers CE (1986) Biochim. Biophys. Acta 884, 326-336.
38. Olson RD, Boerth RC, Gerber JG and Nies AS (1981) Life Sci. 29, 1393-1401.
39. Inaba M, Kobayashi H, Sakurai Y and Johnson RK (1979) Cancer Res. 39, 2200-2203.
40. Scarpa M, Viglino P, Contri D and Rigo A (1984) J. Biol. Chem. 259, 10657-10659.
41. Crane FL, Sun IL, Clark MG, Grebing C and Löw H (1985) Biochim. Biophys. Acta 811, 233-264.
42. Tritton TR and Yee G (1982) Science 217, 248-250.
43. Bredehorst R, Panneerselvam M and Vogel C-W (1987) J. Biol. Chem. 262, 2034-2041.

ERROR IN ENZYMIC STEREOSPECIFICITY AS A PROBE FOR THE OCCURRENCE OF RADICAL INTERMEDIATES

János Rétey
Chair of Biochemistry, Institut of Organic Chemistry, University of Karlsruhe, Federal Republic of Germany

INTRODUCTION

More and more biochemical reactions turn out to be of radical nature. The various energy-trapping and energy-transforming processes as well as light-induced reactions are expectedly so.

There are, however, quite a number of dark reactions of pure organic nature that involve free radical intermediates. Now, the interesting question arises, how the corresponding enzymes generate a radical species in a controlled and reversible manner. One established way to do so is the reversible homolytic cleavage of the cobalt-carbon bond in coenzyme B_{12} [1] (Fig.1). The very hot methylene radical can attack non-activated CH bonds, for instance in methyl groups, and initiate rearrangements. Another not completely elucidated way to generate radicals in enzymes involves iron and a tyrosine residue. Tyrosine radicals have been implicated in the ribonucleotide reductase system of E.coli [2] and in the oxygen evolving system of the photosynthetic apparatus of plants [3].

In the appropriate protein environment the tyrosine radical seems to enjoy longlivity as shown by ESR spectroscopy. However, in the case of very shortliving species ESR spectroscopy might not be appropriate to prove the radical nature of reactions. Here another aspect of radical reactions, their usually low stereoselectivity may come into play.

The steric course of organic chemical reactions provides the most valuable information about their mechanisms. Many of these reactions are concerted what means that bonding interactions remain important during the conversion allowing inference to the geometry of the transition state. Even in the case of nonconcerted processes, the loss or conservation of stereochemical purity are among the best indicators for the existence, nature and symmetry of metas-

373

F. Minisci (ed.), Free Radicals in Synthesis and Biology, 373–390.
© 1989 by Kluwer Academic Publishers.

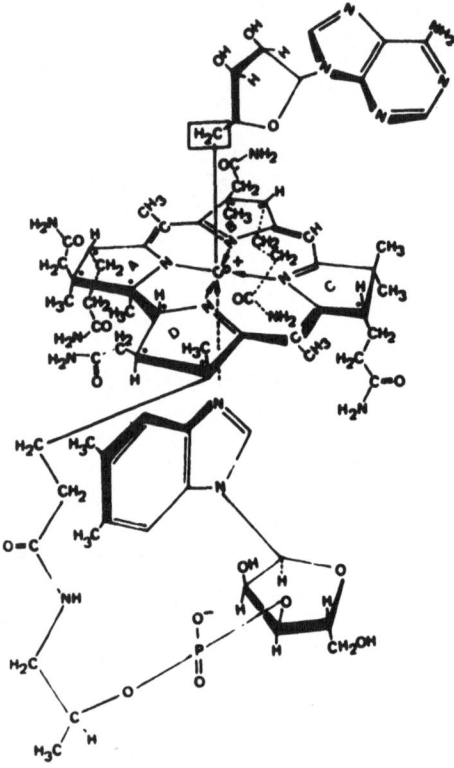

Figure 1.   The Structure of Coenzyme B$_{12}$ (Adenosylcobal-amin).

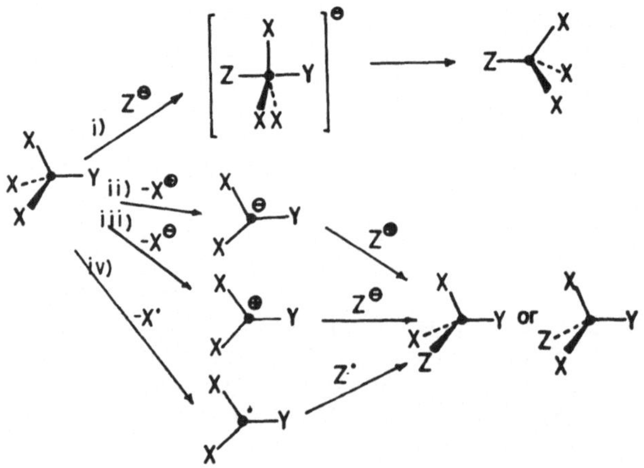

Figure 2.   Some Possibilities for Substitution Reactions at a Tetrahedral Centre Carrying Three Identical Substituents.

table intermediates.

The mechanistic interpretation of the steric course of enzymic reactions is less straightforward. First, very few enzymic reactions are concerted. Second, enzymes are chiral catalysts able to recognize enantiotopic faces or groups in metastable intermediates.

Moreover, such intermediates are in most cases tightly bound to the enzyme and their degrees of freedom may be restricted at the active site. As a result, almost all enzymic reactions are stereospecific. The reasons for this may vary considerably, however. Paradoxically, the best information comes from cases in which an enzyme fails to be stereospecific. It is, however, not easy to catch an enzyme red-handed in such a "sinful" act.

To attempt this we may render the enzyme's job as difficult as possible. For instance, if we consider the substitution at a centre carrying three identical groups, four cases are possible (see Fig.2): i) Concerted substitution or stepwise substitutions, ii) via an anionic, iii) a cationic and iv) a radical intermediate.

Under i) we can cite all S_N2 substitutions occurring with inversion. A few enzymic cases are known using chiral methyl groups [4,5] and many more with chiral phosphorus at the substitution centre [6] although the concertedness of the latter is a matter of debate.

Examples for case ii) are given in the enzymic condensations in which the central atom is carbon, Y and Z are carbonyls and X is hydrogen or its isotopes. With chiral methyl groups it has been shown that all these reactions are stereospecific, since in the carbanion-like intermediate the C-C(Y) bond (see 1 in Fig.3) is in fact a double bond with restricted torsion. As a consequence the two faces of the intermediate are stereoheterotopic, i.e. distinguishable.

Although to my knowledge no example was found that corresponds exactly to case iii), the serine hydroxymethyltransferase reaction may be cited in which formaldehyde or its polarized form is an intermediate. Here torsion about the C-O bond is tantamount to rotation of the whole molecule and this indeed occurs in about 30 % of the cases as it was determined independently by two groups [7-8].

ENZYMIC SUBSTITUTIONS VIA RADICAL INTERMEDIATES

Finally we arrive at the interesting case iv) in which radical intermediates occur. Here free torsion about the C-Y bond can be assumed unless the active site prevents it. Accordingly, the two faces in intermediate 3 are homotopic. Since such faces are indistinguishable even by enzymes, intermediate 3 could be detected by a failure of the enzyme

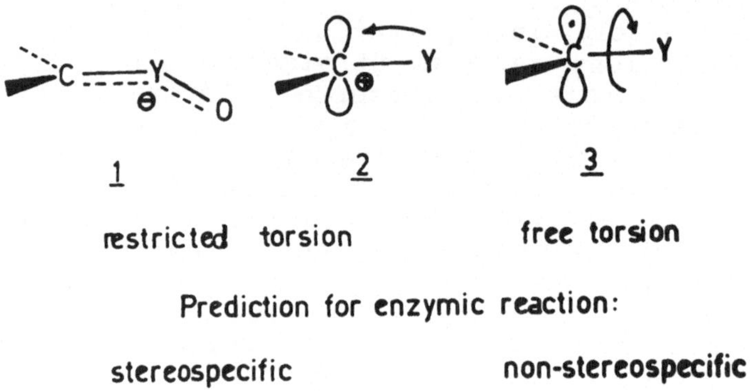

Figure 3. Torsional Freedom of Anionic, Cationic and Radical Intermediates.

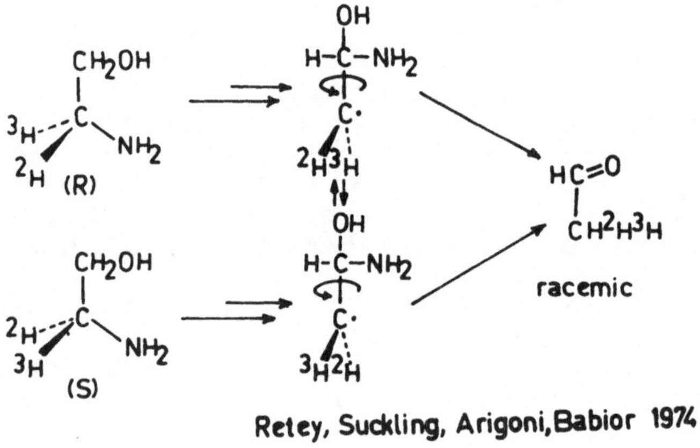

Figure 4. In the Coenzyme B_{12}-dependent Ethanolamine-Ammonia Lyase Reaction the Substitution of the Amino Groups by the Migrating H-atom Occurs with Racemization.

to recognize them.

ETHANOLAMINE-AMMONIA LYASE

An example for such a case was found recently in a coenzyme B_{12} dependent rearrangement [9]. In the conversion catalyzed by ethanolamine-ammonia-lyase the enantiomeric ethanolamines labelled stereospecifically both with deuterium and tritium yielded acetaldehyde with racemic CHDT-groups (Fig.4). It was postulated [9] that a radical intermediate of type 3 with free torsion around the C-C-bond is responsible for the indistinguishability of the two faces.

Further support for this assumption was provided by results obtained with the enantiomeric 2-aminopropanols as substrates [10]. Surprisingly, both react with comparable rates, though the (R)-enantiomer is somewhat less good substrate. Due to the additional methyl group the enzymic reaction was expected to take place in a stereospecific manner, since the intermediate radical exhibits now stereo-heterotopic i.e. distinguishable faces (Fig.5). This turned out to be the case but in a surprising way: While the substitution in the faster reacting (2S)-aminopropanol occurred with retention, inversion was found in the more slowly reacting (2R)-enantiomer.

Moreover in the presence of coenzyme B_{12} tritiated in the cobalt-bound methylene group acetaldehyde and ammonia was converted into tritiated 2-aminopropanol providing evidence for the reversibility of the reaction. Stereochemical analysis revealed that the product was pure (2S)-[2-^{3}H]-2-aminopropanol. When the substrates were stereospecifically deuterated in position 1 a substantial kinetic isotope effect on the reaction rate showed that the H_{Si} atom was the migrating one in both enantiomers of 2-aminopropanol.

Irrespective of whether the starting material was (2S)-, (2R)-2-aminopropanol or propionaldehyde in the presence of [5'-^{3}H]adenosylcobalamin both the product and the substrate were labelled with tritium. Further scrutiny of the recovered tritiated substrates led to another surprise! Starting from (2R)-2-aminopropanol but not from the (2S)enantiomer all labelled substrate molecules were inverted at their chiral centre. In other words, (2S)-[2-^{3}H]-2aminopropanol was formed from either enantiomer.

Such a unique set of stereochemical results puts severe limitations on possible mechanisms and on the stereochemical events (Fig.6). At the enzyme's active site both enantiomers assume a conformation in which the migrating groups (H_{Si} and NH_2) are synclinal. (Note the methyl group and the 2-H atom are interchanged in the two enantiomers.) After abstraction of the H_{Si} atom by the adenosyl radical

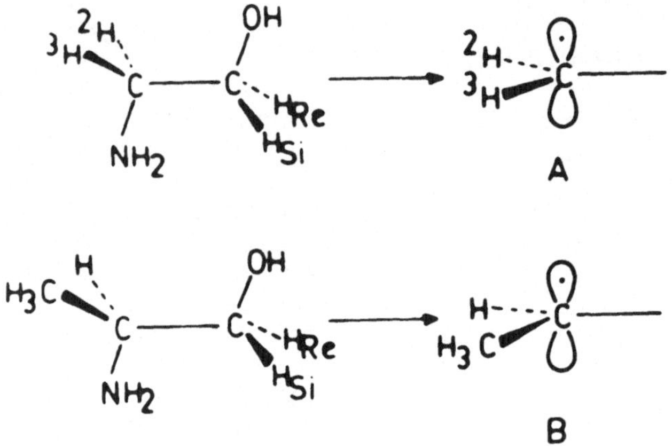

Figure 5. Stereohomotopic (A) and Stereoheterotopic (B) Radical Intermediates Derived from (2S)-2-Amino[2-^{2}H$_1$, ^{3}H]-ethanol and from (2S)-2-Aminopropanol, Respectively.

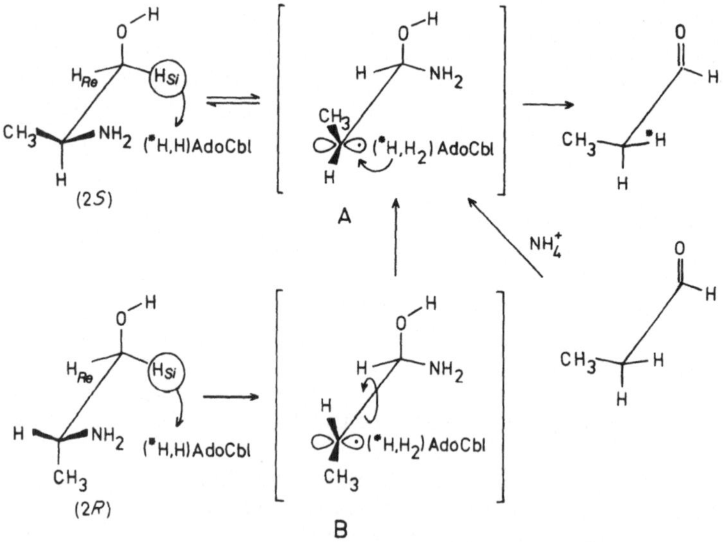

Figure 6. Stereochemical Scheme for the Catalytic Events at the Active Site of Ethanolamine-Ammonia Lyase with the Enantiomeric 2-Aminopropanols.

itself generated by homolytic cleavage of the cobaltcarbon bond - and after migration of the amino group the radical intermediates A and B are formed from (2S)- and (2R)-2-aminopropanol, respectively. Of the two rotamers A seems to be favoured by the active site topology. Moreover, B is longliving enough to allow its conversion into A. All products and also the reformed substrates will be generated from A which explains all stereochemical observations. The racemization observed in the rearrangement of stereospecifically doubly-labelled 2-aminoethanol becomes also obvious, since the two possible rotamers at the active site (corresponding to A and B) are energetically equivalent and indistinguishable.

METHYLMALONYL-CoA MUTASE

In the following I report on the stereochemical extravagancies of another coenzyme B_{12}-dependent enzyme (Fig.7). Methylmalonyl-CoA mutase catalyses a carbon skeleton rearrangement and plays a key role in the propionate metabolism both of animals and of Propionibacterium shermanii (Fig.8). While the major part of our nourishment will be degraded to acetate or acetyl-CoA, a minor but significant portion is converted into propionyl-CoA. Introduction of the latter into the citric acid cycle is only possible by a detour involving carboxylation to methylmalonyl-CoA and its isomerization to succinyl-CoA. In our context it is important to note that the biotin-dependent propionyl-CoA carboxylase produces stereospecifically the (S)-epimer of methylmalonyl-CoA whereas the (R)-epimer is the exclusive substrate for methylmalonyl-CoA mutase. Thus an additional enzyme, methylmalonyl-CoA epimerase is needed to accelerate a reaction which otherwise takes place spontaneously but slowly at neutral pH and in which the intermediacy of the enol form is revealed by the incorporation of solvent protons during epimerization. All the mechanistic work has been done with methylmalonyl-CoA mutase from P.shermanii which is the richest source for the enzyme. This bacterium releases propionic acid as a waste; so the mutase lies in its main metabolic pathway. Until recently it was not possible to obtain methylmalonyl-CoA mutase completely free of epimerase. Even with modern HPLC columns it is difficult to achieve this goal. Presumably a loose complex exists between the two enzymes.

Early stereochemical studies [11,12] revealed that the substitution concomitant with migration of the thioester group takes place with retention (Fig.7). All four methylene hydrogen atoms of the succinyl moiety of succinyl-CoA are stereochemically different and an enzyme is expected to differentiate between them. In the presence of methylmalonyl-CoA mutase three of these H-atoms will be, however,

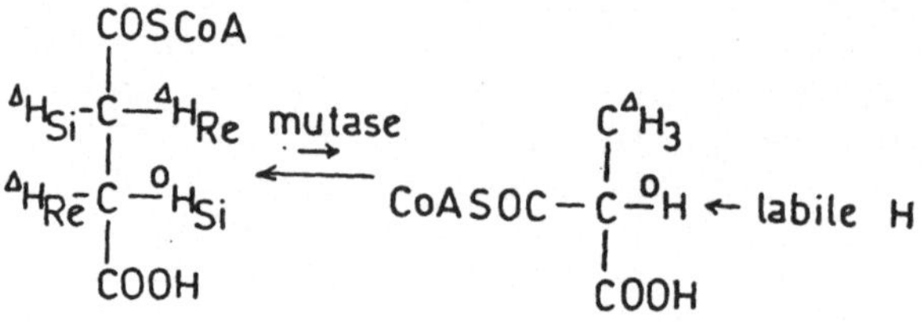

Figure 7. Rearrangement of Methylmalonyl-CoA Catalysed by the Coenzyme B_{12}-dependent Methylmalonyl-CoA Mutase. Specification of Migrating (Δ) and Exchangeable (o) H-atoms.

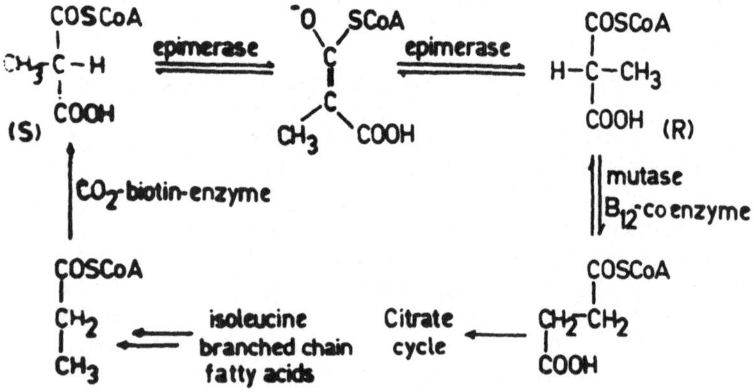

Figure 8. Propionate Metabolism in Higher Organisms.

scrambled, since they become attached to the gyrosymmetric methyl-C-atom of methylmalonyl-CoA. No scrambling is expected, however, between these three H-atoms and the fourth one (marked with a circle in Fig.7) provided that the mutase is strictly stereospecific. Only the latter C-atom will occupy the α-position of methylmalonyl-CoA and becomes exchangeable with solvent protons in the presence of epimerase.

Experiments with the homologous substrate, ethylmalonyl-CoA, indicated for the first time that methylmalonyl-CoA mutase might loose stereochemical control over its substrate at some stage of the reaction. In comparison with the natural substrate the situation in ethylmalonyl-CoA in different insomuch that all three H-atoms attached to the migration centres are distinct (Fig.9). With stereospecifically deuterated substrates it was shown that the substitutions take place mainly with retention at both centres and both in ethylmalonyl-CoA and in methyl-succinyl-CoA migration of H_{Re} is preferred [13]. It was, however, simultaneously revealed, both with deuterium and with tritium as a label, that some H_{Si} migration also occurs in both directions and this "error" is greatly amplified when H_{Re} is substituted by a heavy isotope. In such a situation the heavy isotope will be manoeuvred into the α-position of ethylmalonyl-CoA (see Fig.9) and washed out into the medium by the epimerase. Up to 36% "washing out" of migrating deuterium and about 80% loss of migrating tritium (i.e. from the H_{Re} position) was observed [13].

The significance of these findings may be impaired by the fact that ethylmalonyl-CoA is a reluctant substrate of the mutase, its conversion being some 1000 times slower than that of the natural substrate. We prepared therefore a sample of methylmalonyl-CoA carrying a fully deuterated methyl group [14,15]. In its ^{1}H-NMR spectrum at 500 MHz all methylene protons of the CoA portion appeared as separate triplets and were assigned by double resonance experiments.

In Fig.10 the ^{1}H-NMR spectrum of a reaction mixture is shown in which the conversion of ($^2H_3$)methylmalonyl-CoA to the corresponding deuterated succinyl-CoA was just started with a catalytic amount of coenzyme B_{12}, methylmalonyl-CoA mutase and epimerase. The appearance of a small amount of succinyl-CoA is revealed by two small triplets observed slightly upfield from the large triplets b_M and a_M and marked as b_S and a_S. The triplet a_M and a_S arises from the sulphur-attached methylene group of the CoA portion in methylmalonyl- and succinyl-CoA, respectively, and is ideally suited for monitoring the enzymic conversion. The small broad singlet at 2.6 ppm can be assigned to the single proton of the newly formed trideutero-succinyl-CoA. From the enzymic reaction mixture samples were with-

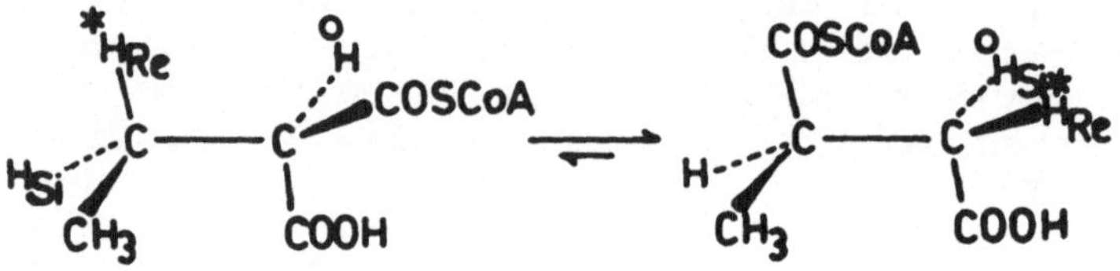

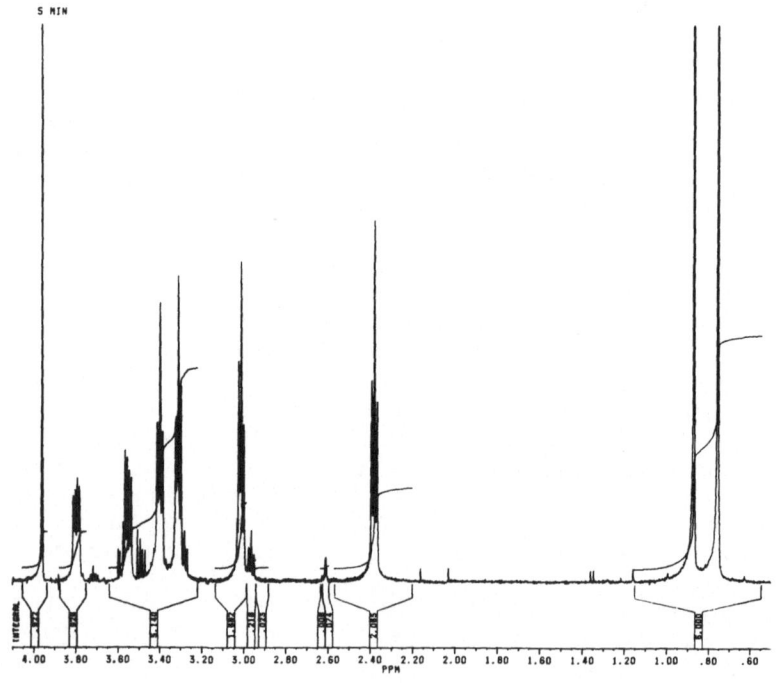

Figure 9.   Enzymic Rearrangement of the Substrate Analogue
Ethylmalonyl-CoA. Specification and Possible Fate of the
H-atoms.

Figure 10.   Partial ^{1}H-NMR Spectrum of (^{2}H$_3$)Methylmalonyl-
CoA 5 min After Starting the Mutase Reaction (see text for
discussion).

drawn at 10 min intervals in which the reaction was terminated by acid. Removal of the denatured protein by centrifugation was followed by lyophilization. Fig.11 shows the
partial ^{1}H-NMR spectra of the samples in deuterium oxide.
A steady increase of the triplet a_S with concomitant decrease of the triplet a_M is a measure for the proceeding
of the reaction. After 60 min equilibrium (methylmalonyl-
CoA/succinyl-CoA = 5/95) is established. The last spectrum
shows downfield from the broadened singlet (the single
proton of the trideutero succinyl-moiety) a narrower singlet
that corresponds to the unlabelled methylene group of
a geminally dideuterated succinyl-CoA. Other minor signals
arise from hydrolytic products of succinyl-CoA. These
NMR spectra reveal that deuterium from migrating positions
of the substrates is lost to the medium even before reaching
equilibrium. In a "mirror image-like" experiment, in which
unlabelled methylmalonyl-CoA was reacted in deuterated
solvent, deuterium "washing-in" into migrating positions
of the substrates was also observed, although this process
was about five times slower (isotope effect!).

Two problems were encountered when interpreting the
results of these experiments. First, an enzyme-catalysed
or spontaneous migration of the CoA-moiety from one carboxyl
of succinate to the other could have been responsible
for the scrambling of "migrating" (Δ) and "exchangeable"
(o) substrate hydrogens. Second, the interpretation of
the ^{1}H-NMR spectra was complicated by a slow spontaneous
hydrolysis of succinyl-CoA under the conditions of the
experiment, which limited the time allowed for monitoring
the scrambling/exchange process and reduced the yield
of products.

In order to exclude the first possibility $(1-^{13}C)$me-
thylmalonyl-CoA was prepared and converted on the mutase
into $(1-^{13}C)$succinyl-CoA. As the ^{13}C-NMR spectrum showed
(Fig.12) little if any $(4-^{13}C)$succinyl-CoA was formed,
although about 20% of the substrate was hydrolysed to
free succinate. To circumvent both the hydrolysis and
the CoA-migration problem the use of an analogue substrate
was ideally suited.

Recently we discovered [16] that methylmalonyl-dethia-
(carba)-CoA is an excellent substrate for methylmalonyl-
CoA mutase. The enzyme seems to tolerate the substitution
of the sulphur atom for a methylene group. Indeed, when
this chemo-enzymically synthesised substrate was converted
on the mutase into the corresponding stable succinyl-
CH_2CoA, its fate could be observed for a much longer period.

Fig.13 shows the ^{1}H-NMR spectrum of $(2-^2H)$methylmalo-
nyl-CH_2CoA in deuterated buffer. The sample is contaminated
by 20% propionyl-CH_2CoA and in addition catalytic amounts
of methylmalonyl-CoA mutase and epimerase. While the concentration of the latter is too low to contribute NMR signals,

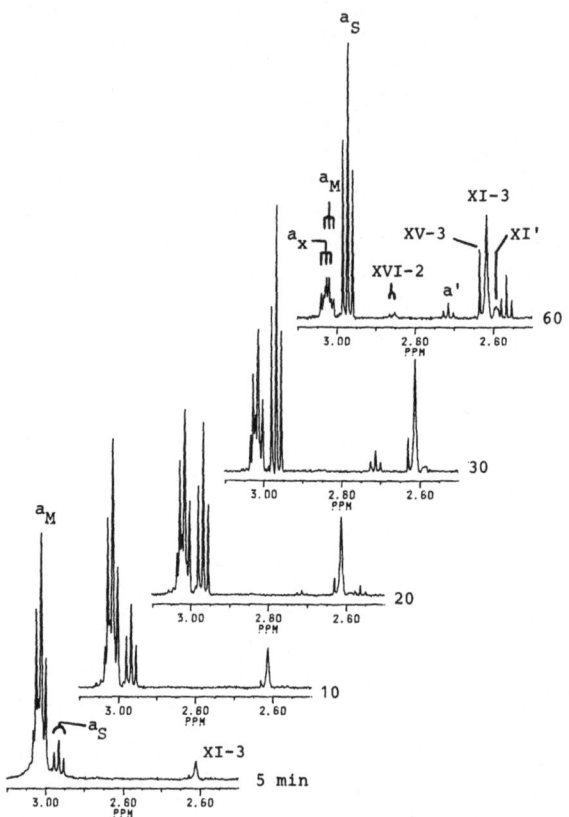

Figure 11. ^{1}H-NMR Monitoring of the Enzyme-catalysed Rearrangement of (^{2}H$_3$)Methylmalonyl-CoA into (2-^{2}H$_2$, 3-^{2}H$_1$)-Succinyl-CoA in Unlabelled Solvent.

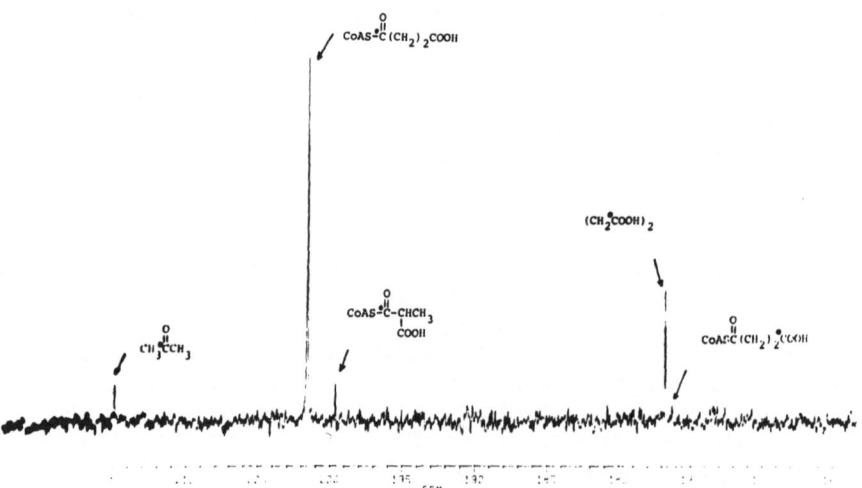

Figure 12. ^{13}C-NMR Spectrum of the Reaction Mixture Resulting from the Conversion of (1-^{13}C)Methylmalonyl-CoA.

385

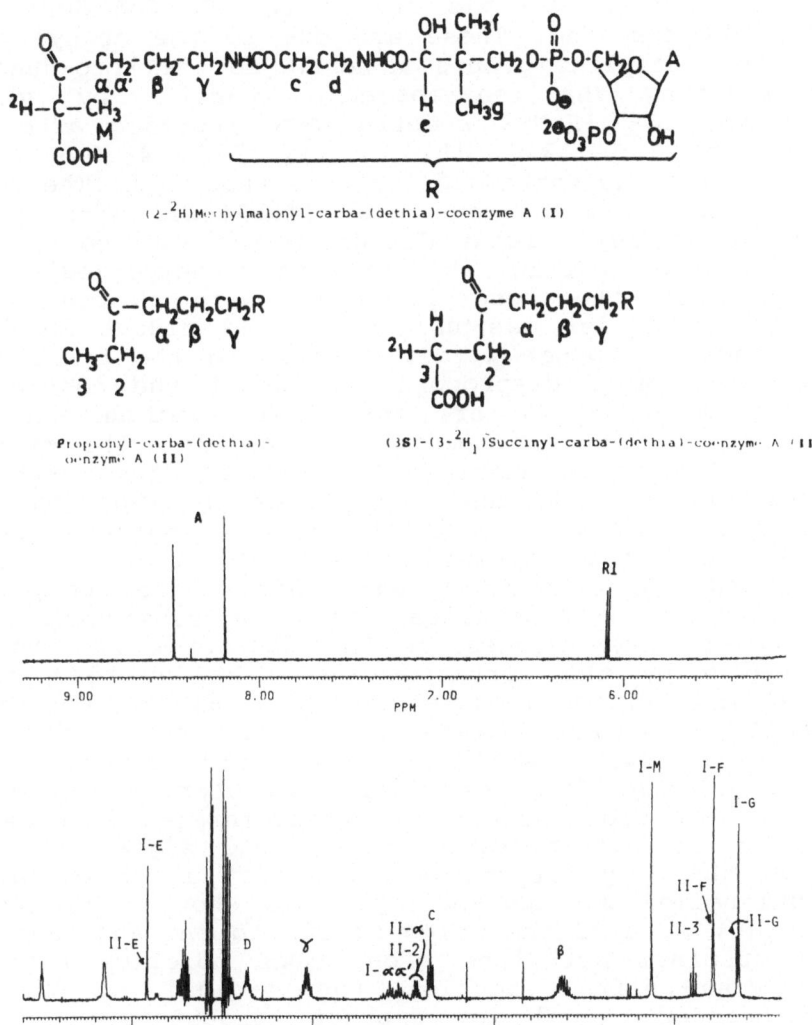

Figure 13. Structure and ^{1}H-NMR Spectrum of $(2-^2H_1)$Methyl-malonyl-CH$_2$CoA (I) and Propional-CH$_2$CoA (II).

the most important signals of the two CH_2-CoA derivatives are clearly resolved. These are due to the methyl groups and to the methylene groups adjacent to the keto-function. After starting the rearrangement reaction by addition of coenzyme B_{12} ^{1}H-NMR spectra were recorded after time intervals of 8.5, 13.8, 19.0, 28.1, 37.5, 48.6 etc. min. The monitoring was continued for more than 17 h. The equilibrium of the rearrangement was established after 28 min and little if any solvent deuterium was incorporated into the substrates during this period. Unexpectedly, only about 60% of the succinyl-CH_2CoA were monodeuterated, while the remainder was unlabelled and dideuterated, ca. 20% of each. In other words, already in the first stages of the reaction a disproportionation of the nonmigrating deuterium occurred! We are forced to conclude that both partial loss of stereochemical control over the migrating hydrogen atom and a multiple H-transfer between substrate and coenzyme are important features of the reaction mechanism. On further incubation the reverse reaction becomes important and the combined effect of imperfect stereocontrol and epimerase-catalysed exchange leads to progressive deuterium incorporation into the substrate. Fig.14 shows three later stages of the deuterium incorporation as revealed by the corresponding ^{1}H-NMR signals of the succinyl moiety. The doublet around 2.65 ppm due to the originally formed monodeuterated species becomes increasingly biased by superposition of the doubly isotope-shifted singlet arising from geminally dideuterated molecules. The vicinally dideuterated and trideuterated species give even more upfield-shifted signals. The signal at lowest field is the only detectable portion of a triplet arising from unlabelled succinyl-moiety. This shows coupling with the triplet at 2.32 ppm which is due to the adjacent unlabelled succinyl methylene group. One observes here also the isotope-shifted doublet for vicinal dideuterated, and a broader signal for another trideuterated species. It can be calculated that after 17 h about 10% tetradeuterated succinyl species must be present. Interestingly they coexist with about 6% unlabelled, 21% monodeuterated, 45% dideuterated and 17% trideuterated species.

A further mechanistically important result was provided by the surprising fact that the contaminating propionyl-CH_2CoA also incorporated deuterium. Fig.15 shows that this incorporation started late and proceeded slowly. At about 2.43 ppm the two methylene groups adjacent to the keto-function give rise to a quartet and a triplet, the former gradually decreasing. Interestingly, the diminished quartet shows no isotope shift which means that both hydrogens of this methylene group are exchanged at once, i.e. little is seen of a monodeuterated intermediate. Dideuteration in the methylene group produces an isotope-

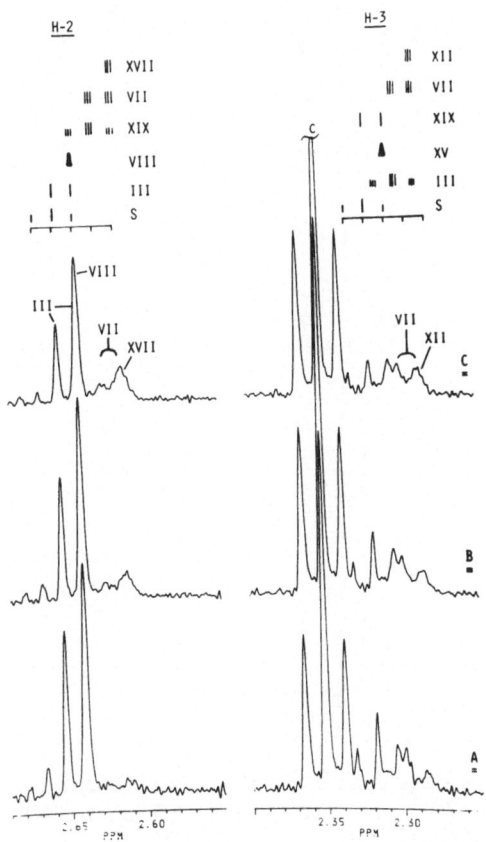

Figure 14. Expansion of the H-2 and H-3 ^{1}H-NMR Signals of the Succinyl-CH_2CoA Species After 1 h 23 min (A), 3 h 53 min (B) and 17 h 23 min (C). Below: Structure and Numbering of the Possible Deuterated Species.

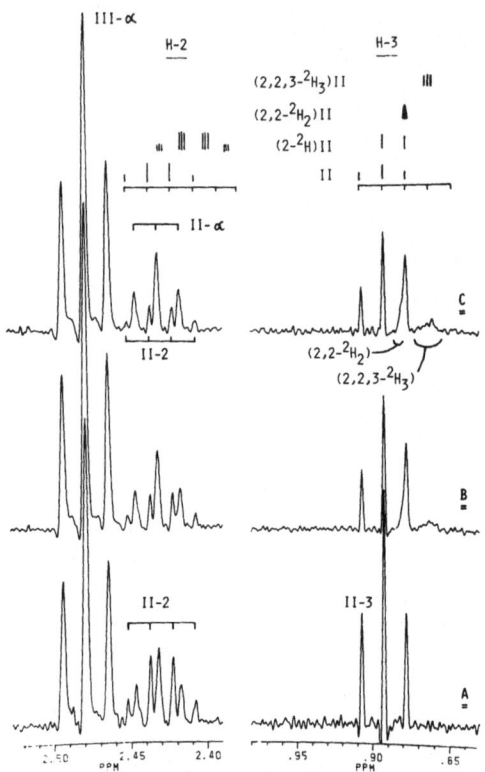

Figure 15. Expansion of the Methyl and Methylene Signals of Propionyl-CH$_2$CoA After 1 h 23 min (A), 12 h 53 min (B) and 17 h 23 min (C) Reaction Time.

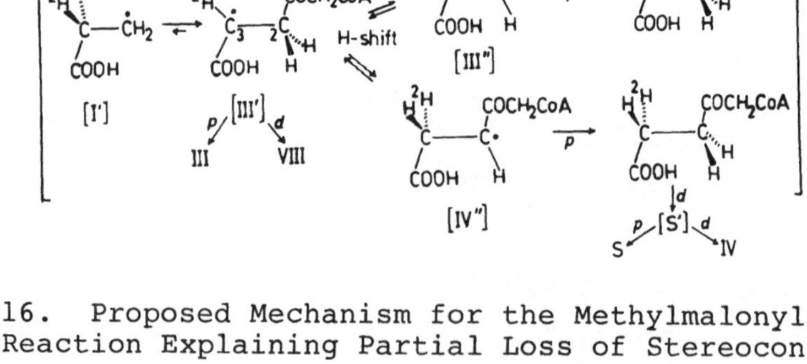

Figure 16. Proposed Mechanism for the Methylmalonyl-CoA Mutase Reaction Explaining Partial Loss of Stereocontrol.

shifted singlet for the methyl group which coincides with the upfield signal of the methyl triplet for unlabelled molecules. At a very late stage deuterium incorporation into the methyl group also occurs as revealed by a signal that are even more isotope-shifted. Such an incorporation occurs only in molecules with fully deuterated methylene group.

It can be concluded that propionyl-CH$_2$CoA binds also to the active site. Since its methyl group should occupy the same site as the methyl group of methylmalonyl-CoA, it is plausible to assume that hydrogen abstraction occurs first from the methyl group, but deuterium backtransfer from the 5'-deoxyadenosine takes place to the adjacent position. This requires a preceding intramolecular 1,2-H-shift in the intermediate radical. A repetition of these reaction series during the same sojourn at the active site, i.e. H-abstraction from methyl, intramolecular 1,2-H-shift and deuterium backtransfer, would explain the observed incorporation pattern. Deuteration in propionyl methyl groups could only take place after a second encounter of previously deuterated substrate and enzyme.

The assumption of an 1,2-H-shift also in the succinyl radical intermediate could explain both the partial loss of stereocontrol over migrating and nonmigrating hydrogen atoms and the previously observed [11,12] high optical purity of the produced monodeuterated succinyl-CoA species. Details of the proposed mechanism are shown in Fig.16. An 1,2-H-shift in the initially formed succinyl radical **III'** would lead to the monodeuterated α -radicals **III''** and **IV''**. In the latter the previously "nonmigrating" deuterium has been manoeuvred into the migrating position and can be abstracted by the adenosine radical. Further stereospecific abstractions and redonations will lead to the observed deuterated species. Thus the intramolecular 1,2-H-shift is identified as the step in which the enzyme loses stereocontrol over its substrate.

ACKNOWLEDGMENTS

I thank my younger colleagues whose names are listed in the references for their important contributions. Financial support from the Deutsche Forschungsgemeinschaft and from the Fonds der Chemischen Industrie is gratefully acknowledged.

390

REFERENCES

[1]   For a review see: R.H. Abeles & D. Dolphin, Accounts.
       Chem.Res. **9**, 114-120 (1976).
[2]   B.M. Sjöberg, P. Reichard, A. Gräslund & A. Ehrenberg,
       J.Biol.Chem. **253**, 6863-6865 (1978).
[3]   R.J. Debus, B.A. Barry, G.T. Babcock & L. McIntosh,
       Proc.Natl.Acad.Sci. USA **85**, 427-430 (1988).
[4]   L. Mascaro,Jr., R. Hörhammer, S. Eisenstein, L.K. Sel-
       lers, K. Mascaro & H.G. Floss, J.Am.Chem.Soc. **99**, 273-
       274 (1977).
[5]   D. Arigoni, cited in H.G. Floss & M.D. Tsai, Adv.En-
       zymol. Vol.**50**, 243-302 (1979).
[6]   For a Review see: P.A. Frey in New Comprehensive Bio-
       chemistry (A. Neuberger and L.L.M. van Deenen, Series
       eds.) Vol.3 Stereochemistry (Ch. Tamm ed.) Elsevier,
       Amsterdam, 202-248 (1982).
[7]   J.F. Biellmann & F. Schuber, Biochem.Biophys.Res.Com-
       mun. **27**, 517-522 (1967)
[8]   C.M. Tatum, P.A. Benkovic, S.J. Benkovic, R. Potts,
       E. Schleicher & H.G. Floss, Biochemistry **16**, 1093-1102
       (1977).
[9]   J. Rétey, C.J. Suckling, D. Arigoni & B.M. Babior,
       J.Biol.Chem. **249**, 6359-6360 (1974)
[10]   P. Diziol, H. Haas, J. Rétey, S.W. Graves & B.M. Ba-
       bior, Eur.J.Biochem. **106**, 211-224 (1980).
[11]   M. Sprecher, M.Y. Clark & D.B. Sprinson, Biochem.Bio-
       phys.Res.Commun. **15**, 581-584 (1964); J.Biol.Chem. **241**,
       872-877 (1966).
[12]   J. Rétey, cited by D. Arigoni & E.L. Eliel in Topics
       in stereochemistry (Eliel & Allinger, eds) Vol.**4**, 127-
       243, J. Wiley & Sons (New York) (1969).
[13]   J. Rétey, E.H. Smith & B. Zagalak, Eur.J.Biochem. **83**,
       437-451 (1978).
[14]   K. Wölfle, M. Michenfelder, A. König, W.E. Hull &
       J. Rétey, Eur.J.Biochem. **156**, 545-554 (1986).
[15]   M. Michenfelder, W.E. Hull & J. Rétey, Eur.J.Biochem.
       168, 659-667 (1987).
[16]   M. Michenfelder & J. Rétey, Angew.Chem. **98**, 337-338;
       Angew.Chem.Internat.Ed.in Engl. **25**, 366-367 (1986)
[17]   W.E. Hull, M. Michenfelder & J. Rétey, Eur.J.Biochem.
       173, 191-201 (1988)

KINETICS OF PEROXYL RADICAL REACTIONS IN MODEL BIOMEMBRANES

L.R.C. Barclay
Department of Chemistry
Mount Allison University
Sackville, New Brunswick
Canada, EOA 3CO

KEYWORDS/ABSTRACT : lipid peroxidation/kinetics mechanism/ketone photosensitized/micelles/phospholipids/membranes/bilayers/rate constants/phenolic antioxidants.

Benzophenone photosensitized oxidation of linoleic acid in SDS micelles involves free radical autoxidation from effects of : antioxidants, singlet oxygen quenchers, and product studies. Kinetic studies indicate the classical rate law $-d[O_2]/dt = k_p/2k_t^{\frac{1}{2}}[R-H]R_i^{\frac{1}{2}}$ applies, where k_p and $2k_t$ are propagation and termination rate constants, RH the substrate, R_i the rate of initiation. This rate law also applies to the autoxidation of mixed phospholipid bilayers. The rate constants, k_p and $2k_t$ and the activity of phenolic antioxidants, are compared in solvents, micelles, and bilayers. Interpretations involve diffusion to and solvation of polar peroxyl radicals in the polar aqueous phase of micelles and bilayers.

1. INTRODUCTION

The autoxidation of organic substrates in bulk homogeneous solution is well known(1-3) to proceed by a free radical chain reaction as indicated in equations [1]-[4]

[1] Initiation: Formation of R• (or ROO•) rate = R_i

[2] Propagation: R• + O_2 $\xrightarrow{\text{fast}}$ ROO•

[3] ROO• + R-H(substrate) $\xrightarrow{k_p}$ ROOH + R•

[4] Termination: ROO• + ROO• $\xrightarrow{2k_t}$ Products + O_2

F. Minisci (ed.), Free Radicals in Synthesis and Biology, 391–406.
© 1989 by Kluwer Academic Publishers.

For systems in which this mechanism applies, the rate of oxygen uptake is first order in concentration of substrate, RH, and one-half order in the rate of chain initiation, Ri, so that equation [5] applies:

$$[5] \quad -d[O_2]/dt = k_p/2k_t^{\frac{1}{2}} [R-H]R_i^{\frac{1}{2}}$$

In quantitative kinetic studies, the rate of free radical chain initiation the R_i, must be known and controlled. For such reactions in solution, this is usually done by using an azo initiator which decomposes at a known rate to form carbon radicals which react rapidly with molecular oxygen to form peroxyl radicals, ROO•. Only those peroxyls which "escape" the solvent cage in which they are formed can initiate chain reactions. Therefore the R_i must be measured; for example, by addition of a phenolic inhibitor known to trap two peroxyls, and by measuring the induction period during which oxidation is inhibited. During this period, equations [6]-[8] apply:

$$[6] \quad ROO\bullet + ArOH \xrightarrow{k_{inh}} ROOH + ArO\bullet$$

$$[7] \quad ROO\bullet + ArO\bullet \xrightarrow{fast} ROOArO$$

[8] $R_i = 2 \times [ArOH]/\tau$, where $\tau =$ inhibition period, and the suppressed oxygen uptake is given by equation [9] :

$$[9] \quad -\frac{d[O_2]}{dt} = \frac{k_p}{k_{inh}} \frac{[RH]R_i}{2[ArOH]}$$

Much useful information has been provided on organic substrates from quantitative kinetic studies; for example, the susceptibility of a substrate to under autoxidation, referred to as its oxidizability, is given by equation [10] :

$$[10] \quad \frac{k_p}{2k_t^{\frac{1}{2}}} = \frac{-d[O_2]/dt}{[RH]R_i^{\frac{1}{2}}}$$

There is now increased interest in peroxidation in particular of biologically sensitive systems, such as polyunsaturated fatty acids (PUFA), because of its relationship to important pathological events related to membrane damage, such as heart disease, cancer, and aging (4-7), and the action of foreign toxic substances such as cigarette smoke (8). Much of the attention into quantitative studies of autoxidation by several groups of researchers has concentrated on: (1) the mechanism and activity of antioxidants, reviewed in references(9-10), (2) the dynamics of free radical initiation(11-12), and (3) the mechanism of lipid peroxidation from trends in hydroperoxide formation(13).

We are currently conducting quantitative kinetic studies in bilayers and micelles(14-17) as model systems expected to mimic free radical reactions in biomembranes. Several basic questions are addressed:

(1) Is the classical kinetic rate law, equation [5], applicable to bilayers and micelles?

(2) What is the mechanism of ketone-photosensitized oxidation of linoleate chains in micelles and phosphatidylcholine (PC) bilayers?

(3) What is the effect of micellar or bilayer aggregation on (a) the products of peroxidation and (b) the oxidizability of PUFA, equation [10]; and what is the effect of peroxidation on bilayer structure and function?

(4) Can absolute rate constants for peroxyl radical, ROO$\cdot$, propagation, k_p, termination, $2k_t$, and inhibition, k_{inh} with phenolic anti-toxidants be determined in bilayers and micelles?

2. MATERIALS AND METHODS

The materials and methods for preparing micellar solutions of linoleic acid, the required antioxidants, alpha-tocopherol, 2,5,7,8-tetramethyl-6-hydroxychroman(PMHC), the quencher, 1,3-diphenylisobenzofuran(DPBF), and the initiator, benzophenone, in 0.50 M sodium dodecyl sulfate(SDS) are as described(17). The azo initiators, azobis-2, 4-dimethylvalero-nitrile(ADVN), and azobis (2-amidinopropane. HCℓ(ABAP) were obtained from Polysciences.

Commercial phosphatidylcholines(PCs), dipalmitoyl PC(DPPC), dimyristoyl PC(DMPC), and dilinoleoyl PC(DLPC) were obtained from Avanti Polar Lipids. 1-Palmitoyl-2-linoleoyl PC(PLPC) was synthesized by known methods(18). DLPC and PLPC were freed of hydroperoxides by reverse phase HPLC before use(17). Mixtures of PC bilayers (DPPC+DLPC) and (DMPC+DLPC) for kinetic studies were prepared by evaporation to films,of mixtures in chloroform by keeping the temperature above the phase transition, as described(17).

Kinetic studies on micellar and bilayer systems used phosphate buffer (pH 7.0), containing 1×10^{-4} M Na$_2$EDTA, which was passed through Chelex 100 to remove traces of heavy metal ions. A custom-made dual-channel pressure transducer system constructed of stainless steel was used for the oxygen uptake studies. One channel was connected to a vessel containing only the buffer used, the other channel was connected to a calibrated reaction vessel containing the sample in buffer. This apparatus automatically corrects for vapour pressure due to the solvent and eliminates "baseline drift". The rotating sector method described earlier(17) was used to measure absolute rate constants. [31]P nmr spectra were measured on 0.1 m mol bilayer dispersions in THAM buffer (pH7.0) on a Nicolet NB FT NMR spectrometer at 146.15 M Hz using two-level broadband decoupling with a 90° pulse of 25 micro s and decoupling power of 10W for data acquisition and 2W during magnetization recovery time, and a delay time of 4 sec between pulses. Peroxidations for ^{31}P NMR studies were also conducted in THAM buffer. Trolox was used to terminate the reaction and the ^{31}P spectra were measured immediately on these samples.

The hydroperoxide oxidation products from linoleic acid in SDS micelles were reduced by Ph_3P, methylated by CH_2N_2 and analyzed by HPLC. The bilayer oxidation products were isolated by Ph_3P reduction followed by transesterification with methanol by known methods(17). The resulting products, the four 9- and 13-hydroxy substituted 9-cis, 11-trans; 9-trans, 11-trans and 10-trans; 12-cis; 10-trans, 12-trans octadecadienoates were identified by cross HPLC comparison with a known reference mixture analyzed in the laboratory of N.A. Porter(19).

3. RESULTS

3.1 Ketone-Photosensitized Oxidation of Linoleate in Micelles Mechanism and Kinetics

The photooxidation of biological systems known as "photodynamic action" (4,20), results in a number of known damaging effects including membrane and protein damage, mutagenesis, and skin cancer. There is great interest in the possible involvement of singlet oxygen in such processes (17,20,21). The many reports of singlet oxygen in micelles and bilayers (see list in Ref.17) usually involve a dye or porphyrin sensitizer and they concentrated on the detection of singlet oxygen in such systems, whereas we are most interested in the fate of an added oxidizable substrate.

Our results from benzophenone-sensitized oxidation of linoleic acid in SDS micelles(17) are entirely consistent with a mechanism involving free radical chain autoxidation (Type I) rather than singlet oxygen reactions (Type II). Known singlet oxygen quenchers, such as sodium azide in the aqueous phase or diphenylisobenzofuran and beta carotene in the micelle phase had no effect on the oxygen uptake. On the other hand phenolic antioxidants including alpha-tocopherol, PMHC and Trolox all acted as antioxidants in this reaction. It is interesting to note that these three antioxidants exhibited different antioxidant activity as shown in different rates of oxygen uptake during the induction periods (see Figure 1). This is the basis for a quantitative comparison of their antioxidant action (vide infra)

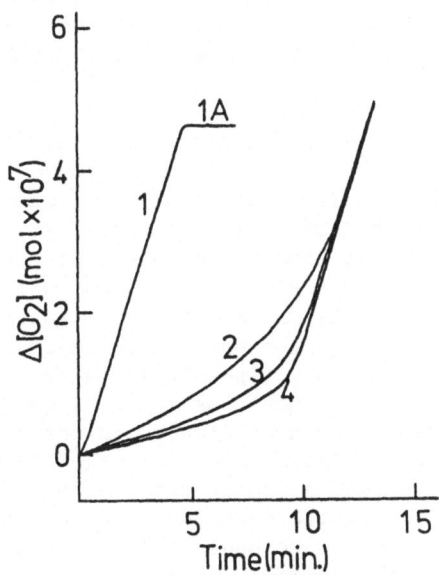

Fig.1.   Autoxidation of linoleic acid (6.43×10^{-5}mol) in 2.00 mL of
0.50 M SDS(pH 7.0) at 30°C, photoinitiated by benzophenone
(2.82×10^{-6}mol): 1,uninhibited reaction (A:lamp off); 2,
inhibited with α-tocopherol (1.67×10^{-8}mol); 3, inhibited with
Trolox (1.66×10^{-8}mol); 4, inhibited with PMHC (1.62×10^{-8}mol).

Alpha-tocopherol and Trolox did not act as oxidation inhibitors
for irradiated micellar solutions of linoleic acid sensitized by
methylene blue.   In contrast to the ketone-sensitized oxidations, these
dye-sensitized ones exhibited a rapid "upsurge" of oxygen uptake on
injection of Trolox, most probably due to rapid attack by singlet
oxygen on the phenolic group(22,23).
    The products from our benzophenone-sensitized oxidations of
linoleic acid in SDS micelles are the known conjugated diene 9- and 13-
hydroperoxides with the cis, trans and trans,trans configurations.   In
addition we find (17) a linear dependence of the cis, trans to trans,
trans ratios with linoleic acid concentration in the micelles similar to
that found for thermal initiation(17) and consistent with the mechanism
of lipid peroxidation proposed by Porter(13).   We also find that the
benzophenone is not consumed in our oxidations.   The result of recycling
the benzophenone provides an important advantage over the usual thermal
initiations for careful quantitative studies;   namely, the rate of free
radical chain initiation remains constant (the initiator is not used up)
and there is no correction to make for nitrogen evolution, which can be
large in bilayers containing high concentrations of thermal initiator(14).
    Quantitative kinetic studies of autoxidation are readily carried
out on benzophenone photoinitiated reactions in micelles.   For example,
we found(17) that the kinetic order in oxygen is unity in linoleic acid
concentration for various light intensities.   In addition this kinetic
order is one-half in the rate of chain initiation (R_i), measured

by varying the light intensity, up to a micellar linoleate concentration of 0.459 M. These results mean that the classical rate law for autoxidation equation [5], is applicable to this system.

3.2. Thermal and Photochemical Peroxidations in Phospholipid Bilayers

We showed recently that the classical rate law, equation [5], also applies to the kinetics of autoxidation in bilayers(17). The kinetic order in oxidizable substrate was found to be <u>unity</u> by conducting oxidation studies on mixed bilayers containing known, variable amounts of an unsaturated PC, DLPC, mixed with a saturated one, DPPC or DMPC as "solvent". Furthermore the same linear and unity kinetic order in [R-H] was found for both a water-soluble (ABAP) and lipid soluble initiator (ADVN), and the kinetic order in R_i was one-half for both types of initiators(17). Product studies on the <u>cis</u>, <u>trans</u> to <u>trans</u>, <u>trans</u> ratios of hydroperoxides formed from the linoleate chains in mixed DLPC+DPPC bilayers showed a linear trend with oxidizable lipid concentration, [DLPC], with the water-soluble initiator, ABAP.

We are currently extending the ketone-sensitized photoinitiation to the kinetics of bilayer peroxidation. Results using benzophenone or a water-soluble derivative, 4-sulphomethylbenzophenone, sodium salt (BP$^-$), indicate oxidation via the same radical chain mechanism of autoxidation in DLPC bilayers as observed for linoleate in micelles. The oxygen uptake for multilamellar DLPC bilayers is one-half order in light intensity showing that the rate varies with $R_i^{1/2}$. Phenolic antioxidants (e.g. Trolox and alpha-tocopherol), act as effective inhibitors consistent with a radical chain mechanism.

3.3 Absolute Rate Constants in Micelles and Bilayers

Our ketone-sensitized initiation has several significant advantages over the usual thermal azo initiation for quantitative kinetic studies. There is no "dark rate" with ketone-photoinitiation, whereas such thermal dark rates can cause large errors in determinations of rate constants with azo initators. Also there are no corrections to make for evolution of nitrogen which can require large corrections in micro-environments such as bilayers(14). In addition the recycling of benzophenone provides a constant R_i.

We used benzophenone-sensitized oxidation and the rotating sector method as modified by Howard and Ingold(24) to determine the absolute rate constants for propagation, k_p, and termination, $2k_t$, for autoxidation of linoleic acid in homogeneous solution and in SDS micelles(17), and recently for oxidation of DLPC bilayers. This method involves measurement of the lifetimes of the reaction chain (τ) given by equation [11] (25).

[11] $\tau = (2k_t R_i)^{-\frac{1}{2}}$

The reaction is photochemically initiated and the rate of oxygen consumption is measured at various rotation speeds of a sectored disc placed between the light source and the reaction flask. The ratio of the reaction rate at intermediate sector speeds, $R\lambda$, to the rate at full light depends on the duration of the light flash, λ, and the lifetime of the reaction chain, equation [12].

$$[12] \quad R_\lambda/R_{steady} = \lambda/\tau$$

From known calibration curves for this dependence(25), τ is evaluated, the R_i is measured using a phenolic inhibitor, and the $2k_t$ readily calculated from equation [11]. The k_p can then be calculated by inserting known values into equation [5]. Actually this assumes the ideal situation of bimolecular chain termination where the rate is exactly proportional to the square root of the light intensity. In most cases our calculations were refined by corrections for some "first order termination"(17,24) which often accompanies the usual second order process. Some typical results from these measurements are summarized in Table I.

TABLE I Absolute Rate Constants for Autoxidation of Linoleic Acid in Solution, in Micelles, and in Phospholipid Bilayers[a] Photoiniated by Benzophenone.

Medium	$2k_t$[b] $M^{-1}s^{-1} \times 10^{-6}$	k_p[b] $M^{-1}s^{-1}$
$(CH_3)_3COH$	17.2±2.7	81.5±7.8
CH_3CN	4.16±0.30	94.0±3.1
SDS micelles (0.5M)	0.352±0.090	36.2±4.8
DLPC bilayers	0.132±0.016	36.1±1.2

[a]Taken in part from reference(17). The same values for $2k_t$ and k_p, within experimental error, were found in $(CH_3)_3COH$ and CH_3CN using an azo initiator.

[b]Values are corrected for the proportion of first order termination (17,24).

3.4 Antioxidant Activities of Phenolic Inhibitors in Micelles and Bilayers

The determination of antioxidant activity involves measurement of the rate of oxygen uptake during inhibition by the antioxidant at a known rate of chain initiation, R_i. Since the k_p is now known in our systems, the various quantities needed to calculate inhibition rate constants,

k_{inh}, from equation [9] are known. For calculations this equation is converted into the integrated forms, equations [11] and [12], (26).

$$[11] \quad \Delta[O_2]_t = - \frac{k_p}{k_{inh}} \times [RH] \, \ln(1 - \frac{R_i \, t}{2[ArOH]_o})$$

$$[12] \quad \Delta[O_2]_t = - \frac{k_p}{k_{inh}} \times [RH] \, \ln(1-t/T)$$

Linear slopes are obtained from plots of $\Delta[O_2]$ versus $-\ln(1-t/T)$ for each of the inhibitors used in solution and in micelles and for Trolox in bilayers, as illustrated in Figure 2 for inhibition periods. The k_{inh} values obtained from the slopes, and using the k_p values given in TABLE I, are given in TABLE II.

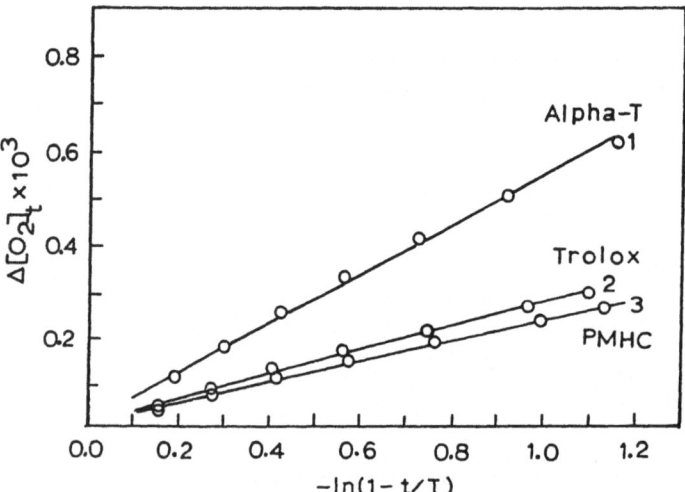

Figure 2. Plots of oxygen uptake versus $-\ln(1-t/T)$. For autoxidation of linoleic acid (6.43×10^{-5}m) in 0.50 M SDS, photoinitiated by Ar_2CO (2.82×10^{-6}m), for 1, α-tocopherol; 2, Trolox, and 3, PMHC. The slope = $k_p[L\overline{H}]/k_{inh}$.

TABLE II   Inhibition Rate Constants for Phenolic Antioxidants in
Solution, in SDS Micelles and in DLPC Bilayers

Medium	Antioxidant	$k_{inh}M^{-1}s^{-1} \times 10^{-4}$ [a]
$(CH_3)_3COH$	α-tocopherol	23 ± 1.2
	PMHC	21 ± 8.0
	Trolox	15 ± 2.0
SDS micelles	PMHC	4 ± 0.09
	Trolox	3.41 ± 0.07
	α-tocopherol	1.66 ± 0.03
DLPC bilayers	Trolox	0.30 ± 0.01
	α-tocopherol	0.245 ± 0.002

[a] Values obtained at 30°C, except DLPC bilayers at 37°C.

4. DISCUSSION

These basic kinetic studies demonstrate that the classical rate law
represented by equation [5] is applicable to the autoxidation of an
oxidizable substrate (e.g. linoleate) sequestered in micelles and to
the linoleate chain when contained bonded within mixed bilayers.   This
conclusion is supported by the data which show a kinetic order of
unity for linoleic acid concentration in micelles and for linoleate
chains in mixtures of DLPC with saturated PCs.   In both systems the
kinetic order of one-half in the rate of free radical chain initiation,
the R_i, supports this conclusion.

The demonstration of the same classical kinetics for benzophenone-
photosensitized initiation is an interesting feature of the oxidation
of linoleate in micelles and in DLPC bilayers.   Product studies and
the effects of phenolic antioxidants support a free radical chain
mechanism (Type I).   The regeneration of benzophenone is of particular
significance for quantitative kinetic studies because this ensures a
constant R_i throughout the study and avoids errors involved with
corrections required with azoinitiators.   The Type I mechanism in
these micellar and bilayer systems is probably a result of a reactive
substrate successfully competing against oxygen for the ketone triplet
sensitizer so that singlet oxygen is not formed(20).   Reactive allylic
hydrogens are known to be effective "quenchers" of benzophenone
triplets by hydrogen abstraction(27) and the doubly allylic hydrogens
in linoleate are readily abstracted.   The mechanism for benzophenone
photoinitiation is outlined in Figure 3.   The characteristic recycling
of benzophenone probably occurs by H-atom transfer from the ketyl
radical to oxygen.

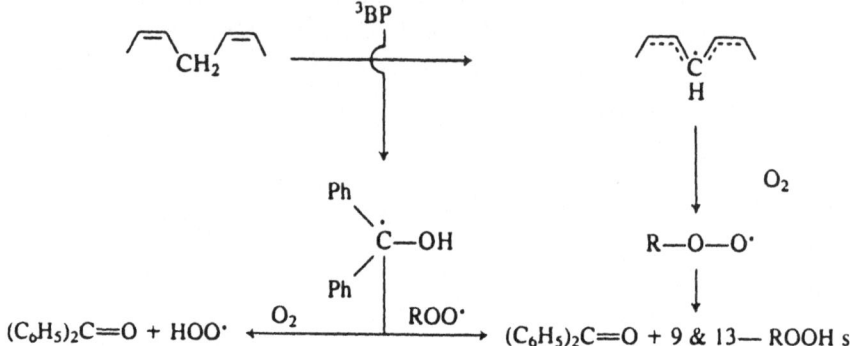

Figure 3. Schematic representation for benzophenone-initiated
oxidation of linoleate, indicating recycling pathways
for benzophenone.

Oxidizability measurements on the linoleate chain obtained in
various media are summarized in Table III.   It is interesting to note
that the susceptibility to autoxidation does not change dramatically
in micelles and in bilayer aggregates from that found in homogeneous
solutions.   An earlier report of a dramatic difference in
oxidizability of PCs in chlorobenzene compared to bilayer aggregates
(14) did not take into account the tendency of PCs to aggregate into
reverse micelles in non protic solvents like chlorobenzene(29).

TABLE III   Oxidizability of Linoleate in Homogeneous Solution,   in
Micelles, and in Bilayers at 30°C.

Medium[a]	Initiator Type	Oxidizability $k_p/2k_t^{\frac{1}{2}}M^{-\frac{1}{2}}s^{-\frac{1}{2}} \times 10^2$	Reference
CH_3CN	BP[b] photo	4.13	17
$O-C_6H_4Cl_2$	thermal azo	4.01	28
C_6H_5Cl	thermal azo	2.30	28
SDS micelles	BP photo	4.42	17
SDS micelles	thermal azo	4.48	16
linoleic acid in DMPC bilayers	thermal azo	2.57	28
DLPC bilayers	thermal azo	2.47[c]	15
	BP photo	3.23[c]	this work

[a] Methyl linoleate was used in organic solvents, linoleic acid
in micelles

[b] BP refers to benzophenone

[c] Calculated per linoleate chain

The absolute rate constants for chain propagation, k_p, and termination, $2k_t$, provide much more specific information on the effects of microenvironments on peroxyl radical dynamics than the rate constant ratios shown in TABLE III. A comparison is made in TABLE IV of the available data on absolute rate constants in SDS and in bilayers with those in solution.

TABLE IV   Comparison of Absolute Rate Constants for Propagation, k_p, and Termination, $2k_t$, of Peroxyl Radicals from Linoleate in Micelles and Bilayers with Values in Solution.

$k_p{}^a$ ratios		$2k_t{}^a$ ratios
SDS/TBA	0.44	0.02
SDS/CH$_3$CN	0.39	0.08
DLPC/TBA	0.44	0.008

^{a}Values of absolute rate constants taken from TABLE I

The rate constants for propagation are reduced by at least one-half in micelles and bilayers compared to values in solution. Much more significant effects are exhibited on the termination rate constants which are <u>dramatically reduced</u> in micelles and especially in bilayers compared to homogeneous solution. It is proposed that these effects on rate constants are due at least in part to the intrinsic polarity of peroxyl radicals(30).

$$R - \overset{..}{\underset{..}{O}} - \overset{..}{\underset{..}{O}} \cdot \quad \longleftrightarrow \quad \overset{(+)}{R - \overset{..}{\underset{.}{O}}} - \overset{(-)}{\overset{..}{\underset{..}{O}}} :$$

Such polar species would diffuse rapidly out of the nonpolar hydrophobic region of a micelle or bilayer where they are formed towards the polar environment of the aqueous buffer. In this polar region, the hydrogen bonding interactions with water would provide strong solvating forces(31) to lower the $2k_t$ substantially and also decrease the reactivity towards hydrogen abstraction. Such diffusion, sometimes termed "The Floating Peroxyl Radical Hypothesis" is shown schematically in Figure 4.

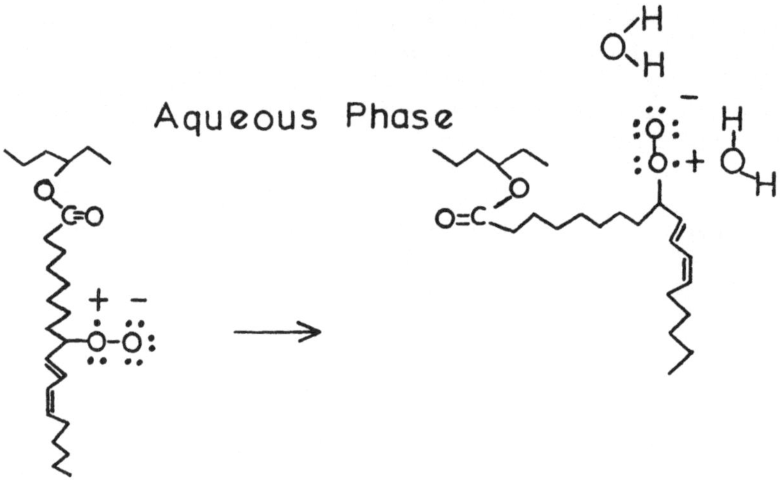

Figure 4.   Schematic for Diffusion of a Polar Peroxyl Radical from
a Hydrophobic to an Aqueous phase.

The diffusion of polar peroxyl radicals out of bilayers could have
very important consequences for the structural integrity of bilayers
in biomembranes.   Some exploratory studies indicate that physical
structural changes can be detected as a result of peroxidation.   For
example, ^{31}P NMR spectra were used to show that unilamellar egg lecithin
liposomes change into multilamellar or larger unilamellar systems as a
result of extensive peroxidation(17)(see Figure 5).   It has recently
been found that starch encapsulated in DLPC liposomes "leaks out" as
a result of peroxidation and this could be a direct consequence of
disruption of bilayer structure.

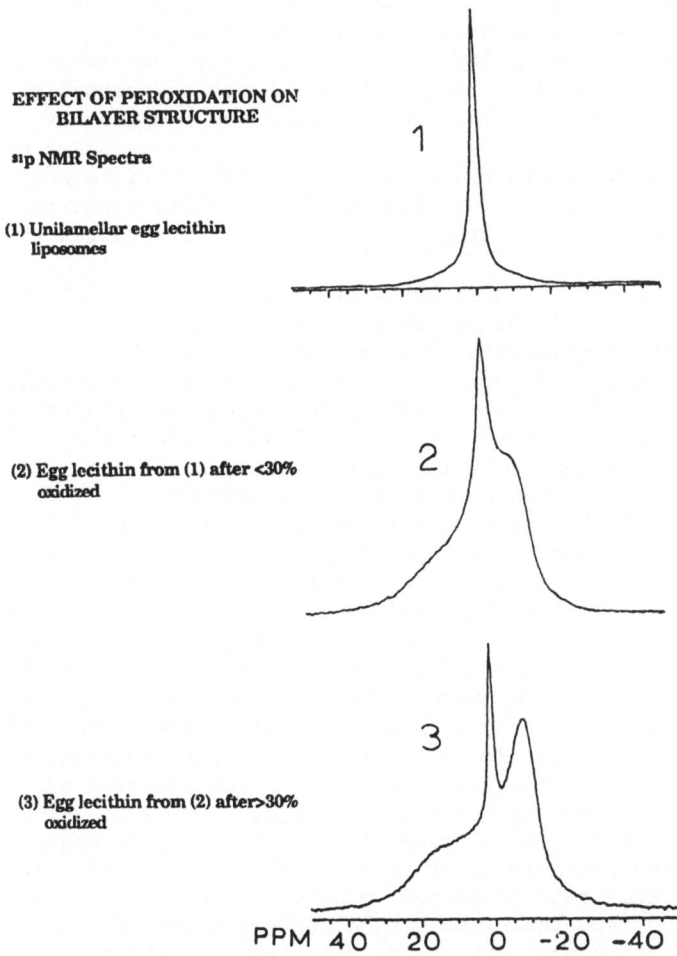

EFFECT OF PEROXIDATION ON BILAYER STRUCTURE

³¹p NMR Spectra

(1) Unilamellar egg lecithin liposomes

(2) Egg lecithin from (1) after <30% oxidized

(3) Egg lecithin from (2) after>30% oxidized

Figure 5. ^{31}P NMR spectra of egg lecithin liposomes, indicating the effect of peroxidation.

Determination of k_{inh} values for antioxidants in micelles and in particular in PC bilayers have important implications for their activities as protectors against biomembrane peroxidation.   Accurate determinations of k_{inh} for various antioxidants in a non-polar solvent have been made by Ingold and co-workers(32).   Our k_{inh} values for alpha-tocopherol, PMHC, and Trolox are similar in a polar organic solvent <u>tert</u>-butyl alcohol (see TABLE II) but as much as an order of magnitude less than in chlorobenzene(32).   There is a significant drop in antioxidant activity of all three antioxidants on changing the medium from <u>tert</u>-butyl alcohol to aqueous SDS micelles, this is especially evident for alpha-tocopherol which is only 1/14 as effective in SDS. A similar effect of SDS micelles on k_{inh} for alpha-tocopherol and other antioxidants has recently been reported by Pryor and co-workers(33). The depressed values are probably due to reduced antioxidant activity of the phenolic hydrogen by damping of the stereoelectronic effect exerted by the lone-pairs on the para ether oxygen(32), due to hydrogen bonding with water in our aqueous systems. In addition, diffusion of lipophilic alpha-tocopherol between micelles may be a limiting factor in its antioxidant activity (17,34).

Alpha-tocopherol is known to be nature's best chain-breaking phenolic antioxidant, at least in homogeneous solution(32). Furthermore it is known to distribute in bilayers with its more hydrophilic chroman group near the lipid-water interface(35) where the phenolic group could be most effective in trapping localized peroxyl radicals (Figure 4). On the other hand, diffusion of alpha-tocopherol between liposomes is retarded by its phytyl side chain(9) and its antioxidant activity when bound to a protein appears to be diffusion limiting(36). The preliminary data in TABLE II for the k_{inh} of alpha-tocopherol and Trolox reveal characteristics which are very important for their antioxidant properties against membrane peroxidation. Their similar k_{inh} values indicate a comparable peroxyl radical trapping ability for the water-soluble Trolox and the lipid-soluble alpha-tocopherol. In addition their activities are both very much reduced compared to those in homogeneous solution. The value for Trolox is only 1/50 while that for alpha-tocopherol only 1/100 of that in tert butyl alcohol. The antioxidant properties of these compounds can be rationalized in terms of the dynamics of the peroxyl radicals and these antioxidants in bilayers. The activity of Trolox when inserted into the aqueous phase of bilayers could be a result of trapping peroxyls at or near the lipid-water interface (Figure 4). The fact that equimolar amounts of Trolox and alpha-tocopherol give the same induction period indicates that Trolox has access to and therefore must diffuse to all layers of the multi-lamellar system used. On the other hand, alpha-tocopherol is already located in the multilamellar system and its drop in activity may be due to the effect of the phytyl side chain in restricting its mobility within bilayers. These conclusions, while speculative in nature, provide a basis for further quantitative kinetic studies in bilayers.

ACKNOWLEDGEMENTS

This research was supported by grants from the Natural Sciences and Engineering Research Council of Canada, the Terry Fox Fund for Cancer Research (New Brunswick), and a NATO Grant for international collaboration in research.

REFERENCES

1. Ingold, K. U. *Acc. Chem. Res.* 2, 1, (1969).
2. Mayo, F. R. *Acc. Chem. Res.* 1, 193 (1968).
3. Howard, J. A. In *Free Radicals*, Vol.II. Edited by J. K. Kochi, Wiley, New York. Chap.12, pp.3-62.
4. Halliwell, B.; Gutteridge, J. M. C. *Free Radicals in Biology and Medicine*. Clarendon Press (1985).
5. Pryor, W. A. 'Cancer and Free Radicals', In *Antimutagenesis and Anticarcinogenesis*, Shankel, D.; Hartman, P.; Kada, T.; Hollander, A. (eds.) Plenum Press, New York (1984).
6. McBrien, D. C. H.; Slater, T. C. (eds.), *Free Radicals, Lipid Peroxidation and Cancer*. Academic Press, New York (1982).
7. Harman, D. 'The Free Radical Theory of Aging' in *Free Radicals in Biology*, Pryor, W. A. (ed.), Vol.V, pp.255-276, Academic Press, New York (1982).
8. Pryor, W. A.; Dooley, M. M.; Church, D. F. *Chem. Biol. Interactions* 57, 271 (1986).
9. Niki, E. *Chem. Phys. Lipids*, 44, 227 (1987).
10. Burton, G. W.; Ingold, K. U. *Acc. Chem. Res.* 19, 194 (1986).
11. Porter, N. A. et al. *J. Am. Chem. Soc.* 106, 813 and 7652 (1984).
12. Winterle, J. S.; Mill, T. *J. Am. Chem. Soc.* 102, 6336 (1980).
13. Porter, N. A. *Acc. Chem. Res.* 19, 262 (1986).
14. Barclay, L. R. C.; Ingold, K. U. *J. Am. Chem. Soc.* 103, 6478 (1981).
15. Barclay, L. R. C.; Locke, S. J.; MacNeil, J. M.; VanKessel, J. S.; Burton, G. W.; Ingold, K. U. *J. Am. Chem. Soc.* 106, 2479 (1984).
16. Barclay, L. R. C.; Locke, S. J.; MacNeil, J. M. *Can. J. Chem.* 63 366 (1985).
17. Barclay, L. R. C. et al. *Can. J. Chem.* 65, 2529 and 2541 (1987).
18. Porter, N. A.; Wolf, R. A.; Weenen, H. *Lipids*, 15, 163 (1980).
19. Porter, N. A.; Weber, B. A.; Weenen, H.; Khan, J. A. *J. Am. Chem. Soc.* 102, 5597 (1980).
20. Foote, C. S. In *Free Radicals in Biology*, Pryor, W. A. (ed.) Vol.II pp.85-133, Academic Press, New York (1976).
21. Straight, R. C.; Spikes, J. D. In 'Singlet O_2'. Vol.IV *Polymers and Biomolecules*, Frimer, A. A. (ed.) CRC Press, Boca Raton, FL. (1985), Chap.2, pp.91-143.
22. Gorman, A. A.; Gould, I. R.; Hamblett, I.; Standen, M. C. *J. Am. Chem. Soc.* 106, 6956 (1984).
23. Clough, R. L.; Yee, B. G.; Foote, C. S. *J. Am. Chem. Soc.* 101, 683 (1979).
24. Howard, J. A.; Ingold, K. U. *Can. J. Chem.* 43, 2729 (1965); 43, 2737 (1965).
25. Burnett, G. M.; Melville, H. W. In *Techniques of Organic Chemistry*. Vol.VIII, Part II. Fries, S. L.; Lewis, E. S.; Weissberger, A. (ed.) Interscience, New York (1963).
26. Burton, G. W.; Ingold, K. U. *J. Am. Chem. Soc.* 103, 6472 (1981).

406

27. Encinas, M. V.; Scaiano, J. C. *J. Am. Chem. Soc.* <u>103</u>, 6393 (1981).
28. Barclay, L. R. C.; Locke, S. J.; MacNeil, J. M.; VanKessel, J. *Can. J. Chem.* <u>63</u>, 2633 (1985).
29. Barclay, L. R. C.; MacNeil, J. M.; VanKessel, J.; Forrest, B. J.; Porter, N. A.; Lehman, L. S.; Smith, K. J.; Ellington, J. C. *J. Am. Chem. Soc.* <u>106</u>, 6740 (1984).
30. Fessenden, R. W.; Hitachi, A.; Nagarajan, V. *J. Phys. Chem.* <u>88</u>, 107 (1984).
31. Rust, F. F.; Youngman, E. A. *J. Org. Chem.* <u>27</u>, 3778 (1962).
32. Burton, G. W.; Doba, T.; Gabe, E. J.; Hughes, L.; Lee, F. L.; Prasad, L.; Ingold, K. U. *J. Am. Chem. Soc.* <u>107</u>, 7053 (1985).
33. Pryor, W. A.; Strickland, T.; Church, D. F. *J. Am. Chem. Soc.* <u>110</u>, 2224 (1988).
34. Castle, L.; Perkins, M. J. *J. Am. Chem. Soc.* <u>108</u>, 6381 (1986).
35. Perly, B.; Smith, I. C. P.; Hughes, L.; Burton, G. W.; Ingold, K.U. *Biochim. Biophys. Acta.* <u>819</u>, 131 (1985).
36. Barclay, L. R. C.; Bailey, A. M. H.; Kong, D. *J. Biol. Chem.* <u>260</u>, 15809 (1985).

REARRANGEMENTS OF OPTICALLY PURE HYDROPEROXIDES

Ned A. PORTER and Patrick H. DUSSAULT
Department of Chemistry
Paul M. Gross Chemical Laboratory
Duke University
Durham, NC 27706
USA

ABSTRACT. The mechanism of autoxidation of lipid materials has been the focus of much investigation. Hydroperoxides are the primary products formed in lipid autoxidation and these hydroperoxide products undergo free radical rearrangements. For example, the hydroperoxides formed in the autoxidation of linoleate rearrange by a mechanism thought to involve formation of the peroxyl radical from the corresponding hydroperoxide and fragmentation of this peroxyl radical to the delocalized carbon radical. Readdition of oxygen to the delocalized radical leads to mixtures of products. Simple allylic hydroperoxides also rearrange by a mechanism involving formation of a peroxyl radical intermediate. The mechanism of this rearrangement has been shown to be concerted. This was demonstrated by the fact that the rearrangement carried out under ^{18}O labelled oxygen does not lead to products that have incorporated labelled oxygen. Furthermore, rearrangement of optically pure allylic hydroperoxides leads to rearrangement products that are also optically pure. A concerted mechanism is consistent with these results. The optically pure allylic hydroperoxides have not been previously prepared and a major focus of our work has been the development of methods for the synthesis of these compounds. We have succeeded in developing methods for the resolution of allylic and dienylic hydroperoxides. Our method involves conversion of the hydroperoxides to perketal derivatives that are formed from the hydroperoxide and α-methyl vinyl ethers derived from (-)-2-phenyl-cyclohexanol. The diastereomeric perketal derivatives can be separated by chromatographic techniques and the perketal protecting group then removed with mild acid. In this way, seven different optically active hydroperoxides have been resolved. In fact, we have yet to encounter a chiral hydroperoxide that cannot be resolved by this approach. Rearrangement of optically pure allylic hydroperoxides occurs with chirality transfer to the new stereocenter formed in the rearrangement.

F. Minisci (ed.), Free Radicals in Synthesis and Biology, 407–421.
© *1989 by Kluwer Academic Publishers.*

1. AUTOXIDATION

Oxygen may act not only as an oxidant to initiate free radical reactions, but it can also act as a substrate for the propagation of these reactions. The spontaneous reaction of molecular oxygen with radicals is commonly referred to as autoxidation. Autoxidation is responsible for the deterioration of many manufactured plastics and rubber goods. Rancidity and spoilage of foodstuffs is a direct result of the autoxidation of fats, which are most susceptible to air oxidation and present, to a large extent, in virtually all foods.

The autoxidative process is commonly represented as consisting of chain initiation, propagation, and termination steps:

Initiation: $\qquad$ $In\bullet + RH \longrightarrow R\bullet + InH$ $\qquad$ (Eq.1)

Propagation: $\qquad$ $R\bullet + O_2 \longrightarrow ROO\bullet$ $\qquad$ (Eq.2)

$$ROO\bullet + RH \overset{k_p}{\longrightarrow} R\bullet + ROOH \qquad \text{(Eq.3)}$$

Termination: $\qquad$ $2ROO\bullet \longrightarrow [ROOOOR] \overset{\text{non radical}}{\longrightarrow} \text{products} + O_2$ $\qquad$ (Eq.4)

The key event in initiation is the formation of $R\bullet$. There are many sources of radical species which may serve to abstract a hydrogen atom from RH. These include nitric oxide (NO), nitrogen dioxide (NO_2), and ozone, all of which are common environmental pollutants and have been shown to initiate autoxidation via hydrogen atom abstraction (1). Another method of generating $R\bullet$ is via thermal or photochemical induced homolytic scission.

$$RH \overset{\text{heat or}}{\underset{\text{UV radiation}}{\longrightarrow}} R\bullet + H\bullet$$

In the initial propagation step of autoxidation (Eq. 2), molecular oxygen adds to $R\bullet$. At partial oxygen pressures above 100 mm Hg, this addition approaches the diffusion controlled limit (ca. 10^9 1/mol-s) (2). This means that the major radical in solution is peroxy radical $ROO\bullet$ and not $R\bullet$. As a result, it is unlikely that any termination reactions involving $R\bullet$ will take place, provided that oxygen is present in sufficient concentration. In the second propagation reaction (Eq. 3), $ROO\bullet$ abstracts hydrogen atom from RH at rate k_p to generate more $R\bullet$ in what is the rate-determining step. For each hydroperoxide product formed, another radical $R\bullet$ is generated. This process could proceed ad infinitum, save for termination reactions that interrupt propagation. One potential termination step shown is that of peroxy radical coupling to ultimately give oxygen and non-radical organic products (Eq. 4) (3,4). Another possible terminating route is that of radical disproportionation.

The rate constant k_p for hydrogen atom abstraction depends primarily on the activation energy of the reaction, i.e., the strength of the C-H bond being broken. Since ROO• is strongly resonance stabilized and a comparatively unreactive radical (2), it is quite selective in abstraction from hydrocarbons and prefers the most weakly bound hydrogen atom.

2. DIENE AUTOXIDATION

The competition of the propagation step illustrated in equation 1 and peroxyl radical β-fragmentation has one important consequence in fatty acid or ester autoxidation; the stereochemistry of products formed in autoxidation is dependent on these competitive pathways. As an example of this, consider the autoxidation of a simple diene fatty acid, linoleic acid (6-10). Four conjugated diene hydroperoxide products are formed, 1-4. Two of these products, 1 and 3, have conjugated diene stereochemistry with the double bond nearest the hydroperoxide being trans and the remote double bond being cis. The two other products have trans/trans diene stereochemistry. Two of the products are substituted at the 9 position, and two at the 13 position of the eighteen carbon chain.

The mechanism for the formation of these four products has been discussed in detail elsewhere (11) and will be described here in outline form. Briefly stated, the mechanism proceeds by H-atom removal from C11 to generate a pentadienyl radical,which couples with oxygen at C9 (recognizing that similar chemistry occurs at C13), giving the peroxyl radical with <u>trans/cis</u> conjugated diene stereochemistry. Entrapment of this radical before β-fragmentation leads to **1**, the kinetic product, the <u>trans/cis</u> compound. If good H-atom donors are not present (R-H or antioxidants, such as tocopherol) then β-fragmentation occurs and subsequent oxygen coupling at C13 of the carbon radical leads ultimately to the thermodynamic products, the 13 <u>trans/trans</u> product **4**. The competition between H-atom transfer to give kinetic product and β-fragmentation to give thermodynamic product thus accounts for the four major products from linoleate autoxidation.

This mechanism is consistent with several experimental observations (8,13). First, higher concentrations of linoleate leads to more <u>trans/cis</u> products. This is because at higher [R-H], the reaction is driven toward kinetic control, since linoleate is an H-atom donor. Second, addition of other good H-atom donors, such as 1,4-cyclohexadiene, causes the product mixture to consist of primarily <u>trans/cis</u> products (kinetic control). Third, product distributions are independent of oxygen pressure between 10 and 1000 mm Hg. Oxygen addition to carbon radicals is not rate limiting at these pressures and β-fragmentation is independent of oxygen concentration.

Evidence for the fragmentation also comes from oxygen labelling studies of Chan et. al., who showed that atmospheric oxygen is incorporated into linoleate hydroperoxides during rearrangement.as shown in Scheme I (7).

Scheme I

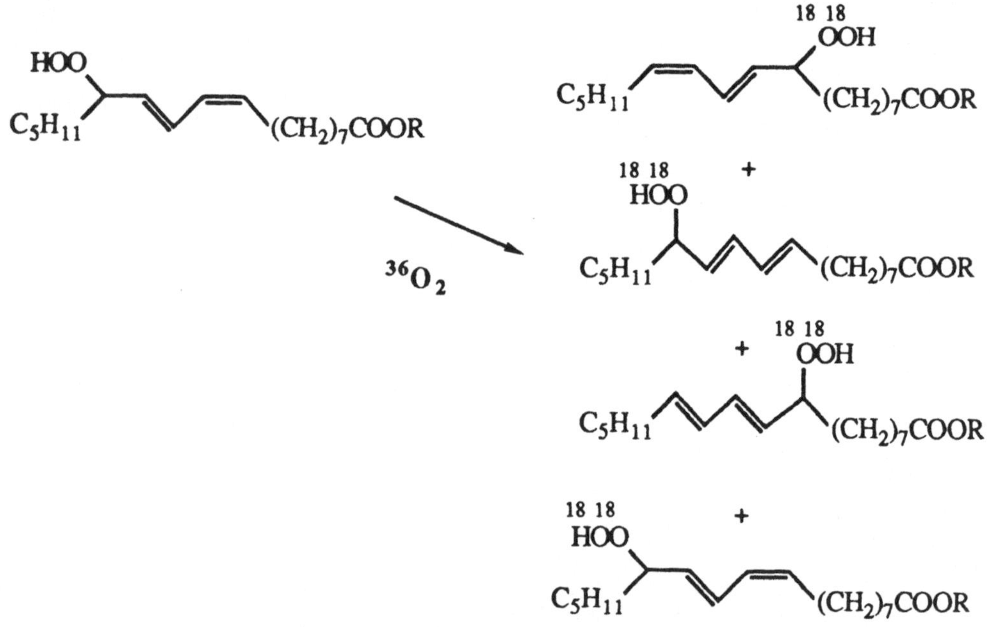

In summary, the mechanisms of autoxidation propagation of diene lipids consist of three steps: (1) H-atom transfer from substrate to peroxy radical; (2) oxygen coupling to intermediate carbon radicals; and (3) β-fragmentation of intermediate peroxyl radical to remove the peroxyl. The kinetics and product distributions of autoxidation can be understood by this fundamental mechanism.

3. MONOENE AUTOXIDATION

The study of monoene oxidation has been much less intensive than work with diene autoxidation. Some studies have been reported on the autoxidation of simple compounds such as *cis* and *trans* hexene and the evidence presented suggests that products like those from diene precursors are formed (14). There is, however, no evidence that oxygen addition to simple allylic radicals is reversible. We have recently provided evidence about the rearrangement of simple allylic peroxyl radicals that suggests, in fact, that this rearrangement occurs by a concerted mechanism.

Allylic hydroperoxides undergo structural rearrangement. This rearrangement has been known since 1957 when Schenck reported that the tertiary C-5 α-hydroperoxide of cholesterol rearranges to its C-7 α-allylic isomer (15). At least three mechanisms for the allylic hydroperoxide rearrangement have been proposed (16-18). These three mechanisms are outlined in Scheme II and involve: (1) formation of a cyclic 5-membered ring peroxide with a free radical at position 4 of the ring. This mechanism amounts to a step-wise reaction pathway with the cyclic radical being a true intermediate in the rearrangement. (2) Formation of a cyclic 5-membered ring transition state that links the two allylic hydroperoxyl radicals. This mechanism is a concerted mechanism in which an authentic reaction intermediate does not exist (3) β-Fragmentation of an allylic peroxyl radical to form molecular oxygen and an allyl carbon radical, which can recombine with oxygen at either end of the radical to give the starting and rearranged peroxyl radicals. Each of these mechanisms involves intermediate peroxyls and consistent with this is the fact that free radical initiators facilitate the reaction and phenolic inhibitors stop the rearrangement .

Several experiments have been carried out to determine the mechanism of the allylic hydroperoxide rearrangement. For example, Brill has attempted to trap the proposed radical intermediate by carrying out the rearrangement under high pressures of O_2 or with allylic systems designed to undergo further molecular rearrangements at the intermediate radical stage. No oxygen entrapment or other evidence for radical intermediate could be presented to support the stepwise mechanism involving a cyclic peroxide radical. Furthermore, when authentic radicals like proposed intermediate are generated, they are found to react by addition of molecular oxygen and cyclic peroxide hydroperoxides (-OOH at C-4) can be isolated. It thus seems reasonable to rule out further consideration of the stepwise rearrangement mechanism involving a true intermediate . The remaining mechanisms could be distinguished by an appropriate experiment involving the use of isotopically labelled oxygen to determine if fragmentation of the peroxyl radical intermediate occurs. Thus, if a β-scission pathway is followed, a rearrangement carried out under $^{36}O_2$ should show incorporation of $^{36}O_2$ into the hydroperoxide products. We have carried out such a study of the allylic hydroperoxides formed from singlet oxygen oxidation of oleic acid.

Scheme II.

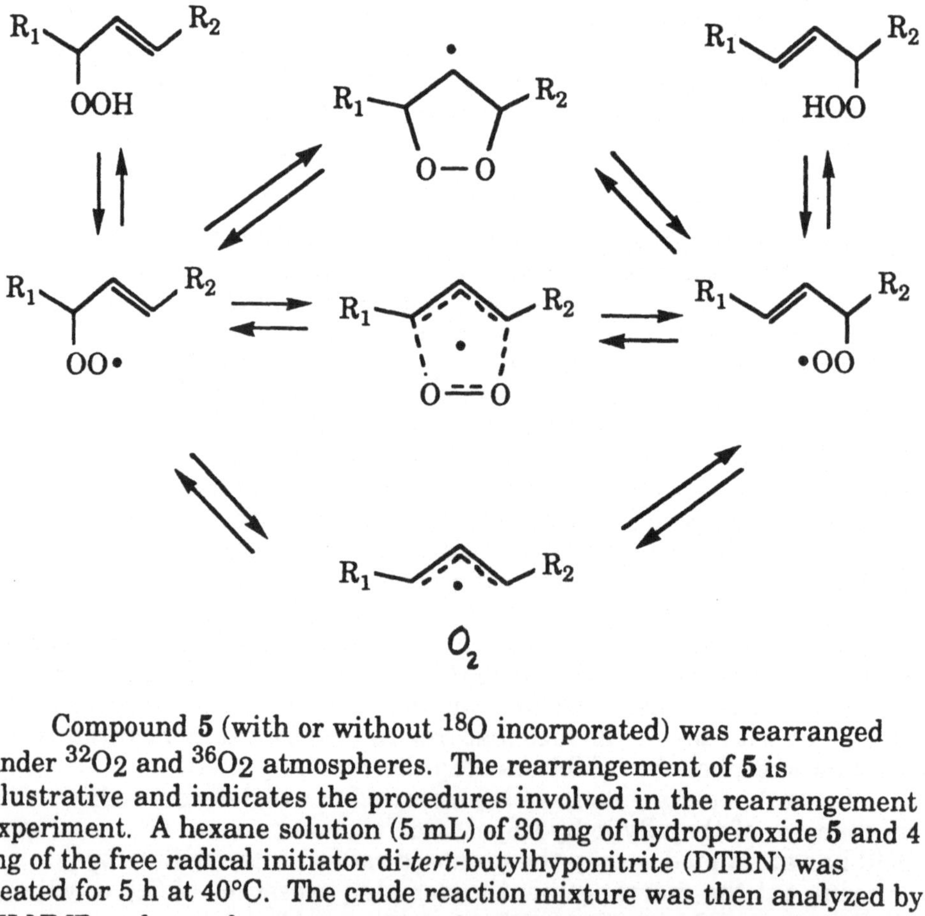

Compound **5** (with or without ^{18}O incorporated) was rearranged under $^{32}O_2$ and $^{36}O_2$ atmospheres. The rearrangement of **5** is illustrative and indicates the procedures involved in the rearrangement experiment. A hexane solution (5 mL) of 30 mg of hydroperoxide **5** and 4 mg of the free radical initiator di-*tert*-butylhyponitrite (DTBN) was heated for 5 h at 40°C. The crude reaction mixture was then analyzed by ^1H NMR and was shown to consist of a 67%:17%:16% mixture of allylic hydroperoxide, allylic alcohol, and α,β-unsaturated ketone. The characteristic NMR resonances used to determine the product ratio for the hydroperoxide are, the proton on the carbon bearing OOH (a quartet at 4.2 δ); for the alcohol, the analogous proton (a quartet at 4.0 δ) and for the α,β-unsaturated ketone; the vinyl proton a to the ketone (a doublet at 6.1 δ). For **5**, the product mixture consisted of hydroperoxides, alcohols, and ketones substituted at the 9-position with a 10-11 *trans* double bond (no rearrangement) and compounds substituted at the 11-position with a 9-10 *trans* double bond (allylic rearrangement). After 5 h at 40°C, the starting material and rearrangement products were present in approximately equal amounts. The allylic alcohols and α,β-unsaturated ketones are presumably formed by termination reactions. The products formed in the rearrangement of **5** are presented in Scheme III. After rearrangement was complete, the crude product mixture was treated with acetyl chloride/pyridine. This treatment converted all allylic hydroperoxides present into the α,β-unsaturated ketones **7** and **8**, while allylic alcohols were converted to acetates. The ketone products **7** and **8** were separated by reverse-phase HPLC. These purified ketones were then analyzed by mass spectrometry.

Scheme III.

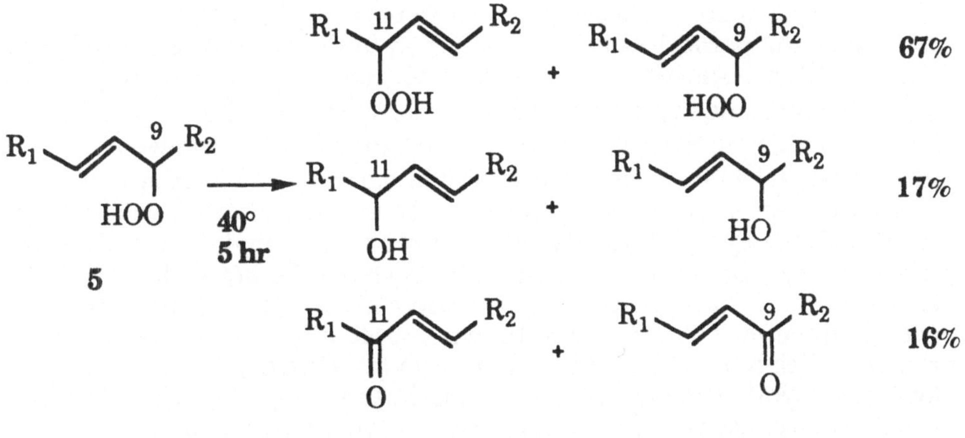

The incorporation of atmospheric oxygen into the hydroperoxide during rearrangement can be ascertained by analysis of the molecular ion of the ketones and by comparison of the M and M+2 (^{18}O) ions. A similar analysis was carried out for rearrangement of the 10-substituted hydroperoxide **6**. The hydroperoxide **6** rearranges to a mixture of 10- and 8-substituted hydroperoxides, alcohols, and ketones analogous to those formed from **5**. The allylic rearrangement of hydroperoxides is catalyzed by UV light and free radical initiators and is inhibited by 2,6-di(*tert*-butyl)-4-methylphenol . One step in chain propagation presumably involves generation of an allylic peroxyl radical from the corresponding hydroperoxide by H-atom transfer. The H-atom abstracting agent may be the initiator radical or another peroxyl radical present in solution. The ROO-H bond is known to be weak and transfer of a hydroperoxyl H to alkoxyl radicals or other peroxyls is a facile process . Once formed, the allylic peroxyl radicals rearrange by an allylic shift. The mechanism of this rearrangement has been the subject of much debate.

Previous studies have ruled out a cyclic intermediate radical from consideration as a reactive intermediate in the rearrangement. The stepwise pathway was first proposed for the cholesterol rearrangements and later for rearrangements of acyclic allylic hydroperoxides. Two other mechanisms are left for consideration. One mechanism involving a cyclic transition state was proposed by Brill after the radical intermediate possiblility was ruled out : Another mechanistic possibility has good precedent in the rearrangements of the dienyl hydroperoxides derived from homoconjugated diene systems, such as linoleic acid. This mechanism, β-scission-recombination involving the allylic radical (Scheme I) is the accepted mechanism for linoleate hydroperoxide rearrangements such as the rearrangement of the 13-linoleate hydroperoxide 3.

The hydroperoxide 3 is suggested to rearrange *via* the corresponding peroxyl radical by β-scission to the stabilized pentadienyl radical. This β-scission-recombination mechanism has been supported by rearrangement of 3 under a $^{36}O_2$ atmosphere. Product hydroperoxides formed by rearrangement in this system have incorporated ^{18}O from the atmosphere in agreement with the fragmentation mechanism. β-Scission of peroxyl radicals has also been substantiated by ESR experiments and benzylic, cumyl, benzhydryl, and trityl peroxyls undergo fragmentation. It should be noted, however, that the peroxyl radical derived from 3 would have a significantly larger driving force for fragmentation than a simple allylic peroxyl (e.g., as shown in Scheme I). Pentadienyl radicals are stabilized by 24-28 kcal/mole, while the allyl radical is stabilized by only 13-14 kcal/mole. Furthermore, concerted rearrangement of dienyl peroxyls would require disruption of diene conjugation, whereas this would not be the case for a simple allyl system. It thus seems likely that the simple allyl systems would be more likely to undergo the concerted radical rearrangement than would dienyl peroxyls.

We have chosen the simple allylic hydroperoxides derived from oleic acid for study because of the importance of these hydroperoxides as primary products in lipid peroxidation and since these acyclic hydroperoxides are prepared by straightforward means. Singlet oxygen oxidation of oleic acid is known to produce only two hydroperoxides and we find that these hydroperoxides (5 and 6) can be readily separated by reverse-phase chromatography. Furthermore, rearrangement of the hydroperoxides readily occurs in organic solvent at 40°C when initiated by DTBN and the product mixture formed is relatively clean. Only allylic hydroperoxides, allylic alcohols, and α,β-unsaturated ketones are present in the rearrangement product mixture. The presence of allylic alcohols and the α,β-unsaturated ketones can be accounted for by Russell termination, as indicated in Scheme IV. Furthermore, the rearrangement product mixture can

Scheme IV.

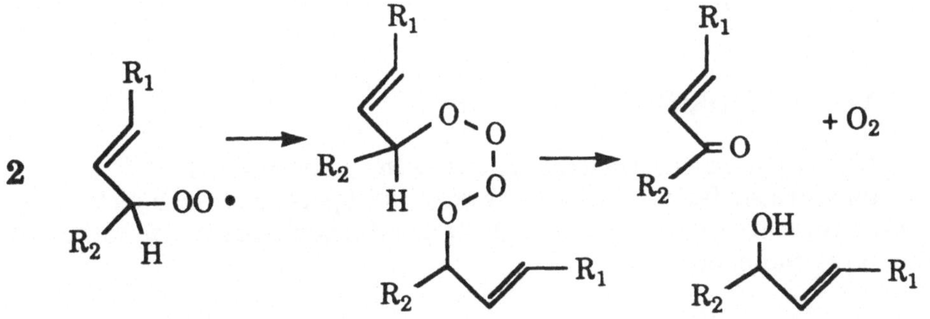

be simplified by conversion of the hydroperoxides to ketones by treatment with acetyl chloride/pyridine.

The results of our rearrangement studies support the concerted radical rearrangement pathway *via* a five-membered ring transition state. Under no circumstance was any significant atmospheric oxygen incorporated into the rearrangement products, nor was atmospheric oxygen detected in the recovered starting hydroperoxide. We conclude that the dienyl peroxyl rearrangements previously investigated differ mechanistically from the rearrangement of simple allyl peroxyls. The dienyl peroxyls rearrange by α β-scission pathway, while the simple allyl peroxyls rearrange by a concerted 3,2-pathway. One might suggest that a caged allyl radical/dioxygen species was responsible for the lack of O_2 incorporation in the allyl peroxyls. If one assumes a 1% uncertainty in the oxygen incorporation data, then the lifetime of the cage radical/O_2 species could be on the order of 10^{-13} sec, and it would be indistinguishable from the activated complex proposed here.

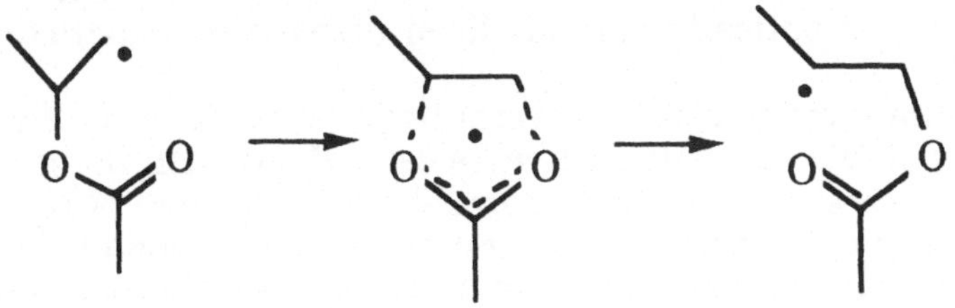

Other examples of concerted 3,2-radical rearrangements have been reported in the literature. For example, β-acetoxyl radicals such as **9** undergo a rearrangement, as described below, and the evidence presented suggests that this rearrangement proceeds *via* a concerted free radical pathway (19-22). It seems likely that this mechanistic option is a general one and a search for other examples of this radical rearrangement seems warranted. Unlike the acetoxyl rearrangement, we note that stereochemical information would be transferred from one end of the allyl peroxyl to the other end of the system in a concerted rearrangement and we are currently exploring the stereochemical implications of the concerted 3,2 allylperoxyl shift, *vide infra*.

4.. HYDROPEROXIDE RESOLUTION

Lipoxygenase enzymes (23) catalyze the conversion of polyunsaturated fatty acids to fatty acid hydroperoxides such as **9**, and these hydroperoxides serve as important intermediates in the formation of diverse compounds of

biological importance. Fatty acid hydroperoxides are precursors to the leukotrienes (24) and the lipoxins for example, and recently a new biochemical pathway involving the conversion of fatty acid hydroperoxides to allene oxides has been reported in flax, corn, and coral (25).
A new stereocenter is generated in the lipoxygenase reaction and fatty acid hydroperoxides isolated from natural sources are essentially one enantiomer if they are formed enzymatically. Nonenzymatic autoxidation of fatty acid substrates also gives fatty acid hydroperoxides formed as racemic mixtures as noted in Section I.

Despite the importance of homochiral unsaturated hydroperoxides in chemistry, biology, and medicine, no general method for the preparation of these labile compounds has been reported. All chemical syntheses described give racemic mixtures and the natural products are thus only available from enzymatic sources while the unnatural enantiomers have not been reported. We have discovered what appears to be a general solution to this problem, the resolution of unsaturated hydroperoxide enantiomers by liquid chromatography of diastereomeric derivatives. This method allows, for the first time, the nonenzymatic preparation of optically pure allylic or dienylic hydroperoxide natural products.

Preliminary screens of several hydroperoxide derivatives of the structure R-OO-CR'XOR", **10**, were carried out with the goal being: a.) ease of conversion of the hydroperoxide to **10**, b.) ease of removal of the protecting group without racemization or destruction of the unsaturated hydroperoxide substructure, and c.) acceptable chromatographic characteristics of the derivative including resolution of diastereomers and stability to the conditions of chromatography. Peracetals, **10** with R'=alkyl or H, X=H are readily formed and stable to chromatography but require harsh conditions for deprotection while perorthoesters, R'=H, aryl or alkyl and X=OR", and peraminals are unstable to chromatography. Perketals **10** with R'=X=alkyl are readily prepared from hydroperoxides , they are stable to normal or reverse-phase chromatography and they can be deprotected under very mild acid conditions.

We have had the most success with resolution utilizing the vinyl ether **12** , with R=isopropy or phenyl prepared as shown in Scheme V. Standard

Scheme V.

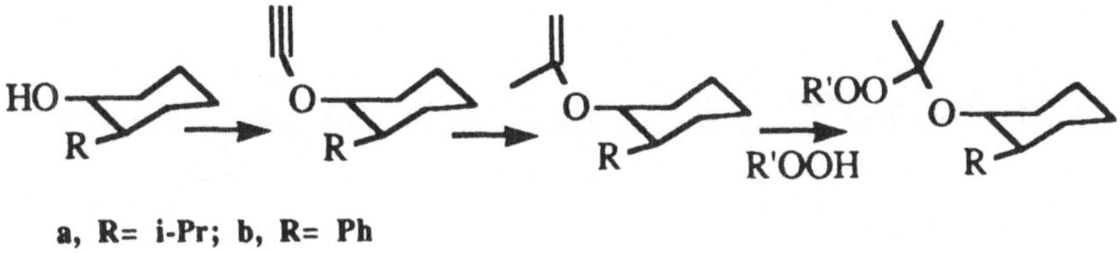

a, R= i-Pr; b, R= Ph

| 11 | 12 | 13 |

procedures for vinyl ether synthesis directly from the alcohol were unsucessful but the path through the acetylenic ether gave **11** in good yield for the two steps. The perketal **3** could be prepared from 2 and hydroperoxide with pyridinium p-toluenesulfonate catalyst in yields in excess of 90%. Hydroperoxides **14-16** (R_1=H, R_2=Me) were converted to the perketals **13a** and **14-18** (R_1=H, R_2=Me) were reacted with **12b** to give the corresponding perketals **13b**. In each case examined, the perketals were readily formed, were stable to chromatography under appropriate conditions and the perketal protecting group could be removed with acetic acid/ THF/ water. Although the perketals are completely stable to normal-phase chromatography, some decomposition occurs in reverse-phase solvents such as acetonitrile/ water or methanol/ water unless 0.01% Et3N is added .

Separation of the menthol derivatives **13a** was acceptable on analytical reverse-phase columns but troublesome on a preparative scale. In contrast to the menthol auxiliary, the diastereomeric phenylcyclohexanol-derived perketals **13b** generally gave baseline separation on normal or reverse-phase chromatography and hundreds of mg quantities of the diastereomers could be isolated with isomeric purities of 96% or better. After recovery of hydroperoxide from theperketal by hydrolysis with acetic acid/ THF/ water, the optical purity of the hydroperoxide was assayed by rederivatization with **12b** and analytical chromatography or by comparison of the specific rotation of the resolved enantiomers with a known natural product. Alternatively,

14

15

16

17

18

reduction of the hydroperoxide to the corresponding alcohol and conversion of the alcohol to its Mosher ester followed by spectroscopic or chromatographic analysis was used to confirm the purity of the hydroperoxides. No evidence for racemization during deprotection was observed.

The hydroperoxides **15** and **17** were of particular interest since **15** ($R_1=R_2=H$) is a product of the soybean lipoxygenase reaction of linoleic acid and Koshino *et. al.* have recently isolated the alcohol derived from **15** (R_2= Me, $[\alpha]_D$=-2.14° in EtOH) from the timothy plant fungus <u>Epichloe typhina</u> (26). The natural product **17** ($R_1=R_2=H$) and its enantiomer could be prepared from the corresponding perketals by LiOH hydrolysis of the methyl ester and neutralization to give the deprotected free acid hydroperoxide which was identical to the lipoxygenase product in every respect. The enzymatic formation of **15** is unexpected since all known lipoxygenase enzymes require at least diene substrates for activity and **7** is only a monene. The hydroperoxide **15** could also be prepared in >97% optical purity and the derived alcohol with (R_2= Me) had $[\alpha]_D$= -4.2° in EtOH, indicating that the material isolated from <u>Epichloe typhina</u> (26) has a substantial amount of the enantiomer present. It seems likely that **15** is not an enzymatic product but is derived from singlet oxygen and is enriched in one enantiomer by virtue of the fact that it is formed in a chiral environment or is selectively isolated from that environment.

The use of chiral perketals derivatives to resolve hydroperoxides may have widespread application. We have yet to study a hydroperoxide (nine total) that could not be resolved by chromatography of one of the perketal derivatives. Furthermore, the diastereomeric perketals described here give better chromatographic resolution than the corresponding alcohol Mosher esters. Thus, perketals may be the derivative of choice for the assay of enantiomeric purity of hydroperoxides formed enzymatically. We are currently exploring the utility and limitations of this approach to optically pure hydroperoxides.

With optically pure hydroperoxides such as **15** (9-OOH) or **16** (10-OOH) available, the stereochemistry of allylic rearrangement can be investigated. Starting with the **9R**- compound,**15**,the rearrangement gives only one 11 substituted hydroperoxide which we assign the **11-S** configuration. We thus suggest that a concerted rearrangement *via* a transition state as shown in Scheme VI.

Scheme VI.

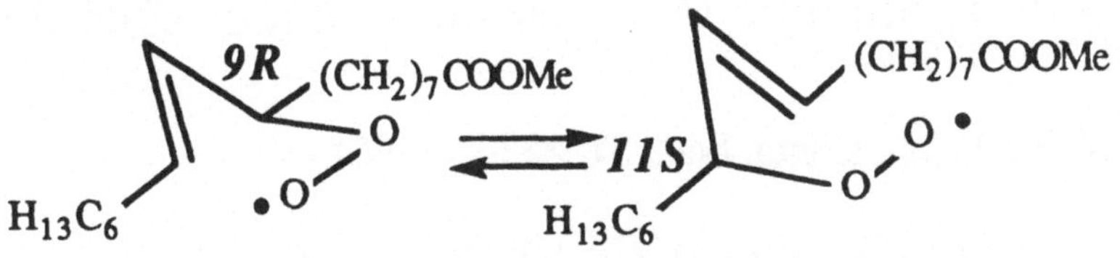

REFERENCES

1. A. P. Autor,In: The Biology and Chemistry of Active Oxygen (J. V. Bannister, W. H. Bannister, Eds), pp. 139-145. Elsevier, New York, (1984)

2. K. U. Ingold, Acc. Chem. Res., **2**, 1-9 (1969)

3. J. A. Howard, Free radicals (J. K. Kochi, Ed.), pp. 3-62. Wiley, New York, (1973)

4. G. A. Russell, J. Am. Chem. Soc., **79**, 3871-3877 (1957)

5. J. A. Howard, K. U. Ingold, and M. Symonds, Can J. Chem., **45**, 1017-1022 (1968)

6. F. Haslbeck, and W. Grosch, J. Food Biochem., **9**, 1-3 (1985)

7. H. W. S. Chan, V. K. Newby, and G. Levett, J. Chem. Soc., Chem. Commun., 82-83 (1978)

8. N. A. Porter, B. A. Weber, H. Weenen, and J. A. Khan, J. Am. Chem. Soc., **102**, 5597-5601 (1980)

420

9. Yamamoto, S. Haga, E. Niki, and Y. Kamiya, Bull. Chem. Soc..
    Jpn,**57**, 1260-1264 (1984)

10. Y. Yamamoto, E. Niki, and Y. Kamiya, Lipids, **17**, 870-877 (1982)

11. N. A. Porter, Acc. Chem. Res., **19**, 262-268 (1986)

12. E. Bascetta, F. D. Gunstone, and J. C. Walton, J. Chem. Soc..
    Perkin Trans. II, 603-613 (1983)

13. N. A. Porter, L. S. Lehman, B. A. Weber, and K. J. Smith, J. Am.
    Chem. Soc., **103**, 6447-6455 (1981)

14. Frankel, E. N., Garwood, R. F., Khambay, B. P. S., Moss, G. P. O.,
    Weedon, B. C. L., J. Chem. Soc. Perkin Trans. I, 2233, (1984)

15. Schenck, G.D., Neumiller, O. A., Eisfield, W., Angew. Chem. Int.
    Ed., **70**, 595, (1958)

16. Brill; W. F., J. Am. Chem. Soc., **87,** 3286, (1965)

17. Brill, W. F., J. Chem. Soc.. Perkin Trans. II, 621,(1984)

18. Porter, N., Zuraw, P., J. Chem. Soc.. Chem. Comm., 1472, (1985)

19. Korth, H.-G., Heinrich, T., Sustmann, R., J. Am. Chem. Soc.,
    103, 4483, (1981)

20. Barclay, L. R. C., Griller, D., Ingold, K. U., J. Am. Chem. Soc.,
    104, 4399, (1982)

21. Saebo, S., Beckwith, A. L. J., Radom, L., J. Am. Chem. Soc.,
    106, 5119, (1984)

22. Barclay, L. R. C., Lusztyk, J., Ingold, K. U., J. Am. Chem. Soc.,
    106, 1793, (1984)

23. Galliard, T.; Chan, H. W.-S, The Biochemistry of Plants,
    Lipoxygenases, Galliard, T.; Chan, H. W.-S., Ed. Academic Press,
    Inc., 1980, pg. 131.

24. (a) Serhan, C. N.; Hamberg, M.; Samuelsson, B.; Morris, J.;
    Wishka, D. G., Proc. Natl. Acad. Sci.. USA, **83**, 1983, (1986]); (b)
    Adams, J.; Fitzsimmons, B. J.; Girard, Y.; Leblanc, Y.; Evans, J.
    F.; Rokach, J., J. Am. Chem. Soc., **107**, 464, (1985)

25. (a) Corey, E. J.; d'Alarcao, M.; Matsuda, S. P. T.; Lansbury, Jr., P.
    T., J. Am. Chem. Soc., **109**, 289, (1987]); (b) Hamberg, M.,

421

Biochimica et Biophysica Acta, **920**, 76, (1987); (c) Brash, A. R.;
Baertschi, S. W.; Ingram, C. D.; Harris, T. M., Journal of Biological
Chemistry, **262**, 15829, (1987); (d) Gardner, H. W.; Kleiman, R.;
Christianson, D. D.; Weisleder, D., Lipids,**10**, 602, (1975)

26. (a) Koshino, H.; Togiya, S.; Yoshihara, T.; Sakamura, S.;
     Shimanuki, T.; Sato, T.; Tajimi, A., Tetrahedron Lett, **28**, 73, (1987);
     (b) See also Phillips, N. J.; Lynn, D. G.; Lynn, W. S., Ninth Cotton
     Dust Research Conference Proceedings, National Cotton Dust
     Council, Memphis, TN, 91 (1985)

IN VIVO DETECTION OF FREE RADICAL METABOLITES

Kirk R. Maples, Kathryn T. Knecht, and Ronald P. Mason
Laboratory of Molecular Biophysics
National Institute of Environmental Health Sciences
National Institutes of Health
P.O. Box 12233
Research Triangle Park, NC   27709   USA

ABSTRACT.
     In the last two decades since the introduction of the ESR spin trapping technique the ESR study of free radical formation in vitro due to xenobiotic metabolism has been significantly enhanced. Although this ESR technique has proven useful in vitro, it has only recently been applied to drug metabolism in vivo. Spin trapping has successfully been employed to study the metabolism of halocarbons and hydrazines in perfused organs and in whole animals. The results of these current studies are detailed herein. In addition, the merits and pitfalls associated with the in vivo application of spin trapping are discussed.

INTRODUCTION

     The technique of spin trapping involves the addition of a reactive free radical across the double bond of a diamagnetic compound, the spin trap, to form a more persistent free radical, the radical adduct. This technique allows the indirect detection of primary free radicals that cannot be directly observed by conventional ESR due to low steady-state concentrations or to very short radical relaxation times, which lead to very broad lines (Janzen, 1980).
     When choosing a particular spin trap for in vivo use, the qualities of stability and sensitivity must be weighed. There are four spin traps commonly used in biological systems: tert-nitrosobutane or 2-methyl 2-nitrosopropane (MNP),

$$(CH_3)_3CN=O + R^\cdot \quad ------> \quad (CH_3)_3C\underset{\underset{R}{|}}{N}-O^\cdot$$

MNP

F. Minisci (ed.), Free Radicals in Synthesis and Biology, 423–436.

phenyl-<u>tert</u>-butylnitrone (PBN),

$$\text{C}_6\text{H}_5-\overset{\text{H}}{\underset{}{\text{C}}}=\overset{+}{\text{N}}(\text{O}^-)-\text{C}(\text{CH}_3)_3 \quad \xrightarrow{\text{R}^\cdot} \quad \text{C}_6\text{H}_5-\overset{\text{R}}{\underset{\text{H}}{\text{C}}}-\overset{\text{O}^\cdot}{\underset{}{\text{N}}}-\text{C}(\text{CH}_3)_3$$

(α)-4-pyridyl-1-oxide N-<u>tert</u>-butylnitrone (4-POBN),

$$\text{O}-\text{N}\langle\text{pyridyl}\rangle-\text{CH}=\overset{+}{\text{N}}(\text{O}^-)-\text{C}(\text{CH}_3)_3 \quad \xrightarrow{\text{R}^\cdot} \quad \text{O}-\text{N}\langle\text{pyridyl}\rangle-\overset{\text{R}}{\underset{\text{H}}{\text{C}}}-\overset{\text{O}^\cdot}{\underset{}{\text{N}}}-\text{C}(\text{CH}_3)_3$$

and 5,5-dimethyl-1-pyrroline N-oxide (DMPO)

$$\xrightarrow{\text{R}^\cdot}$$

For all four spin traps, the resulting radical adducts are nitroxide
free radicals, which are relatively stable free radical species.

MNP is a nitroso compound. With this class of spin trap, the
reactive free radical adds directly to the nitrogen atom of the spin
trap, usually giving rise to hyperfine splittings distinctive for that
radical and thus greatly facilitating radical identification. However,
this advantage of nitroso compounds is accompanied by photochemical and
thermal instability (Perkins, 1980). MNP, for instance, decomposes
thermally and photochemically to di-<u>t</u>-butyl nitroxide, a persistent
radical that produces an often-observed three-line spectrum.

PBN, 4-POBN, and DMPO are nitrone spin traps. Nitrone compounds
are much more stable than nitroso compounds, but yield far less spectral
information about the primary radical trapped.

In most cases, radical adducts of nitrone spin traps produce
six-line ESR spectra. The hyperfine splittings arise from the nitrogen
and β-hydrogen of the spin trap rather than from atoms of the primary
radical. Identification of the trapped radical species therefore
depends on a careful comparison of hyperfine splitting constant values
with those of reference nitroxides analyzed under exactly the same
experimental conditions, or on spectral changes produced by isotopic
labeling of compounds.

When designing a spin trapping experiment, several additional considerations must be addressed, as delineated by Perkins (1980) in his
excellent review of spin trapping.

(a) Will the spin trap participate in reactions other than those with the reactive radicals generated in the experiment? Can these alternative reactions yield nitroxides that will appear as radical-adduct imposters?

(b) How readily can the spectrum be interpreted and the structure of R* be determined?

(c) How fast is the trapping reaction, and how stable are the radical adducts formed?

(d) Does the appearance of a radical adduct signify a major reaction pathway, or can it be a minor side reaction?

To these questions perhaps a fifth should be added:

(e) Is there isotopically labeled (^{13}C, ^{2}H, ^{17}O, ^{15}N, etc.) material available for proof of structure? If not, is there an independent synthesis for the radical adduct (Mottley and Mason, 1988)?

Such questions are especially important for in vivo work where a multitude of enzymes may produce an array of radical adducts of various stabilities and distributions.

In addition to the review articles referenced above there are several good reviews of spin trapping (Evans, 1979; Finkelstein et al., 1980; Buettner, 1982; Kalyanaraman, 1982; Mason, 1982; Mason, 1984; Rosen and Finkelstein, 1985; Green et al., 1985; Mason and Mottley, 1987), which address biological applications of this technique.

IN VIVO BIOLOGICAL CONSEQUENCES OF SPIN TRAP ADMINISTRATION

Comparatively little spin trapping has been done in living animals. Production of a radical adduct stable enough to be detected in biological samples is a major difficulty, but other factors must also be considered. When using the spin trapping technique in whole animals, spin traps may interfere with the experimental system by inhibiting or stimulating enzymes, or by producing toxicity. The latter possibility has not seemed to affect in vivo work to date, although this issue has not been directly addressed in detail.

Spin Traps as Enzyme Inhibitors

Surprisingly, despite the high concentrations of spin traps used in vitro and the known interactions of these compounds with various enzyme systems, the inhibition of enzymes by spin traps has not been a serious problem. PBN, 4-POBN (Augusto et al., 1982; Cheeseman et al., 1985), and DMPO (Augusto et al., 1982) have all been shown to interact with microsomal cytochrome P-450. PBN showed a type I binding spectrum indicating interaction with the substrate-binding site of the cytochrome (Cheeseman et al., 1985), while 4-POBN (Cheeseman et al., 1985) and DMPO (Augusto et al., 1982) showed a type II spectrum, which results from coordination of the N-oxide to the heme iron of the cytochrome. PBN significantly inhibits the cytochrome P-450-catalyzed formation of carbon-centered free radicals from alkylhydrazines (Ortiz de Montellano

et al., 1982), and carbon tetrachloride (Albano et al., 1982). Both PBN
and 4-POBN inhibit P-450 mixed function oxidase catalyzed aminopyrine
N-demethylation, aniline p-hydroxylation, and ethoxycoumarin
O-deethylation (Cheeseman et al., 1985). In addition, at low
concentrations PBN stimulated aniline hydroxylase activity (Cheeseman et
al., 1985). This stimulatory effect has been observed for other
compounds such as 2,2'-bipyridine, acetone, and various nitrogen
heterocycles (Cinti, 1978; Murray and Ryan, 1982) but the mechanism is
not known. However, neither PBN, 4-POBN nor DMPO affects the
NADPH-cytochrome C (P-450) reductase activity of rat liver microsomes
(Kalyanaraman et al., 1980; Cheeseman et al., 1985). At 30 mM, DMPO has
also been found without effect on microsomal aminopyrine demethylation
or aniline hydroxylation (Cederbaum and Cohen, 1980). In addition, DMPO
does not inhibit xanthine oxidase activity as measured by the oxidation
of xanthine to uric acid (Finkelstein et al., 1979).

Mason et al. (1980) found that MNP inhibited the biooxygenation of
arachidonic acid by prostaglandin synthase. The enzyme-bound carbon-
centered arachidonic acid intermediate was proposed to react with the
spin trap, instead of with oxygen, forming characteristic arachidonic
acid radical adducts. DMPO does not inhibit prostaglandin formation,
nor does it form a detectable concentration of the DMPO-carbon-centered
arachidonic acid radical adduct (Mottley et al., 1982a). The inhibition
of prostaglandin synthase and soybean lipoxygenase (Aoshima et al.,
1977) by spin traps appears to depend on the rate of trapping of the
carbon-centered radical intermediate.

These effects on enzyme activity must be considered when spin traps
are used in the presence of these or other enzymes, in vitro or in vivo.
Quantitative interpretation of data should be made cautiously since the
spin trap may be affecting results by increasing or decreasing enzyme
activity, either directly or by reacting with substrate radical metabolites.

Spin Traps as Enzyme Substrates

Radical adducts are nitroxides, and most nitroxides can be reduced
to their corresponding (ESR-invisible) hydroxylamines by ascorbate
(Eriksson et al., 1987), or by NADPH-cytochrome c reductase (Stier and
Sackman, 1973) or cytochrome P-450 (Rosen and Rauckman, 1977). In fact,
radical adducts will generally be reduced at some rate in any microsomal
incubation containing NADPH, especially under anaerobic conditions, and
thus similar reactions are likely to occur in vivo as well.

Nitroso spin traps can also be reduced, even under fairly mild conditions. MNP is reduced to tert-butyl hydronitroxide by the superoxide
formed in NADPH-supplimented rat liver microsomal incubations
(Kalyanaraman et al., 1979a). The enzymatic formation of hydronitroxides via the one-electron reduction of nitroso compounds by flavoproteins
has not been specifically reported, but enzymatic nitroso reduction is

known (Becker and Sternson, 1980; Horie et al., 1980) and enzymatic hydronitroxide formation is therefore to be expected. Phenyl hydronitroxide is formed by the reduction of another nitroso spin trap, nitrosobenzene, with ascorbate, epinephrine (Mottley et al., 1981), or isoproterenol (Sridhar, 1981). Spectra of hydronitroxides have been misidentified in some cases as radical adducts (Mottley et al., 1981; Kalyanaraman et al., 1979). Similar reactions in vivo may thus result in artifactual ESR signals.

PBN increases endogenous microsomal NADPH oxidation and oxygen consumption as if it were a substrate of cytochrome P-450 (Wolf et al., 1980), consistent with its inhibitory activity mentioned above. However, neither the oxidation products nor the effect of this oxidation on in vivo distribution and availability of the spin trap is known.

Mutagenicity of Spin Traps

The mutagenicity of many spin traps has been determined using Salmonella typhimurium histidine (-) mutants in the Ames test (Hampton et al., 1981). This work showed that none of the eighteen spin traps tested were very potent mutagens and the commonly-used spin traps, DMPO, PBN, and MNP, are not mutagenic at the levels tested.

IN VIVO SPIN TRAPPING APPLICATIONS

Halocarbons

In vivo spin trapping was developed using the prototypical hepatotoxicant carbon tetrachloride, for which a free radical mechanism of toxicity had long been established. As early as 1961, the reductively dehalogenated trichloromethyl radical had been proposed as a reactive intermediate (Butler, 1961). However, spin trapping provided the first direct proof of this radical's existence in vivo. In 1979, Lai et al. administered carbon tetrachloride and PBN to rats and found a radical adduct in organic extracts of the liver. Its identity was assigned by comparing experimental splitting constants with those of PBN/·CCl$_3$ generated in a microsomal system. More definitive identification of this species as the PBN/trichloromethyl radical adduct was later made using ^{13}C carbon tetrachloride, which results in an additional splitting in the radical adduct spectrum (Poyer et al., 1980; Albano et al., 1982).

A carbon tetrachloride-derived radical adduct from a living animal was first detected in rat urine by Connor et al., who discovered the novel radical adduct metabolite PBN/·CO$_2^-$ (Connor et al., 1986). Use of ^{13}C labeling showed that this adduct was carbon tetrachloride-derived. This adduct was also detected in the perfusate of carbon tetrachloride- and PBN-treated perfused liver, and in the urine after bromotrichloromethane administration in vivo (LaCagnin et al., 1988). The lysis of the hepatocytes in the perfused liver as measured by the release of lactate dehydrogenase occurs after the detection of the PBN/'CO_2^-. The concentration of this radical adduct at the start of lysis is statistically correlated with the time required for lactate dehydrogenase release. Since correlation does not imply causation; PBN/'CO_2^- can only be considered a marker for the reaction species responsible for membrane damage. PBN did not protect the liver against CCl_4-induced lysis. Presumably because of the very low rates of radical trapping by PBN, it could not prevent the membrane damaging free radical reactions from occurring. Both PBN/'CCl_3 and PBN/'CO_2^- were detected in the bile of living rats treated intraperitoneally with PBN and intragastically with carbon tetrachloride. Either hypoxia or phenobarbital pretreatment was required for the detection of PBN/'CO_2^-. Both treatments also increased the biliary concentration of PBN/'CCl_3 (Knecht and Mason, 1988) and increased the hepatotoxicity of CCl_4.

Carbon-centered radical adducts from other halogenated hydrocarbons have also been detected in the organic extracts of livers from treated animals. Of special interest have been studies with the volatile anesthetic halothane, which produces hepatitis in humans. Under hypoxic conditions, halothane produces liver damage in animals and free radicals that can be trapped by PBN (Poyer et al., 1981). The trapped radical species has not yet been identified. Although metabolite profiles show both debromination and defluorination are involved in metabolic activation, reductive defluorination is unlikely on chemical grounds due to the relative strength of the carbon-fluorine bond. Reductive debromination is probably responsible for the reported carbon-centered free radical formation (Poyer et al., 1981; Plummer et al., 1982; Fujii et al., 1984). Chloroform, iodoform, bromoform, and bromodichloromethane are other compounds that have been found to form free radical adducts in vivo (Albano et al., 1982). CCl_3Br produces the same metabolites as does carbon tetrachloride, but is metabolized more readily, since the C-Br bond is weaker than the C-Cl bond.

Halocarbon free radical formation may result in lipid peroxidation, and it is logical to expect that these carbon-centered radicals might also be trapped by PBN. Identification of lipid adducts is inherently difficult, since ^{13}C labeling is unavailable and chromatography or mass spectroscopy of these heterogeneous biological products is a formidable task. Nevertheless, several studies of lipid radical adducts have been reported. Lipid peroxidation is after all an important toxicological process, and lipid radical adducts may be the only evidence of initiating radical species that do not form stable adducts themselves. For instance, 3-methylindole has been metabolized in vitro to form a nitrogen-centered free radical that was spin trapped with PBN. In vivo, however, using organic extracts of lungs from goats treated with 3-methylindole and PBN, only a carbon-centered lipid radical adduct was detected. Radical adduct detection varied inversely with GSH, as was suggested with experiments with the GSH precursor cysteine and the GSH-depleting agent diethylmaleate (Kubow et al., 1984). In a similar manner, carbon-centered lipid-derived radical adducts have been extracted from the brain, spleen, liver, and heart, but not the kidney, of rats dosed with PBN and exposed to a non-lethal burst of gamma-irradiation (Lai et al., 1986).

Although carbon tetrachloride is associated with lipid peroxidation, no conclusive evidence of lipid-derived radical adduct formation upon treatment of whole animals with carbon tetrachloride and PBN has been reported. Lipid-derived radical adducts have been detected in vitro (Kalyanaraman et al., 1979b; Poyer et al., 1980; McCay et al., 1984) and studies with the PBN analog $(CH_3O)_3PBN$ and carbon tetrachloride have demonstrated a [^{13}C]-invariant spectrum in extracts of rat liver (McCay et al., 1984). However, O-demethylation of the spin trap precludes a simple interpretation of these studies. An ethanol and high fat diet has also been used to produce this same radical adduct in extracts of hearts and livers from animals treated in vivo (Reinke et al., 1988). Administration of carbon tetrachloride and PBN to gerbils has resulted in detectable free radicals in organic extracts of liver, kidney, heart, lung, testis, brain, and blood, with signal intensity decreasing in the above list. In extracts of liver PBN/˙CCl$_3$ was detected, but no assignment of the radical adduct identity in other tissues could be made (Ahmed et al., 1987).

Hydrazines

The reaction of oxyhemoglobin with phenylhydrazine and hydrazine-based drugs within red blood cells induces a series of processes which lead to premature destruction of the cell and results in hemolytic anemia. Considerable evidence obtained from in vitro ESR investigations implicates free radicals in the events contributing to red blood cell hemolysis. The following reaction mechanism has been proposed for the reaction of phenylhydrazine with oxyhemoglobin:

$$PhNHNH_2 + HbFe(II)O_2 \longrightarrow Ph\overset{\cdot}{N}NH_2 + HbFe(II)O_2H$$

$$HbFe(II)O_2H \overset{H^+}{\longrightarrow} HbFe(III) + H_2O_2$$

$$Ph\overset{\cdot}{N}NH_2 + O_2 \longrightarrow PhN{=}NH + H^+ + O_2^{\overline{\cdot}}$$

$$2\ Ph\overset{\cdot}{N}NH_2 \longrightarrow PhN{=}NH + PhNHNH_2$$

$$PhN{=}NH + O_2 \longrightarrow PhN{=}N^{\cdot} + H^+ + O_2^{\overline{\cdot}}$$

$$PhN{=}N^{\cdot} \longrightarrow Ph^{\cdot} + N_2$$

$$Ph^{\cdot} + heme\ Fe \longrightarrow heme\ Fe\text{-}Ph \longrightarrow N\text{-}phenyl\ heme$$

This mechanism incorporates the following experimental observations of the phenylhydrazine-oxyhemoglobin reaction: nitrogen gas evolution (Nizet, 1946), benzene formation (Beaven and White, 1954), hydrogen peroxide production (Cohen and Hochstein, 1964), and N-phenylated heme formation (Ortiz de Montellano et al., 1981). In addition, in vitro ESR experiments have conclusively proven the formation of the superoxide anion radical (Goldberg et al., 1979), the phenyl radical (Hill and Thornalley, 1981), and the phenylhydrazyl radical (Smith and Maples, 1985). These in vitro studies have recently been thoroughly reviewed (Mottley and Mason, 1988) and, therefore, will not be further discussed here.

Although ESR has been successfully applied to the study of the reaction of oxyhemoglobin with hydrazine-based drugs in vitro, ESR has only recently been applied to the analogous in vivo investigations. In the perfused liver, PBN trapped the same radical from hydrazine, acetylhydrazine, and isoniazid (Sinha, 1987). The hyperfine splitting constants of this species in benzene (a^N = 14.2 G and a^H_β = 3.2 G) are the same as those of PBN/$_N^{\cdot}$COCH$_3$. The liver perfusion of iproniazid formed another species (a^N = 14.6 G and a^H_β = 2.7 G), which was assigned to the PBN/isopropyl radical adduct (Sinha, 1987). In the absence of any hydrazine a third species, present in the fluosal-43 perfusate, was assigned as a lipid-derived radical adduct by Sinha (1987). The marked difference in the hyperfine splitting constants reported for the PBN/$^{\cdot}$COCH$_3$ and PBN/isopropyl radical adducts by Albano and Tomasi (1987) and Sinha (1987) may be due to the solvent effect of chloroform and benzene, respectively.

In the whole animal, Maples et al. (1988a) detected the formation of an immobilized radical adduct ($2a^N_{zz}$ = 63.6 G and a^H_{zz} = 9.5 G) in the blood of rats which received an intraperitoneal injection of DMPO followed by an intragastric dose of phenylhydrazine. This immobilized radical adduct was detected when phenylhydrazine was administered at a dose of only 1 mg/kg. Hydrazine itself gives a weaker spectrum of the same species. The immobilized radical adduct co-chromatographs with oxyhemoglobin and could be detected in vitro using purified rat hemoglobin, phenylhydrazine, and DMPO. The sulfhydryl reagents, iodoacetamide, maleimide, and N-ethylmaleimide all inhibit phenylhydrazine dependent radical adduct formation when whole rat blood is treated in vitro (Maples et al., 1988a). This sulfhydryl-dependent radical adduct has been assigned to a DMPO/hemoglobin thiyl radical adduct. This is the first report of the formation of a radical adduct of a biological macromolecule free radical formed as a consequence of free radical metabolism. In addition, PBN could replace DMPO in vivo to yield the PBN/hemoglobin thiyl radical adduct, $2a^N_{zz}$ = 61.6 G (Maples et al. (1988b). Maples et al. (1988a) were also able to detect a weak signal from the DMPO/phenyl radical adduct in chloroform extracts of whole blood. In subsequent work, Maples et al. (1988b) examined the formation

of the DMPO/hemoglobin thiyl radical adduct in the blood of rats
following the administration of hydrazine-based drugs. The drugs examined were hydralazine, iproniazid, isoniazid, and phenelzine. Of the
four drugs, only iproniazid and phenelzine were able to induce
DMPO/hemoglobin thiyl radical adduct formation in vivo, whereas only
hydralazine and phenelzine were able to form this adduct in vitro.
Maples et al. (1988b) were able to decrease the in vivo iproniazid-
induced radical adduct formation by pretreating the rats with bis-para-
nitrophenylphosphate, an arylamidase inhibitor. These results support
the argument that iproniazid is hydrolyzed in the liver to a more reactive metabolite, most likely isopropyl hydrazine, which is subsequently
released into the blood stream. In contrast, phenylhydrazine and phenelzine react directly with red blood cells to yield the DMPO/hemoglobin
thiyl radical adduct. As hydralazine did not yield this adduct in vivo,
Maples et al. (1988b) propose that hydralazine is metabolized in vivo
into a less reactive compound, possibly via acetylation. Likewise, isoniazid is rapidly acetylated in vivo. Maples et al. (1988b) showed that
if acetylhydrazine, the metabolite which would result from the action of
arylamidase enzymes on acetylated isoniazid, was released into the blood
stream, it would not be very effective in yielding the DMPO/hemoglobin
thiyl radical adduct.

CONCLUSION

In conclusion, the detection and identification of free radical
metabolites produced in vivo should be more useful to the understanding,
and even to the prediction of toxicities than traditional analytical
techniques, such as HPLC, which are inherently limited to the detection
of stable metabolites.

REFERENCES

Ahmad, F.F., D.L. Cowan, and A.Y. Sun. Detection of free radical formation in various tissues after acute carbon tetrachloride administration in the gerbil. Life Sciences 41: 2469-2475 (1987).

Albano, E., K.A.K. Lott, T.F. Slater, A. Stier, M.C.R. Symons, and A.
Tomasi. Spin-trapping studies on the free-radical products formed by
metabolic activation of carbon tetrachloride in rat liver microsomal
fractions isolated hepatocytes and in vivo in the rat. Biochem. J. 204:
593-603 (1982).

Aoshima, H., T. Kajiwara, T. Hatanaka, and H. Hatano. Electron spin
resonance studies on the lipoxygenase reaction by spin trapping and spin
labelling methods. J. Biochem. 82: 1559-1565 (1977).

Augusto, O., H.S. Beilan, and P.R. Ortiz de Montellano. The catalytic
mechanism of cytochrome P-450. Spin trapping evidence for one-electron
substrate oxidation. J. Biol. Chem. 257: 11288-11295 (1982).

432

Beaven, G.H., and J.C. White. Oxidation of phenylhydrazines in the presence of oxyhaemoglobin and the origin of Heinz bodies in erythrocytes.
Nature 173: 389-391 (1954).

Becker, A.R., and L.A. Sternson. Nonenzymatic reduction of nitrosobenzene to phenylhydroxylamine by NAD(P)H. Bioorg. Chem. 9: 305-312
(1980).

Buettner, G.R., 1982, 'The spin trapping of superoxide and hydroxyl
radicals', in: Superoxide Dismutase Volume II (L.W. Oberely, ed.) CRC
Press, Boca Raton, Florida, pp. 63-81.

Butler, T.C. Reduction of carbon tetrachloride in vivo and reduction of
carbon tetrachloride and chloroform in vitro by tissues and tissue
constituents. J. Pharmacol. Exp. Ther. 134: 311-319 (1961).

Cederbaum, A.I., and G. Cohen. Inhibition of the microsomal oxidation of
ethanol and 1-butanol by the free-radical, spin-trapping agent
5,5-dimethyl-1-pyrroline-1-oxide. Arch. Biochem. Biophys. 204: 397-403
(1980).

Cheeseman, K.H., E.F. Albano, A. Tomasi, M.U. Dianzani, and T.F. Slater.
The effects of the spin traps PBN and 4-POBN on microsomal drug metabolism and hepatocyte viability. Life Chem. Reports 3: 259-264 (1985).

Cinti, D.L. Agents activating the liver microsomal mixed function oxidase system. Pharmac. Therap:A 2: 727-749 (1978).

Cohen, G. and P. Hochstein. Generation of hydrogen peroxide in erythrocytes by hemolytic agents. Biochemistry 3: 895-900 (1964).

Connor, H.D., R.G. Thurman, M.D. Galizi, and R.P. Mason. The formation
of a novel free radical metabolite from CCl_4 in the perfused rat liver
and in vivo. J. Biol. Chem. 261: 4542-4548 (1986).

Eriksson, U.G., R.C. Brasch, and T.N. Tozer. Nonenzymatic bioreduction
in rat liver and kidney of nitroxyl spin labels, potential contrast
agents in magnetic resonance imaging. Drug Metab. Disp. 15: 155-160
(1987).

Finkelstein, E., G.M. Rosen, E.J. Rauckman, and J. Paxton. Spin trapping
of superoxide. Mol. Pharmacol. 16: 676-685 (1979).

Finkelstein, E., G.M. Rosen, and E.J. Rauckman. Spin trapping of
superoxide and hydroxyl radical: practical aspects. Arch. Biochem.
Biophys. 200: 1-16 (1980).

Fujii, K., M. Morio, H. Kikuchi, S. Ishihara, M. Okida, and F. Ficor.
In vivo spin-trap study on anaerobic dehalogenation of halothane. Life
Sci. 35: 463-468 (1984).

Goldberg, B., A. Stern, J. Peisach, and W.E. Blumberg. The detection of superoxide anion from the reaction of oxyhemoglobin and phenylhydrazine using EPR spectroscopy. Experientia 35: 488-489 (1979).

Green, M.J., H.A.O. Hill, and D.G. Tew. Applications of spin-trapping to biological systems. Biochem. Soc. Trans. 13: 600-603 (1985).

Hampton, M.J., R.A. Floyd, E.G. Janzen, and R.V. Shetty. Mutagenicity of free-radical spin-trapping compounds. Mut. Res. 91: 279-283 (1981).

Horie, S., T. Watanabe, and Y. Ogura. Studies on the enzymatic reduction of C-nitroso compounds. J. Biochem. 88: 847-857 (1980).

Janzen, E.G., 1980, 'A critical review of spin trapping in biological systems', in: Free Radicals in Biology Volume IV (W.A. Pryor, ed.) Academic Press, New York, pp. 115-154.

Kalyanaraman, B., E. Perez-Reyes, and R.P. Mason. The reduction of nitroso-spin traps in chemical and biological systems. A cautionary note. Tetrahedron Lett. 1979: 4809-4812 (1979a).

Kalyanaraman, B., R.P. Mason, E. Perez-Reyes, C.F.Chignell, C.R. Wolf, and R.M. Philpot. Characterization of the free radical formed in aerobic microsomal incubations containing carbon tetrachloride and NADPH. Biochem. Biophys. Res. Commun. 89: 1065-1072 (1979b).

Kalyanaraman, B., E. Perez-Reyes, and R.P. Mason. Spin-trapping and direct electron spin resonance investigations of the redox metabolism of quinone anticancer drugs. Biochim. Biophys. Acta 630: 119-130 (1980).

Kalyanaraman, B., 1982, 'Detection of toxic free radicals in biology and medicine', in: Rev. Biochem. Toxic., Volume 4 (E. Hodgson, J.R. Bend, and R.M. Philpot, eds.) Elsevier Biomedical, New York, pp. 73-139.

Knecht, K.T., and R.P. Mason. In vivo radical trapping and biliary secretion of radical adducts of carbon-tetrachloride-derived free radical metabolites. Drug. Metab. Disp., in press (1988).

Kubow, S., E.G. Janzen, and T.M. Bray. Spin-trapping of free radicals formed during in vitro and in vivo metabolism of 3-methylindole. J. Biol. Chem. 259: 4447-4451 (1984).

LaCagnin, L.B., H.D. Connor, R.P. Mason, and R.G. Thurman. The carbon dioxide anion radical adduct in the perfused rat liver: Relationship to halocarbon-induced toxicity. Mol. Pharmacol. 33: 351-357 (1988).

Lai, E.K., P.B. McCay, T. Noguchi, and K.-L. Fong. In vivo spin-trapping of trichloromethyl radicals formed from CCl_4. Biochem. Pharm. 28: 2231-2235 (1979).

Lai, E.K., C. Crossley, R. Sridhar, H.P. Misra, E.G. Janzen, and P.B. McCay. In vivo spin-trapping of free radicals generated in brain, spleen, and liver during γ radiation of mice. Arch. Biochem. Biophys. 244: 156-160 (1986).

Maples, K.R., S.J. Jordan, and R.P. Mason. In vivo rat hemoglobin thiyl free radical formation following phenylhydrazine administration. Mol. Pharmacol. 33: 344-350 (1988a).

Maples, K.R., S.J. Jordan, and R.P. Mason. In vivo rat hemoglobin thiyl free radical formation following the administration of phenylhydrazine and hydrazine-based drugs. Drug Metab. Dispos., in press (1988b).

Mason, R.P., B. Kalyanaraman, B.E. Tainer, and T.E. Eling. A carbon-centered free radical intermediate in the prostaglandin synthetase oxidation of arachidonic acid. J. Biol. Chem. 255: 5019-5022 (1980).

Mason, R.P., 1982, 'Free-radical intermediates in the metabolism of toxic chemicals', in: Free Radicals in Biology, Volume V (W.A. Pryor, ed.), Academic Press, New York, pp. 161-222.

Mason, R.P., 1984, 'Spin trapping free radical metabolites of toxic chemicals', in: Spin Labeling in Pharmacology, (J.L. Holtzman, ed.) Academic Press, New York, pp. 87-129.

Mason, R.P., and C. Mottley, 1987, 'Spin trapping free radical metabolites of inorganic chemicals', in: Electron Spin Resonance, Volume 10B (M.C.R. Symons, senior reporter), Royal Society of Chemistry, London, pp. 185-197.

McCay, P.B., E.K. Lai, J.L. Poyer, C.M. DuBose, and E.G. Janzen. Oxygen- and carbon-centered free radical formation during carbon tetrachloride metabolism. J. Biol. Chem. 259: 2135-2143 (1984).

Mottley, C., B. Kalyanaraman, and R.P. Mason. Spin trapping artifacts due to the reduction of nitroso spin traps. FEBS Lett. 130: 12-14 (1981).

Mottley, C., T.B. Trice, and R.P. Mason. Direct detection of the sulfur trioxide radical anion during the horseradish peroxidase-hydrogen peroxide oxidation of sulfite (aqueous sulfur dioxide). Mol. Pharmacol. 22: 732-737 (1982).

Mottley, C., and R.P. Mason, 1988, 'Nitroxide radical adducts in biology: chemistry, applications, and pitfalls', in: Spin Labeling. Theory and Applications, Volume III (L.J. Berliner, ed.) Academic Press, New York.

Murray, M., and A.J. Ryan. Inhibition and enhancement of mixed-function oxidases by nitrogen heterocycles. Biochem. Pharmacol. 31: 3002-3005 (1982).

Nizet, A. Une reaction de la phenylhydrazine avec l'oxyhemoglobine. Application au dosage de l'hemoglobine dans le sang. Bull. Soc. Chim. Biol. 28: 527-530 (1946).

Ortiz de Montellano, P.R., and K.L. Kunze. Formation of N-phenylheme in the hemolytic reaction of phenylhydrazine with hemoglobin. J. Am. Chem. Soc. 103: 6534-6536 (1981).

Ortiz de Montellano, P.R., O. Augusto, F. Viola, and K.L. Kunze. Carbon radicals in the metabolism of alkyl hydrazines. J. Biol. Chem. 258: 8623-8629 (1983).

Perkins, M.J., Spin trapping, Advan. Phys. Org. Chem. 17: 1-64 (1980).

Plummer, J.L., A.L.J. Beckwith, F.N. Bastin, J.F. Adams, M.J. Cousins, and P. Hall. Free radical formation in vivo and hepatotoxicity due to anesthesia with halothane. Anesthesiology 57: 160-166 (1982).

Poyer, J.L., P.B. McCay, C.C. Weddle, P.E. Downs. In vivo spin-trapping of free radicals formed during halothane metabolism. Biochem. Pharm. 30: 1517-1519 (1981).

Poyer, J.L., P.B. McCay, E.K. Lai, E.G. Janzen, and E.R. Davis. Confirmation of assignment of the trichloromethyl radical spin adduct detected by spin trapping during ^{13}C-carbon tetrachloride metabolism in vitro and in vivo. Biochem. Biophys. Res. Commun. 94: 1154-1160 (1980).

Reinke, L.A., E.K. Lai, C.M. DuBose, and P.B. McCay. Reactive free radical generation in vivo in the heart and liver of ethanol-fed rats: Correlation with in vitro radical formation. Proc. Natl. Acad. Sci. USA, in press (1988).

Rosen, G.M., and E.J. Rauckman. Formation and reduction of a nitroxide radical by liver microsomes. Biochem. Pharmacol. 26: 675-678 (1977).

Rosen, G.M., and E. Finkelstein, 1985, Use of spin traps in biological systems, Adv. Free Rad. Biol. Med. 1: 345-375.

Sinha, B.K. Activation of hydrazine derivatives to free radicals in the perfused liver: a spin trapping study. Biochim. Biophys. Acta 924: 261-269 (1987).

Smith, P., and K.R. Maples. EPR study of the oxidation of phenylhydrazine initiated by the titanous chloride/hydrogen peroxide reaction and by oxyhemoglobin. J. Magn. Reson. 65: 491-496 (1985).

Sridhar, R., 1981, 'Accelerated oxygen consumption by catecholamines in the presence of aromatic nitro and nitroso compounds. Implications for neurotoxicity of nitro compounds', in: Oxygen and Oxy-radicals in

Chemistry and Biology (M.A.J. Rodgers and E.L. Powers, eds.), Academic Press, New York, pp. 363-365.

Stier, A., and E. Sackmann. Spin labels as enzyme substrates. Heterogeneous lipid distribution in liver microsomal membranes. Biochim. Biophys. Acta 311: 400-408 (1973).

Tomasi, A., E. Albano, F. Biasi, T.F. Slater, V. Vannini, and M.U. Dianzani. Activation of chloroform and related trihalomethanes to free radical intermediates in isolated hepatocytes and in the rat in vivo as detected by the ESR-spin trapping technique. Chem.-Biol. Interactions 55: 303-316 (1985).

Wolf, C.R., W.G. Harrelson, Jr., W.M. Nastainczyk, R.M. Philpot, B. Kalyanaraman, and R.P. Mason. Metabolism of carbon tetrachloride in hepatic microsomes and reconstituted monooxygenase systems and its relationship to lipid peroxidation. Mol. Pharmacol. 18: 553-558 (1980).

THE INVOLVEMENT OF FREE RADICALS IN THE METABOLISM OF DRUGS

V. MALATESTA

R. & D. Farmitalia-Carlo Erba
Via dei Gracchi 35,
20146 Milano, Italy

ABSTRACT. The cytotoxycity of the anthracycline antitumour drugs results from multiple biochemical alterations of the cellular organization at several levels. The expression of the cytotoxic effects is subsequent to the activation of the drug through: reduction, metal-ion complexation and irradiation. A correlation between the cardiotoxicity, the main side-effect displayed by these drugs, the reductive deglycosidation and the metal-ion catalyzed side-chain degradation is proposed.

F. Minisci (ed.), Free Radicals in Synthesis and Biology, 437–460.
© 1989 by Kluwer Academic Publishers.

INTRODUCTION

The anthracyclines are natural glycosides currently used in the clinical practice for the treatment of various malignancies such as cancer of the ovary, breast, lung, stomach and less common tumours including lymphoma, sarcoma and leukaemia. The anthraquinone moiety of the structure is responsible for the lipophylicity, the redox and metal-ion chelating properties of these molecules, whereas the hydrophylic sugar moiety S, which is generally the aminosugar L-daunosamine, is thought to be essential for the drug transport. The doxorubicin (adriamycin) and daunorubicin (daunomycin) are the best known representatives of these large group of quinonoid xenobiotics.

R = OH ; R' = O   adriamycin
R = H ; R' = O   daunomycin

There is no general consensus regarding the anthracyclines mode of action despite the many efforts and the over 2000 papers published to date. It is however becoming evident that the cytotoxicity of anthracyclines results from multiple biochemical alterations at several levels of the cellular organization. It has been documented that doxorubicin inhibits microsomal (1) and mitochondrial (2) electron transport enzymes; associates to cardiolipin rich membranes (3, 4); is engaged in a dynamic intercalative interaction with DNA (5, 6); alkylates proteins (7); induces DNA strand scission (8) and enhances the peroxidation of microsomal (9), mitochondrial (10), intact cells (11) and nuclear membranes(12). In general the one or two-electron *reduction* of anthracyclines, the *chelation* of transition metal ions and subsequent reaction with oxygen leads to generation of reduced active oxygen species which are ultimately responsible for the cell kill. Cytotoxic oxygen activated species may, in principle, originate also *via* a *photodynamic* mechanism with the anthracyclines behaving as photosensitizers.

REDUCTIVE ACTIVATION

The doxorubicin and other clinically useful anthracyclines have been demonstrated to undergo enzymatic one or two-electron reduction *in vitro* and *in vivo*. The reduction is usually accomplished by enzymes such as xanthine oxidase, cytochrome P450 reductase, cytochrome b5 reductase, NADH reductase and a not yet well identified class of enzymes generally called doxorubicin reductases. The monoelectronic reduction product is the semiquinone, which can either be reoxidized back to the parent glycoside by reaction with oxygen or undergo, when certain substitution pattern requirements are met, an intramolecular rearreangement (deglycosidation) (13) to a C7 centered free radical which may, by reaction with DNA, RNA, proteins etc. lead to the alkylation of these biomolecules(14, 15). *Scheme 1* .

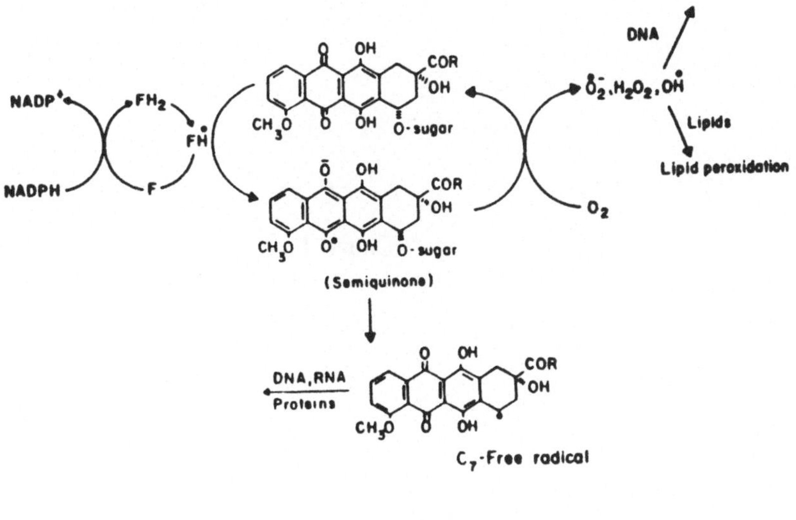

Scheme 1

We have found (13, 16) that the 6-deoxyanthracyclines are unable to undergo the sugar linkage splitting reaction when reduced either electrochemically or enzymatically. Indeed while daunomycin , 4-demethoxy-11-deoxydaunomycin,10(R)-methoxy-daunomycin (*Chart 1*) lead, upon reduction, to the formation of the corresponding 7-deoxyaglycone through an E.C.E. (Electrochemical-Chemical,-Electrochemical) mechanism and two reduction peaks are observed, the first being attributed to the reduction of the glycoside and the second to that of the deglycosidation product (7-deoxyaglycone), the 10(S)-methoxy-daunomycin and the 4-demethoxy-6-deoxy-daunomycin are characterized by a *single* reduction peak in either cyclic voltammetry or differential pulse polarography (13). *Figures 1 , 2.*

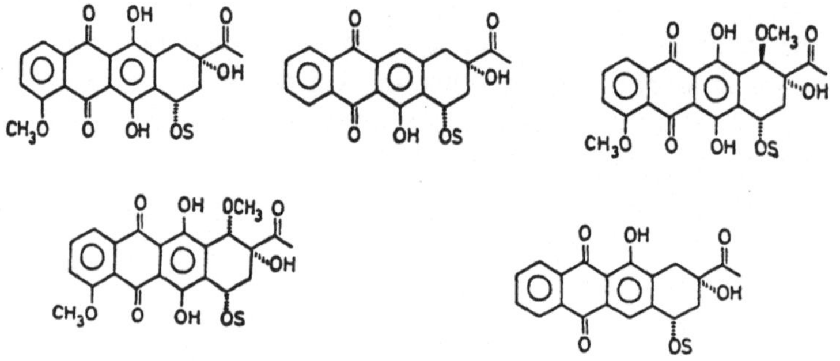

Chart 1

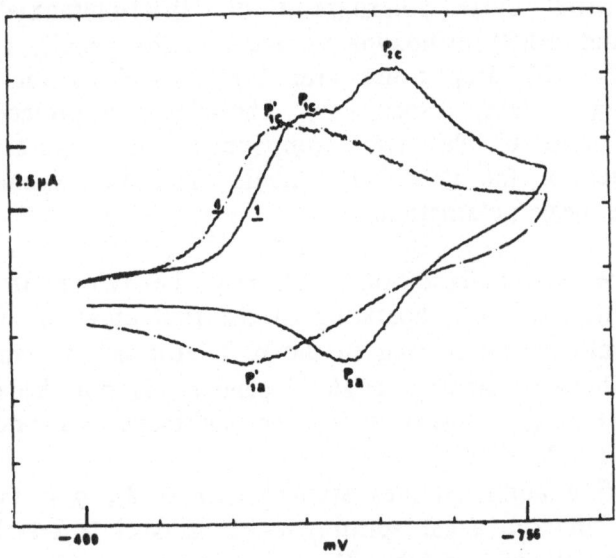

Fig. 1 . Cyclic voltammograms of 6-deoxycarmynomycin **4** and daunomycin **1** (see Chart 2) , 8.10^{-5} M, pH=7.5 at 37.5 °C, v= 0.1V/s

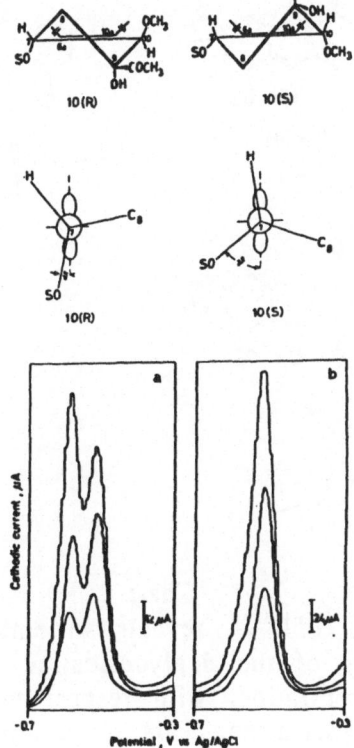

Fig.2. Differential pulse polarograms of 10(R)-methoxydaunomycin (left panel) and 10(S)-methoxydaunomycin (right panel), 1.9.10-5 M, v=1.33 mV/s. The drop times are: 1.4, 3, and 6s for the lower, middle and top curves, respectively. The Newman projections along the C7-C6a bond of the two compounds are characterized by different values of the Ø dihedral angle. The 6-deoxyanthracyclines give a single peak polarogram.

We have rationalized this different behaviour in terms of *stereoelectronic effect* operating in the first class of anthracylines having the cyclohexenic A ring in the α conformation, whereas such an effect , which is dihedral angle dependent, is not, because of the larger dihedral angle, observed for the anthracyclines preferring the β conformation (13) *Figure 2.*

Of the five anthracyclines studied, (Chart 2), 4 is the only one that does not undergo sugar splitting, whereas such a reaction is very efficient in the case of 5 *(Chart 2) .*

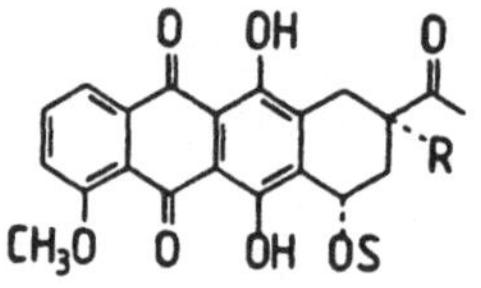

1 : R = OH
2 : R = H

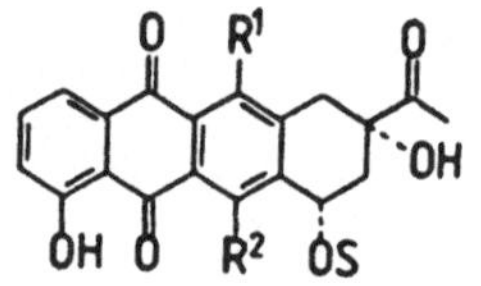

3 : R^1 = H ; R^2 = OH
4 : R^1 = OH ; R^2 = H

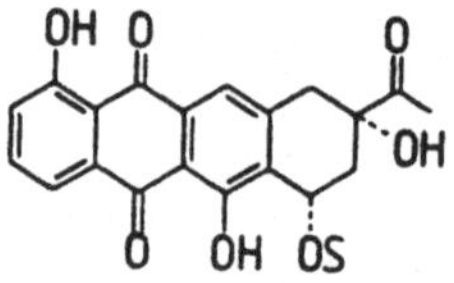

5

S =

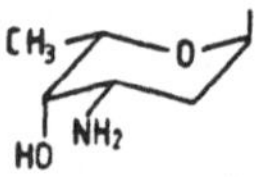

Chart 2

An inspection of *Table 1* reveals an empirical correspondence between the efficiency of the deglycosidation reaction of 5 and its relatively higher concentration, with respect to *e.g.* 1. required for inhibition of colony formation. This observation seems to indicate that somehow an efficient deglycosidation leads to a lesser availability of the parent anthracycline and hence to a lower toxicity. A similar trend seems to be displayed by 2. On the contrary 4 behaves as an efficient electron-shuttle between the enzyme and molecular oxygen thus generating highly toxic reduced oxygen species (*Scheme 1*). The same type of *rationale* seems to emerge by comparing the values of the optimal doses for antitumour activity on Gross leukaemia or mammary carcinoma, required for 1 and 4 and their analogues (*Table 2*).

Table 1. Colony-forming ability of HeLa cell cultures after exposition to the drugs for 24 hrs

Compound	Concentration(ng/mL) required for 50% inhibition
1	15
2	1200
3	14
4	3.7
5	250

If the cardiotoxicity, the most serious side effect displayed by these drugs, is linked to mitochondrial membrane lipid peroxidation(17), we anticipate that the anthracyclines which are able to preserve,after reduction, their structural integrity, will function as efficient oxygen redox cycling catalysts (*Scheme 1*) and induce severe cardiac injury at relatively lower doses. The results reported in *Table 2*, seem to support this mechanistic picture.

Table 2. Antitumour activity and cardiotoxicity of 6-deoxycarminomycin (4), its doxorubicin analogue (6) and of daunomycin(1) and doxorubicin

| Compound | Gross Leukaemia[a] | | Mammary[b] | Cardiotoxicity[c] | |
	O.D.(mg/Kg)	T/C%	carcinoma	dose (mg/kg)	AGL[d]
1	15	183		8	0.5
				12	2
4	2.5	175			
doxorubicin	13	208	6	6	1.7
			7.5	7	3.2
6	5	217	1.8	1.4	1.2
			2.3	1.8	3.2
			3	2.3	6.0

a) CH3 mice received 2.10^6 leukaemia cells iv on day O. The drugs were injected iv on day 1.

b) CH3 female mice were inoculated sc with 20.10^6 cells from a third generation transplanted of mammary carcinoma oh CH3 mice. The drugs were administered iv once a weel, starting when tumours were palpable.

c) cardiotoxicity on CH3 female mice treated iv once a week for 4 weeks

d) average grade of ventricular lesions

We have recently found (18) that a reduced intermediate may originate during the irradiation of anthracyclines at 308nm with an excimer laser (XeCl), probably through a charge transfer mechanism between the excited and the ground state molecule. The main product formed is the 7-deoxyaglycone when the anthracycline concentration is $> 10^{-3}$M. At lower concentration ($< 10^{-5}$sM) the fully aromatized compound predominates Scheme 2.

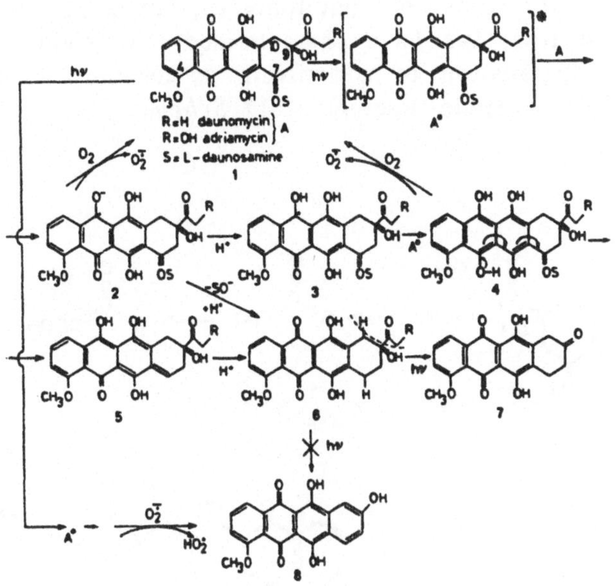

Scheme 2

METAL-ION COMPLEXATION

The anthracyclines are characterized by a very high affinity for Fe(III) and Cu(II) ions (19, 20). Although the extracellular concentration of free Fe(III) and Cu(II) ions is exceedingly small (ca. 10^{-12}M) it has been reported that *iron pools* may form when the bound metal ion is released from e.g. ferritin by reaction with the adriamycin semiquinone (23) or from transferrin by a drop of the cellular pH to 5.5-6.5 (24), and concentrations close to 10^{-5}M can be achieved. As for the copper, the typical plasma concentration is ca. 10^{-5}M (25) with most of the Cu(II) ions being "locked up" in structures such as those of thyrosinase, SOD, amineoxidase, cytochrome c oxidase, etc. However the metal ion concentrations tend to increase during infections, epilepsy, arthritic diseases, cancer, etc. It is then meaningful to consider the interaction of these metal-ions with the anthracyclines.The Fe(III) complex of adriamycin is, at

446

variance with that of daunomycin, capable of reducing the coordinated iron. We have characterized the product of the iron(III) catalyzed degradation of adriamycin as deriving from the oxidation of the hydroxyketone side-chain(21) *Scheme 3*.

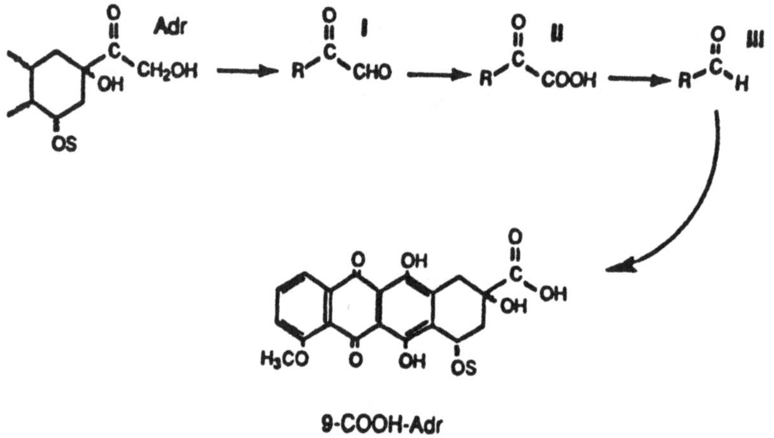

9-COOH-Adr

Scheme 3

The formation of this product, the 9-carboxyderivative, is modulated by the availability of the sugar NH2 group. We have reported (22) that the conformational equilibrium about the C7-07 bond is different in the case of adriamycin and 4'-epi-adriamycin because of the presence in the latter compound of H-bonds between the C4'-OH group and the C5,C6 hydroxyquinone functionality. If this conformational preference if verified also in the case of the Fe(III) complex, then the sugar NH2 is prevented from participating in a *further* coordination of the Fe(III) chelated on the side-chain. In the 3'-N-acetyl-adriamycin,although the conformational equilibrium should not differ from that of adriamycin, the electronic availability on the nitrogen of the NH-acetyl group is reduced and no additional participation in the Fe(III) chelation is expected *Chart 3*.

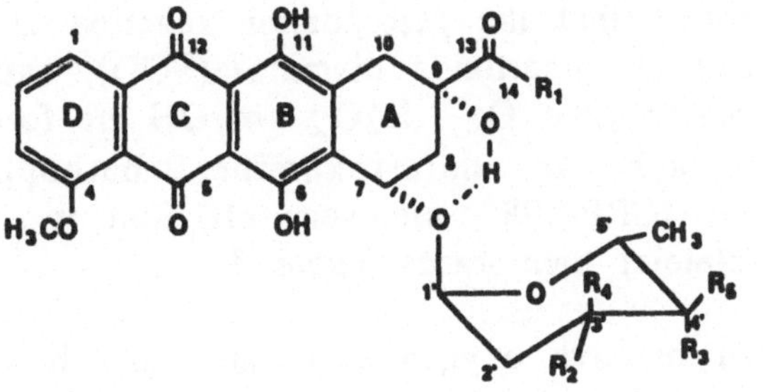

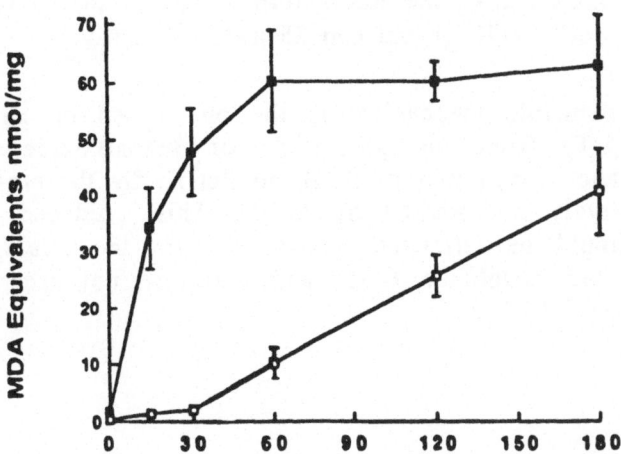

Anthracycline	R_1	R_2	R_3	R_4	R_5
Adriamycin	CH_2OH	NH_2	OH	H	H
4'-epi-Adr	CH_2OH	NH_2	H	H	OH
Daunorubicin	CH_3	NH_2	OH	H	H
3'-N-acetyl-Adr	CH_2OH	$NHCOCH_3$	OH	H	H

Chart 3

We have found that the amount of malondialdehyde (MDA) formed in the presence of peroxidizeable substrates such as platelet membranes, parallels the amount of reducible Fe(III) and it reaches a *plateau* after 1h in the case of adriamycin but the reaction is rather sluggish with daunomycin *Figure 3*.

Fig.3. Peroxidation of platelet membranes by Fe(III) complexes of adriamycin (solid squares) and daunomycin (open squares). The complexes (100 μM anthracycline and 25 μM Fe(III)), were reacted with 120 μg/ml of membrane protein.

It is worthnoting that the peroxidation reaction is not inhibited by scavengers such as superoxide dismutase(SOD), catalase,mannitol and this implies that no *free* O_2^- ,H_2O_2 or $\cdot OH$ are formed. However iron chelators such as deferioxamine, bathophenanthroline sulphonate(BPS) or ICRF-198 are very efficient in quenching the peroxydation of platelet membranes *Table 3.*

Table 3. Effect of reactive oxygen scavengers and iron chelators on lipid peroxidation by ferric complexes of adriamycin and daunorubicin[a]

Addition	Adr		Dnr	
	nmol/mg	%	nmol/mg	%
None	43.6	100	6.32	100
SOD (200 U/ml)	43.1	99	7.12	113
Catalase (400 U/ml)	45.5	104	7.21	114
Mannitol (10 mM)	44.6	102	8.00	128
Deferioxamine(20mM)	0	0	0	0
BPS (20 mM)	0.8	2	0.54	9
ICRF-198 (20 mM)	3.7	8	0.85	13

a)Results were subtracted of the MDA formed (2) in the presence of membranes and $FeCl_3$ (110 µg/ml and 25 µM, respectively)

The most plausible mechanism is the one based on an initial charge transfer(CT) from the phenoxide or ketoalkoxide ion to chelated Fe(III) and reoxydation of Fe(II) to Fe(III) by the *molecular oxygen* reductively coordinated by Fe(II). This reduced oxygen *crypto*species should be ultimately responsible for the *site-specific oxydation* of the membrane lipids and therefore not accesible to scavengers *Scheme 4.*

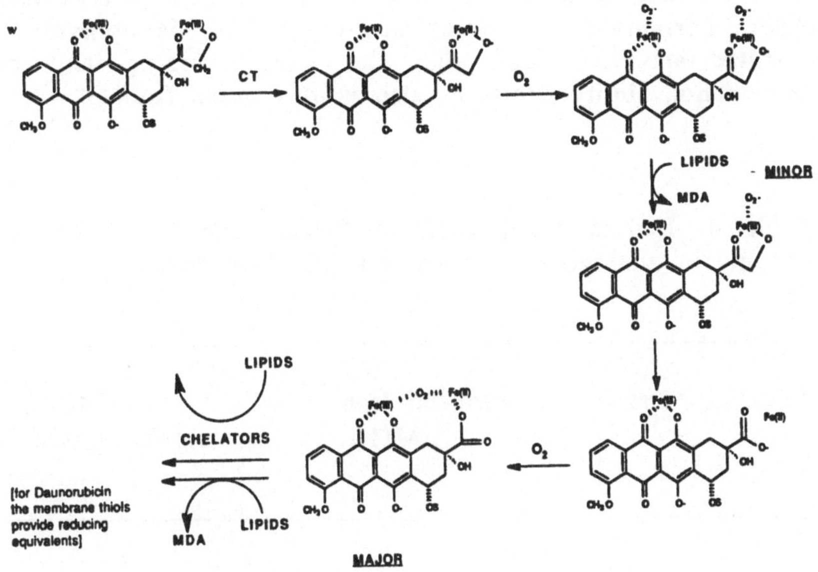

Scheme 4

Finally the participation of the sugar NH_2 in the Fe(III) coordination may affect the metal ion reduction potential through a *spin-dependent* mechanism and explain the reduced ability of 4'-epìadriamycin and 3'N-acetyladriamycin iron complexes to induce lipid membrane peroxidation*Scheme 5*.

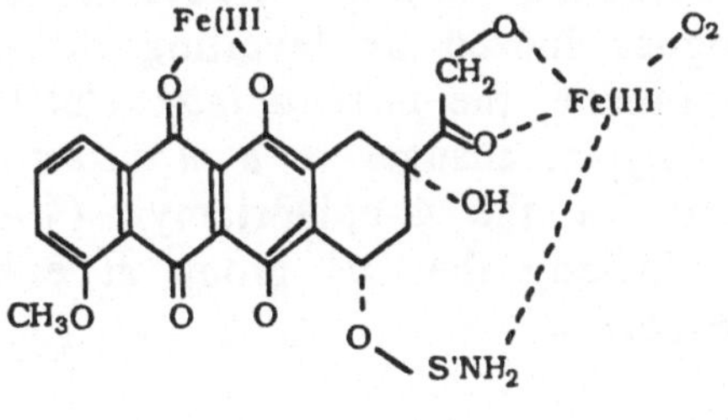

$$Fe(III)_{LS} \underset{}{\overset{K_A}{\rightleftarrows} } Fe(III)_{HS} \underset{}{\overset{K_B}{\rightleftarrows} } Fe(II)_{HS}$$

Scheme 5

It is interesting to notice that the cardiotoxicity decreases in the order adriamycin > 4'-epiadriamycin > daunomycin > 3'N-acetyladriamycin and that the iron reducing and membrane peroxidation ability decreases also in this order *Table 4*.

Table 4. Reduction of chelated Fe(III) and lipid peroxidation by adriamycin analogues modified on the daunosamine.

Anthracycline	Reduced iron µM/1h	MDA nmol/mgprotein/1h
Adr	12.9	49.9
4'-epi-Adr	9.5	39.0
3'N-acetyl-Adr	5.3	6.1

The copper(II) ion behaves as a powerful paramagnetic probe being able to evidence minor structural differences in the anthracycline analogues Indeed by inverting the chirality of the C4' center of the daunosamine, the *polynuclear* Cu(II) complex observed in the case of adriamycin, changes to a *mononuclear* structure This structure is verified for the 4'-epiadriamyn (4'-epidoxorubicin) and also the analogues lacking the OH group at either the C6 or C11 positions.(26, 27) *Figure 4*.

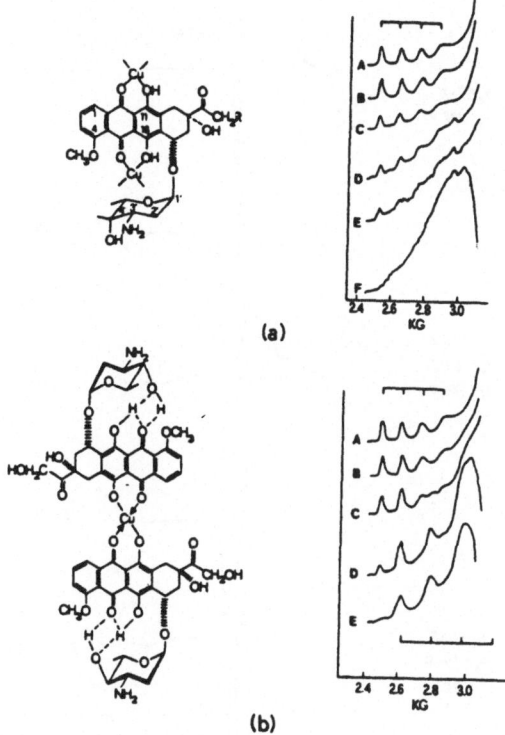

(a)

(b)

Fig.4. (a) g region of the e.s.r. spectra of 10^{-3}M frozen (-150°C) aqueous solution of Cu(II)/adriamycin,r =1, at pH: A,5.0; B, 5.5; C, 6.0; D,6.5; E, 6.8; F, 7.0; (b) spectra at the reported pH values, of the Cu(II)/4'-epiadriamycin complex.

The structure of the copper complexes is also dependent upon the anthracycline to Cu(II) ratio r , *Figure 5,* with the metal ion being also able to evidence the *self-aggregation* of the anthracycline when r =5-10 (28) In the latter case a different type of *mononuclear* complex having the following structure is postulated *Figure 6.*

452

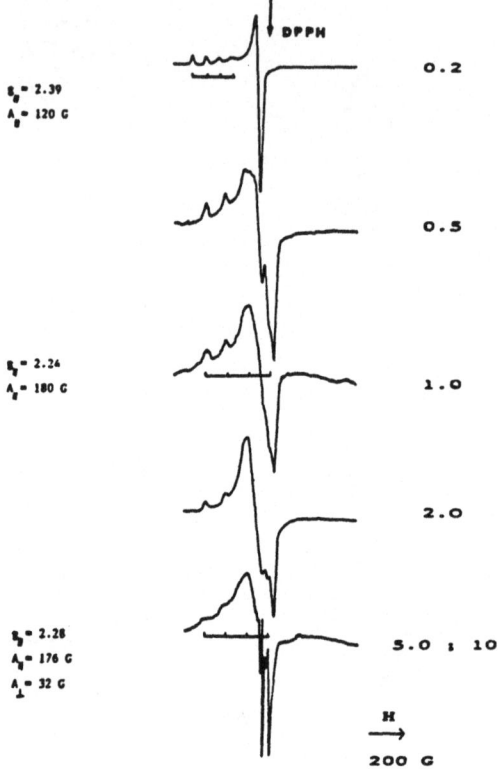

$g_0 = 2.39$
$A_0 = 120$ G

$g_1 = 2.24$
$A_0 = 180$ G

$g_1 = 2.28$
$A_1 = 176$ G
$A_\perp = 32$ G

DPPH

0.2

0.5

1.0

2.0

5.0 : 10

H
$\longrightarrow$
200 G

Fig.5. r dependence of e.s.r. spectra of Cu(II)/4'epi-adriamycin at pH=6.5.

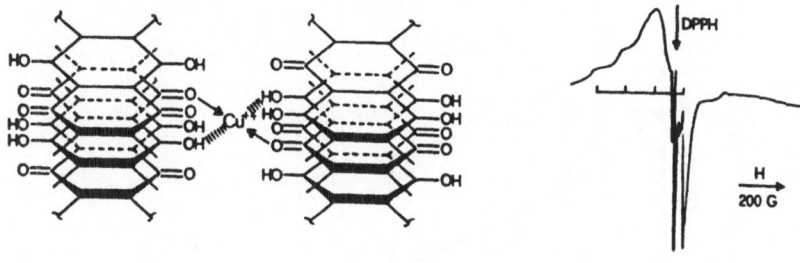

Fig.6. Proposed structure and e.s.r. spectrum of the Cu(II)/4'-epiadriamycin at r =10; pH=6.5 and -150°C.

A similar ESR spectrum is also observed by interaction of adriamycin with SOD(29). Another important feature of these complexes is their easier formation in the presence of *molecular oxygen* and the pH value (ca.6) at which they form that is generally lower than the pKa of the anthracycline phenolic OH (pKa>10) (28).

From preliminary experiments on rat hearts, it appears that the anthracycline-Cu(II) complexes have a lower cardiotoxicity than the corresponding parent *free-ligand* as shown by the smaller change of the heart contractile force, observed after separate administration of the drug or its complex (30) *Figure 7*.

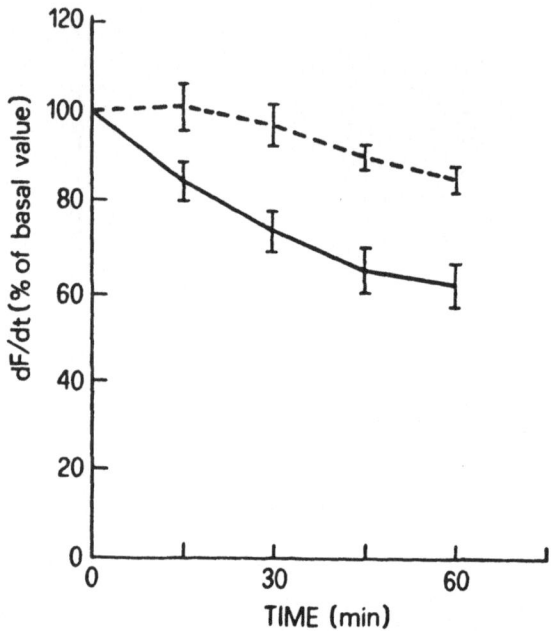

Fig.7. Contractile force % change induced by adriamycin (solid line) and Cu(II)/adriamycin complex (broken line).

This difference may result from a *dismutase type activity* of the copper complexes. This suggestion is not unreasonable as their redox potentials (ca.-150mV vs.Ag/AgCl) fall in between those of the O_2^-/O_2 and O_2^-/H_2O_2 couples and are slightly more cathodic than those of the natural superoxide dismutases(31).

PHOTOACTIVATION

Free radicals have been reported to form during irradiation of daunomycin or doxorubicin with UV or visible light. Spectroscopic ESR evidences for the anthracyclines mediated formation of O_2^-, ·OH and carbon centered radicals have been obtained through spin trapping experiments(32). Biological effects such as severe inflammation and skin ulceration of the parts of the body exposed to light(33), light-sensitized viral and bacterial death(34, 35), photodegradation of DNA(36), peptides or pyrimidine bases(37) following treatment of the experimental systems with anthracyclines have also been observed.

Although the question as to whether *singlet oxygen* is formed as result of a photosensitizing action by anthracycline has been addressed in the past, no photophysical studies have been,so far, carried out on this class of drugs. We have recently completed a series of measurements of the *singlet state, triplet state* lifetimes and quantum yields on three representative anthracyclines: adriamycin,daunomycin and 5-imminodaunomycin (5ID) (38) *Table 5*.

Table 5. Photophysical parameters , singlet oxygen quenching rate constants and oxydation potentials of anthracyclines.

| Anthracycline | 1_τ | 3_τ | Φ_T | Φ_Δ | kT^a | kR^b | kQ^b | Eox^c |
	ns	ns			$M^{-1}s^{-1}$ $x10^{-8}$	$M^{-1}s^{-1}$ $x10^{-7}$	$M^{-1}s^{-1}$ $x10^{-9}$	mV
Daunomycin	1	1.8	0.23	0.02	1	1.6	1.1	250
Adriamycin	1	1.7	0.23	0.02	1	1.5	0.9	320
5ID	0.6	1.6		0.02	1	1.5	3.2	310

a) for singlet oxygen generated photochemically at 22°C (38)
b) for singlet oxygen generated thermally at 35°C (39)
c) vs SCE, carbon glass electrode, pH 7.8 at 35°C (39)

The *singlet oxygen* quantum yields are exceedingly low(ca.0.02) and its quenching by the anthracyclines extremely fast and close to the diffusional limit. In a separate study(39), by generating the *singlet oxygen* thermally by decomposition at 35°C of 3-(4-methyl-1-naphtyl)propionic acid 1,4-endoperoxide, we have succeded in measuring its chemical kR and physical kQ quenching rate constants. As for the quenching mechanism, we believe that because of the anthracycline triplet energy values i.e. 36-42 Kcal/mol, an energy transfer from the *singlet oxygen* to *ground state* anthracycline is predicted to be highly endothermic and therefore not likely to occur. We favour the charge-transfer mechanism leading to a solvent-separated radical ion pair for which we can calculate, from the knowledge of the oxidation potentials of the three anthracyclines considered in our study and through the Rehm-Weller equation a ΔG=-7-8Kcal/mol, corresponding to the transfer of a full electron. It would then seem that although *singlet oxygen* is formed through the photosensitizing action of the anthracyclines,this activated species is promptly and efficiently deactivated by reaction with the parent drug molecule yielding superoxide anion O_2^- and its dismutation species. Based on these results one can conclude that the anthracyclines should indeed be very poor photosensitizers. However recent experiments (40) carried out on LoVo colon carcinoma *(Figure 8)* and/or FRTL-5 thyroid cell cultures treated with 5-imminodaunomycin and then irradiated at 350nm (XeF excimer laser) or at 590nm (argon pumped dye laser, Rhodamine 6G) show that these cells are severely damaged at the cellular membrane level and the % survival, in the case of the thyroid cells, is ca.2 after 30min. irradiation.

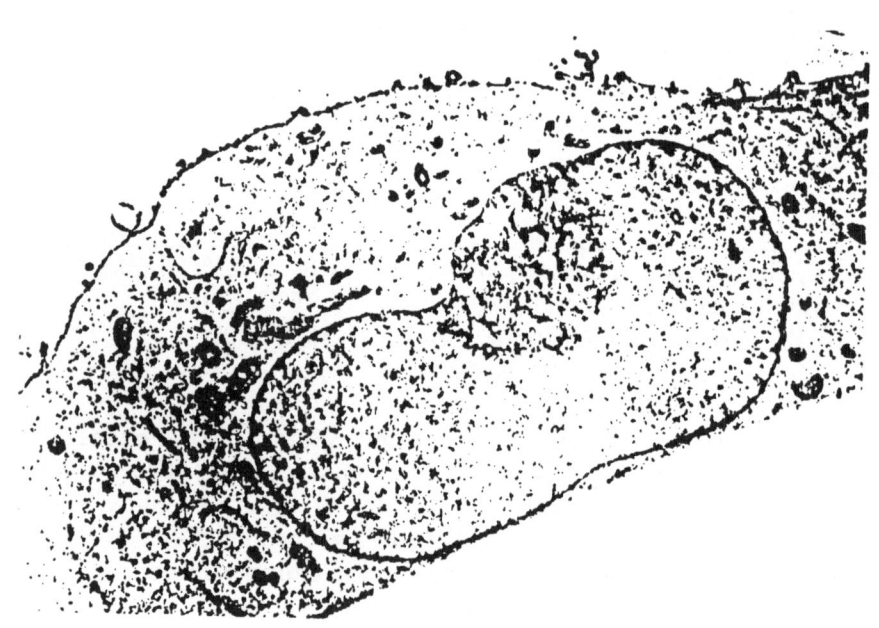

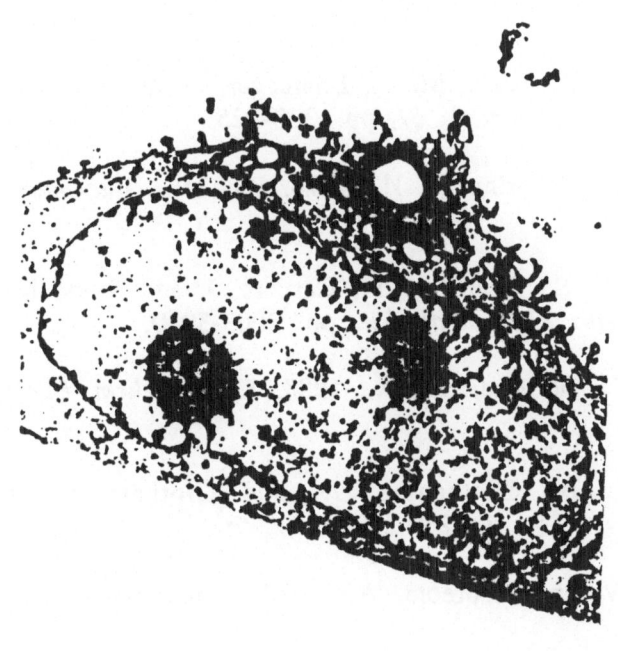

Fig.8. Top, Electron microscopy of LoVo cells treated with 5ID before and (bottom) after irradiation for 5 min (40Hz) at 350nm(20 mJ/p).

We believe that this is a clear demonstration of how difficult a task is the extrapolation of *bulk solution chemistry* results to the much more complex biological systems whose compartmentalized and sophisticated organization may result in a suppression or enhancement of the reactivity of species identified in the more simplistic aqueous systems.

AKNOWLEDGEMENTS

The financial support by the Italian NRC/Special Project Oncology is gratefully aknowledged (Grants no. 84.00690.44; 85.02262.44; 86.00495.44). The precious help of Dr. D. Martini in the preparation of the manuscript is appreciated.

REFERENCES

1. Mimnaugh, E. G., Trush, M. A., Bhatnagar, M. and Gram, T.E.: 1981, *Toxicol. Appl.Pharmacol. 61,* pp. 313-325.

2. Mailer, K. and Petering, D. H.: 1976, *Biochem. Pharmacol. 25*, pp. 2085-2089.

3. Goormaghtigh, E., Chatelain, P., Caspers, J. and Ruysschaert, J. M.: 1980, *Biochem. Pharmacol. 29*, pp. 3003-3010.

4. Malatesta, V. and Andreoni, A.: 1988, *J. Photochem. Photobiol.* in press

5. Chaires, J. B., Dattagupta, A. and Crothers, D. M.: 1985, *Biochemistry,* 24, pp. 260-267.

6. Malatesta, V. and Andreoni, A.: 1988, *Photochem. Photobiol.* in press

7. Sinha, B. K., Trush, M. A., Kennedy, K. A. and Mimnaugh, E.G.: 1984, *Cancer Res.* 44, pp.2892-2896.

8. Someya, A. and Tanaka, N.: 1979, *J. Antibiot.* 32, pp. 839-845.

9. Mimnaugh, E. G., Trush, M. A., Bhatnagar, M. and Garm, T. E.: 1984, *Biochem. Pharmacol.* 34, pp. 847-856.

10. Kharasch, E. D. and Novak, R. F.: 1983, *J. Pharmacol. Exp. Ther.* 226, pp. 500-506.

11. Babson, J. R., Abell, N. S. and Reed, D. J.: 1981, *Biochem. Pharmacol.* 30, pp. 2299-2304.

12. Mimnaugh, E. G., Kennedy, K. A., Trush, M. A. and Sinha, B. K.: 1985, *Cancer Res.* 45, pp. 3296-3304.

13. Malatesta, V., Penco, S., Sacchi, N., Valentini, L., Vigevani, A., and Arcamone, F.: 1984, *Can. J. Chem.* 62, pp. 2845-2850.

14. Pan, S., Pedersen, L. and Bachur, N. R.: 1981, *Molec. Pharmacol.* 19, pp. 184-186.

15. Sinha, B. K. and Gregory, J. L.: 1981, *Biochem. Pharmacol.* **130**, pp. 2626-2631.

16. Penco, S., Malatesta, V., Barchielli, G., Sacchi, N., Bordoni, T., Bellini, O. and Arcamone, F.: *Molec. Pharmacol.* submitted

17. Doroshow, J. H. and Davies, J. A.: 1986, *J. Biol. Chem.* **261**, pp. 3068-3074.

18. Malatesta, V.: 1988, *Med. Biol. Envir.* **16**, pp. 77-82.

19. May, P. M , Williams, G. N. and Williams, D. R.: 1980, *Eur. J. Cancer* **16**, pp. 1275-1276.

20. Beraldo, H., Garnier-Suillerot, A. and Tosi, L.: 1983, *Inorg. Chem.* **22**, pp. 4117-4124.

21. Gianni, L., Viganò, L., Lanzi, C., Niggeler, M. and Malatesta, V.: 1988, *Cancer Res.* in press

22. Malatesta, V., Tosi, C. and Fusco, R.; 1986, *Chem. Phys. Lett.* **128**, pp. 565-568.

23. Thomas, G. E. and Aust, S. D.: 1986, *Archiv. Biochem. Biophys.* **248**, pp. 684-689.

24. Halliwell, B., Gutteridge, J. M. and Blake, D.: 1985, *Philos. Trans. R. SOC. London, Series B* **311**, pp.659-671.

25. Phillips, D. R. and Carlyle, G. A.: 1981, *Biochem. Pharmacol.* **30**, pp. 2021-2024.

26. Malatesta, V., Morazzoni, F., Gervasini, A. and Arcamone, F.: 1985, *Anti Cancer Drug Design* **1**, pp. 53-57.

27. Morazzoni, F., Gervasini, A. and Malatesta, V.: 1987, *Inorg. Chim. Acta* **136**, pp. 111-115.

28. Malatesta, V., Gervasini, A. and Moarazzoni, F.: 1987, *Inorg. Chim. Acta* **136**, pp. 81-85.

29. Malatesta, V., Morazzoni, F., Pellicciari-Bollini, L. and Scotti, R.: 1987, *J. Chem. Soc. Faraday Trans. 1* **83**, pp. 3669-3673.

30. Malatesta, V., Morazzoni, F., Piccinini, F. and Monti, E.: in preparation

31. Malatesta, V. and Sacchi, N.: unpublished results

32. Li, A. S. W. and Chignell, F.: 1987, *Photochem. Photobiol.* **45**, pp. 565-570.

33. Reilly, J. J., Neifeld, J. P. and Rosenberg, S. A.: 1977, *Cancer* **40**, pp. 2053-2056.

34. Sanfilippo, A. G. , Schioppacassi, G., Movillo, E. and Ghione, M.: 1968, *G. Microbiol.* **16**, pp. 49-54.

35. Verini, M. A., Casazza, A. M , Fioretti, A. , Rodenghi, F. anf Ghione, M.: 1968, *G. Microbiol.* **16,** pp. 55-64.

36. Gray, P. J., Phillips, D. R. and Wedd, A.G.: 1982, *Photochem. Photobiol.* **36**, pp. 49-57.

37. Carmichael, A. J. and Riesz, P.: 1985, *Archiv. Biochem. Biophys.* **236**, pp. 433-444.

38. Andreoni, A., Land, T., Malatesta, V., McLean, A. and Truscott, T. G.: *J. Chem. Soc.* submitted

39. Manitto, P., Speranza, G. and Malatesta, V.: *J. Am. Chem. Soc.* submitted

40. Andreoni, A., Colasanti, A., Kisslinger, A. M. and Malatesta, V. in preparation

THE FORMATION OF FLUORESCENT PIGMENTS DUE TO RADICALS IN BIOLOGICAL TISSUES

J.F. Koster[a], A. Montfoort[b] and H. Esterbauer[c]
[a]Dept. of Biochemistry I, [b]Dept. of Pathology,
Medical Faculty, Erasmus University Rotterdam, P.O.Box 1738,
3000 DR Rotterdam, The Netherlands
[c]Institute of Biochemistry, University of Graz,
Schubertstrasse 1, A-8010 Graz, Austria.

ABSTRACT. In the ageing process fluorescent pigments (lipofuscin, aged pigments) are formed. There is a lot of evidence that these pigments are the results of an on-going process, lipid peroxidation. Lipid peroxidation is a process in which unsaturated fatty acids are broken down, resulting in a tremendous amount of various aldehydes. Probably these aldehydes play a role in the formation of the fluorescent pigments. It is shown that one of the main aldehydes, 4-hydroxynonenal is important in this process, which is stimulated by iron. Also the possible involvement of NADPH dependent enzymatic reaction is shown. The inhibition of fluorescent pigment formation by reduced scavengers indicate that radicals are involved in this formation.

INTRODUCTION

Although, there are many contradictory data concerning the process of ageing, the formation of fluorescent pigments (lipofucsin, aged pigments) is well accepted marker about the age of the tissue. However, the mechanism by which these pigments are formed is still unknown and it is rather doubtful if these pigments are harmful for the cell. During the course of time, there is a lot of evidence obtained that the so-called process of lipid peroxidation is involved (1,2,3,4). Lipid peroxidation is a process in which the polyunsaturated fatty acids (PUFA) are broken down to a tremendous variety of products, of which 4-hydroxynonenal is one of the main products (5). From these products the unsaturated aldehydes are involved in the genesis of fluorescent pigments. The process of lipid peroxidation is initiated by radicals and especially oxygen radicals, in this mechanism the presence of a transit metal is necessary. From physiological point of view, iron is the most likely candidate.

The oxygen radicals are most likely formed in any oxidative process. During the normal mitochondrial oxidation most of the oxygen is reduced to water, but still oxygen is reduced to superoxide anion (O_2^-). Under pathological conditions in which cell injury occurs

461

F. Minisci (ed.), Free Radicals in Synthesis and Biology, 461–468.
© *1989 by Kluwer Academic Publishers.*

it is quite well-known that PUFA are broken down.

It should be mentioned that the cell possesses a defence system against these radicals as well enzymatic as non-enzymatic, but this is out of scope.

One of the most well-known products of lipid peroxidation is malondialdehyde. It is proposed that the fluorescent pigments are formed from the reaction between malondialdehyde and the amino group of the phospholipids (phosphatidylethanolamine, phosphatidylserine) or proteins. These reactions reveal fluorescent Schiff's bases with the 1-amino-3-amino propane structure (6). The latter is debated by Kikugawa et al. (7,8) who propose that dihydropyridine derivatives are the molecular species responsible for the fluorescence.

In vitro it is shown that biological samples (cell organelles or intact cells) in which lipid peroxidation is induced also fluorescent pigments are formed (3,4). If the lipid peroxidation is inhibited the formation of fluorescent pigments is also impaired. Furthermore, the maxima of emission and extinction of these pigments are similar to the maxima observed for the fluorescent pigment isolated from biological tissues (2,3).

The question remains whether these fluoroscent pigments are formed during the cascade of lipid peroxidation or whether the products of lipid peroxidation (especially the aldehydes) are responsible.

As already mentioned there is great doubt that the fluorescent pigments are the result of the reaction between malondialdehyde and amino groups from either phospholipid or protein. Shimasaki et al. (9) have suggested that radicals are involved in the formation of fluorescent pigments. Although this cannot be excluded their suggestion is based on the finding that thiobarbituric acid-reactive substances obtained from membrane are not reactive with the amine compounds. However, it is known that this barbituric acid-reactive substance is mainly malondialdehyde, while the 4-hydro alkenales do not react with thiobarbituric acid (10). This is rather important, because one of the 4-hydroxy alkenales is 4-hydroxynonenal which hardly reacts with thiobarbituric acid and which is one of the main products of lipid peroxidation. Therefore, we started an investigation on the possible involvement of 4-hydroxynonenal in the formation of fluorescent pigments (11). For this purpose rat-liver microsomes and rat-liver mitochondria, isolated according conventional methods, were used. For model studies L-α-phosphatidylethanol, L-α-phosphatidylcholine, L-α-phosphatidylinositol and L-α-phosphatidylserine were used.

METHODS AND MATERIALS

Most of the methodology used can be found in ref. 11.

RESULTS

Incubation of 4-hydroxynonenal with various phospholipids reveals that 4-hydroxynonenal reacts well with phosphatidylethanolamine and phosphatidylserine as is to be expected. Phosphatidylinositol, however,

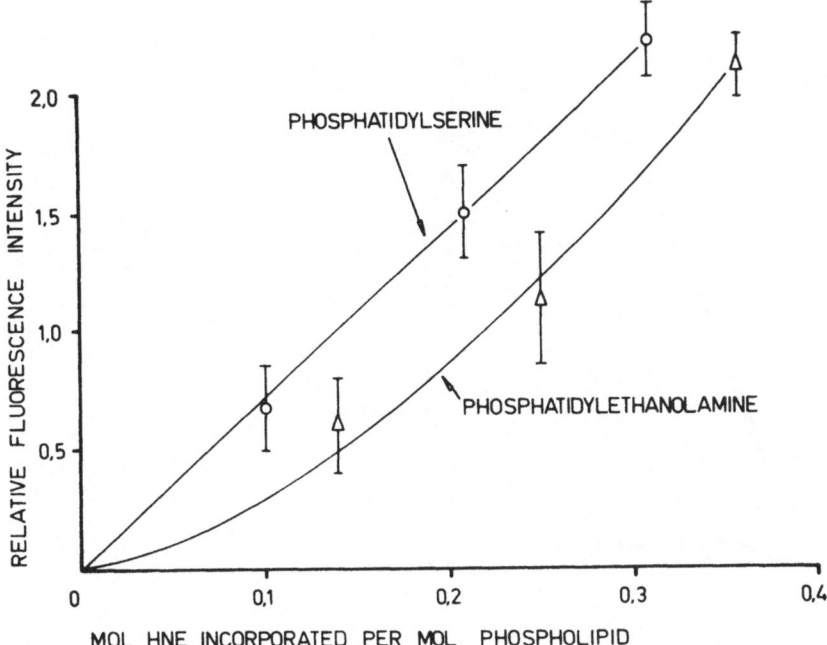

Figure 1. The dose-dependent fluorescent increment of phospholipids induced by 4-hydroxynonenal.
o-o, phosphatidylserine; Δ-Δ, phosphatidylethanolamine. Phospholipid concentration 1 mg/ml, [4-hydroxynonenal] 1 mM in 0.1 M Tris-HCl pH 7.4 at 37°C.

reveals much less, while phosphatidylcholine not at all. These findings are in accordance with the fluorescence as shown in Fig. 1.

Phosphatidylethanolamine and phosphatidylserine lead to the formation of fluorescence with the same characteristics as generated by microsomes or mitochondria (see below) (excitation 360 nm/emission 430 nm).

The addition of ADP iron does not change the pattern and the amount of fluorescence pigments.

Incubation of 4-hydroxynonenal with cell organelles (rat-liver microsomes and mitochondria) gives the following results. Microsomes incubated with 4-hydroxynonenal show a time-dependent increase in fluorescence, which is dependent on the amount of 4-hydroxynonenal added (Fig. 2b). As mentioned, the addition of ADP-iron to phospholipids does not alter the results. With microsomes, however, an increase in fluorescence is observed (Fig. 2a). Also the addition of NADPH to the incubation does result in an enhanced fluorescence (Fig. 2a). This latter is probably due to some enzymatic reaction, because boiled microsomes (5°, 100°C) do not form fluorescent material in the presence of NADPH, but still do with ADP iron.

464

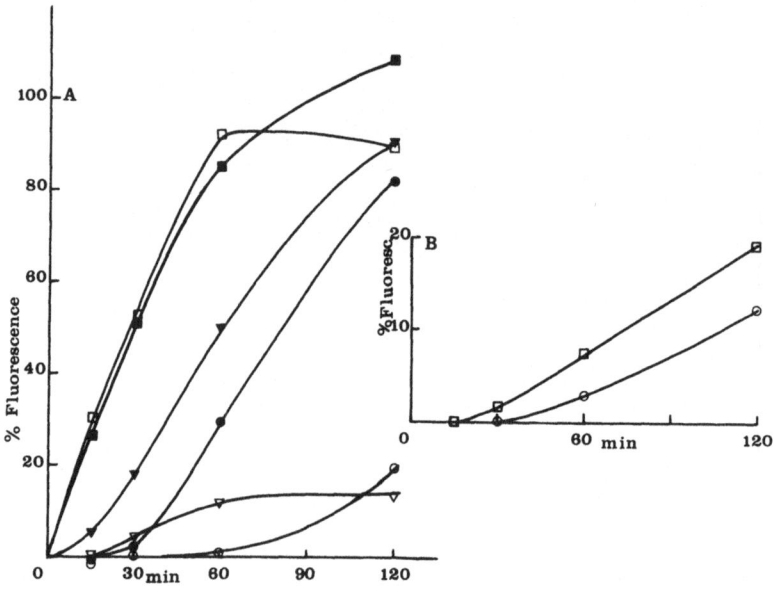

Figure 2. Time-dependency of the formation of fluorescent chromolipids in microsomes in the presence of ADP-iron, NADPH or 4-hydroxynonenal. (a) ●,o, ADP-iron; ▼,▽, NADPH; ■,□, ADP-iron plus NADPH. Open symbols indicate the absence, and closed symbols the presence, of 4-hydroxynonenal (0.77 mM). (b) o, 0.77 mM 4-hydroxynonenal; □, 1.54 mM 4-hydroxynonenal in the absence of ADP-iron and NADPH. The amount of microsomes added was 0.89 mg of protein/ml.

Mitochondria behave quite similar as microsomes. Incubation of mitochondrial with 4-hydroxynonenal leads to a small concentration and time-dependent increase of fluorescence (Fig. 3b). The addition of ADP iron enhances the fluorescence drastically, but the addition of NADPH does not stimulate the chromolipid formation (Fig. 3a). This latter contrasts the finding with microsomes and strengthens the possibility that with microsomes some enzymatic reaction is involved (aldehyde-dehydrogenase?).

The role of ADP iron is also shown in Fig. 4. This figure shows clearly that iron is involved. The addition of the ADP iron chelator desferral results in an inhibition of fluorescent material formation. Also if the chelator is added 30 min or 60 min after the incubation with ADP iron, the subsequent formation of fluorescent material is decreased significantly.

Additional support for the possible involvement of radicals in the formation of fluorescent material without concurrent lipid peroxidation is found in the experiments with radical scavengers. Fig. 5 shows that

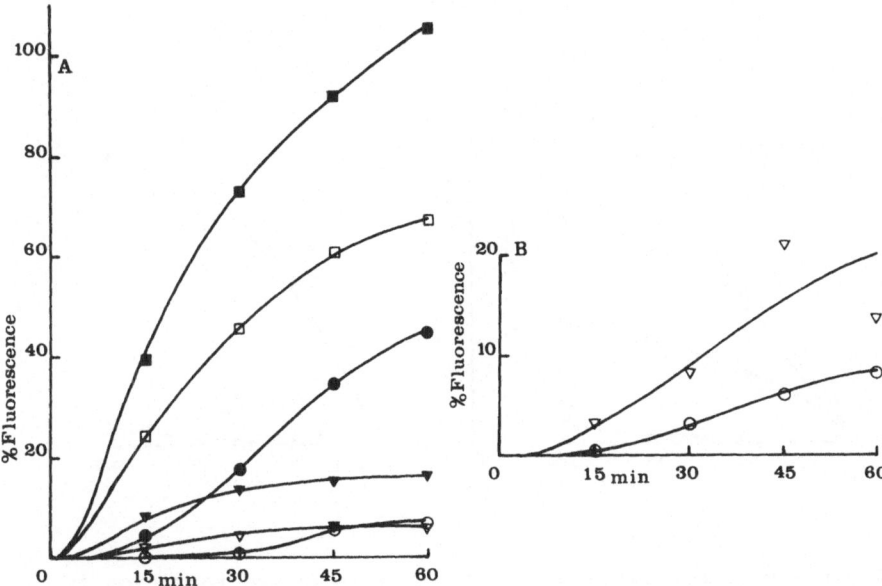

Figure 3. Time-dependency of the formation of fluorescent chromolipids in mitochondria in the presence of NADPH, ADP-iron or 4-hydroxynonenal. (a) ●,o, ADP-iron; ▼,▽, NADPH; ■,□, ADP-iron plus NADPH. Open symbols indicate the presence, and closed symbols the presence, of 4-hydroxynonenal (0.77 mM). (b) o, 0.77 mM 4-hydroxynonenal; ▽, 1.54 mM 4-hydroxynonenal in the absence of ADP-iron and NADPH. The amount of mitochondria added was 4.37 mg/ml.

the scavenger cyanidanol inhibits the formation of chromolipids, also the antioxidant butylated hydroxy toluene and the hydroxyl scavenger thiourea block the formation of fluorescent chromolipids.

To compare the capacity of microsomes and mitochondria to generate fluorescent material, the amount of fluorescence formed under various conditions are summarized in Table I. It shows that microsomes do form more fluorescent chromolipid based on the amount of protein.

DISCUSSION

From the data presented it can be concluded that it is most likely that 4-hydroxynonenal is involved in the fluorescent pigment formation. The involvement of the other lipid peroxidation product malondialdehyde has fluorescent characteristics of excitation at 400 nm and emission at 475 nm (2). These differences are too large to be neglected.

466

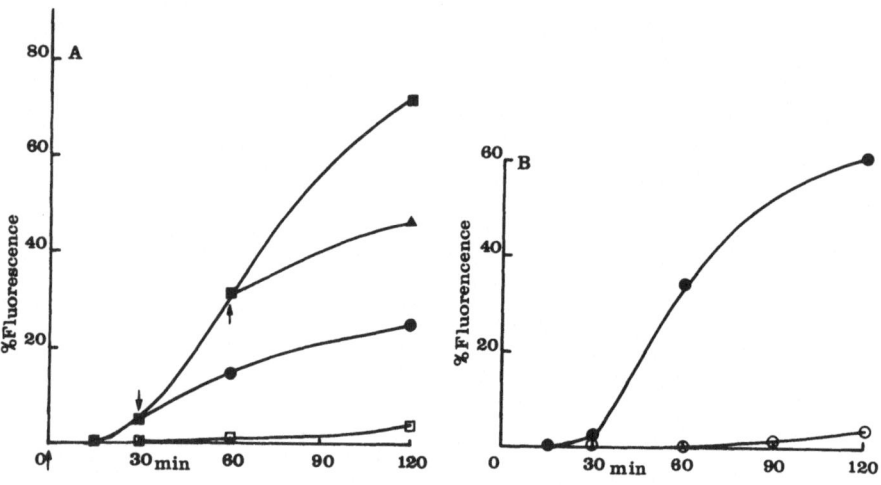

Figure 4. Effect of desferral on the ADP-iron-enhanced formation of
fluorescent chromolipids by 4-hydroxynonenal in microsomes and
mitochondria.
The arrows indicate the time of addition of desferral (1.5 mM). (a)
Microsomes; ■, control; □, desferral added at zero time. o,△,
desferrral added 30 min (●) and at 60 min (▲) after starting the
reaction. (b) Mitochondria: ●, control; o, desferral added at zero
time. Amounts added; microsomes, 0.73 mg/ml; mitochondria, 3.38 mg/ml;
4-hydroxynonenal, 0.74 mM.

Furthermore, chromolipids extracted from biological samples in which
lipid peroxidation (3,4) was provoked show the same spectral
characteristics, as the chromolipids obtained with 4-hydroxynonenal and
the natural occurring lipofucsin.
　　　Also it should be realized that during the lipid peroxidation much
4-hydroxynonenal is formed (5). From physiological point the role of
iron is quite intriguing, because the amount of free iron in biological
tissue is very small. Most of the iron is stored in the protein
ferritin. It is known that reductive equivalents are necessary to
mobilize iron from ferritin. O_2^- is able to mobilize iron (12) as is
NADPH (13). The exact mechanism by which iron is mobilized from
ferritin is still obscure, but it is shown that ferritin can be the
physiological iron donor for lipid peroxidation (13).

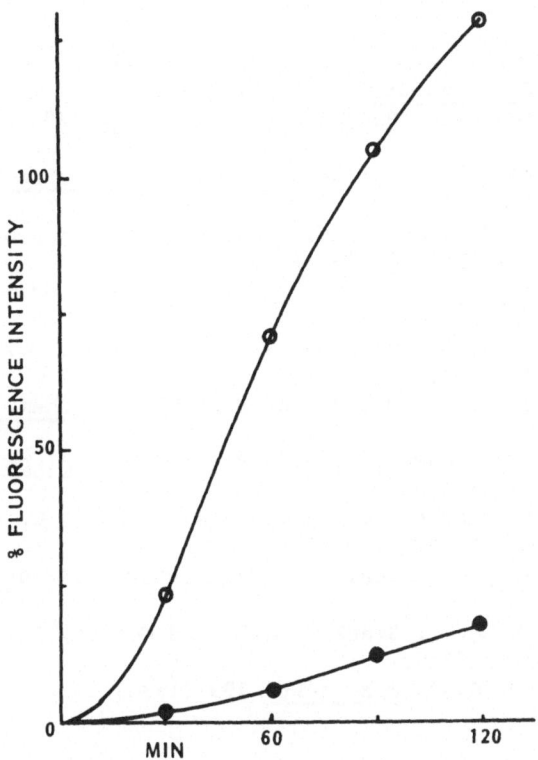

Figure 5. The effect of cyanidamol (50 μM) on the fluorescent pigment formation by 4-hydroxynonenal (0.77 mM) and microsomes (1 mg/ml). o-o, 4-hydroxynonenal + microsomes in the absence and ●-●, in the presence of 50 μM cyanidanol.

Table I. Effect of NADPH, ADP-iron and 4-hydroxynonenal (HNE) on the formation of fluorescent products by microsomes and mitochondria.

Microsomes or mitochondria were incubated in the presence of the indicated additions for 120 min and the fluorescence of the $CHCl_3/CH_3OH$ extractable lipid material was measured in % relative to a quinine standard (= 100%)

| | % fluorescence | |
Supplements	Microsomes	Mitochondria
+ NADPH	14+2	8+2
+ ADP/Fe^{3+}	18+2	8+3
+ NADPH + ADP/Fe^{3+}	100+5	65+3
+ HNE	12+2	8+2
+ NADPH + NHE	90+3	18+3
+ ADP/Fe^{3+} + HNE	85+2	50+4
+ NADPH + ADP/Fe^{3+} + HNE	105+5	105+6

Microsomes (0.9 mg/ml), mitochondria (4.4 mg/ml), [4-hydroxynonenal] = 0.77 mM. The values are means ± S.D. from three experiments.

REFERENCES

1. Dillard, C.J. and Tappel, A.L., Lipids **16** (1971) 715.
2. Tappel, A.L., Adv. Exp. Med.Biol. **97** (1978) 111.
3. Koster, J.F. and Slee, R.G., Biochim. Biophys. Acta **670** (1980) 489.
4. Koster, J.F., Slee, R.G. and Van Berkel, Th.J.C., Biochim. Biophys. Acta **710** (1982) 230.
5. Poli, G., Dianzini, M.U., Cheeseman, J.H., Slater, T.F., Levy, J., and Esterbauer, H., Biochem. J. **227** (1985) 629.
6. Tappel, A.L. in Free Radicals in Biology (Pryor, W.A. ed.) 1980, Vol. 4, p. 2, Acad. Press, New York.
7. Kikugawa, K., Maruijama, F., Machida, Y., and Kurachi, T., Chem. Pharm. Bull. **29** (1981) 1423.
8. Kikugawa, K., Machida, Y., Kida, M. and Kurechi, T., Chem. Pharm. Bull. **29** (1981) 3003.
9. Shimasaki, H., Ueta, N., Mowri, H.-O. and Inove, K., Biochim. Biophys. Acta **792** (1984) 123.
10. Esterbauer, H., Cheeseman, K.H., Dianzini, M.U., Pali, G. and Slater, T.F., Biochem. J. **208** (1982) 129.
11. Esterbauer, H., Koller, E., Slee, R.G. and Koster, J.F. Biochem. J. **239** (1986) 405.
12. Biemond, P., Van Eijk, H.G., Swaak, A.J.G. and Koster, J.F. J. Clin. Invest. **73** (1984) 1576.
13. Koster, J.F. and Slee, R.G., FEBS lett. **199** (1986) 185.

THE CHEMISTRY/BIOCHEMISTRY INTERFACE IN FREE-RADICAL REACTIONS

Robert Louw* and Ned A. Porter

It was a excellent idea to have a general discussion session, on the conferences' Thursday. In so far as necessary, participants have made acquaintance then, and are well conditioned by the advanced chemistry and novel results already presented.

Invited to present a short introductory statement, Professor Minisci recalled that, up to some or twenty years ago, free radicals were mainly the domain of physical-organic chemists, synthesis and biological reactions via free radicals being exotic subjects. Next to explosive developments in these directions, physical-organic chemistry continues to play a key role, and therefore, scientists of all three areas had been invited to this NATO-workshop.

Indeed, looking back to the fifties and sixties, structure (often by ESR) and rate measurements of combination/disproportionation, addition (also: polymerization) and abstraction (for example, in chlorination) were the usual subjects of free radical chemists. Complex "in vitro" phenomena, such as hydrocarbon pyrolysis (cracking!), or combustion, though practiced on a huge scale, were little understood on a quantitative molecular basis. These areas awaited technical developments: computers to handle large sets of contributing reactions, and modern instrumental methods enabling chemical physicists and physical chemists to determine, or refine thermochemical and kinetic parameters.

* I am indebted to D. Schulte-Frohlinde and N.A. Porter for their willingness to make opening statements in the discussion session, and to J.F. Koster for taking down major points. Ned Porter is also thanked for preparing a written report at my request; I am pleased to acknowledge his contribution by appointing him as a co-author.

F. Minisci (ed.), Free Radicals in Synthesis and Biology, 469–474.
© 1989 by Kluwer Academic Publishers.

Better understanding of chemical reactivity and rates both
in vitro and in vivo, and extension to organometallic
derivatives, have certainly been instrumental in the success
of radicals in synthesis, in making fine chemicals, by well-
tuned transfer and cyclization reactions. It is interesting
to see, that a number of chemists, already renowned for
their contributions to the "even", have become adepts in the
"odd"; in any case, they have shown that, if properly
trained, free radicals can react with great selectivity.
Meanwhile, the importance of radical reactions in biological
systems has become clear, and there is a growing research
effort (mostly by biomedical groups) to uncover the role of
radicals in the chemistry of life - its maintenance,
protection, challenge, or destruction.

D. Schulte-Frohlinde stressed the rapid progress in
methods and techniques, both as to in vivo and in vitro
research. He pointed at the major role of free radicals in
the chemistry of respiration; certainly a convincing
example, involving redox reactions, set in by the most
abundant free radical species, triplet atmospheric dioxygen.
He also reminded that a single cell is already an amazingly
complex system, wherein many chemistries go on at the same
time. Cells may contain some 10^{-2} M of glutathione, and to
regular organic chemists this is a rather high concentration
of a free radical scavenger. This example underscores that
in vitro chemical studies pertinent to biological processes
should be done under biologically relevant conditions. He
also stated that most free radical (organic) chemists know
too little about molecular biology and the methods and
approaches in that area; he suggested these collegues to
obtain a good book and to read it up.

Physical-organic chemists usually deal with processes
in homogeneous solution. However, much of what happens in
cells, tissues, and organs are heterogenous, site specific
reactions. For example, an iron ion meeting a H_2O_2 molecule
promotes formation of •OH. But if the ion is bound at, or
close to, DNA, •OH generated there has a good chance to
attack DNA. It is very difficult to prevent •OH from doing
so, by adding a free-radical scavenger, unless this compound
has a metal-binding ability.
In other words, radicals in biology often are not free.
Whatever the challenges of single cell-chemistry, there is a
tremendous added complexity in tissues, organs, and on top,
a wide variety of species - plants, and animals up and
including man. For the chemists amongst the latter species
there is one important consolation: Reaction mechanisms are
basically the same - in test tubes, and in life.

N.A. Porter put forward that one of the interfaces that

is developing between free radical chemistry and biology
involves the study of enzyme reaction mechanisms. Twenty
years ago, a proposal that an enzyme operates by a free
radical mechanism would have been unusual and would probably
have brought a skeptical response. Today, however, several
enzymatic systems have been investigated that suggest the
possibility that free radical intermediates are involved in
an enzyme reaction mechanism. It would appear that free
radical chemists can contribute to the understanding of
these enzyme mediated processes and can, at the same time,
learn new free radical chemistry from nature.

One example of a free radical enzyme mechanism is
prostaglandin synthase, an enzyme that incorporates
molecules of oxygen into a polysaturated fatty acid and, in
the process, converts the acyclic precursor into a bicyclic
peroxide-hydroperoxide structure, PGG. PGG is a key
intermediate in the formation of a multitude of compounds
that have important biological functions that range from
initiating blood platelet aggregation to causing smooth
muscle relaxation.

The conversion of polyunsaturated fatty acids to PGG is
chemically interesting since the achiral precursor is
converted into a bicyclic product that has five chiral
centers. This enzymatic conversion can be modelled by free
radical reactions in nonenzymatic systems. In fact,
autoxidation of polyunsaturated fatty acids follows a course
that is very similar to the prostaglandin enzymatic
conversion. Autoxidation of unsaturated fatty acids gives a
multitude of products that result from random H atom
abstraction and random addition of oxygen to intermediate
carbon radicals while the enzyme gives only one of the many
possible isomeric products. Apparently the enzyme serves to
initiate the free radical process, control stereochemistry
of the conversion and deliver oxygen at the regiochemically
correct position. Thus, free radical intermediates are tamed
by the enzyme.

Free radical chemists have contributed to this problem
by outlining the fundamental chemistry that occurs in this
process. The enzymatic conversion appears to be nothing more
than a radical H atom abstraction, addition of oxygen to
intermediate carbon radicals and cyclization of intermediate
carbon or peroxyl free radicals. Free radical chemists can
also learn from this example, however. How does the enzyme
tame the free radical? What factors are important in the
control of stereochemistry and regiochemistry in enzymatic
free radical processes? These are important questions which
represent a real challenge to our understanding of free
radical enzyme processes.

A second example of free radical intermediates in
enzyme reaction mechanisms is the enzyme cytochrome P-450.
One of the functions of this enzyme is to rid the system of

xenobiotic materials and one of the chemical pathways used
to do this involves the oxidation of carbon-hydrogen bonds
to give alcohols. The activation of a carbon-hydrogen bond
is chemically difficult and free radicals are among the few
reactive intermediates that are capable of this difficult
task. The chemistry of cytochrome P-450 can be modelled by
free radical reactions in nonenzymatic systems. The critical
intermediate is apparently an iron-oxo species that
abstracts hydrogen from an alkyl group and delivers OH to
the intermediate carbon radical species. Hydrogen atom
abstraction from hydrocarbon precursors is, of course, well
known in free radical chemistry but this process usually
occurs randomly, with all C-H bonds being attacked, and with
loss of stereochemistry in the products. Apparently the
enzyme serves to initiate the free radical process, control
stereochemistry of the conversion and deliver OH at the
correct position. Thus, free radical intermediates are tamed
by the enzyme.

Many other examples of free radical enzyme mechanisms
are being investigated today and the lessons from the two
examples cited above appear to be general. It seems
reasonable to suggest that nature uses free radical
reactions for enzymatic conversions when it is appropriate
to do so. Conversions involving the introduction of oxygen
into a molecule and the rupture of a strong C-H bond are
difficult for the other reactive intermediates but easy for
free radical species. It seems likely that many new
enzymatic pathways will be discovered where free radicals
play an important role and the free radical chemist will
have an opportunity to contribute to this developing field.
Fruitful collaborations can develop and, as has already been
mentioned , the chemist must be willing to learn new
techniques and approaches if he is to make a significant
contribution to the problem.

Further comment should be made about enzymes taming
free radical reactivity. The chemo, regio, and stereoselectivity exhibited by enzymes utilizing free radical
intermediates is undoubtedly due to the nature of the
enzyme-substrate-transition state complex. Forces that cause
substrate enzyme binding are nothing more than weak
intermolecular forces that have been studied for decades.
Thus, the hydrophobic effect, electrostatic effects, and
other intermolecular forces must be responsible for how free
radical reactivity is tamed. It seems appropriate to
endeavor to begin to understand how these weak
intermolecular forces can alter free radical reactivity.
This is surely a difficult problem but one that will provide
insight into how enzymes operate and allow new selective
free radical chemistry.

The understanding of how weak intermolecular forces can
be used to influence chemical reactivity of free radicals

leads to the conclusion that free radical chemistry needs to be pursued in more complex systems. The free radical chemistry of peroxidation in cell membranes has been discussed in several of the talks at this meeting and the need to understand this chemistry in aggregates and other organized media such as micelles, lipid bilayers, and liquid crystals is an example of a productive interface between free radical chemistry and biology. In these more complex systems, many tools may be needed to address the problems presented. In fact, the more complex the system, the less likely it seems that one single tool or technique will provide sufficient information for conclusive experiments. Chemists must be willing to use a variety of techniques to provide useful unambiguous data in these complex systems.

One other point should be mentioned. Many of the tools that are currently being used in free radical chemistry and biology were developed by individual researchers studying the details of free radical mechanism. Free radical clocks are frequently used today in the study of free radical mechanisms in chemistry, biochemistry, and biology. The fundamental idea of the free radical clock was developed as a simple chemical mechanistic probe in the last decade and it was only after this fundamental work that the idea was accepted and applied to the study of biochemical reaction mechanisms. The point is, that chemists should not give up chemistry in order to work on biochemically relevant problems. It is important to learn more about the biochemical systems of interest and accept technology from the biochemists and molecular biologists. Without doing this, the chemist may not understand the real problems in the field. On the other hand, the chemist can frequently reduce problems encountered to simpler ones than those found in biological systems and the chemist has an advantage in providing important insights into mechanism in this respect. Other examples of fundamental chemistry leading the biochemistry or biology can be cited but the message is the same. Fundamental new insight about mechanisms in free radical chemistry may have a broad impact on the chemical-biological interface. The study of the free radical mechanisms at the fundamental level should therefore not be abandoned by individual investigators.

In a lively discussion the participants generally agreed upon the sketched "states of the arts", expected scientific and technical developments, and opportunities and needs for cooperation between the relevant disciplines.

For studying life processes, it would be beneficial to bring enough knowledge and skills together to master such complex subjects. Physical-organic chemists could bring in the sophistication and quantification sometimes lacking in biology. It was also stressed that better contacts be laid

with biophysicists, considering the increased impact of
physical methods and techniques - in spectroscopy,
optoacoustics etc. Although the identification of some
radical does not mean it is important in a particular
chemical process, improved techniques for direct detection
of radicals in their living environments are welcome.
Anyhow, recognition of essential steps and crucial
intermediates are the primary goals of the science of
chemistry, also in the complexity of biological systems.

*It was an honour and pleasure to chair the discussion
session. Finally, I am quite convinced to speak on behalf of
all participants in thanking Prof. F. Minisci and his group
for the organization, the invitation, and the hearty care
Italian style. Aided by this book, we will remember a very
stimulating conference.*